Technical Calculus
with Analytic Geometry
Second Edition

Other books by the author

Essentials of Basic Mathematics, Second Edition (with Boyd & Plotkin)

Essentials of Algebra (with Boyd & Plotkin)

Introduction to Technical Mathematics, Second Edition

Basic Technical Mathematics, Third Edition

Basic Technical Mathematics with Calculus, Third Edition

Basic Technical Mathematics with Calculus, Third Edition,
Metric Version

Modules in Technical Mathematics
   Introductory Topics from Arithmetic
   Introduction to Algebra
   Introduction to Trigonometry
   Introduction to Geometry

Plane Trigonometry (with Edmond)

Mathematics: A Developmental Approach

Technical Calculus
with Analytic Geometry
Second Edition

Allyn J. Washington

Dutchess Community College
Poughkeepsie, New York

The Benjamin/Cummings Publishing Company
Menlo Park, California • Reading, Massachusetts
London • Amsterdam • Don Mills, Ontario • Sydney

Sponsoring editor: Susan A. Newman
Production: Greg Hubit

Copyright © 1980, 1966 by The Benjamin/Cummings Publishing Company, Inc. Philippines
Copyright 1980, 1966 by The Benjamin/Cummings Publishing Company, Inc.

Library of Congress Cataloging in Publication Data

Washington, Allyn J
  Technical calculus with analytic geometry.

  Includes index.
  1. Calculus.   2. Geometry, Analytic.   I. Title.
OA303.W38   1980   515'.15   79-17595

ISBN 0-8053-9519-9
GHIJKLM-HA-8987654

The Benjamin/Cummings Publishing Company, Inc.
2727 Sand Hill Road
Menlo Park, California 94025

Preface

Scope of the Book

This book is intended primarily for students taking technical programs at two-year and four-year colleges and technical institutes. The emphasis in this text is on elementary topics in calculus and certain selected topics from the more advanced areas. These have been selected due to their application in the technologies, with an emphasis on electrical and mechanical technologies. The topics are developed in a nonrigorous and intuitive manner, with stress being placed on the interpretation and applications of the material presented.

It is assumed that students using this text will have had courses covering the basic topics of algebra and trigonometry. However, several sections on algebra and trigonometry are included for review and reference purposes. These are the topics essential for proper development and understanding of the calculus.

The general topics covered in the text are basic analytic geometry, differentiation and integration of algebraic and elementary transcendental functions, an introduction to partial derivatives and double integrals, expansion of functions in series, and differential equations.

The analytic geometry is developed primarily for use in the calculus, although numerous direct applications of curves such as the conic sections are discussed in order to show the importance of graphical methods. A chapter on polar coordinates is included to demonstrate the use of different coordinate systems. Also, empirical curve fitting is included to show how a curve may be fitted to data. In the chapter dealing with partial derivatives and double integrals, a section is devoted to solid analytic geometry. In addition to these specific topics from analytic geometry, graphical techniques and interpretations are included throughout the text. The graphs of the trigonometric, inverse trigonometric, exponential and logarithmic functions as well as methods of graphical differentiation and integration are discussed.

The calculus will give the student a mathematical understanding of many topics which arise in other courses. Numerous applications from all fields of technology are included to show the wide application of calculus. Among these are many definitions of physical quantities, the relations among displacement, velocity and acceleration, areas and volumes, centroids and moments of inertia, work and liquid pressure.

The chapter on partial derivatives and double integrals is included in order to introduce these topics in curricula for which they may be deemed necessary. Their applications are shown for such fields as electronics, thermodynamics, and mechanics.

The chapter on expansion of functions in series will give an understanding of the development of numerical tables and some other uses of this method of expressing functions. Also, the section on Fourier series is appropriate for curricula in electrical technology.

Differential equations provide one of the richest areas of mathematics for applications in technical areas, especially electrical and mechanical technologies. Here a mathematical basis can be established for many concepts and formulas which previously the student has had to accept without basis. For this reason considerable space is devoted to differential equations.

Laplace transforms are introduced to show the methods and terminology for electrical and mechanical technology students. Numerical methods of solving differential equations are appropriate for use with electronic calculators and computers. Introduction of such topics will make the technician more effective in modern industry.

New Features

This second edition of *Technical Calculus with Analytic Geometry* includes all the basic features of the first edition. However, most sections have been rewritten to some degree to include additional explanatory material, examples, and exercises. Specifically, among the new features of the second edition are the following: (1) New sections have been added and other sections of the first edition have been split to provide more detailed coverage. Thus, Section 2–9 is a separate section on differentiation of implicit functions, Section 4–8 is a new section on Simpson's rule, Section 5–6 is a separate section on work by a variable force, and Section 5–7 is a new section on force due to liquid pressure. (2) Continuity is now briefly introduced in Section 2–2 with limits. (3) It is assumed that the students will use a scientific calculator for most of their calculating work, and a number of problems are specifically designed for solution using the calculator. These problems may be used in developing certain basic concepts such as a limit. (4) There are more problems using metric units, and a summary of the metric system (SI) is included in Appendix B. (5) Important formulas have been set off so that they are easily located and utilized. (6) There are now over 3000 exercises, an increase of over 40%. Each exercise group is designated by the number of the section which it follows. Exercises are generally grouped such that there is an even-numbered exercise equivalent to each odd-numbered exercise. (7) Many new examples are included, and more detail is included in examples which appeared in the first edition. There are now over 500 worked examples, an increase of over 15%.

Other features are: (1) The many examples included in this text are often used advantageously to introduce concepts, as well as to clarify and illustrate points made in the text. (2) Those topics which experience has shown to be more difficult for the student have been developed in more detail, with many examples. (3) The order of coverage can be changed in several places without loss of continuity. Any omissions or changes in order will, of course, depend on the type of course and the completeness required. (4) Review exercises are included after each chapter. These may be used either for additional problems or for review assignments. (5) The answers to all the odd-numbered exercises are given at the back of the book. Included are answers to graphical problems and other types which are not always included in textbooks.

Acknowledgments

I wish to acknowledge the comments and suggestions given to me by many of those who used the first edition of this text. Among these, I wish to thank the members of the Mathematics Department of Dutchess Community College. In particular I wish to thank Carolyn Edmond for her help in checking most of the answers. Also, the valuable suggestions by William D. Brower of New Jersey Institute of Technology, Harry Wilson of California State Polytechnic University at Pomona, and Kenneth Tekel of Essex County College in their reviews of the manuscript are most appreciated. Finally, the assistance and cooperation of The Benjamin/Cummings staff during the production of this text are also greatly appreciated.

A. J. W.

Contents

Chapter 8   Methods of Integration   269

Chapter 9   Introduction to Partial Derivatives and Double Integrals   302

Chapter 10   Polar Coordinates   335

Chapter 11   Empirical Curve Fitting   350

Chapter 12   Expansion of Functions in Series   376

Appendix C   Tables   466

Answers to Odd-Numbered Exercises   481

Index   511

1

Plane Analytic Geometry

1–1 Introduction

The problems which can be solved by the methods of algebra and trigonometry are numerous. There are, however, a great many problems which arise in the various fields of technology which require for their solution methods beyond those available from algebra and trigonometry. The use of these traditional topics remains of definite importance, but it is necessary to develop additional methods of analyzing and solving problems.

One very important type of problem which arises involves the rate of change of one quantity with respect to another. Examples of rates of change are velocity (the rate of change of distance with respect to time), the rate of change of the length of a metal rod with respect to temperature, the rate of change of light intensity with respect to the distance from the source, the rate of change of electric current with respect to time, and many other similar physical situations. The methods of **differential calculus**, which we shall start developing in Chapter 2, will enable us to solve problems involving these quantities.

Another principal type of problem which calculus allows us to solve is that of finding a function when its rate of change is known. This is **integral calculus**, a study of which starts in Chapter 4. One of the principal applications comes from electricity, where current is the time rate of change of electric charge. Although apparently unrelated, integral calculus also leads to the solution of a great many other problems, which include plane areas, volumes, and the physical concepts of work and pressure.

Before starting a study of the calculus, we will first develop some of the basic concepts of **analytic geometry**. This branch of mathematics deals with the relationship between algebra and geometry. The definitions and methods developed here, along with certain basic curves and their characteristics will enable us to more readily develop the calculus. Also, many technical applications of analytic geometry itself are illustrated.

1—2 Rectangular Coordinates

From algebra we recall that *a variable is a quantity which may take on admissible values during a given discussion. A constant is a quantity which remains fixed during the discussion.* Generally letters toward the end of the alphabet are used to denote variables, and letters near the beginning of the alphabet denote constants.

One of the most valuable ways of representing an equation in two variables is by means of a graph of the equation. By using graphs we are able to obtain a "picture" of the equation, and this picture enables us to learn a great deal about the equation.

To achieve this graphical representation we use the fact that we may represent numbers by points on a line. Since two variables are involved, it is necessary to use a line for each. This is done most conveniently by placing the lines perpendicular to each other.

Thus *we place one line horizontally and label it the x-axis. The other line we place vertically and label it the y-axis. The point of intersection is called the* **origin** *and is designated by 0. This is called the* **rectangular coordinate system.**

Starting from the origin, equal intervals are marked off along each of the axes, and the **integers** (*the numbers 1, 2, 3, . . . which represent whole quantities*) are placed at these positions. The **positive integers** are placed on the x-axis to the right of the origin, and the **negative integers** are placed to the left. On the y-axis, the positive integers are placed above the origin and the negative integers are placed below. *The four parts into which the plane is divided are called* **quadrants** (see Fig. 1–1).

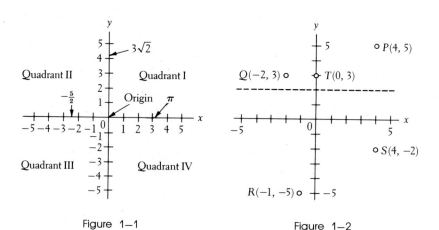

Figure 1—1   Figure 1—2

The points between the integers represent either the **rational numbers** (*numbers which are represented by dividing one integer by another*) or the **irrational numbers** (*those numbers which cannot be represented by dividing one integer by another; π and $\sqrt{2}$ are such numbers*). In this way all **real numbers,** which include the rational numbers (the integers are rational) and irrational numbers, may be represented as points on the axes. We shall use real numbers throughout this text unless otherwise noted. We will occasionally refer to and use **imaginary numbers,** *which is the name given to square roots of negative numbers.*

Example A

The number 7 is an integer, rational (since $7 = \frac{7}{1}$), and real (the real numbers include all the rational numbers); 3π is irrational and real; $\sqrt{5}$ is irrational and real; $\frac{1}{8}$ is rational and real; $7\sqrt{-1}$ is imaginary; $\frac{6}{3}$ is rational and real (it is an integer when we use the symbol "2" to represent it); $\pi/6$ is irrational and real; $\sqrt{-3}/2$ is imaginary.

A point P in the plane is designated by the pair of numbers (x, y). *The x-value, called the* **abscissa** *is the perpendicular distance of P from the y-axis. The y-value, called the* **ordinate,** *is the perpendicular distance of P from the x-axis. The values x and y together, written as (x, y) are the* **coordinates** *of P.*

Example B

The positions of points $P(4, 5)$, $Q(-2, 3)$, $R(-1, -5)$, $S(4, -2)$ and $T(0, 3)$ are shown in Fig. 1–2. Note that this representation allows for *one point for any pair of values (x, y).*

Example C

Where are all the points whose ordinates are 2? All such points are two units above the x-axis; thus the answer can be stated as "on a line 2 units above the x-axis." Note the dashed line in Fig. 1–2.

Another important mathematical concept we use in dealing with numbers is the **absolute value** of a number. By definition, *the absolute value of a positive number is the number itself, and the absolute value of a negative number is the corresponding positive number (obtained by changing its sign).* We may interpret the absolute value as being the number of units a given number is from the origin, regardless of direction. The absolute value is designated by | | placed around the number.

Example D

The absolute value of 6 is 6, and the absolute value of -7 is 7. We designate this by $|6| = 6$ and $|-7| = 7$.

Other examples are $|-\pi| = \pi$, $|\frac{7}{5}| = \frac{7}{5}$, and $|-\sqrt{2}| = \sqrt{2}$.

On the x-axis, *if a first number is to the right of a second number, the first number is said to be* **greater than** *the second. If the first number is to the left of the second, it is* **less than** *the second.* On the y-axis, if one number is above a second, the first number is greater than the second. "Greater than" is designated by $>$, and "less than" is designated by $<$. *These are called* **signs of inequality.**

Example E

$$6 > 3, 8 > -1, 5 < 9, 0 > -4, -2 > -4, -1 < 0$$

In Exercises 1 and 2 designate the given numbers as integers, rational, irrational, real, or imaginary. (More than one designation may be correct.)

1. (a) $\frac{5}{4}$　　(b) $-\pi$　　(c) $\sqrt{-4}$　　(d) $6\sqrt{-1}$

2. (a) 3　　(b) $6 + \sqrt{2}$　　(c) $\frac{1}{3}\sqrt{7}$　　(d) $\frac{7}{3}$

In Exercises 3 and 4 find the absolute values of the given numbers.

3. (a) 3　　(b) $\frac{7}{2}$　　(c) -4　　4. (a) $-(\pi/18)$　　(b) x　(if $x < 0$)

In Exercises 5 and 6 insert the correct inequality signs ($>$ or $<$) between the numbers of the given pairs of numbers.

5. (a) 6　　8　　(b) 7　　-5　　(c) π　　-1

6. (a) -4　　-3　　(b) $-\sqrt{2}$　　-9　　(c) 0.2　　0.6

In Exercises 7 through 10 plot the given points.

7. $A(2, 7)$, $B(-1, -2)$, $C(-4, 2)$　　8. $A(3, \frac{1}{2})$, $B(-6, 0)$, $C(-\frac{5}{2}, -5)$

9. $A(1, 3)$, $B(-1, -1)$, $C(6, 13)$ Join these points by straight-line segments. What conclusion can you make?

10. $A(-3, -2)$, $B(-3, 5)$, $C(3, 5)$ Join these points by straight-line segments. What conclusion can you make?

In Exercises 11 through 22 locate the indicated points, and answer the questions.

11. Where are all the points whose abscissas are 1?

12. Where are all the points whose ordinates are -3?

13. What is the abscissa of all points on the y-axis?

14. What is the ordinate of all points on the x-axis?

15. Where are all the points (x, y) for which $x > 0$?

16. Locate all points (x, y) for which $|x| < 1$.

17. Locate all points (x, y) for which $x < 0$ and $y > 1$.

18. Locate all points (x, y) for which $x > 0$ and $y < 0$.

19. In which quadrants is the ratio y/x positive?

20. In which quadrants is the ratio y/x negative?

21. Three vertices of a rectangle are $(5, 2)$, $(-1, 2)$ and $(-1, 4)$. What are the coordinates of the fourth vertex?

22. Two vertices of an equilateral triangle are $(7, 1)$ and $(2, 1)$. What is the abscissa of the third vertex?

In Exercises 23 and 24 shade in the appropriate regions for the given inequalities. See Appendix B for the use of units of measurement, and the symbols used to represent them.

23. After 2 s, the current in a certain circuit is less than 3 A. Using t to represent time and i to represent the current, this statement may be written as "for $t > 2$ s, $i < 3$ A." Using the x-axis for values of t and the y-axis for values of i, shade in the region for the above statement, assuming that $i > 0$.

24. Less than 3 ft from a light source the illuminance is greater than 8 lx (lux). Using x for the distance and y for the illuminance, shade in the appropriate region.

1–3 The Graph of an Equation.

The graph of an equation is the set of all points whose coordinates (x, y) satisfy the equation relating x and y. In order to find sets of coordinates for the purpose of constructing the graph, we assume certain values of one of the variables, usually x, and then determine the values of the other variable by use of the equation.

Since there is no limit to the possible number of points which can be chosen, we normally select a few values of x, obtain the corresponding values of y, and plot these points. The points are then connected by a *smooth* curve (not short straight lines from one point to the next), and are normally connected from left to right.

Example A

Graph the equation $y = 3x - 5$.

We now let x take on various values and determine the corresponding values of y. Note that once we choose a given value of x, we have no choice about the corresponding y-value, as it is determined by the equation. If $x = 0$ we find that $y = -5$. This means that the point $(0, -5)$ is on the graph of the equation $y = 3x - 5$. Choosing another value of x, for example, 1, we find that $y = -2$. This means that the point $(1, -2)$ is on the graph of the equation. Continuing to choose a few other values of x, we tabulate the results, as shown in Fig. 1–3. It is best to arrange the table so that the values of x increase; then there is no doubt how they are to be connected, for they are connected in the order shown. Finally, we connect the points as shown in Fig. 1–3, and see that the graph of the equation $y = 3x - 5$ is a straight line.

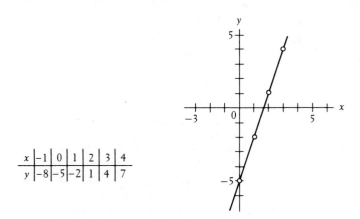

x	-1	0	1	2	3	4
y	-8	-5	-2	1	4	7

Figure 1–3

There are some special points with regard to certain curves which should be noted. Each of the next four examples illustrates and explains a particular point to be considered when plotting the graph of an equation.

Example B

Graph the equation $y = x - x^2$.

First we determine the values in the table as shown with Fig. 1–4. We must be careful with negative values of x. For $x = -1$, we have $y = (-1) - (-1)^2 = -1 - (+1) = -1 - 1 = -2$. Once all the values have been found and plotted, we note that $y = 0$ for both $x = 0$ and $x = 1$. The question arises—what happens between these values? Trying $x = \frac{1}{2}$, we find that $y = \frac{1}{4}$. Using this point completes the necessary information. Note that in plotting these graphs we do not stop the graph with the last point determined, but indicate that the curve continues.

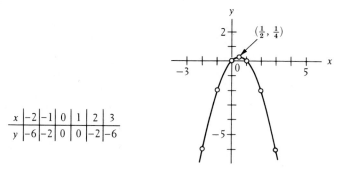

x	-2	-1	0	1	2	3
y	-6	-2	0	0	-2	-6

Figure 1–4

Example C

Graph the equation $y = 2x^2 - 4$.

We first tabulate the values as shown. In constructing the table we note that the same value of y appears for both $+3$ and -3. Also the same value of y appears for $+2$ and for -2, and again for $+1$ and for -1. Since $(-x)^2 = x^2$ and $(+x)^2 = x^2$ we can understand why this happens. The graph of this equation (see Fig. 1–5) is seen to be **symmetrical** to the y-axis. This can be thought of as meaning that the right half of the curve is the reflection of the left half, and conversely. As in this case, *any time $-x$ can replace x in an equation without changing the equation, its graph will be symmetrical to the y-axis.* In the same way, *if $-y$ can replace y in an equation without changing the equation, its graph is symmetrical to the x-axis.* Many of the curves to be discussed in the coming sections will demonstrate the property of symmetry.

Example D

Graph the equation $y = 1 + \frac{1}{x}$.

In finding points on this graph (see Fig. 1–6) we note that y is not defined if we try $x = 0$, since *division by zero is not defined.* Although x cannot *equal* zero, it can take on any value near zero. Thus, by choosing values of x between -1 and $+1$ we obtain the points necessary to determine the shape of the graph.

x	-3	-2	-1	0	1	2	3
y	14	4	-2	-4	-2	4	14

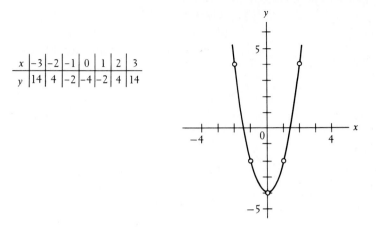

Figure 1–5

In examining the possible values of x more closely, we note that as the values of x get closer and closer to zero the values of y become numerically larger and larger. That is, as x *approaches* zero, y becomes numerically unbounded. In this case we say that $x = 0$ (the y-axis) is an **asymptote** of the curve. More generally, *an asymptote is a line which the graph of an equation approaches.* We can find the asymptotes by determining those values of x which lead to division by zero. In this way we determine those asymptotes which are vertical, although we must realize that the graph of any given equation may have asymptotes which are horizontal or oblique. We shall not consider these other types at this time. Asymptotes are of importance in a number of graphs. In trigonometry the graph of $y = \tan x$ has vertical asymptotes for $x = \frac{1}{2}\pi, \frac{3}{2}\pi, \ldots$, and we shall find asymptotes of importance when discussing the hyperbola.

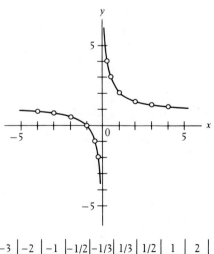

x	-4	-3	-2	-1	-1/2	-1/3	1/3	1/2	1	2	3	4
y	3/4	2/3	1/2	0	-1	-2	4	3	2	3/2	4/3	5/4

Figure 1–6

Example E

Graph the equation $y = \sqrt{x + 1}$.

When finding the points for the graph, we may not let x equal any negative value less than -1. All such values lead to imaginary values for y, and *only real values of the variables are permissible*. Also, *since we have the positive square root indicated, all values of y are positive or zero* (see Fig. 1–7). Note that the graph starts at the point $(-1, 0)$.

Graphs and graphical methods will be used throughout this text. It will be seen that a great deal of information can be learned about an equation by analyzing its graph. Also, graphs can often demonstrate a given property more clearly than can other descriptions. The following example illustrates an application of graphing an equation along with conclusions that can easily be seen from the graph.

Example F

The power of a certain voltage source in terms of the load resistance is given by

$$P = \frac{100R}{(0.5 + R)^2}$$

where P is measured in watts and R in ohms. Plot the graph, using the y-direction for power and the x-direction for resistance.

Since a negative value for resistance has no physical significance, we need plot P for positive values of R only. The following table of values is obtained.

R	0	0.25	0.50	1.0	2.0	3.0	4.0	5.0	10.0
P	0.0	44.4	50.0	44.4	32.0	24.5	19.8	16.5	9.1

The values 0.25 and 0.50 are used for R when it is found that P is less for $R = 2$ than for $R = 1$. In this way a smoother, more useful curve is obtained (see Fig. 1–8).

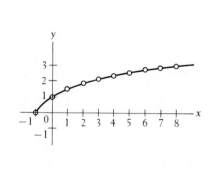

x	y
-1	0
0	1
1	1.4
2	1.7
3	2
4	2.2
5	2.4
6	2.6
7	2.8
8	3

Figure 1–7

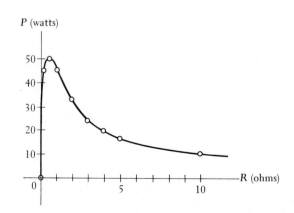

Figure 1–8

From the graph we see that P changes quite rapidly for small values of R with a peak value of about 50 W when $R = 0.50\ \Omega$ (actually this is exact for this equation). Also, as R becomes larger, P becomes smaller, but much less rapidly. This shows some of the conclusions that can be drawn from a graph.

Equations of a particular type have graphs which are of a specific form, and many of them have been named. We noted that the graph of the function in Example A is a **straight line**. A **parabola** is illustrated in Example B and in Example C. The graph of the function in Example D is a **hyperbola**. Other types of graphs are found in the exercises, and a more detailed analysis of several of these curves is found later in this chapter. The use of graphs is extensive in mathematics and in nearly all areas of application.

Exercises 1–3

In Exercises 1 through 32 graph the given equations.

1. $y = 3x$ 2. $y = -2x$ 3. $y = 2x - 4$

4. $y = 3x + 5$ 5. $y = 7 - 2x$ 6. $y = 5 - 3x$

7. $y = \frac{1}{2}x - 2$ 8. $y = 6 - \frac{1}{3}x$ 9. $y = x^2$

10. $y = -2x^2$ 11. $y = 3 - x^2$ 12. $y = x^2 - 3$

13. $y = \frac{1}{2}x^2 + 2$ 14. $y = 2x^2 + 1$ 15. $y = x^2 + 2x$

16. $y = 2x - x^2$ 17. $y = x^2 - 3x + 1$ 18. $y = 2 + 3x + x^2$

19. $y = x^3$ 20. $y = -2x^3$ 21. $y = x^3 - x^2$

22. $y = 3x - x^3$ 23. $y = x^4 - \tfrac{1}{2}x^2$ 24. $y = x^3 - x^4$

25. $y = \dfrac{1}{x}$ 26. $y = \dfrac{1}{x + 2}$ 27. $y = \dfrac{1}{x^2}$

28. $y = \dfrac{1}{x^2 + 1}$ 29. $y = \sqrt{x}$ 30. $y = \sqrt{4 - x}$

31. $y = \sqrt{16 - x^2}$ 32. $y = \sqrt{x^2 - 16}$

In Exercises 33 and 34 determine whether or not the graphs of the equations of the indicated exercises have symmetry to the x-axis or to the y-axis.

33. (a) Exercise 12   (b) Exercise 16   (c) Exercise 29

34. (a) Exercise 14   (b) Exercise 17   (c) Exercise 32

In Exercises 35 and 36 determine whether or not the graphs of the equations of the indicated exercises have vertical asymptotes.

35. (a) Exercise 19   (b) Exercise 25   (c) Exercise 28

36. (a) Exercise 22   (b) Exercise 27   (c) Exercise 26

In Exercises 37 through 44 graph the given equations. In each case plot the first mentioned variable as the ordinate and the second variable as the abscissa.

37. The velocity v (in feet per second) of an object under the influence of gravity, in terms of time t (in seconds), is given by $v = 100 - 32t$. If the object strikes the ground after 4 s, graph v vs. t.

38. If $1000 is placed in an account earning 6% simple interest, the amount A in the account after t years is given by the equation $A = 1000(1 + 0.06t)$. If the money is withdrawn from the account after 6 years, plot A vs. t.

39. The surface area of a cube is given by $A = 6e^2$, where e is the side of the cube. Plot A vs. e.

40. The energy in the electric field around an inductor is given by $E = \frac{1}{2}LI^2$, where I is the current in the inductor and L is the inductance. Plot E vs. I (a) if $L = 1$ unit, (b) if $L = 0.1$ unit, and (c) if the E-axis is marked off in units of L.

41. The illuminance (in lumens per square meter) of a certain source of light in terms of the distance (in meters) from the source is given by $I = 400/r^2$. Plot I vs. r.

42. The heat capacity (in joules per kilogram) of an organic liquid is related to the temperature (in degrees Celsius) by the equation $c_p = 2320 + 4.73T$ for the temperature range of $-40°C$ to $120°C$. Graphically show that this equation does not satisfy experimental data above $120°C$. Plot the graph of the equation and the following data points.

c_p	3030	3220	3350
T	140	160	200

43. If a rectangular tract of land has a perimeter of 600 m, its area in terms of its width is $A = 300w - w^2$. Plot A vs. w.

44. The deflection y of a beam at a horizontal distance x from one end is given by $y = -k(x^4 - 30x^2 + 1000x)$, where k is a constant. Plot the deflection (in terms of k) vs. x (in feet) if there are 10 ft between the end supports of the beam.

1–4 Some Basic Definitions

In Section 1–1 it was noted that analytic geometry deals with the relationship between algebra and geometry. In the sections which follow we will see that proper recognition of certain important equations will lead immediately to the graph of that equation. In this section we will develop certain basic concepts which are needed to establish the proper relationships between equation and curve.

The first of these concepts involves the distance between any two points in the coordinate plane. If these two points lie on a line parallel to the x-axis, the **directed distance** from the first point $A(x_1, y)$ to the second point $B(x_2, y)$ is denoted as $\overline{AB}$ and is defined as $x_2 - x_1$. We can see that $\overline{AB}$ *is positive if B is to the right of A, and it is negative if B is to the left of A.* Similarly, the directed distance between two points $C(x, y_1)$ and $D(x, y_2)$ on a line parallel to the y-axis is $y_2 - y_1$. We see that $\overline{CD}$ *is positive if D is above C, and it is negative if D is below C.*

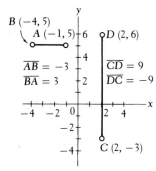

Figure 1–9

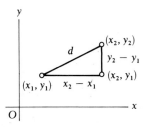

Figure 1–10

Example A

The line segment joining $A(-1, 5)$ and $B(-4, 5)$ in Fig. 1–9 is parallel to the x-axis. Therefore, the directed distance $\overline{AB} = -4 - (-1) = -3$, and the directed distance $\overline{BA} = -1 - (-4) = 3$.

Also in Fig. 1–9, the line segment joining $C(2, -3)$ and $D(2, 6)$ is parallel to the y-axis. The directed distance $\overline{CD} = 6 - (-3) = 9$, and the directed distance $\overline{DC} = -3 - 6 = -9$.

We now wish to find the length of a line segment joining any two points in the plane. If these points are on a line which is not parallel to either of the axes (Fig. 1–10), we must use the Pythagorean theorem to find the distance between them. By making a right triangle with the line segment joining the two points as the hypotenuse, and line segments parallel to the axes as the legs, we have the formula which gives the distance between any two points in the plane. This formula, called the **distance formula**, is

$$d = \sqrt{(x_2 - x_1)^2 + (y_2 - y_1)^2} \tag{1-1}$$

Here we choose the positive square root since we are concerned only with the magnitude of the length of the line segment.

Example B

The distance between $(3, -1)$ and $(-2, -5)$ is given by

$$d = \sqrt{[(-2) - 3]^2 + [(-5) - (-1)]^2}$$
$$= \sqrt{(-5)^2 + (-4)^2} = \sqrt{25 + 16} = \sqrt{41}$$

It makes no difference which point is chosen as (x_1, y_1) and which is chosen as (x_2, y_2), since the difference in the x-coordinates (and y-coordinates) is squared. We also obtain $\sqrt{41}$ if we set up the distance as

$$d = \sqrt{[3 - (-2)]^2 + [(-1) - (-5)]^2}$$

Another important quantity which is defined for a line is its **slope**. *The slope gives a measure of the direction of a line, and is defined as the vertical directed distance from one point to another on the same straight line, divided by the horizontal directed distance from the first point to the second.* Thus, using the letter m to represent slope, we have

$$m = \frac{y_2 - y_1}{x_2 - x_1} \tag{1-2}$$

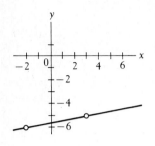

Figure 1–11

Example C
The slope of the line joining $(3, -5)$ and $(-2, -6)$ is

$$m = \frac{-6 - (-5)}{-2 - 3} = \frac{-6 + 5}{-5} = \frac{1}{5}$$

See Fig. 1–11. Again we may interpret either of the points as (x_1, y_1) and the other as (x_2, y_2). We can also obtain the slope of this same line from

$$m = \frac{-5 - (-6)}{3 - (-2)} = \frac{1}{5}$$

The larger the numerical value of the slope of a line, the more nearly vertical is the line. Also, a line rising to the right has a positive slope, and a line falling to the right has a negative slope.

Example D
The line through the two points in Example C has a positive slope, which is numerically small. From Fig. 1–11 it can be seen that the line rises slightly to the right.

The line joining $(3, 8)$ and $(4, -6)$ has a slope of -14. This line falls sharply to the right (see Fig. 1–12).

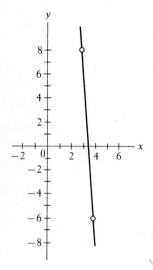

Figure 1–12

From the definition of slope, we may conclude that the slope of a line parallel to the y-axis cannot be defined (the x-coordinates of any two points on the line would be the same, which would then necessitate division by zero). This, however, does not prove to be of any trouble.

If a given line is extended indefinitely in either direction, it must cross the x-axis at some point unless it is parallel to the x-axis. *The angle measured from the x-axis in a positive direction to the line is called the **inclination** of the line* (see Fig. 1–13). The inclination of a line parallel to the x-axis is defined to be zero. An alternative definition of slope, in terms of the inclination, is

$$m = \tan \alpha, \qquad 0° \leq \alpha < 180° \tag{1-3}$$

where α is the inclination. This can be seen from the fact that the slope can be defined in terms of any two points on the line. Thus, if we choose as one of these points that point where the line crosses the x-axis and any other point, we see from the definition of the tangent of an angle that Eq. (1–3) is in agreement with Eq. (1–2).

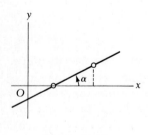

Figure 1–13

Example E
The slope of a line with an inclination of $45°$ is $m = \tan 45° = 1.000$. The slope of a line having an inclination of $120°$ is $m = \tan 120° = -\sqrt{3} = -1.732$. See Fig. 1–14.

Any two parallel lines crossing the x-axis will have the same inclination. Therefore, *the slopes of parallel lines are equal.* This can be stated as

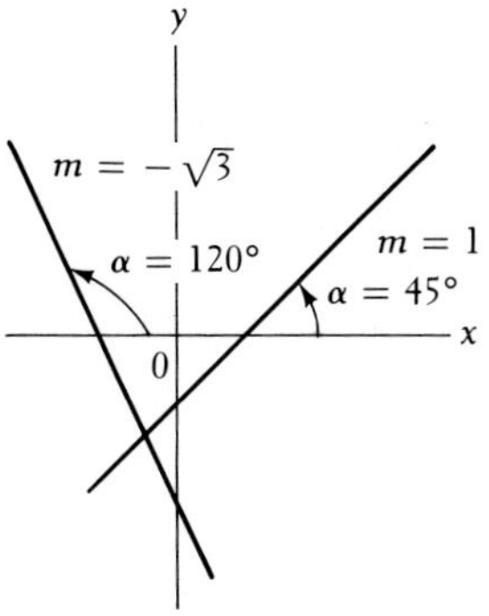

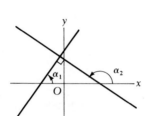

Figure 1–14

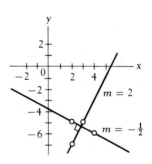

Figure 1–15

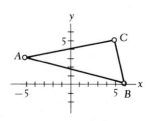

Figure 1–16

$$m_1 = m_2 \qquad \text{for } \| \text{ lines} \tag{1–4}$$

If two lines are perpendicular, this means that there must be 90° between their inclinations (Fig. 1–15). The relation between their inclinations is

$$\alpha_2 = \alpha_1 + 90°$$

which can be written as

$$90° - \alpha_2 = -\alpha_1$$

Taking the tangent of each of the angles in this last relation, we have

$$\tan (90° - \alpha_2) = \tan (-\alpha_1)$$

or

$$\cot \alpha_2 = -\tan \alpha_1$$

since a function of the complement of an angle is equal to the cofunction of that angle, and since $\tan (-\alpha) = -\tan \alpha$. But $\cot \alpha = 1/\tan \alpha$, which means that $1/\tan \alpha_2 = -\tan \alpha_1$. Using the inclination definition of slope, we may write, as the relation between slopes of perpendicular lines,

$$m_2 = -\frac{1}{m_1} \qquad \text{or} \qquad m_1 m_2 = -1 \qquad \text{for } \perp \text{ lines} \tag{1–5}$$

Example F

The line through $(3, -5)$ and $(2, -7)$ has a slope of 2. The line through $(4, -6)$ and $(2, -5)$ has a slope of $-\frac{1}{2}$. These lines are perpendicular. See Fig. 1–16.

Using the formulas for distance and slope, we can show certain basic geometric relations. The following examples illustrate the use of the formulas, and thus show the use of algebra in solving problems which are basically geometric. This illustrates the methods of analytic geometry.

Example G

Show that line segments joining $A(-5, 3)$, $B(6, 0)$, and $C(5, 5)$ form a right triangle (see Fig. 1–17).

If these points are vertices of a right triangle, the slopes of two of the sides must be negative reciprocals. This would show perpendicularity. Thus, we find the slopes of the three lines to be

$$m_{AB} = \frac{3 - 0}{-5 - 6} = -\frac{3}{11}, \qquad m_{AC} = \frac{3 - 5}{-5 - 5} = \frac{1}{5}, \qquad m_{BC} = \frac{0 - 5}{6 - 5} = -5$$

We see that the slopes of AC and BC are negative reciprocals, which means that $AC \perp BC$. From this we can conclude that the triangle is a right triangle.

Figure 1–17

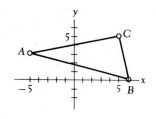

Figure 1–17

Example H
Find the area of the triangle in Example G.

Since the right angle is at C, the legs of the triangle are AC and BC. The area is one-half the product of the lengths of the legs of a right triangle. The lengths of the legs are

$$d_{AC} = \sqrt{(-5-5)^2 + (3-5)^2} = \sqrt{104} = 2\sqrt{26}$$

and

$$d_{BC} = \sqrt{(6-5)^2 + (0-5)^2} = \sqrt{26}$$

Therefore, the area is

$$A = \frac{1}{2}(2\sqrt{26})(\sqrt{26}) = 26$$

Exercises 1–4

In Exercises 1 through 8 find the distances between the given pairs of points.

1. $(3, 8)$ and $(-1, -2)$ 2. $(-1, 3)$ and $(-8, -4)$
3. $(4, -5)$ and $(4, -8)$ 4. $(-3, 7)$ and $(2, 10)$
5. $(-1, 0)$ and $(5, -7)$ 6. $(15, -1)$ and $(-11, 1)$
7. $(-4, -3)$ and $(3, -3)$ 8. $(-2, 5)$ and $(-2, -2)$

In Exercises 9 through 16 find the slopes of the lines through the points in Exercises 1 through 8.

In Exercises 17 through 20 find the slopes of the lines with the given inclinations.

17. $30°$ 18. $60°$ 19. $150°$ 20. $135°$

In Exercises 21 through 24 find the inclinations of the lines with the given slopes.

21. 0.3640 22. 0.8243 23. -6.691 24. -1.428

In Exercises 25 through 28 determine whether the lines through the two pairs of points are parallel or perpendicular.

25. $(6, -1)$ and $(4, 3)$, and $(-5, 2)$ and $(-7, 6)$
26. $(-3, 9)$ and $(4, 4)$, and $(9, -1)$ and $(4, -8)$
27. $(-1, -4)$ and $(2, 3)$, and $(-5, 2)$ and $(-19, 8)$
28. $(-1, -2)$ and $(3, 6)$, and $(2, -6)$ and $(5, 0)$

In Exercises 29 through 32 determine the value of k.

29. The distance between $(-1, 3)$ and $(11, k)$ is 13.
30. The distance between $(k, 0)$ and $(0, 2k)$ is 10.

31. The points $(6, -1)$, $(3, k)$, and $(-3, -7)$ are all on the same line.
32. The points in Exercise 31 are the vertices of a right triangle, with the right angle at $(3, k)$.

In Exercises 33 through 36 show that the given points are vertices of the indicated geometric figures.

33. Show that the points $(2, 3)$, $(4, 9)$, and $(-2, 7)$ are the vertices of an isosceles triangle.
34. Show that $(-1, 3)$, $(3, 5)$, and $(5, 1)$ are the vertices of a right triangle.
35. Show that $(3, 2)$, $(7, 3)$, $(-1, -3)$, and $(3, -2)$ are the vertices of a parallelogram.
36. Show that $(-5, 6)$, $(0, 8)$, $(-3, 1)$, and $(2, 3)$ are the vertices of a square.

In Exercises 37 and 38 find the indicated areas.

37. Find the area of the triangle of Exercise 34.
38. Find the area of the square of Exercise 36.

1–5 The Straight Line

Using the definition of slope, we can derive the equation which always represents a straight line. This is another basic method of analytic geometry. That is, equations of a particular form can be shown to represent a particular type of curve. When we recognize the form of the equation, we know the kind of curve it represents. This can be of great assistance in sketching the graph.

A straight line can be defined as a "curve" with constant slope. By this we mean that for any two different points chosen on a given line, if the slope is calculated, the same value is always found. Thus, if we consider one point (x_1, y_1) on a line to be fixed (Fig. 1–18), and another point $P(x, y)$ which can *represent* any other point on the line, we have

$$m = \frac{y - y_1}{x - x_1}$$

which can be written as

$$y - y_1 = m(x - x_1) \tag{1–6}$$

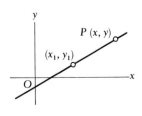

Figure 1–18

Equation (1–6) is known as the **point-slope form** of the equation of a straight line. It is useful when we know the slope of a line and some point through which the line passes. Direct substitution can then give us the equation of the line. Such information is often available about a given line.

Example A

Find the equation of the line which passes through $(-4, 1)$ with a slope of -2.

By using Eq. (1–6), we find that

$$y - 1 = (-2)(x + 4)$$

which can be simplified to

$$y + 2x + 7 = 0$$

This line is shown in Fig. 1–19.

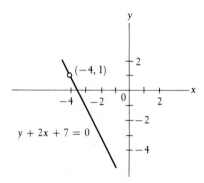

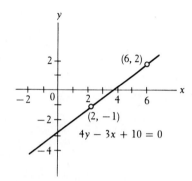

Figure 1–19   Figure 1–20

Example B

Find the equation of the line through $(2, -1)$ and $(6, 2)$.

We first find the slope of the line through these points:

$$m = \frac{2 + 1}{6 - 2} = \frac{3}{4}$$

Then, by using either of the two known points and Eq. (1–6), we can find the equation of the line:

$$y + 1 = \frac{3}{4}(x - 2)$$

or

$$4y + 4 = 3x - 6$$

or

$$4y - 3x + 10 = 0$$

This line is shown in Fig. 1–20.

Equation (1–6) can be used for any line except one parallel to the y-axis. Such a line has an undefined slope. However, it does have the property that all points have the same *x*-coordinate, regardless of the *y*-coordinate. We represent a line parallel to the *y*-axis as

$$x = a \qquad \text{(1–7)}$$

A line parallel to the *x*-axis has a slope of 0. From Eq. (1–6), we can find its equation to be $y = y_1$. To keep the same form as Eq. (1–7), we normally write this as

$$y = b \qquad \text{(1–8)}$$

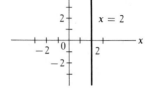

Figure 1–21

Example C
The line $x = 2$ is a line parallel to the *y*-axis and two units to the right of it. This line is shown in Fig. 1–21.

Example D
The line $y = -4$ is a line parallel to the *x*-axis and 4 units below it. This line is shown in Fig. 1–22.

If we choose the special point $(0, b)$, which is the *y*-intercept of the line, as the point to use in Eq. (1–6), we have

$$y - b = m(x - 0)$$

or

$$y = mx + b \qquad \text{(1–9)}$$

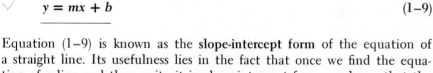

Figure 1–22

Equation (1–9) is known as the **slope-intercept form** of the equation of a straight line. Its usefulness lies in the fact that once we find the equation of a line and then write it in slope-intercept form, we know that the slope of the line is the coefficient of the *x*-term and that it crosses the *y*-axis with the coordinate indicated by the constant term.

Example E
Find the slope and the *y*-intercept of the straight line whose equation is $2y + 4x - 5 = 0$.

We write this equation in slope-intercept form:

$$2y = -4x + 5$$

$$y = -2x + \frac{5}{2}$$

Since the coefficient of *x* in this form is -2, the slope of the line is -2. The constant on the right is $\frac{5}{2}$, which means that the *y*-intercept is $\frac{5}{2}$.

From Eqs. (1–6) and (1–9), and from the examples of this section, we see that the equation of the straight line has certain characteristics: we have a term in y, a term in x, and a constant term if we simplify as much as possible. This form is represented by the equation

$$Ax + By + C = 0 \qquad (1–10)$$

which is known as the **general form** of the equation of the straight line. We have seen this form before in Section 1–3. Now we have shown why it represents a straight line.

Example F

What are the intercepts of the line $3x - 4y - 12 = 0$?

We know that the intercepts of a curve are those points where it crosses each of the axes. At the point where any curve crosses the x-axis, the y-coordinate must be 0. Thus, by letting $y = 0$ in the equation above, we solve for x and obtain $x = 4$. Hence one intercept is $(4, 0)$. By letting $x = 0$, we find the other intercept as $(0, -3)$.

In many physical situations a linear relationship exists between variables. A few examples of situations where such relationships exist are between (1) the distance traveled by an object and the elapsed time, when the velocity is constant, (2) the amount a spring stretches and the force applied, (3) the change in electric resistance and the change in temperature, (4) the force applied to an object and the resulting acceleration, and (5) the pressure at a certain point within a liquid and the depth of the point. The following example illustrates the use of a straight line in dealing with an applied problem.

Example G

Under the condition of constant acceleration, the velocity of an object varies linearly with the time. If after 1 s a certain object has a velocity of 40 ft/s, and 3 s later it has a velocity of 55 ft/s, find the equation relating the velocity and time, and graph this equation. From the graph determine the initial velocity (the velocity when $t = 0$) and the velocity after 6 s.

If we treat the velocity v as the dependent variable, and the time t as the independent variable, the slope of the straight line is

$$m = \frac{v_2 - v_1}{t_2 - t_1}$$

Using the information given in the problem, we have

$$m = \frac{55 - 40}{4 - 1} = 5$$

Then, using the point-slope form of the equation of a straight line, we have $v - 40 = 5(t - 1)$, or $v = 5t + 35$, which is the required equation (see Fig. 1–23). For purposes of graphing the line, the values given

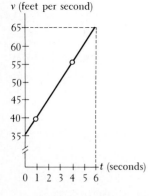

Figure 1–23

are sufficient. Of course, there is no need to include negative values of t, since these have no physical meaning. From the graph we see that the line crosses the v-axis at 35. This means that the initial velocity is 35 ft/s. Also, when $t = 6$ we see that $v = 65$ ft/s.

Exercises 1–5

In Exercises 1 through 16 find the equation of each of the lines with the given properties.

1. Passes through $(-3, 8)$ with a slope of 4
2. Passes through $(-2, -1)$ with a slope of -2
3. Passes through $(-2, -5)$ and $(4, 2)$
4. Passes through $(-3, -5)$ and $(-2, 3)$
5. Passes through $(1, 3)$ and has an inclination of $45°$
6. Has a y-intercept of -2 and an inclination of $120°$
7. Passes through $(6, -3)$ and is parallel to the x-axis
8. Passes through $(-4, -2)$ and is perpendicular to the x-axis
9. Is parallel to the y-axis and is 3 units to the left of it
10. Is parallel to the x-axis and is 5 units below it
11. Has an x-intercept of 4 and a y-intercept of -6
12. Has an x-intercept of -3 and a slope of 2
13. Is perpendicular to a line with a slope of 3 and passes through $(1, -2)$
14. Is perpendicular to a line with a slope of -4 and has a y-intercept of 3
15. Is perpendicular to the line joining $(4, 2)$ and $(3, -5)$ and passes through $(4, 2)$
16. Is parallel to the line $2y - 6x - 5 = 0$ and passes through $(-4, -5)$

In Exercises 17 through 20 draw the lines with the given equations.

17. $4x - y = 8$ 18. $2x - 3y - 6 = 0$
19. $3x + 5y - 10 = 0$ 20. $4y = 6x - 9$

In Exercises 21 through 24 reduce the equations to slope-intercept form and determine the slope and y-intercept.

21. $3x - 2y - 1 = 0$ 22. $4x + 2y - 5 = 0$
23. $5x - 2y + 5 = 0$ 24. $6x - 3y - 4 = 0$

In Exercises 25 through 28 determine the value of k.

25. What is the value of k if the lines $4x - ky = 6$ and $6x + 3y + 2 = 0$ are to be parallel?
26. What must k equal in Exercise 25, if the given lines are to be perpendicular?
27. What must be the value of k if the lines $3x - y = 9$ and $kx + 3y = 5$ are to be perpendicular?
28. What must k equal in Exercise 27, if the given lines are to be parallel?

In Exercises 29 and 30 show that the given lines are parallel. In Exercises 31 and 32 show that the given lines are perpendicular.

29. $3x - 2y + 5 = 0$ and $4y = 6x - 1$
30. $3y - 2x = 4$ and $6x - 9y = 5$

31. $6x - 3y - 2 = 0$ and $x + 2y - 4 = 0$

32. $4x - y + 2 = 0$ and $2x + 8y - 1 = 0$

In Exercises 33 through 36 find the equations of the given lines.

33. The line which has an x-intercept of 4 and a y-intercept the same as the line $2y - 3x - 4 = 0$

34. The line which is perpendicular to the line $8x + 2y - 3 = 0$ and has the same x-intercept as this line

35. The line with a slope of -3 which also passes through the intersection of the lines $5x - y = 6$ and $x + y = 12$

36. The line which passes through the point of intersection of $2x + y - 3 = 0$ and $x - y - 3 = 0$ and through the point $(4, -3)$

In Exercises 37 through 50 some applications and methods involving straight lines are shown.

37. The average velocity of an object is defined as the change in displacement s divided by the corresponding change in time t. Find the equation relating the displacement s and time t for an object for which the average velocity is 50 m/s and $s = 10$ m when $t = 0$ s.

38. The acceleration of an object is defined as the change in velocity v divided by the corresponding change in time t. Find the equation relating the velocity v and time t for an object for which the acceleration is 20 ft/s² and $v = 5.0$ ft/s when $t = 0$ s.

39. Within certain limits, the amount which a spring stretches varies linearly with the amount of force applied. If a spring whose natural length is 15 in. stretches 2.0 in. when 3.0 lb of force are applied, find the equation relating the length of the spring and the applied force.

40. The electric resistance of a certain resistor increases by 0.005 Ω for every increase of 1°C. Given that its resistance is 2.000 Ω at 0°C, find the equation relating the resistance and temperature. From the equation find the resistance when the temperature is 50°C.

41. The amount of heat required to raise the temperature of water varies linearly with the increase in temperature. However, a certain quantity of heat is required to change ice (at 0°C) into water without a change in temperature. An experiment is performed with 10 g of ice at 0°C. It is found that ice requires 4.19 kJ to change it into water at 20°C. Another 1.26 kJ is required to warm the water to 50°C. How many kilojoules are required to melt the ice at 0°C into water at 0°C?

42. The pressure at a point below the surface of a body of water varies linearly with the depth of the water. If the pressure at a depth of 10.0 m is 199 kPa and the pressure at a depth of 30.0 m is 395 kPa, what is the pressure at the surface? (This is the atmospheric pressure of the air above the water.)

43. A survey of the traffic on a particular highway showed that the number of cars passing a particular point each minute varied linearly from 6:30 AM to 8:30 AM on workday mornings. The study showed that an average of 45 cars passed the point in one minute at 7 AM, and that 115 cars passed in one minute at 8 AM. If n is the number of cars passing the point in one minute and t is the number of minutes after 6:30 AM, find the equation relating n and t, and graph the equation. From the graph, determine n at 6:30 AM and at 8:30 AM.

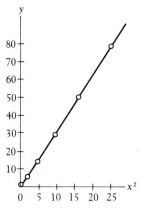

Figure 1–24

44. The total fixed cost for a company to operate a certain plant is \$400 per day. It also costs \$2 for each unit produced in the plant. Find the equation relating the total cost C of operating the plant and the number of units n produced each day. Graph the equation, assuming that $n \le 300$.

45. Sometimes it is desirable to treat a nonlinear function as a linear function. For example, if $y = 2 + 3x^2$, this can be plotted as a straight line by plotting y vs. x^2. A table of values for this graph is

x	0	1	2	3	4	5
x^2	0	1	4	9	16	25
y	2	5	14	29	50	77

and the graph appears in Fig. 1–24. Using this method, sketch the function $y = 1 + \sqrt{x}$. That is, plot y vs. $\sqrt{x}$.

46. Using the method of Exercise 45, sketch the function $C = 10/E$ as a linear function. This is the equation between the voltage across a capacitor and the capacitance, assuming the charge on the capacitor is constant.

47. Another method of treating a nonlinear function as a linear one is to use logarithms (see Section 7–1). Curves plotted on logarithmic and semi-logarithmic paper often come out as straight lines. Since distances corresponding to log y and log x are plotted automatically if the graph is logarithmic in the respective directions, the graphs of many nonlinear equations are linear when plotted on this paper. The graph of $y = ax^n$ is straight when plotted on a logarithmic paper, since log y = log a + n log x is in the form of a straight line. The variables are log y and log x; the slope can be found from (log y − log a)/log x = n, and the intercept is a. (To get the slope from the graph, it is necessary to measure vertical and horizontal distances between two points. The log y intercept is found where log $x = 0$, and this occurs when $x = 1$.) Plot $y = 3x^4$ on logarithmic paper to verify this analysis.

48. A function of the form $y = a(b^x)$ is a straight line on semilogarithmic paper, since log y = log a + x log b is in the form of a straight line. The variables are log y and x, the slope is log b, and the intercept is a. [To get the slope from the graph, we must calculate (log y − log a)/x for some set of values x and y. The intercept is read directly off the graph where $x = 0$.] Plot $y = 3(2^x)$ on semilogarithmic paper to verify this analysis.

49. If experimental data are plotted on logarithmic paper and the points lie on a straight line, it is possible to determine the function (see Exercise 47). The following data come from an experiment to determine the functional relationship between the tension in a string and the velocity of a wave in the string. From the graph on logarithmic paper, determine v as a function of T.

v (meters per second)	16.0	32.0	45.3	55.4	64.0
T (newtons)	0.100	0.400	0.800	1.20	1.60

50. If experimental data are plotted on semilogarithmic paper, and the points lie on a straight line, it is possible to determine the function (see Exercise 48). The following data come from an experiment designed to determine the relationship between the voltage across an inductor and the time, after the switch is opened. Determine v as a function of t.

v (volts)	40	15	5.6	2.2	0.8
t (milliseconds)	0.0	20	40	60	80

1–6 The Circle

We have found that we can obtain a general equation which represents a straight line by considering a fixed point on the line and then a general point $P(x, y)$ which can represent any other point on the same line. Mathematically we can state this as "the line is the **locus** of a point $P(x, y)$ which *moves* along the line." That is, the point $P(x, y)$ can be considered as a variable point which moves along the line.

In this way we can define a number of important curves. *The **circle** is defined as the locus of a point $P(x, y)$ which moves so that it is always equidistant from a fixed point. This fixed distance we call the **radius**, and the fixed point is called the **center** of the circle.* Thus, using this definition, calling the fixed point (h, k) and the radius r, we have

$$\sqrt{(x - h)^2 + (y - k)^2} = r$$

or, by squaring both sides, we have

$$(x - h)^2 + (y - k)^2 = r^2 \qquad (1\text{–}11)$$

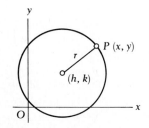

Figure 1–25

Equation (1–11) is called the **standard equation** of a circle with center at (h, k) and radius r (Fig. 1–25).

Example A

The equation $(x - 1)^2 + (y + 2)^2 = 16$ represents a circle with center at $(1, -2)$ and a radius of 4. We determine these values by considering the equation of this circle to be in the form of Eq. (1–11) as $(x - 1)^2 + (y - (-2))^2 = 4^2$. Note carefully the way in which we find the y-coordinate of the center. This circle is shown in Fig. 1–26.

If the center of the circle is at the origin, which means that the coordinates of the center are $(0, 0)$, the equation of the circle becomes

$$x^2 + y^2 = r^2 \qquad (1\text{–}12)$$

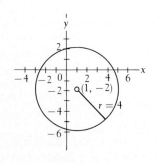

Figure 1–26

A circle of this type clearly exhibits the important property of symmetry, which we first discussed in Section 1–3. We can see that if we replace y by $-y$ the equation is unchanged, since $x^2 + (-y)^2 = r^2$ can be simplified to $x^2 + y^2 = r^2$. Also, the equation remains unchanged if we replace x by $-x$. Therefore, it is symmetrical to the x-axis and to the y-axis.

This type of circle is symmetrical to the origin as well as being symmetrical to both axes. The meaning of symmetry to the origin is that the origin is the midpoint of any two points (x, y) and $(-x, -y)$ which are

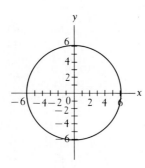

Figure 1–27

on the curve. Thus, if *−x can replace x, and −y replace y at the same time, without changing the equation, the graph of the equation is symmetrical to the origin.*

Example B

The equation of the circle with its center at the origin and with a radius of 6 is $x^2 + y^2 = 36$.

The symmetry of this circle can be shown analytically by the substitutions mentioned above. Replacing x by $−x$, we obtain $(−x)^2 + y^2 = 36$. Since $(−x)^2 = x^2$, this equation can be rewritten as $x^2 + y^2 = 36$. Since this substitution did not change the equation, the graph is symmetrical to the y-axis.

Replacing y by $−y$, we obtain $x^2 + (−y)^2 = 36$, which is the same as $x^2 + y^2 = 36$. This means that the curve is symmetrical to the x-axis.

Replacing x by $−x$, and simultaneously replacing y by $−y$, we obtain $(−x)^2 + (−y)^2 = 36$, which is the same as $x^2 + y^2 = 36$. This means that the curve is symmetrical to the origin. The circle is shown in Fig. 1–27.

Example C

Find the equation of the circle with center at (2, 1) and which passes through (4, 8).

In Eq. (1–11), we can determine the equation if we can find h, k, and r for this circle. From the given information, $h = 2$ and $k = 1$. To find r, we use the fact that *all points on the circle must satisfy the equation of the circle.* The point (4, 8) must satisfy Eq. (1–11), with $h = 2$ and $k = 1$. Thus, $(4 − 2)^2 + (8 − 1)^2 = r^2$. From this relation we find $r^2 = 53$. The equation of the circle is

$$(x − 2)^2 + (y − 1)^2 = 53$$

This circle is shown in Fig. 1–28.

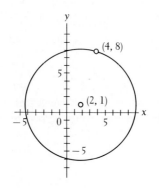

Figure 1–28

If we multiply out each of the terms in Eq. (1–11), we may combine the resulting terms to obtain

$$x^2 − 2hx + h^2 + y^2 − 2ky + k^2 = r^2$$
$$x^2 + y^2 − 2hx − 2ky + (h^2 + k^2 − r^2) = 0 \qquad (1–13)$$

Since each of h, k, and r is constant for any given circle, the coefficients of x and y and the term within parentheses in Eq. (1–13) are constants. Equation (1–13) can then be written as

$$x^2 + y^2 + Dx + Ey + F = 0 \qquad (1–14)$$

Equation (1–14) is called the **general equation** of the circle. It tells us that any equation which can be written in that form will represent a circle.

Example D

Find the center and radius of the circle

$$x^2 + y^2 - 6x + 8y - 24 = 0$$

We can find this information if we write the given equation in standard form. To do so, we must complete the square in the x-terms and also in the y-terms. This is done by first writing the equation in the form

$$(x^2 - 6x \quad) + (y^2 + 8y \quad) = 24$$

To complete the square of the x-terms, we take half of 6, which is 3, square it, and add the result, 9, to each side of the equation. In the same way, we complete the square of the y-terms by adding 16 to each side of the equation, which gives

$$(x^2 - 6x + 9) + (y^2 + 8y + 16) = 24 + 9 + 16$$
$$(x - 3)^2 + (y + 4)^2 = 49$$

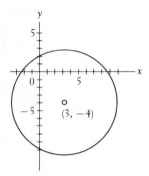

Figure 1—29

Thus, the center is $(3, -4)$, and the radius is 7 (see Fig. 1–29).

Example E

The equation $3x^2 + 3y^2 + 6x - 20 = 0$ can be seen to represent a circle by writing it in general form. This is done by dividing through by 3. In this way we have

$$x^2 + y^2 + 2x - \frac{20}{3} = 0$$

To determine the center and radius for this circle, we write the equation in standard form and then complete the necessary squares. This leads to

$$(x^2 + 2x + 1) + y^2 = \frac{20}{3} + 1$$

$$(x + 1)^2 + y^2 = \frac{23}{3}$$

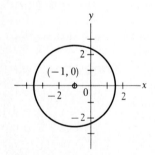

Figure 1—30

Thus, the center is $(-1, 0)$; the radius is $\sqrt{23/3} = 2.77$ (see Fig. 1–30). We can also see that this circle is symmetrical to the x-axis, but it is not symmetrical to the y-axis or to the origin. If we replace y by $-y$, the equation does not change, but if we replace x by $-x$, the 6x term in the original equation becomes negative, and the equation *does* change.

Exercises 1—6

In Exercises 1 through 4 determine the center and radius of each circle.

1. $(x - 2)^2 + (y - 1)^2 = 25$ 2. $(x - 3)^2 + (y + 4)^2 = 49$
3. $(x + 1)^2 + y^2 = 4$ 4. $x^2 + (y + 6)^2 = 64$

In Exercises 5 through 16 find the equation of each of the circles from the given information.

5. Center at $(0, 0)$, radius 3   6. Center at $(0, 0)$, radius 1
7. Center at $(2, 2)$, radius 4   8. Center at $(0, 2)$, radius 2

9. Center at $(-2, 5)$, radius $\sqrt{5}$ 10. Center at $(-3, -5)$, radius 8

11. Center at $(2, 1)$, passes through $(4, -1)$

12. Center at $(-1, 4)$, passes through $(-2, 3)$

13. Center $(-3, 5)$, tangent to the x-axis

14. Tangent to both axes, radius 4, and in the second quadrant

15. Center on the line $5x = 2y$, radius 5, tangent to the x-axis

16. The points $(3, 8)$ and $(-3, 0)$ are the ends of a diameter.

In Exercises 17 through 24 determine the center and radius of each circle. Sketch each circle.

17. $x^2 + y^2 - 25 = 0$ 18. $x^2 + y^2 - 9 = 0$

19. $x^2 + y^2 - 2x - 8 = 0$ 20. $x^2 + y^2 - 4x - 6y - 12 = 0$

21. $x^2 + y^2 + 8x - 10y - 8 = 0$ 22. $x^2 + y^2 + 8x + 6y = 0$

23. $2x^2 + 2y^2 - 4x - 8y - 1 = 0$ 24. $3x^2 + 3y^2 - 12x + 4 = 0$

In Exercises 25 through 28 determine whether the circles with the given equations are symmetrical to either axis or the origin.

25. $x^2 + y^2 = 100$ 26. $x^2 + y^2 - 4x - 5 = 0$

27. $x^2 + y^2 + 8y - 9 = 0$ 28. $x^2 + y^2 - 2x + 4y - 3 = 0$

In Exercises 29 through 36 find the indicated quantities.

29. Determine where the circle $x^2 - 6x + y^2 - 7 = 0$ crosses the x-axis.

30. Find the points of intersection of the circle $x^2 + y^2 - x - 3y = 0$ and the line $y = x - 1$.

31. Find the locus of a point $P(x, y)$ which moves so that its distance from $(2, 4)$ is twice its distance from $(0, 0)$. Describe the locus.

32. Find the equation of the locus of a point $P(x, y)$ which moves so that the line joining it and $(2, 0)$ is always perpendicular to the line joining it and $(-2, 0)$. Describe the locus.

33. If an electrically charged particle enters a magnetic field of flux density B with a velocity v at right angles to the field, the path of the particle is a circle. The radius of the path is given by $R = mv/Bq$, where m is the mass of the particle and q is its charge. If a proton ($m = 1.67 \times 10^{-27}$ kg, $q = 1.60 \times 10^{-19}$ C) enters a magnetic field of flux density 1.50 T (tesla: $V \cdot s/m^2$) with a velocity of 2.40×10^8 m/s, find the equation of the path of the proton. (This is important for the construction of a cyclotron.)

34. The relationship of the impedance Z, resistance R, capacitive reactance X_C, and inductive reactance X_L in an electric circuit is $Z = \sqrt{(X_L - X_C)^2 + R^2}$. Sketch the graph of R vs. X_L if Z and X_C are constant.

35. A cylindrical oil tank, 4.00 ft in diameter, is on its side. A circular hole, with center 18.0 in. below the center of the end of the tank, has been drilled in the end of the tank. If the radius of the hole is 3.00 in., what is the equation of the circle of the end of the tank and the equation of the circle of the hole? Use the center of the end of the tank as the origin.

36. A draftsman is drawing a friction drive in which two circular disks are in contact with each other. They are represented by circles in his drawing. The first has a radius of 10.0 cm and the second has a radius of 12.0 cm. What is the equation of each circle if the origin is at the center of the first circle and the x-axis passes through the center of the second circle?

1—7 The Parabola

Another important curve is the parabola. We have come across this curve several times in the past. In this section we shall find the form of the equation of the parabola and see that this is a familiar form.

The **parabola** *is defined as the locus of a point* $P(x, y)$ *which moves so that it is always equidistant from a given line and a given point. The given line is called the* **directrix**, *and the given point is called the* **focus.** The line through the focus which is perpendicular to the directrix is called the **axis** of the parabola. The point midway between the directrix and focus is the **vertex** of the parabola. Using the definition, we shall find the equation of the parabola for which the focus is the point $(p, 0)$ and the directrix is the line $x = -p$. By choosing the focus and directrix in this manner, we shall be able to find a general representation of the equation of a parabola with its vertex at the origin.

According to the definition of the parabola, the distance from a point $P(x, y)$ on the parabola to the focus $(p, 0)$ must equal the distance from $P(x, y)$ to the directrix $x = -p$. The distance from P to the focus can be found by use of the distance formula. The distance from P to the directrix is the perpendicular distance, and this can be found as the distance between two points on a line parallel to the x-axis. These distances are indicated in Fig. 1–31.

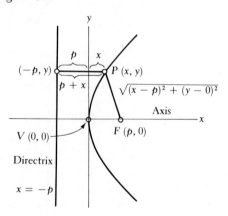

Figure 1–31

Thus, we have

$$\sqrt{(x - p)^2 + (y - 0)^2} = x + p$$

Squaring both sides of this equation, we have

$$(x - p)^2 + y^2 = (x + p)^2$$

or

$$x^2 - 2px + p^2 + y^2 = x^2 + 2px + p^2$$

Simplifying, we obtain

$$y^2 = 4px \qquad\qquad (1\text{--}15)$$

Equation (1–15) is called the **standard form** of the equation of a parabola with its axis along the x-axis and the vertex at the origin. Its symmetry to the x-axis can be proven since $(-y)^2 = 4px$ is the same as $y^2 = 4px$.

Example A
Find the coordinates of the focus, the equation of the directrix, and sketch the graph of the parabola $y^2 = 12x$.

From the form of the equation, we know that the vertex is at the origin (Fig. 1–32). The coefficient 12 tells us that the focus is (3, 0), since $p = \frac{12}{4}$ [note that the coefficient of x in Eq. (1–15) is $4p$]. Also, this means that the directrix is the line $x = -3$.

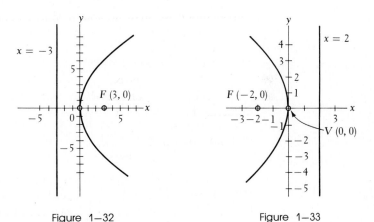

Figure 1–32   Figure 1–33

Example B
If the focus is to the left of the origin, with the directrix an equal distance to the right, the coefficient of the x-term will be negative. This tells us that the parabola opens to the left, rather than to the right, as is the case when the focus is to the right of the origin. For example, the parabola $y^2 = -8x$ has its vertex at the origin, its focus at $(-2, 0)$, and the line $x = 2$ as its directrix. This is consistent with Eq. (1–15), in that $4p = -8$, or $p = -2$. The parabola opens to the left as shown in Fig. 1–33.

If we chose the focus as the point $(0, p)$ and the directrix as the line $y = -p$, we would find that the resulting equation is

$$x^2 = 4py \tag{1-16}$$

This is the standard form of the equation of a parabola with the y-axis as its axis and the vertex at the origin. Its symmetry to the y-axis can be proven since $(-x)^2 = 4py$ is the same as $x^2 = 4py$. We note that the difference between this equation and Eq. (1–15) is that x is squared and y appears to the first power in Eq. (1–16), rather than the reverse, as in Eq. (1–15).

Example C

The parabola $x^2 = 4y$ has its vertex at the origin, focus at the point $(0, 1)$, and its directrix the line $y = -1$. The graph is shown in Fig. 1–34, and we see in this case that the parabola opens up.

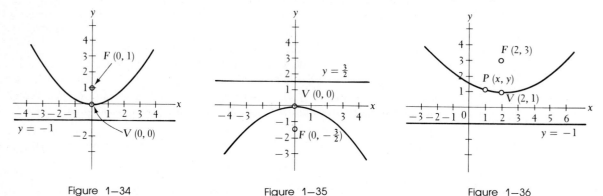

Figure 1–34   Figure 1–35   Figure 1–36

Example D

The parabola $x^2 = -6y$ has its vertex at the origin, its focus at the point $(0, -\frac{3}{2})$, and its directrix the line $y = \frac{3}{2}$. This parabola opens down, as Fig. 1–35 shows.

Equations (1–15) and (1–16) give us the general form of the equation of a parabola with its vertex at the origin and its focus on one of the coordinate axes. The following example illustrates the use of the definition of the parabola to find the equation of a parabola which has its vertex at a point other than at the origin.

Example E

Using the definition of the parabola, determine the equation of the parabola with its focus at $(2, 3)$ and its directrix the line $y = -1$ (see Fig. 1–36).

Choosing a general point $P(x, y)$ on the parabola, and equating the distances from this point to $(2, 3)$ and to the line $y = -1$, we have

$$\sqrt{(x - 2)^2 + (y - 3)^2} = y + 1$$

Squaring both sides of this equation and simplifying, we have

$$(x - 2)^2 + (y - 3)^2 = (y + 1)^2$$
$$x^2 - 4x + 4 + y^2 - 6y + 9 = y^2 + 2y + 1$$

or

$$8y = 12 - 4x + x^2$$

When we see it in this form, we note that this type of equation appeared numerous times earlier in Section 1–3. The x-term and the constant (12 in this case) are characteristic of a parabola which does not have its vertex at the origin if the directrix is parallel to the x-axis.

Thus we can conclude that a parabola is characterized by the presence of the square of either (but not both) x or y, and a first-power term in the other. In this way we can recognize a parabola by inspecting the equation.

Exercises 1—7

In Exercises 1 through 12 determine the coordinates of the focus and the equation of the directrix of each of the given parabolas. Sketch each curve.

1. $y^2 = 4x$ 2. $y^2 = 16x$ 3. $y^2 = -4x$ 4. $y^2 = -16x$

5. $x^2 = 8y$ 6. $x^2 = 10y$ 7. $x^2 = -4y$ 8. $x^2 = -12y$

9. $y^2 = 2x$ 10. $x^2 = 14y$ 11. $y = x^2$ 12. $x = 3y^2$

In Exercises 13 through 20 find the equations of the parabolas satisfying the given conditions.

13. Focus $(3, 0)$, directrix $x = -3$ 14. Focus $(-2, 0)$, directrix $x = 2$

15. Focus $(0, 4)$, vertex $(0, 0)$ 16. Focus $(-3, 0)$, vertex $(0, 0)$

17. Vertex $(0, 0)$, directrix $y = -1$ 18. Vertex $(0, 0)$, directrix $y = \frac{1}{2}$

19. Vertex at $(0, 0)$, axis along the y-axis, passes through $(-1, 8)$

20. Vertex at $(0, 0)$, axis along the x-axis, passes through $(2, -1)$

In Exercises 21 through 24 use the definition of the parabola to find the equations of the parabolas satisfying the given conditions. Sketch each curve.

21. Focus $(6, 1)$, directrix $x = 0$ 22. Focus $(1, -4)$, directrix $x = 2$

23. Focus $(1, 1)$, vertex $(1, 3)$ 24. Vertex $(-2, -4)$, directrix $x = 3$

In Exercises 25 through 32 find the indicated quantities.

25. In calculus it can be shown that a light ray emanating from the focus of a parabola will be reflected off parallel to the axis of the parabola. Suppose that a light ray from the focus of a parabolic reflector described by the equation $y^2 = 8x$ strikes the reflector at the point $(2, 4)$. Along what line does the incident ray move, and along what line does the reflected ray move?

26. The rate of development of heat H (measured in watts) in a resistor of resistance R (measured in ohms) of an electric circuit is given by $H = Ri^2$, where i is the current (measured in amperes) in the resistor. Sketch the graph of H versus i, if $R = 6.0 \, \Omega$.

27. Under certain conditions, a cable which hangs between two supports can be closely approximated as being parabolic. Assuming that this cable hangs in the shape of a parabola, find its equation if a point 10 ft horizontally from its lowest point is 1.0 ft above its lowest point. Choose the lowest point as the origin of the coordinate system.

28. A wire is fastened 36.0 ft up on each of two telephone poles which are 200 ft apart. Halfway between the poles the wire is 30.0 ft above the ground. Assuming the wire is parabolic, find the height of the wire 50.0 ft from either pole.

29. The period T (measured in seconds) of the oscillation for resonance in an electric circuit is given by $T = 2\pi\sqrt{LC}$, where L is the inductance (in henrys) and C is the capacitance (in farads). Sketch the graph of the period versus capacitance for a constant inductance of 1 H. Assume values of C from 1 μF to 250 μF for this is a common range of values.

30. The velocity v of a jet of water flowing from an opening in the side of a certain container is given by $v = 8\sqrt{h}$, where h is the depth of the opening. Sketch a graph of v (in feet per second) vs. h (in feet).

31. A small island is 4 km from a straight shoreline. A ship channel is equidistant between the island and the shoreline. Write an equation for the channel.

32. Under certain circumstances, the maximum power P in an electric circuit varies as the square of the voltage of the source E_0 and inversely as the internal resistance R_i of the source. If 10 W is the maximum power for a source of 2.0 V and internal resistance of 0.10 Ω, sketch the graph of P versus E_0, if R_i remains constant.

1–8 The Ellipse

The next important curve we shall discuss is the ellipse. *The ellipse is defined as the locus of a point P(x, y) which moves so that the sum of its distances from two fixed points is constant.* These two fixed points are called the **foci** of the ellipse. Letting this fixed sum equal $2a$ and the foci be the points $(c, 0)$ and $(-c, 0)$, we have (see Fig. 1–37) from the definition of the ellipse

$$\sqrt{(x - c)^2 + y^2} + \sqrt{(x + c)^2 + y^2} = 2a$$

When solving equations involving radicals, we find that we should remove one of the radicals to the right side, and then square each side. Thus we have the following steps:

$$\sqrt{(x + c)^2 + y^2} = 2a - \sqrt{(x - c)^2 + y^2}$$
$$(x + c)^2 + y^2 = 4a^2 - 4a\sqrt{(x - c)^2 + y^2} + (\sqrt{(x - c)^2 + y^2})^2$$
$$x^2 + 2cx + c^2 + y^2 = 4a^2 - 4a\sqrt{(x - c)^2 + y^2} + x^2 - 2cx + c^2 + y^2$$
$$4a\sqrt{(x - c)^2 + y^2} = 4a^2 - 4cx$$
$$a\sqrt{(x - c)^2 + y^2} = a^2 - cx$$
$$a^2(x^2 - 2cx + c^2 + y^2) = a^4 - 2a^2cx + c^2x^2$$
$$(a^2 - c^2)x^2 + a^2y^2 = a^2(a^2 - c^2)$$

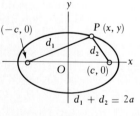

Figure 1–37

At this point we define (and the usefulness of this definition will be shown presently) $a^2 - c^2 = b^2$. This substitution gives us

$$b^2x^2 + a^2y^2 = a^2b^2$$

Dividing through by a^2b^2, we have

$$\frac{x^2}{a^2} + \frac{y^2}{b^2} = 1 \qquad (1\text{--}17)$$

If we let $y = 0$, we find that the x-intercepts are $(-a, 0)$ and $(a, 0)$. We see that the distance $2a$, originally chosen as the sum of the distances in the derivation of Eq. (1–17), is the distance between the x-intercepts. *These two points are known as the* **vertices** *of the ellipse, and the line between them is known as the* **major axis** *of the ellipse. Thus, a is called the* **semi-major axis.**

If we now let $x = 0$, we find the y-intercepts to be $(0, -b)$ and $(0, b)$. *The line joining these two intercepts is called the* **minor axis** *of the ellipse* (Fig. 1–38) *which means b is called the* **semi-minor axis.** The point $(0, b)$ is on the ellipse, and is also equidistant from $(-c, 0)$ and $(c, 0)$. Since the sum of the distances from these points to $(0, b)$ equals $2a$, the distance from $(c, 0)$ to $(0, b)$ must be a. Thus, we have a right triangle formed by line segments of lengths a, b, and c, with a as the hypotenuse. From this we have the relation

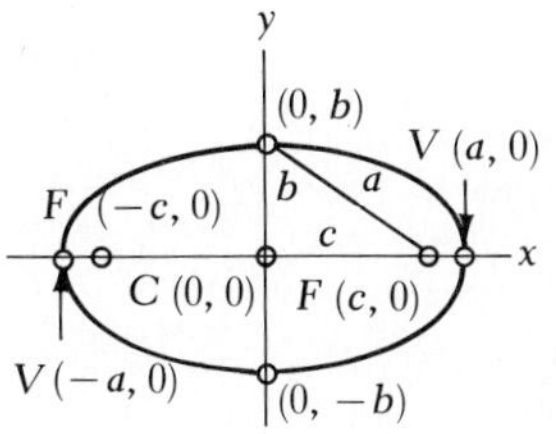

Figure 1–38

$$a^2 = b^2 + c^2 \qquad (1\text{--}18)$$

between the distances a, b, and c. This equation also shows us why b was defined as it was in the derivation.

Equation (1–17) is called the **standard equation** of the ellipse with its major axis along the x-axis and its center at the origin.

If we choose points on the y-axis as the foci, the standard equation of the ellipse, with its center at the origin and its major axis along the y-axis, is

$$\frac{y^2}{a^2} + \frac{x^2}{b^2} = 1 \qquad (1\text{--}19)$$

In this case the vertices are $(0, a)$ and $(0, -a)$, the foci are $(0, c)$ and $(0, -c)$, and the ends of the minor axis are $(b, 0)$ and $(-b, 0)$.

The symmetry of the ellipses given by Eqs. (1–17) and (1–19) to each of the axes and the origin can be proven since each of x and y can be replaced by its negative in these equations without changing the equations.

Example A
The ellipse

$$\frac{x^2}{25} + \frac{y^2}{9} = 1$$

has vertices at $(5, 0)$ and $(-5, 0)$. Its minor axis extends from $(0, 3)$ to $(0, -3)$, as we see in Fig. 1–39. This information is directly obtainable from this form of the equation, since the 25 tells us that $a^2 = 25$, or $a = 5$. In the same way we have $b = 3$. Since we know both a^2 and b^2, we can find c^2 from the relation $c^2 = a^2 - b^2$. Thus, $c^2 = 16$, or $c = 4$. This in turn tells us that the foci of this ellipse are at $(4, 0)$ and $(-4, 0)$.

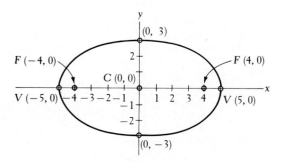

Figure 1–39

Example B
The ellipse

$$\frac{x^2}{4} + \frac{y^2}{9} = 1$$

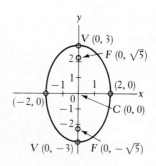

Figure 1–40

has vertices at $(0, 3)$ and $(0, -3)$. The minor axis extends from $(2, 0)$ to $(-2, 0)$. This we can find directly from the equation, since $a^2 = 9$ and $b^2 = 4$. One might ask how we chose $a^2 = 9$ in this example and $a^2 = 25$ in the preceding example. Since $a^2 = b^2 + c^2$, a is always larger than b. Thus, we can tell which axis (x or y) the major axis is along by seeing which number (in the denominator) is larger, when the equation is written in standard form. The larger one stands for a^2. In this example, the foci are $(0, \sqrt{5})$ and $(0, -\sqrt{5})$ (see Fig. 1–40).

Example C
Find the coordinates of the vertices, the ends of the minor axis, and the foci of the ellipse $4x^2 + 16y^2 = 64$.

This equation must be put in standard form first, which we do by dividing each term by 64. When this is done, we obtain

$$\frac{x^2}{16} + \frac{y^2}{4} = 1$$

Thus, $a = 4$, $b = 2$, and $c = 2\sqrt{3}$. The vertices are at $(4, 0)$ and $(-4, 0)$. The ends of the minor axis are $(0, 2)$ and $(0, -2)$, and the foci are $(2\sqrt{3}, 0)$ and $(-2\sqrt{3}, 0)$ (see Fig. 1–41).

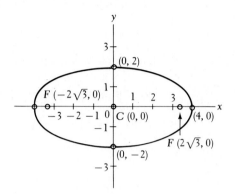

Figure 1–41

Example D

Find the equation of the ellipse for which the center is at the origin, the major axis is along the y-axis, the minor axis is 6 units long, and there are 8 units between foci.

Directly we know that $2b = 6$, or $b = 3$. Also $2c = 8$, or $c = 4$. Thus, we find that $a = 5$. Since the major axis is along the y-axis, we have the equation

$$\frac{y^2}{25} + \frac{x^2}{9} = 1$$

This ellipse is shown in Fig. 1–42.

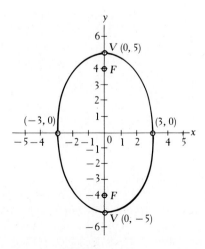

Figure 1–42

Equations (1–17) and (1–19) give us the standard form of the equation of an ellipse with its center at the origin and its foci on one of the coordinate axes. The following example illustrates the use of the definition of the ellipse to find the equation of an ellipse with its center at a point other than the origin.

Example E

Using the definition, find the equation of the ellipse with foci at $(1, 3)$ and $(9, 3)$, with a major axis of 10.

Using the same method as in the derivation of Eq. (1–17), we have the following steps. (Remember, the sum of distances in the definition equals the length of the major axis.)

$$\sqrt{(x - 1)^2 + (y - 3)^2} + \sqrt{(x - 9)^2 + (y - 3)^2} = 10$$
$$\sqrt{(x - 1)^2 + (y - 3)^2} = 10 - \sqrt{(x - 9)^2 + (y - 3)^2}$$
$$(x - 1)^2 + (y - 3)^2 = 100 - 20\sqrt{(x - 9)^2 + (y - 3)^2} + (\sqrt{(x - 9)^2 + (y - 3)^2})^2$$
$$x^2 - 2x + 1 + y^2 - 6y + 9 = 100 - 20\sqrt{(x - 9)^2 + (y - 3)^2} + x^2 - 18x + 81 + y^2 - 6y + 9$$
$$20\sqrt{(x - 9)^2 + (y - 3)^2} = 180 - 16x$$
$$5\sqrt{(x - 9)^2 + (y - 3)^2} = 45 - 4x$$
$$25(x^2 - 18x + 81 + y^2 - 6y + 9) = 2025 - 360x + 16x^2$$
$$25x^2 - 450x + 2250 + 25y^2 - 150y = 2025 - 360x + 16x^2$$
$$9x^2 - 90x + 25y^2 - 150y + 225 = 0$$

The additional x- and y-terms are characteristic of the equation of an ellipse whose center is not at the origin (see Fig. 1–43).

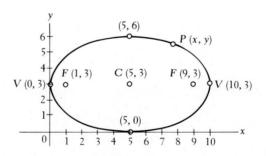

Figure 1–43

We can conclude that the equation of an ellipse is characterized by the presence of both an x^2- and a y^2-term, having different coefficients (in value but not in sign). The difference between the equation of an ellipse and that of a circle is that the coefficients of the squared terms in the equation of the circle are the same, whereas those of the ellipse differ.

Exercises 1–8

In Exercises 1 through 12 find the coordinates of the vertices and foci of the given ellipses. Sketch each curve.

1. $\dfrac{x^2}{4} + \dfrac{y^2}{1} = 1$

2. $\dfrac{x^2}{100} + \dfrac{y^2}{64} = 1$

3. $\dfrac{x^2}{25} + \dfrac{y^2}{36} = 1$

4. $\dfrac{x^2}{49} + \dfrac{y^2}{81} = 1$

5. $4x^2 + 9y^2 = 36$

6. $x^2 + 36y^2 = 144$

7. $49x^2 + 4y^2 = 196$

8. $25x^2 + y^2 = 25$

9. $8x^2 + y^2 = 16$

10. $2x^2 + 3y^2 = 6$

11. $4x^2 + 25y^2 = 25$

12. $9x^2 + 4y^2 = 9$

In Exercises 13 through 20 find the equations of the ellipses satisfying the given conditions. The center of each is at the origin.

13. Vertex $(15, 0)$, focus $(9, 0)$

14. Minor axis 8, vertex $(0, -5)$

15. Focus $(0, 2)$, major axis 6

16. Semi-minor axis 2, focus $(3, 0)$

17. Vertex $(8, 0)$, passes through $(2, 3)$

18. Focus $(0, 2)$, passes through $(-1, \sqrt{3})$

19. Passes through $(2, 2)$ and $(1, 4)$

20. Passes through $(-2, 2)$ and $(1, \sqrt{6})$

In Exercises 21 through 24 find the equations of the ellipses with the given properties by use of the definition of an ellipse.

21. Foci at $(-2, 1)$ and $(4, 1)$, a major axis of 10

22. Foci at $(-3, -2)$ and $(-3, 8)$, major axis 26

23. Vertices at $(1, 5)$ and $(1, -1)$, foci at $(1, 4)$ and $(1, 0)$

24. Vertices at $(-2, 1)$ and $(-2, 5)$, foci at $(-2, 2)$ and $(-2, 4)$

In Exercises 25 through 32 solve the given problems.

25. Show that the ellipse $2x^2 + 3y^2 - 8x - 4 = 0$ is symmetrical to the x-axis.

26. Show that the ellipse $5x^2 + y^2 - 3y - 7 = 0$ is symmetrical to the y-axis.

27. The arch of a bridge across a stream is in the form of half an ellipse above the water level. If the span of the arch at water level is 100 ft and the maximum height of the arch above water level is 30 ft, what is the equation of the arch? Choose the origin of the coordinate system at the most convenient point.

28. An elliptical gear (Fig. 1–44) which rotates about its center is kept continually in mesh with a circular gear which is free to move horizontally. If the equation of the ellipse of the gear (with the origin of the coordinate system at its center) in its present position is $3x^2 + 7y^2 = 20$, how far does the center of the circular gear move going from one extreme position to the other? (Assume the units are centimeters.)

Figure 1–44

29. An artificial satellite of the earth has a minimum altitude of 500 mi and a maximum altitude of 2000 mi. If the path of the satellite about the earth is an ellipse with the center of the earth at one focus, what is the equation of its path? (Assume the radius of the earth is 4000 mi.)

30. An electric current is caused to flow in a loop of wire rotating in a magnetic field. In a study of this phenomenon, a piece of wire is cut into two pieces,

one of which is bent into a circle and the other into a square. If the sum of areas of the circle and square is always π units, find the relation between the radius r of the circle and the side x of the square. Sketch the graph of r versus x.

31. A vertical pipe 6.0 in. in diameter is to pass through a roof inclined at 45°. What are the dimensions of the elliptical hole which must be cut in the roof for the pipe?

32. The ends of a horizontal tank 20.0 ft long are ellipses, which can be described by the equation $9x^2 + 20y^2 = 180$, where x and y are measured in feet. The area of an ellipse is $A = \pi ab$. Find the volume of the tank.

1–9 The Hyperbola

The final curve we shall discuss in detail is the hyperbola. *The **hyperbola** is defined as the locus of a point $P(x, y)$ which moves so that the difference of the distances from two fixed points is a constant. These fixed points are the **foci** of the hyperbola.* Assuming that the foci of the hyperbola are the points $(c, 0)$ and $(-c, 0)$ (see Fig. 1–45) and the constant difference is $2a$, we have

$$\sqrt{(x + c)^2 + y^2} - \sqrt{(x - c)^2 + y^2} = 2a$$

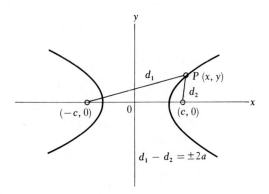

Figure 1–45

Following this same procedure as in the preceding section, we find the equation of the hyperbola to be

$$\frac{x^2}{a^2} - \frac{y^2}{b^2} = 1 \qquad \text{sideways} \tag{1–20}$$

When we derive this equation, we have a definition of the relation between a, b, and c which is different from that for the ellipse. This relation is

$$c^2 = a^2 + b^2 \tag{1–21}$$

An analysis of Eq. (1–20) will reveal the significance of *a* and *b*. First, by letting $y = 0$, we find that the *x*-intercepts are $(a, 0)$ and $(-a, 0)$, just as they are for the ellipse. These points are called the **vertices** of the hyperbola. By letting $x = 0$, we find that we have imaginary solutions for *y*, which means that there are no points on the curve which correspond to a value of $x = 0$.

To find the significance of *b*, we shall solve Eq. (1–20) for *y* in a particular form. First we have

$$\frac{y^2}{b^2} = \frac{x^2}{a^2} - 1$$

$$= \frac{x^2}{a^2} - \frac{a^2 x^2}{a^2 x^2}$$

$$= \frac{x^2}{a^2}\left(1 - \frac{a^2}{x^2}\right)$$

Multiplying through by b^2 and then taking the square root of each side, we have

$$y^2 = \frac{b^2 x^2}{a^2}\left(1 - \frac{a^2}{x^2}\right)$$

$$y = \pm\frac{bx}{a}\sqrt{1 - \frac{a^2}{x^2}} \tag{1–22}$$

We note that, if large values of *x* are assumed in Eq. (1–22), the quantity under the radical becomes approximately 1. In fact, the larger *x* becomes, the nearer 1 this expression becomes, since the x^2 in the denominator of a^2/x^2 makes this term nearly zero. Thus, for large values of *x*, Eq. (1–22) is approximately

$$y = \pm\frac{bx}{a} \tag{1–23}$$

Equation (1–23) can be seen to represent the equations for two straight lines, each of which passes through the origin. One has a slope of b/a and the other a slope of $-b/a$. *These lines are the* **asymptotes of the hyperbola.** The concept of an asymptote was introduced in Section 1–3. The fact that the hyperbola *approaches* the asymptote as *x* becomes large without bound (*approaches* infinity) is designated by

$$y \to \frac{bx}{a} \text{ as } x \to \infty$$

Since straight lines are easily sketched, the easiest way to sketch a hyperbola is to draw its asymptotes and then to draw the hyperbola so that it comes closer and closer to these lines as *x* becomes larger numerically. To draw in the asymptotes, the usual procedure is to first draw a small rectangle, $2a$ by $2b$, with the origin in the center. Then straight lines are drawn through opposite vertices. These lines are the asymptotes (see Fig. 1–46). Thus we see that the significance of the value of b lies in the slope of the asymptotes of the hyperbola.

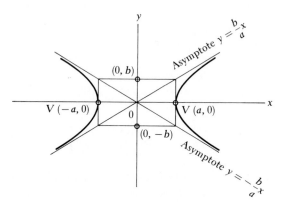

Figure 1–46

Equation (1–20) is called the **standard equation** of the hyperbola with its center at the origin. *It has a* **transverse axis** *of length 2a along the x-axis and a* **conjugate axis** *of length 2b along the y-axis.* This means that a represents the length of the semi-transverse axis, and b represents the length of the semi-conjugate axis. The relation between a, b, and c is given in Eq. (1–21).

If the transverse axis is along the y-axis and the conjugate axis is along the x-axis, the equation of a hyperbola with its center at the origin is

up and down

$$\frac{y^2}{a^2} - \frac{x^2}{b^2} = 1 \tag{1–24}$$

The symmetry of the hyperbolas given by Eqs. (1–20) and (1–24) to each of the axes and to the origin can be proven since each of x and y can be replaced by its negative in these equations without changing the equations.

Example A

The hyperbola

$$\frac{x^2}{16} - \frac{y^2}{9} = 1$$

has vertices at $(4, 0)$ and $(-4, 0)$. Its transverse axis extends from one vertex to the other. Its conjugate axis extends from $(0, 3)$ to $(0, -3)$.

Since $c^2 = a^2 + b^2$, we find that $c = 5$, which means the foci are the points $(5, 0)$ and $(-5, 0)$. Drawing in the rectangle and then the asymptotes (Fig. 1–47), we draw the hyperbola from each vertex toward the asymptotes. Thus, the curve is sketched.

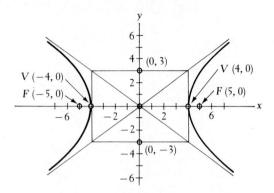

Figure 1–47

Example B

The hyperbola

$$\frac{y^2}{4} - \frac{x^2}{16} = 1$$

has vertices at $(0, 2)$ and $(0, -2)$. Its conjugate axis extends from $(4, 0)$ to $(-4, 0)$. The foci are $(0, 2\sqrt{5})$ and $(0, -2\sqrt{5})$. Since, with 1 on the right side of the equation, the y^2-term is the positive term, this hyperbola is in the form of Eq. (1–24). [Example A illustrates a hyperbola of the type of Eq. (1–20), since the x^2-term is the positive term.] Since $2a$ extends along the y-axis, we see that the equations of the asymptotes are $y = \pm(a/b)x$. This is not a contradiction of Eq. (1–23), but an extension of it for the case of a hyperbola with its transverse axis along the y-axis. The ratio a/b simply expresses the slope of the asymptote (see Fig. 1–48).

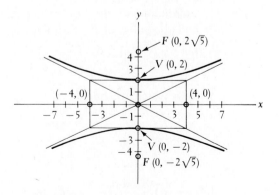

Figure 1–48

Example C

Determine the coordinates of the vertices and foci of the hyperbola $4x^2 - 9y^2 = 36$.

First, by dividing through by 36, we can put this equation in standard form. Thus, we have

$$\frac{x^2}{9} - \frac{y^2}{4} = 1$$

The transverse axis is along the x-axis with vertices at $(3, 0)$ and $(-3, 0)$, since $c^2 = a^2 + b^2$ and $c = \sqrt{13}$. This means that the foci are at $(\sqrt{13}, 0)$ and $(-\sqrt{13}, 0)$ (see Fig. 1–49).

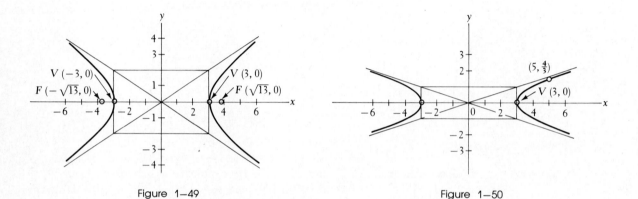

Figure 1–49   Figure 1–50

Example D

Find the equation of the hyperbola with its center at the origin, a vertex at $(3, 0)$, and which passes through $(5, \frac{4}{3})$.

When we know that the hyperbola has its center at the origin and a vertex at $(3, 0)$, we know that the standard form of its equation is given by Eq. (1–20). This also tells us that $a = 3$. By using the fact that the coordinates of the point $(5, \frac{4}{3})$ must satisfy the equation, we are able to find the value of b^2. Substituting, we have

$$\frac{25}{9} - \frac{(16/9)}{b^2} = 1$$

from which we find $b^2 = 1$. Therefore, the equation of the hyperbola is

$$\frac{x^2}{9} - \frac{y^2}{1} = 1$$

or

$$x^2 - 9y^2 = 9$$

This hyperbola is shown in Fig. 1–50.

x	y
-8	$-\frac{1}{2}$
-4	-1
-1	-4
$-\frac{1}{2}$	-8
$\frac{1}{2}$	8
1	4
4	1
8	$\frac{1}{2}$

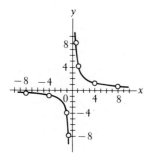

Figure 1–51

Equations (1–20) and (1–24) give us the standard form of the equation of the hyperbola with its center at the origin and its foci on one of the coordinate axes. There is one other important equation form which respresents a hyperbola, and that is

$$xy = c \qquad\qquad (1\text{–}25)$$

The asymptotes of this hyperbola are the coordinate axes, and the foci are on the line $y = x$, or on the line $y = -x$, if c is negative. This hyperbola is symmetrical to the origin, for if $-x$ replaces x, and $-y$ replaces y at the same time, we have $(-x)(-y) = c$, or $xy = c$. The equation remains unchanged. However, if either $-x$ replaces x, or if $-y$ replaces y, but not both, the sign on the left is changed. Therefore, it is not symmetrical to either axis. The c here represents a constant and is not related to the focus. The following example illustrates this type of hyperbola.

Example E
Plot the graph of the equation $xy = 4$.

We find the values of the table at the left, and then plot the appropriate points. Here it is permissible to use a limited number of points, since we know that the equation represents a hyperbola (Fig. 1–51). Thus, using $y = 4/x$, we obtain the values in the table.

We conclude that the equation of a hyperbola is characterized by the presence of both an x^2- and a y^2-term, having different signs, or by the presence of an xy-term with no squared terms.

Exercises 1–9

In Exercises 1 through 12 find the coordinates of the vertices and the foci of the given hyperbolas. Sketch each curve.

1. $\dfrac{x^2}{25} - \dfrac{y^2}{144} = 1$ 2. $\dfrac{x^2}{16} - \dfrac{y^2}{4} = 1$ 3. $\dfrac{y^2}{9} - \dfrac{x^2}{1} = 1$

4. $\dfrac{y^2}{2} - \dfrac{x^2}{2} = 1$ 5. $2x^2 - y^2 = 4$ 6. $3x^2 - y^2 = 9$

7. $2y^2 - 5x^2 = 10$ 8. $3y^2 - 2x^2 = 6$ 9. $4x^2 - y^2 + 4 = 0$

10. $9x^2 - y^2 - 9 = 0$ 11. $4x^2 - 9y^2 = 16$ 12. $y^2 - 9x^2 = 25$

In Exercises 13 through 20 find the equations of the hyperbolas satisfying the given conditions. The center of each is at the origin.

13. Vertex $(3, 0)$, focus $(5, 0)$ 14. Vertex $(0, 1)$, focus $(0, \sqrt{3})$

15. Conjugate axis $= 12$, vertex $(0, 10)$ 16. Focus $(8, 0)$, transverse axis $= 4$

17. Passes through $(2, 3)$, focus $(2, 0)$ 18. Passes through $(8, \sqrt{3})$, vertex $(4, 0)$

19. Passes through $(5, 4)$ and $(3, \frac{4}{5}\sqrt{5})$ 20. Passes through $(1, 2)$ and $(2, 2\sqrt{2})$

In Exercises 21 through 24 sketch the graphs of the hyperbolas given.

21. $xy = 2$ 22. $xy = 10$ 23. $xy = -2$ 24. $xy = -4$

In Exercises 25 through 28 find the equations of the hyperbolas with the given properties by use of the definition of the hyperbola.

25. Foci at (1, 2) and (11, 2), with a transverse axis of 8
26. Vertices (−2, 4) and (−2, −2), with conjugate axis of 4
27. Center at (2, 0), vertex at (3, 0), conjugate axis of 6
28. Center at (1, −1), focus at (1, 4), vertex at (1, 2)

In Exercises 29 through 34 solve the given problems.

29. Sketch the graph of impedance versus resistance if the reactance $X_L - X_C$ of a given electric circuit is constant at 60 Ω (see Exercise 34 of Section 1–6).

30. One statement of Boyle's law is that the product of the pressure and volume, for constant temperature, remains a constant for a perfect gas. If one set of values for a perfect gas under the condition of constant temperature is that the pressure is 300 kPa for a volume of 8.0 L, sketch a graph of pressure versus volume.

31. The relationship between the frequency f, wavelength λ, and the velocity v of a wave is given by $v = f\lambda$. The velocity of light is a constant, being 3.0×10^{10} cm/s. The visible spectrum ranges in wavelength from about 4.0×10^{-5} cm (violet) to about 7.0×10^{-5} cm (red). Sketch a graph of frequency f (in Hertz) as a function of wavelength for the visible spectrum.

32. Wavelengths of gamma rays vary from about 10^{-8} cm to about 10^{-13} cm, but the gamma rays have the same velocity as light. On logarithmic paper, plot the graph of frequency versus wavelength for gamma rays. What type of curve results when this type of hyperbola is plotted on logarithmic paper?

33. An electronic instrument located at point P records the sound of a rifle shot and the impact of the bullet striking the target at the same instant. Show that P lies on a branch of a hyperbola.

34. Two concentric hyperbolas are called conjugate hyperbolas if the transverse and conjugate axes of one are respectively the conjugate and transverse axes of the other. What is the equation of the hyperbola conjugate to the hyperbola of Exercise 14?

1–10 Translation of Axes

Until now, except by direct use of the definition, the equations considered for the parabola, the ellipse, and the hyperbola have been restricted to the particular cases in which the vertex of the parabola is at the origin and the center of the ellipse or hyperbola is at the origin. In this section we shall consider, without specific use of the definition, the equations of these curves for the cases in which the axis of the curve is parallel to one of the coordinate axes. This is done by **translation of axes.**

We choose a point (h, k) in the xy-coordinate plane and let this point be the origin of another coordinate system, the $x'y'$-coordinate system. The x'-axis is parallel to the x-axis and the y'-axis is parallel to the y-axis.

Every point in the plane now has two sets of coordinates associated with it, (x, y) and (x', y'). From Fig. 1–52 we see that

$$x = x' + h \quad \text{and} \quad y = y' + k \tag{1–26}$$

Equations (20–26) can also be written in the form

$$x' = x - h \quad \text{and} \quad y' = y - k \tag{1–27}$$

The following examples illustrate the use of Eqs. (1–27) in the analysis of equations of the parabola, ellipse, and hyperbola.

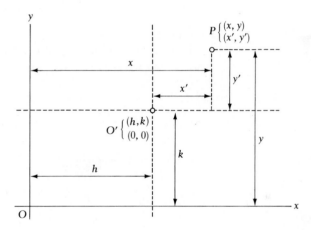

Figure 1–52

Example A
Describe the locus of the equation

$$\frac{(x - 3)^2}{25} + \frac{(y + 2)^2}{9} = 1$$

In this equation, given that $h = 3$ and $k = -2$, we have $x' = x - 3$ and $y' = y + 2$. In terms of x' and y', the equation is

$$\frac{(x')^2}{25} + \frac{(y')^2}{9} = 1$$

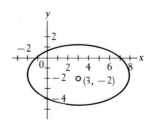

Figure 1–53

We recognize this equation as that of an ellipse (see Fig. 1–53) with a semi-major axis of 5 and a semi-minor axis of 3. The center of the ellipse is at $(3, -2)$ since this was the choice of h and k to make the equation fit a standard form.

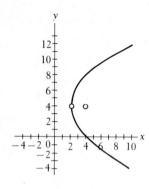

Figure 1–54

Example B

Find the equation of the parabola with vertex at $(2, 4)$ and focus at $(4, 4)$.

If we let the origin of the $x'y'$-coordinate system be the point $(2, 4)$, the point $(4, 4)$ would be the point $(2, 0)$ in the $x'y'$-system. This means that $p = 2$ and $4p = 8$ (Fig. 1–54). In the $x'y'$-system, the equation is

$$(y')^2 = 8(x')$$

Since $(2, 4)$ is the origin of the $x'y'$-system, this means that $h = 2$ and $k = 4$. Using Eq. (1–27), we have

$$(y - 4)^2 = 8(x - 2)$$

as the equation of the parabola in the xy-coordinate system. If this equation is multiplied out, and like terms are combined, we obtain

$$y^2 - 8x - 8y + 32 = 0$$

Example C

Find the center of the hyperbola $2x^2 - y^2 - 4x - 4y - 4 = 0$.

To analyze this curve, we first complete the square in the x-terms and in the y-terms. This will allow us to recognize properly the choice of h and k.

$$2x^2 - 4x - y^2 - 4y = 4$$
$$2(x^2 - 2x \quad) - (y^2 + 4y \quad) = 4$$
$$2(x^2 - 2x + 1) - (y^2 + 4y + 4) = 4 + 2 - 4$$
$$2(x - 1)^2 - (y + 2)^2 = 2$$

$$\frac{(x - 1)^2}{1} - \frac{(y + 2)^2}{2} = 1$$

Thus, if we let $h = 1$ and $k = -2$, the equation in the $x'y'$-system becomes

$$\frac{(x')^2}{1} - \frac{(y')^2}{2} = 1$$

This means that the center is at $(1, -2)$, since this point corresponds to the origin of the $x'y'$-coordinate system (see Fig. 1–55).

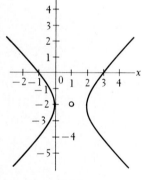

Figure 1–55

Example D

Find the vertex of the parabola $2x^2 - 12x - 3y + 15 = 0$.

First, we complete the square in the x terms, placing all other resulting terms on the right. By factoring the resulting expression on the right, we may determine the values of h and k.

$$2x^2 - 12x = 3y - 15$$
$$2(x^2 - 6x \quad) = 3y - 15$$
$$2(x^2 - 6x + 9) = 3y - 15 + 18$$
$$2(x - 3)^2 = 3(y + 1)$$

$$(x - 3)^2 = \frac{3}{2}(y + 1)$$

$$x'^2 = \frac{3}{2}y'$$

Therefore, the vertex is at $(3, -1)$, since this point corresponds to the origin of the $x'y'$-coordinate system (see Fig. 1–56).

Example E

Glass beakers are to be made with a height of 3 in. Express the surface area of the beakers in terms of the radius of the base. Sketch the graph of area versus radius.

The total surface area of a beaker is the sum of the area of the base and the lateral surface area of the side. In general, this surface area S in terms of the radius r of the base and the height h of the side is

$$S = \pi r^2 + 2\pi r h$$

Since h is constant at 3 in., we have

$$S = \pi r^2 + 6\pi r$$

which is the desired relationship.

For the purposes of sketching the graph of S and r, we now complete the square of the r terms:

$$S = \pi(r^2 + 6r)$$
$$S + 9\pi = \pi(r + 6r + 9)$$
$$S + 9\pi = \pi(r + 3)^2$$
$$(r + 3)^2 = \frac{1}{\pi}(S + 9\pi)$$

We note that this equation represents a parabola with vertex at $(-3, -9\pi)$ for its coordinates (r, S). Since $4p = \frac{1}{\pi}$, $p = \frac{1}{4\pi}$ and the focus of this parabola is at $(-3, \frac{1}{4\pi} - 9\pi)$. Only positive values for S and r have meaning, and therefore the part of the graph for negative r is shown as a dashed curve (see Fig. 1–57).

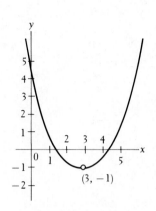

Figure 1–56

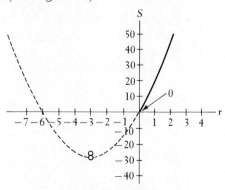

Figure 1–57

Exercises 1—10

In Exercises 1 through 8 describe the locus of each of the given equations. Identify the type of curve and its center (vertex if it is a parabola). Sketch each curve.

1. $(y - 2)^2 = 4(x + 1)$

2. $\dfrac{(x + 4)^2}{4} + \dfrac{(y - 1)^2}{1} = 1$

3. $\dfrac{(x - 1)^2}{4} - \dfrac{(y - 2)^2}{9} = 1$

4. $(y + 5)^2 = -8(x - 2)$

5. $\dfrac{(x + 1)^2}{1} + \dfrac{y^2}{9} = 1$

6. $\dfrac{(y - 4)^2}{16} - \dfrac{(x + 2)^2}{4} = 1$

7. $(x + 3)^2 = -12(y - 1)$

8. $\dfrac{x^2}{16} + \dfrac{(y + 1)^2}{1} = 1$

In Exercises 9 through 16 find the equation of each of the curves described by the given information.

9. Parabola: vertex $(-1, 3)$, $p = 4$, axis parallel to x-axis
10. Parabola: vertex $(2, -1)$, directrix $y = 3$
11. Parabola: vertex $(-3, 2)$, focus $(-3, 3)$
12. Parabola: focus $(2, 4)$, directrix $x = 6$
13. Ellipse: center $(-2, 2)$, focus $(-5, 2)$, vertex $(-7, 2)$
14. Ellipse: center $(0, 3)$, focus $(12, 3)$, major axis 26 units
15. Hyperbola: vertices $(2, 1)$ and $(-4, 1)$, focus $(-6, 1)$
16. Hyperbola: center $(1, -4,)$ focus $(1, 1)$, transverse axis 8 units

In Exercises 17 through 24 determine the center (or vertex if the curve is a parabola) of the given curves. Sketch each curve.

17. $x^2 + 2x - 4y - 3 = 0$
18. $y^2 - 2x - 2y - 9 = 0$
19. $4x^2 + 9y^2 + 24x = 0$
20. $2x^2 + 9y^2 + 8x - 72y + 134 = 0$
21. $9x^2 - y^2 + 8y - 7 = 0$
22. $5x^2 - 4y^2 + 20x + 8y = 4$
23. $2x^2 - 4x = 9y - 2$
24. $4x^2 + 16x = y - 20$

In Exercises 25 through 28 find the required equations.

25. Find the equation of the hyperbola with asymptotes $x - y = -1$ and $x + y = -3$, and vertex $(3, -1)$.

26. The circle $x^2 + y^2 + 4x - 5 = 0$ passes through the foci and the ends of the minor axis of an ellipse which has its major axis along the x-axis. Find the equation of the ellipse.

27. A first parabola has its vertex at the focus of a second parabola, and its focus at the vertex of the second parabola. If the equation of the second parabola is $y^2 = 4x$, find the equation of the first parabola.

28. Verify each of the following equations as being the standard form as indicated. Parabola, vertex at (h, k), axis parallel to the x-axis:

$$(y - k)^2 = 4p(x - h)$$

Parabola, vertex at (h, k), axis parallel to the y-axis:

$$(x - h)^2 = 4p(y - k)$$

Ellipse, center at (h, k), major axis parallel to the x-axis:

$$\frac{(x - h)^2}{a^2} + \frac{(y - k)^2}{b^2} = 1$$

Ellipse, center at (h, k), major axis parallel to the y-axis:

$$\frac{(y - k)^2}{a^2} + \frac{(x - h)^2}{b^2} = 1$$

Hyperbola, center at (h, k), transverse axis parallel to the x-axis:

$$\frac{(x - h)^2}{a^2} - \frac{(y - k)^2}{b^2} = 1$$

Hyperbola, center at (h, k), transverse axis parallel to the y-axis:

$$\frac{(y - k)^2}{a^2} - \frac{(x - h)^2}{b^2} = 1$$

In Exercises 29 through 32 solve the given problems.

29. The power supplied to a circuit by a battery with a voltage E and an internal resistance r is given by $P = EI - rI^2$, where P is the power (in watts) and I is the current (in amperes). Sketch the graph of P vs. I for a 6.0 V battery with an internal resistance of 0.30 Ω.

30. A calculator company determined that its total income I from the sale of a particular type of calculator is given by $I = 100x - x^2$, where x is the selling price of the calculator. Sketch a graph of I vs. x.

31. The planet Pluto moves about the sun in an elliptical orbit, with the sun at one focus. The closest that Pluto approaches the sun is 2.8 billion miles, and the farthest it gets from the sun is 4.6 billion miles. If the sun is at the origin of a coordinate system and the other focus is on the positive x-axis, what is the equation of the path of Pluto?

32. A rectangular tract of land is to have a perimeter of 800 m. Express the area in terms of its width and sketch the graph.

1—11 The Second-Degree Equation

The equations of the circle, parabola, ellipse, and hyperbola are all special cases of the same general equation. In this section we shall discuss this general equation and how to identify the particular form it takes when it represents a specific type of curve.

Each of these curves can be represented by a **second-degree equation** of the form

$$Ax^2 + Bxy + Cy^2 + Dx + Ey + F = 0 \qquad (1\text{--}28)$$

The coefficients of the second-degree terms determine the type of curve which results. Recalling the discussions of the general forms of the equations of the circle, parabola, ellipse, and hyperbola from the previous sections of this chapter, we have the following results.

Equation (1–28) represents the indicated curve for the given conditions for A, B, and C.

(1) If $A = C$, $B = 0$, a circle.
(2) If $A \neq C$ (but they have the same sign), $B = 0$, an ellipse.
(3) If A and C have different signs, $B = 0$, a hyperbola.
(4) If $A = 0$, $C = 0$, $B \neq 0$, a hyperbola.
(5) If either $A = 0$ or $C = 0$ (but not both), $B = 0$, a parabola.
(Special cases, such as a single point or no real locus, can also result.)

Another conclusion about Eq. (1–28) is that, if either $D \neq 0$ or $E \neq 0$ (or both), the center (or vertex of a parabola) of the curve is not at the origin. If $B \neq 0$, the axis of the curve has been rotated. We have considered only one such case (the hyperbola $xy = c$) in this chapter.

Example A

The equation $3x^2 = 6x - y^2 + 3$ represents an ellipse. Before we analyze the equation, we should put it in the form of Eq. (1–28). For the given equation, this form is

$$3x^2 + y^2 - 6x - 3 = 0$$

Here we see that $B = 0$ and $A \neq C$. Therefore, it is an ellipse. The $-6x$ term indicates that the center of the ellipse is not at the origin.

Example B

The equation $2(x + 3)^2 = y^2 + 2x^2$ represents a parabola. Putting it in the form of Eq. (1–28), we have

$$2(x^2 + 6x + 9) = y^2 + 2x^2$$
$$2x^2 + 12x + 18 = y^2 + 2x^2$$
$$y^2 - 12x - 18 = 0$$

We now note that $A = 0$, $B = 0$, and $C \neq 0$. This indicates that the equation represents a parabola. Here, the -18 term indicates that the vertex is not at the origin.

Example C

Identify the curve represented by the equation $2x^2 + 12x = y^2 - 14$. Determine the appropriate important quantities associated with the curve, and sketch the graph.

Writing this equation in the form of Eq. (1–28), we have

$$2x^2 - y^2 + 12x + 14 = 0$$

In this form we identify the equation as representing a hyperbola, since A and C have different signs, and $B = 0$. We now write it in the standard form of a hyperbola.

$$2x^2 + 12x - y^2 = -14$$
$$2(x^2 + 6x \quad) - y^2 = -14$$
$$2(x^2 + 6x + 9) - y^2 = -14 + 18$$
$$2(x + 3)^2 - y^2 = 4$$

$$\frac{(x + 3)^2}{2} - \frac{y^2}{4} = 1$$

$$\frac{x'^2}{2} - \frac{y'^2}{4} = 1$$

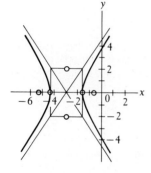

Thus, we see that the center (h, k) of the hyperbola is the point $(-3, 0)$. Also, $a = \sqrt{2}$ and $b = 2$. This means that the vertices are $(-3 + \sqrt{2}, 0)$ and $(-3 - \sqrt{2}, 0)$, and the conjugate axis extends from $(-3, 2)$ to $(-3, -2)$. Also, $c^2 = 2 + 4 = 6$, which means that $c = \sqrt{6}$. The foci are $(-3 + \sqrt{6}, 0)$ and $(-3 - \sqrt{6}, 0)$. The graph is shown in Fig. 1–58.

Figure 1–58

Example D

Identify the curve represented by the equation $4y^2 - 23 = 4(4x + 3y)$. Determine the appropriate important quantities associated with the curve, and sketch the graph.

Writing this equation in the form of Eq. (1–28), we have

$$4y^2 - 23 = 16x + 12y$$
$$4y^2 - 16x - 12y - 23 = 0$$

Therefore, we recognize the equation as representing a parabola, since $A = 0$ and $B = 0$. Now writing the equation in the standard form of a parabola, we have

$$4y^2 - 12y = 16x + 23$$
$$4(y^2 - 3y \quad) = 16x + 23$$

$$4\left(y^2 - 3y + \frac{9}{4}\right) = 16x + 23 + 9$$

$$4\left(y - \frac{3}{2}\right)^2 = 16(x + 2)$$

$$\left(y - \frac{3}{2}\right)^2 = 4(x + 2)$$

$$y'^2 = 4x'$$

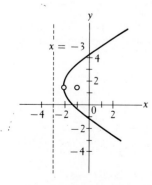

Figure 1–59

We now note that the vertex is the point $(-2, \frac{3}{2})$ and that $p = 1$. Also, it is symmetric to the x' axis. Therefore, the focus is $(-1, \frac{3}{2})$ and the directrix is $x = -3$. The graph is shown in Fig. 1–59.

The circle, parabola, ellipse, and hyperbola are usually referred to as **conic sections.** If a plane is passed through a cone, the intersection of the plane and the cone results in one of these curves, the curve formed depends on the angle of the plane with respect to the axis of the cone. This is indicated in Fig. 1–60.

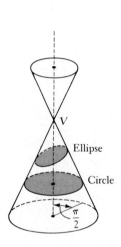
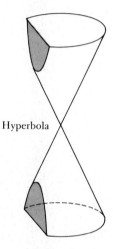
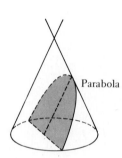

Figure 1–60

Exercises 1–11

In Exercises 1 through 16 identify each of the equations as representing either a circle, parabola, ellipse, or hyperbola.

1. $2x^2 + 2y^2 - 3y - 1 = 0$
2. $x^2 - 2y^2 - 3x - 1 = 0$
3. $2x^2 - x - y = 1$
4. $2x^2 + 4y^2 - y - 2x = 4$
5. $x^2 = y^2 - 1$
6. $3x^2 = 2y - 4y^2$
7. $x^2 = y - y^2$
8. $y = 3 - 6x^2$
9. $x(y + 3x) = x^2 + xy - y^2 + 1$
10. $x(2 - x) = y^2$
11. $2xy + x - 3y = 6$
12. $(y + 1)^2 = x^2 + y^2 - 1$
13. $2x(x - y) = y(3 - y - 2x)$
14. $2x^2 = x(x - 1) + 4y^2$
15. $y(3 - 2y) = 2(x^2 - y^2)$
16. $4x(x - 1) = 2x^2 - 2y^2 + 3$

In Exercises 17 through 24 identify the curve represented by each of the given equations. Determine the appropriate important quantities associated with the curve, and sketch the graph.

17. $x^2 = 8(y - x - 2)$
18. $x^2 = 6x - 4y^2 - 1$
19. $y^2 = 2(x^2 - 2x - 2y)$
20. $4x^2 + 4 = 9 - 8x - 4y^2$
21. $y^2 + 42 = 2x(10 - x)$
22. $x^2 - 4y = y^2 + 4(1 - x)$
23. $4(y^2 - 4x - 2) = 5(4y - 5)$
24. $2(2x^2 - y) = 8 - y^2$

In Exercises 25 through 28 set up the necessary equation and then determine the type of curve it represents.

25. For a given alternating-current circuit the resistance and capacitive reactance are constant. What type of curve is represented by the equation relating impedance and inductive reactance (see Exercise 34 of Section 1–6)?

26. A room is 8.0 ft high, and the length is 6.5 ft longer than the width. Express the volume V of the (rectangular) room in terms of the width w. What type of curve is represented by the equation?

27. The sides of a rectangle are $2x$ and y, and the digaonal is $x + 5$. What type of curve is represented by the equation relating x and y?

28. The legs of a right triangle are x and y and the hypotenuse is $x + 2$. What type of curve is represented by the equation relating x and y?

1–12 Review Exercises for Chapter 1

In Exercises 1 through 12 find the equation of the indicated curve subject to the given conditions. Sketch each curve.

1. Straight line: passes through $(1, -7)$ with a slope of 4
2. Straight line: passes through $(-1, 5)$ and $(-2, -3)$
3. Straight line: perpendicular to $3x - 2y + 8 = 0$ and has a y-intercept of -1
4. Straight line: parallel to $2x - 5y + 1 = 0$ and has an x-intercept of 2
5. Circle: center at $(1, -2)$, passes through $(4, -3)$
6. Circle: tangent to the line $x = 3$, center at $(5, 1)$
7. Parabola: focus $(3, 0)$, vertex $(0, 0)$
8. Parabola: directrix $y = -5$, vertex $(0, 0)$
9. Ellipse: vertex $(10, 0)$, focus $(8, 0)$, center $(0, 0)$
10. Ellipse: center $(0, 0)$, passes through $(0, 3)$ and $(2, 1)$
11. Hyperbola: vertex $(0, 13)$, center $(0, 0)$, conjugate axis of 24
12. Hyperbola: foci $(0, 10)$ and $(0, -10)$, vertex $(0, 8)$

In Exercises 13 through 24 find the indicated quantities for each of the given equations. Sketch each curve.

13. $x^2 + y^2 + 6x - 7 = 0$, center and radius
14. $x^2 + y^2 - 4x + 2y - 20 = 0$, center and radius
15. $x^2 = -20y$, focus and directrix
16. $y^2 = 24x$, focus and directrix
17. $16x^2 + y^2 = 16$, vertices and foci
18. $2y^2 - 9x^2 = 18$, vertices and foci
19. $2x^2 - 5y^2 = 8$, vertices and foci
20. $2x^2 + 25y^2 = 50$, vertices and foci
21. $x^2 - 8x - 4y - 16 = 0$, vertex and focus
22. $y^2 - 4x + 4y + 24 = 0$, vertex and directrix
23. $4x^2 + y^2 - 16x + 2y + 13 = 0$, center
24. $x^2 - 2y^2 + 4x + 4y + 6 = 0$, center

In Exercises 25 through 48 solve the given problems.

25. Find the points of intersection of the ellipses $25x^2 + 4y^2 = 100$ and $4x^2 + 9y^2 = 36$.

26. Find the points of intersection of the hyperbola $y^2 - x^2 = 1$ and the ellipse $x^2 + 25y^2 = 25$.

27. In two ways show that the line segments joining $(-3, 11)$, $(2, -1)$, and $(14, 4)$ form a right triangle.

28. Show that the altitudes of the triangle with vertices $(2, -4)$, $(3, -1)$, and $(-2, 5)$ meet at a single point.

29. By means of the definition of a parabola, find the equation of the parabola with focus at $(3, 1)$ and directrix the line $y = -3$. Find the same equation by the method of translation of axes.

30. Repeat the instructions of Exercise 29 for the parabola with focus at $(0, 2)$ and directrix the line $y = 4$.

31. In an electric circuit the voltage V equals the product of the current I and the resistance R. Sketch the graph of V (in volts) vs. I (in amperes) for a circuit in which $R = 3 \, \Omega$.

32. An airplane touches down when landing at 100 mi/h. Its velocity v while coming to a stop is given by $v = 100 - 20000t$, where t is the time in hours. Sketch the graph of v vs. t.

33. In a certain electric circuit, two resistors having resistances R_1 and R_2 respectively, act as a voltage divider. The relation between them is $\alpha = R_1/(R_1 + R_2)$, where α is a constant less than 1. Sketch the curve of R_1 vs. R_2 (both in ohms), if $\alpha = \frac{1}{2}$.

34. Let C and F denote corresponding Celsius and Fahrenheit temperature readings. If the equation relating the two is linear, determine this equation given that $C = 0$ when $F = 32$ and $C = 100$ when $F = 212$.

35. The arch of a small bridge across a stream is parabolic. If, at water level, the span of the arch is 80 ft and the maximum height of the span above water level is 20 ft, what is the equation of the arch? Choose the most convenient point for the origin of the coordinate system.

36. If a source of light is placed at the focus of a parabolic reflector, the reflected rays are parallel. Where is the focus of a parabolic reflector which is 8 cm across and 6 cm deep?

37. Sketch the curve of the total surface area of a right circular cylinder as a function of its radius, if the height is always 10 units.

38. At very low temperatures certain metals have an electric resistance of zero. This phenomenon is called superconductivity. A magnetic field also affects the superconductivity. A certain level of magnetic field, the threshold field, is related to the temperature T by

$$\frac{H_T}{H_0} = 1 - \left(\frac{T}{T_0}\right)^2$$

where H_0 and T_0 are specifically defined values of magnetic field and temperature. Sketch H_T/H_0 vs. T/T_0.

39. The vertical position of a projectile is given by $y = 120t - 16t^2$, and its hozizontal position is given by $x = 60t$. By eliminating the time t, determine the path of the projectile. Sketch this path for the length of time it would need to strike the ground, assuming level terrain.

40. The rate r in grams per second at which a substance is formed in a chemical reaction is given by $r = 10.0 - 0.010t^2$. Sketch the graph of r vs. t.

41. The inside of the top of an arch is a semi-ellipse with a major axis (width) of 26 ft and a minor axis of 10 ft. The arch is 7 ft thick at all points. Is the outside of the arch a portion of an ellipse? (*Hint:* Check the equation of the outer "ellipse" 1 ft to the right of center.)

42. The earth moves about the sun in an elliptical path. If the closest the earth gets to the sun is 90.5 million miles, and the farthest it gets from the sun is 93.5 million miles, find the equation of the path of the earth. The sun is at one focus of the ellipse. Assume the major axis to be along the x-axis and that the center of the ellipse is at the origin.

43. Soon after reaching the vicinity of the moon, Apollo 11 (the first spacecraft to land a man on the moon) went into an elliptical lunar orbit. The closest the craft was to the moon in this orbit was 70 mi, and the farthest it was from the moon was 190 mi. What was the equation of the path if the moon was at one of the foci of the ellipse? Assume the major axis to be along the x-axis and that the center of the ellipse is at the origin. The radius of the moon is 1080 mi.

44. A machine-part designer wishes to make a model for an elliptical cam by placing two pins in her design board, putting a loop of string over the pins, and marking off the outline by keeping the string taut. (Note that she is using the definition of the ellipse.) If the cam is to measure 10 cm by 6 cm, how long should the loop of string be and how far apart should the pins be?

45. Under certain conditions the work W done on a wire by increasing the tension from T_1 to T_2 is given by $W = k(T_2^2 - T_1^2)$, where k is a constant depending on the properties of the wire. If the work is constant for various sets of initial and final tensions, sketch a graph of T_2 vs. T_1.

46. In order to study the relationship between velocity and pressure of a stream of water, a pipe is constructed such that its cross-section is a hyperbola. If the pipe is 10 ft long, 1 ft in diameter at the narrow point (middle) and 2 ft in diameter at the ends, what is the equation of the cross-section of the pipe? (In physics it is shown that where the velocity of a fluid is greatest, the pressure is the least.)

47. A hallway 16 ft wide has a ceiling whose cross-section is a semi-ellipse. The ceiling is 10 ft high at the walls and 14 ft high at the center. Find the height of the ceiling 4 ft from each wall.

48. A 60-ft rope passes over a pulley 10 ft above the ground and an object on the ground is attached at one end. A man holds the other end at a level of 4 ft above the ground. If the man walks away from the pulley, express the height of the object above the ground in terms of the distance the man is from directly below the object. Sketch the graph of distance vs. height.

2

The Derivative

2—1 Algebraic Functions

In Chapter 1 we discussed a number of equations and their associated graphs. In each case discussed, the equation expressed a relationship between the two variables. *If the equation is solved for one of the variables, say y, in terms of the other, x, then we say that y is a* **function** *of x. Here y is called the* **dependent variable** *and x is called the* **independent variable.**

Example A
The equation of a straight line written in slope-intercept form, $y = mx + b$, expresses y, the dependent variable, as a function of x, the independent variable.

Relationships which can be expressed as functions exist between various quantities in all fields of technology. An experiment to determine whether or not a relationship exists between the distance an object falls and the time of fall should indicate (approximately) that $s = 16t^2$, where s is the distance in feet and t is the time in seconds. Electrical measurements of current and voltage with respect to a particular resistor would show that $V = kI$, where V is the voltage, I is the current, and k is a constant.

Example B
The power developed in a certain resistor by a current I is given by $P = 4I^2$. Here P is a function of I. The dependent variable is P, and the independent variable is I.

As we mentioned in Section 1–1 the rate of change of a quantity is of importance in many physical situations. Since relationships between the appropriate quantities can be expressed as functions, and through differential calculus we can determine the rate of change of a function, we find it necessary to review various operations on functions.

1 to 1 correspondence

We shall now make our definition of a function more explicit. *If two variables are so related that for each value of the first variable a single value of the second variable can be determined, the second variable is a function of the first variable.*

There are many ways to express functions. Formulas, such as those mentioned above, define functions. Other ways to express functions are by means of tables, charts, and graphs.

Example C

We can demonstrate an important aspect of a function by inspecting the graphs shown in Figures 2–1(a) and 2–1(b).

In Fig. 2–1(a) there is only one value ot y for each value of x. In Fig. 2–1(b) we can see that there are two values of y for some values of x. Therefore, the graph in Fig. 2–1(a) defines a function, whereas the graph in Fig. 2–1(b) does *not* define a function. There must be only one value of the dependent variable for each value of the independent variable for there to be a function.

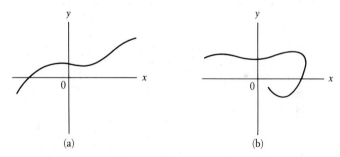

(a)   (b)

Figure 2–1

For convenience of notation, the phrase "function of x" is written as $f(x)$. This, in turn, is a simplification of the statement that "y is a function of x," in that we may now write it as $y = f(x)$. One of the most important uses of this notation is to designate the value of the function for a particular value of the independent variable. That is, for the expression "the value of the function $f(x)$ when $x = a$" we may write $f(a)$.

Example D

If $y = 6x^3 - 5x$, we may say that y is a function of x, where the function $f(x)$ is $6x^3 - 5x$. We may also write $f(x) = 6x^3 - 5x$. It is common to write functions in this manner, as well as in the form $y = 6x^3 - 5x$. Also, when $x = -2$, we find that $f(x) = -38$, or $f(-2) = -38$.

The functional notation $f(x)$ is very useful in another respect. This notation emphasizes the significance of a function. The function itself may be looked upon as a set of instructions. These instructions tell us how to obtain the value of the dependent variable for a particular value of the independent variable, even if this is expressed in symbols. For example, the function $f(x) = x^2 - 3x$ tells us to "square the independent variable x, multiply x by 3, and finally subtract the second result from the first." An analogy would be a computer which was programmed so that when a number x was introduced into the program, it would square the number and then subtract 3 times the value of the number. In diagram form, this can be represented as in Fig. 2–2.

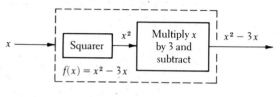

Figure 2–2

Example E

As we noted, the distance s (in feet) which an object falls in t seconds is given by $s = 16t^2$. A firm determines that the profits from a particular sale should be $p = 16x^2$, where x is the side of the square display area. In the first statement s is a function of t, and in the second p is a function of x. However, *the function is the same in both cases*: to evaluate the dependent variable, we square the value of the independent variable and multiply the result by 16. *Even though the letters are different, the operation is the same.*

Numerous functions are restricted as to the values of the independent variable which are permissible. *Values of the independent variable which would lead to imaginary values for the function or which lead to division by zero are not included in the permissible values.* Also, if the function is related to a physical situation, the values may be restricted due to their physical significance. It is also possible that a given function is defined only for particular values of the independent variable.

Example F

The function $f(u) = \dfrac{u}{u-4}$ is not defined if $u = 4$ since this would require division by zero. Therefore, the values of u are restricted to values other than 4.

The function $g(s) = \sqrt{3-s}$ is not defined for real numbers if s is greater than 3, since such values would result in imaginary values for $g(s)$. Thus, the values of s are restricted to values equal to or less than 3.

Also note the use of $g(s)$ to represent the function. The use of letters other than f is common.

Example G
If the formula in Example B is solved for I, we have $I = \sqrt{P}/2$. In this case the current is a function of the power. This could be stated as $g(P) = \sqrt{P}/2$. Even though P is the independent variable in this case it is restricted to values greater than or equal to zero. Negative values of P would lead to imaginary values of I. Such values would not have significance either mathematically or from the physical point of view.

transcendental Funtions
exponential, log, trig, inverse trig

The functions which are studied in mathematics fall into two general classes: **algebraic functions** and **transcendental functions.** In this chapter we shall be concerned with the calculus associated with algebraic functions. Transcendental functions, such as the trigonometric functions and the logarithmic function, will be taken up in later chapters.

Algebraic functions are numerous, but three are of concern to us here: **Polynomials,** which are of the form

$$P(x) = a_0 x^n + a_1 x^{n-1} + \cdots + a_{n-1} x + a_n \qquad (2\text{--}1)$$

The a's are constants and n is an integer.
Rational fractions, which are of the form

$$F(x) = \frac{f(x)}{g(x)} \qquad (2\text{--}2)$$

where $f(x)$ and $g(x)$ are polynomials.
Irrational functions of the form

$$I(x) = \sqrt[n]{F(x)} \qquad (2\text{--}3)$$

which is the root of a rational function. Combinations of these functions may also be encountered.

Example H
If $f(x) = 2x^2 - 3$ and $g(x) = \sqrt{x + 2}$, then

$$g[f(x)] = \sqrt{(2x^2 - 3) + 2}$$
$$= \sqrt{2x^2 - 1}$$

Here we replace x in the function $g(x)$ with the expression for $f(x)$. This is the meaning of $g[f(x)]$, following functional notation.

Since we will be dealing with exponents and radicals, we shall now review their basic properties. The basic laws of exponents are given on the following page.

$$a^m \cdot a^n = a^{m+n} \tag{2-4}$$

$$\frac{a^m}{a^n} = a^{m-n}, \quad a \neq 0 \tag{2-5}$$

$$(a^m)^n = a^{mn} \tag{2-6}$$

$$(ab)^n = a^n b^n, \quad \left(\frac{a}{b}\right)^n = \frac{a^n}{b^n} \text{ if } b \neq 0 \tag{2-7}$$

$$a^0 = 1 \tag{2-8}$$

$$a^{-n} = \frac{1}{a^n} \tag{2-9}$$

We shall often find it convenient to express radicals in the form of fractional exponents. The relationship between fractional exponents and radicals, along with the basic operations with radicals, are listed below.

$$a^{m/n} = \sqrt[n]{a^m} \quad (m \text{ and } n \text{ integers}) \tag{2-10}$$

$$\sqrt[n]{a^n} = a \tag{2-11}$$

$$\sqrt[m]{\sqrt[n]{a}} = \sqrt[mn]{a} \tag{2-12}$$

$$\sqrt[n]{a} \cdot \sqrt[n]{b} = \sqrt[n]{ab} \tag{2-13}$$

$$\frac{\sqrt[n]{a}}{\sqrt[n]{b}} = \sqrt[n]{\frac{a}{b}}, \quad b \neq 0 \tag{2-14}$$

It should be noted that the above properties hold in general if $a > 0$ and $b > 0$. However, they are not necessarily valid for negative values. The following examples illustrate the use of these laws of exponents and radicals with functions.

Example I

For the function $f(x) = 2x^2 - 1$, we have:

(a) $(2x^2 - 1)^2(2x^2 - 1)^5 = (2x^2 - 1)^{2+5} = (2x^2 - 1)^7$

(b) $[(2x^2 - 1)^3]^4 = (2x^2 - 1)^{3 \cdot 4} = (2x^2 - 1)^{12}$

(c) $(2x^2 - 1)^{-2} = 1/(2x^2 - 1)^2$

(d) $(2x^2 - 1)^{2/3} = \sqrt[3]{(2x^2 - 1)^2}$

(e) $\sqrt[4]{\sqrt{2x^2 - 1}} = \sqrt[4 \cdot 2]{2x^2 - 1} = \sqrt[8]{2x^2 - 1}$

Example J

Simplify the expression for the function

$$f(x) = \frac{(3x^2 - 1)^{1/3}(2x) - (2x^3)(3x^2 - 1)^{-2/3}}{(3x^2 - 1)^{2/3}}$$

We multiply the numerator and denominator of this expression by

$(3x^2 - 1)^{2/3}$ to clear the fractional exponents in the numerator. Applying the laws of exponents, we then proceed with the simplification.

$$f(x) = \frac{(3x^2 - 1)^{1/3}(2x) - (2x^3)(3x^2 - 1)^{-2/3}}{(3x^2 - 1)^{2/3}} \cdot \frac{(3x^2 - 1)^{2/3}}{(3x^2 - 1)^{2/3}}$$

$$= \frac{(3x^2 - 1)^{1/3}(3x^2 - 1)^{2/3}(2x) - (2x^3)(3x^2 - 1)^{-2/3}(3x^2 - 1)^{2/3}}{[(3x^2 - 1)^{2/3}]^2}$$

$$= \frac{(3x^2 - 1)(2x) - (2x^3)(3x^2 - 1)^0}{(3x^2 - 1)^{4/3}}$$

$$= \frac{6x^3 - 2x - 2x^3}{(3x^2 - 1)^{4/3}} = \frac{4x^3 - 2x}{(3x^2 - 1)^{4/3}}$$

Algebraic simplifications such as this one are found commonly in calculus problems.

Exercises 2–1

In Exercises 1 through 8 determine the required functions.

1. Express the area A of a square as a function of its side s.
2. Express the circumference c of a circle as a function of its radius r.
3. Express the volume V of a right circular cone of height 8 as a function of the radius r of the base.
4. Express the perimeter p of a square as a function of its area A.
5. A certain article costs \$3 to produce and distribute. Express the profit p made by selling 100 of these articles at a price of c dollars each.
6. A taxi fare is 55¢ plus 10¢ for every $\frac{1}{5}$ mile traveled. Express the fare F as a function of the distance s traveled.
7. A particular factory uses two kinds of fuel. To keep air pollutants below a specified level, it is determined that the number of tons n_1 of the first fuel and the number of tons n_2 of the second fuel which may be used weekly is determined by the equation $6n_1 + 9n_2 = 12{,}000$. Express n_2 as a function of n_1.
8. The distance p of an object from a particular lens is related to the distance q of the image from the lens by the equation $p = 5q/(q - 5)$. Express q as a function of p.

In Exercises 9 through 16 find the indicated values of the given functions.

9. $f(x) = x^2 - 9x$. Find $f(3)$ and $f(-5)$.
10. $f(x) = 2x^3 - 7x$. Find $f(1)$ and $f(\frac{1}{2})$.
11. $g(x) = 6\sqrt{x} - 3$. Find $g(9)$ and $g(\frac{1}{4})$.
12. $H(t) = t^4 - t + 3$. Find $H(0)$ and $H(-2)$.
13. $f(x) = ax^2 - a^2x$. Find $f(a)$ and $f(2a)$.
14. $K(s) = 3s^2 - s + 6$. Find $K(2s)$ and $K(s - 1)$.
15. $T(t) = 5t + 7$. Find $T(-t)$ and $T[T(t)]$.
16. $X(f) = \dfrac{f + 5}{1 - f}$. Find $X(f - 2)$ and $X[X(-f)]$.

In Exercises 17 through 20 use the functions $F(x) = \sqrt{x - 1}$ and $G(x) = x^2 + 4$.

17. Find $F[G(x)]$.   **18.** Find $[F(x)]^6$.

19. Find $F(x)/G(x)$.   **20.** Find $\{G[F(x)]\}^2$.

In Exercises 21 through 24 sketch the graph of the given functions. Algebraically each can be shown to be a familiar form. However, be careful to show only the appropriate portion.

21. $f(x) = 6 - 3x$ **22.** $f(x) = 4x^2 - 8$

23. $f(x) = \sqrt{9 - x^2}$ **24.** $f(x) = \sqrt{x^2 - 9}$

In Exercises 25 through 32 perform the indicated operations and simplify.

25. $\sqrt{\sqrt[3]{x^6 + 1}}$ **26.** $(\sqrt{x} - 1)(\sqrt{x^2 + x + 1})$

27. $\dfrac{(4x - 5)^4}{(4x - 5)^{1/2}}$ **28.** $\dfrac{3(x^2 + 1)^0}{(x^2 + 4)^{-1/2}}$

29. $(2x + 1)^{1/2} + (x + 3)(2x + 1)^{-1/2}$ **30.** $(3x - 1)^{-2/3}(1 - x) - (3x - 1)^{1/3}$

31. $\dfrac{(x^2 + 1)^{1/2} - x^2(x^2 + 1)^{-1/2}}{x^2 + 1}$ **32.** $\dfrac{(1 - 2x^2)^{1/4}(2x) + 4x^3(1 - 2x^2)^{-3/4}}{(1 - 2x^2)^{1/2}}$

In Exercises 33 through 36 answer the given questions involving functions.

33. Does the table in Fig. 2–3(a) define y as a function of x? x as a function of y?

34. Does the graph in Fig. 2–3(b) define y as a function of x? x as a function of y?

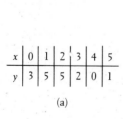

x	0	1	2	3	4	5
y	3	5	5	2	0	1

(a)

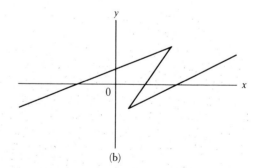

(b)

Figure 2–3

35. If $Y(y) = (y + 1)/(y - 1)$, are there any restrictions on the values of the independent variable?

36. If $F(z) = \sqrt{z - 1}$, are there any restrictions on the values of the independent variable?

In Exercises 37 through 42 solve the given problems involving functions.

37. The voltage of a certain thermocouple as a function of the temperature is given by $E = 2.8T + 0.006T^2$. Find the voltage when the temperature is $T + 100$.

38. The electrical resistance of a certain ammeter as a function of the resistance of the coil of the meter is

$$R = \frac{10R_c}{10 + R_c}$$

Find the function which represents the resistance of the meter if the resistance of the coil is doubled.

39. A piece of wire 60 in. long is cut into two pieces, one of which is bent into a circle and the other into a square. Express the total area of the two figures as a function of the perimeter of the square.

40. A window is in the form of a rectangle surmounted by a semicircle. If the perimeter of the window is 20 ft, express the area of the window as a function of the vertical side of the rectangle.

41. Express the perimeter of a rectangular field in terms of one of its dimensions if the area of the field is A.

42. The power of an electric circuit varies jointly as the resistance and the square of the current. Given that the power is 10 W when the current is 0.5 A and the resistance is 40 Ω, express the power as a function of the current for this 40-Ω resistor.

2–2 Limits

Another important concept for the development of differential calculus is that of a **limit**. Reasoning similar to that which is necessary for establishing the meaning of a limit was encountered when we discussed the asymptotes of a hyperbola. We shall now introduce the necessary terminology in order to establish the meaning of a limit.

In order to help understand and develop the concept of a limit, we consider briefly the **continuity** of a function. *For a function to be* **continuous at a point,** *the function must exist at the point, and any small change in x produces only a small change in f(x).* In fact, the change in $f(x)$ can be made as small as we wish by restricting the change in x sufficiently, if the function is continuous. Also, *a function is said to be* **continuous over an interval** *if it is continuous at each point in the interval.*

Example A

The function $f(x) = 3x^2$ is continuous for all values of x. That is, $f(x)$ is defined for all values of x, and a small change in x for any given value produces only a small change in $f(x)$. If we choose $x = 2$ and then let x change by 0.1, 0.01, and so on, we obtain the values in the following table.

x	2	2.1	2.01	2.001
$f(x)$	12	13.23	12.1203	12.012003

We can also see that the change in $f(x)$ is made smaller by the smaller changes in x. This shows that $f(x)$ is continuous at $x = 2$. Since this type of result would be obtained for any other x we may choose, we see that $f(x)$ is continuous for all values, and therefore it is continuous over the interval of all values of x.

Example B

The function $f(x) = \frac{1}{x-2}$ is not continuous at $x = 2$. When we substitute 2 for x in the function, we have division by zero. This means that the function is not defined. Therefore, the condition that the function must exist is not satisfied.

Considering continuity from a graphical point of view, a function which is continuous over an interval will have no "breaks" in its graph over that interval. If a function is not continuous, either the graph of the function has a break in it or it does not exist for the values of x.

Example C

The graph of the function $f(x) = 3x^2$, which we determined to be continuous at all values of x in Example A, is shown in Fig. 2–4. We see that there are no breaks in the graph.

The graph of $f(x) = \frac{1}{x-2}$, which we determined to be not continuous at $x = 2$ in Example B, is shown in Fig. 2–5. We see that there is a break in the graph for $x = 2$.

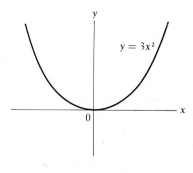

Figure 2–4

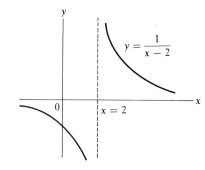

Figure 2–5

Example D

The function represented by the graph in Fig. 2–6 is continuous in the interval for which $x > 1$. The graph does not exist for values $x < 1$.

The function represented by the graph in Fig. 2–7 is not continuous at $x = 1$. The function is defined (by the solid circle point) for $x = 1$ (the open circle point indicates that point is not part of the graph). However, a small change from 1 may result in a change of at least 3 in $f(x)$, regardless of how small the change in x is made. Therefore, the small change condition is not satisfied.

The function represented by the graph in Fig. 2–8 is not continuous for $x = -2$. The open circle shows that point is not part of the graph, and therefore $f(x)$ is not defined for $x = -2$.

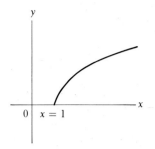

Figure 2–6

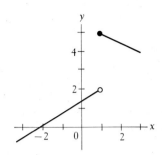

Figure 2–7

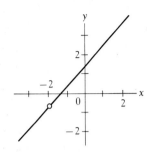

Figure 2–8

In our earlier discussions of the asymptotes of a hyperbola, we used the symbol → which means "approaches." *When we say that x → 2, we mean that x may take on any value as close to 2 as desired, but it also distinctly means that x cannot be set equal to 2.*

Example E
Let us consider the behavior of the function $f(x) = 2x + 1$ as $x \rightarrow 2$.

Since we are not to use $x = 2$, we set up tables to determine values of $f(x)$, as x gets close to 2.

x	1.000	1.500	1.900	1.990	1.999
$f(x)$	3.000	4.000	4.800	4.980	4.998

x	3.000	2.500	2.100	2.010	2.001
$f(x)$	7.000	6.000	5.200	5.020	5.002

We can see that $f(x)$ approaches 5, as x approaches 2, from above 2 and from below 2.

In Example E, since $f(x) \rightarrow 5$ as $x \rightarrow 2$, the number 5 is called the limit of $f(x)$ as $x \rightarrow 2$. This leads us to the meaning of the limit of a function. In general, *the* **limit of a function** $f(x)$ *is that value which the function approaches as x approaches a given value a.* This is written as

$$\lim_{x \to a} f(x) = L \tag{2–15}$$

where L is the value of the limit of the function. Remember, in approaching a, x may come as arbitrarily close as desired to a, but it may not equal a.

An important conclusion can be drawn from the limit in Example E. The function $f(x)$ is a continuous function, and $f(2)$ equals the value of the limit as $x \rightarrow 2$. In general, it is true that *the limit of $f(x)$ as $x \rightarrow a$ is the same as $f(a)$, if the function is continuous at $x = a$.*

Although we can evaluate the limit for a continuous function as $x \rightarrow a$ by evaluating $f(a)$, it is possible that a function is not continuous at $x = a$, and that the limit exists and can be determined. Thus, we must be able to determine the value of a limit without finding $f(a)$. The following example illustrates the evaluation of such a limit.

Example F
Find

$$\lim_{x \to 2} \frac{2x^2 - 3x - 2}{x - 2}$$

We note immediately that the function is not continuous at $x = 2$, for division by zero is indicated. Thus, we cannot evaluate the limit by substituting $x = 2$ into the function. By setting up tables, we determine the value which $f(x)$ approaches, as x approaches 2.

x	1.000	1.500	1.900	1.990	1.999
$f(x)$	3.000	4.000	4.800	4.980	4.998

x	3.000	2.500	2.100	2.010	2.001
$f(x)$	7.000	6.000	5.200	5.020	5.002

We see that the values obtained are identical to those in Example E. Since $f(x) \to 5$ as $x \to 2$, we have

$$\lim_{x \to 2} \frac{2x^2 - 3x - 2}{x - 2} = 5$$

Thus, the limit exists at $x = 2$, although the function does not exist at $x = 2$.

The reason that the functions in Examples E and F have the same limit is shown in the following example.

Example G
The function $\frac{2x^2 - 3x - 2}{x - 2}$ in Example F is the same as the function $2x + 1$ in Example E, except when $x = 2$. By factoring the numerator of the function of Example F, we have

$$\frac{2x^2 - 3x - 2}{x - 2} = \frac{(2x + 1)(x - 2)}{x - 2} = 2x + 1$$

The cancellation here is valid, as long as x does not equal 2, for we have division by zero at $x = 2$. Also, in finding the limit as $x \to 2$, we do not use the value $x = 2$. Therefore,

$$\lim_{x \to 2} \frac{2x^2 - 3x - 2}{x - 2} = \lim_{x \to 2} (2x + 1) = 5$$

The limits of the two functions are equal, since, again, in finding the limit, we do not let $x = 2$. The graphs of the two functions are shown

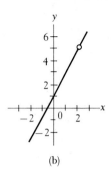

(a)

(b)

Figure 2–9

in Fig. 2–9 (a) and (b). We can see from the graphs that the limits are the same, although one of the functions is not continuous.

The limit of the function in Example F was determined by calculating values near $x = 2$ and by means of an algebraic change in the function. This illustrates that limits may be found through the meaning and definition, and through other procedures when the function is not continuous. The following examples further illustrate the evaluation of limits.

substitution

Example H
Find $\lim\limits_{x \to 4} (x^2 - 7)$.

Since the function $x^2 - 7$ is continuous at $x = 4$, we may evaluate this limit by substitution. For $f(x) = x^2 - 7$, we have $f(4) = 9$. This means that

$$\lim_{x \to 4} (x^2 - 7) = 9$$

Example I
Find

factor

$$\lim_{x \to 2} \left(\frac{x^2 - 4}{x - 2} \right)$$

Since

$$\frac{x^2 - 4}{x - 2} = \frac{(x - 2)(x + 2)}{x - 2} = x + 2$$

is valid as long as $x \neq 2$, we find that

$$\lim_{x \to 2} \left(\frac{x^2 - 4}{x - 2} \right) = \lim_{x \to 2} (x + 2) = 4$$

Again, we do not have to concern ourselves with the fact that the cancellation is not valid for $x = 2$. In finding the limit we do not consider the value of $f(x)$ at $x = 2$.

Example J

Find $\lim\limits_{x \to 0} (x\sqrt{x-3})$.

We see that $x\sqrt{x-3} = 0$ if $x = 0$, but this function does not have real values for values of x less than 3 other than $x = 0$. *Since x cannot approach* 0, *f(x) does not approach* 0 *and the limit does not exist.* The point of this example is that even if $f(a)$ exists, we cannot evaluate the limit by finding $f(a)$, unless $f(x)$ is continuous at $x = a$. Here, $f(a)$ exists but the limit does not exist.

Limits as x approaches infinity are also of importance. When we write $x \to \infty$ we know that we are to consider values of x which are becoming large without bound. Consider the following examples.

Example K

Find

divide by highest exponent

$$\lim_{x \to \infty} \left(3 + \frac{1}{x^2}\right)$$

As x becomes larger and larger, $1/x^2$ becomes smaller and smaller and approaches zero. This means that $f(x) \to 3$ as $x \to \infty$. Thus,

$$\lim_{x \to \infty} \left(3 + \frac{1}{x^2}\right) = 3$$

Example L

Find

$$\lim_{x \to \infty} \frac{x^2 + 1}{2x^2 + 3}$$

We note that as $x \to \infty$ both the numerator and denominator become large without bound. However, if we divide numerator and denominator by x^2, the function becomes

$$\frac{1 + \dfrac{1}{x^2}}{2 + \dfrac{3}{x^2}}$$

we see that $1/x^2$ and $3/x^2$ both approach zero as $x \to \infty$. This means that the numerator approaches 1 and the denominator approaches 2. Thus,

$$\lim_{x \to \infty} \frac{x^2 + 1}{2x^2 + 3} = \lim_{x \to \infty} \frac{1 + \dfrac{1}{x^2}}{2 + \dfrac{3}{x^2}} = \frac{1}{2}$$

The definitions and development of continuity and of a limit presented in this section are not mathematically completely rigorous. However, the development is consistent with a more rigorous development, and the concept of a limit is the principal concern.

Exercises 2–2

In Exercises 1 through 4 determine the values of x for which the function is continuous. If the function is not continuous, determine the reason.

1. $f(x) = 3x - 2$

2. $f(x) = 9 - x^2$

3. $f(x) = \dfrac{1}{x + 3}$

4. $f(x) = \dfrac{2}{x^2 - x}$

In Exercises 5 through 8 determine the values of x for which the function, as represented by the graph in Fig. 2–10, is continuous. If the function is not continuous, determine the reason.

5.
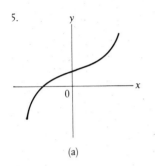
(a)

6.
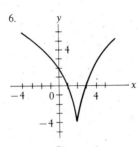
(b)

7.
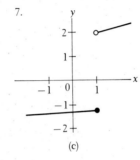
(c)

8.
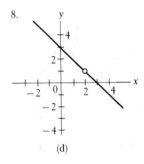
(d)

Figure 2–10

In Exercises 9 through 12 evaluate the given functions for the given values of x. Do not change the form of the functions. Then find the indicated limits.

9. Evaluate $f(x) = 3x - 2$ for the following values of x: 4.000, 3.500, 3.100, 3.010, 3.001 and for 2.000, 2.500, 2.900, 2.990, 2.999. Find

$$\lim_{x \to 3} (3x - 2)$$

10. Evaluate $f(x) = 2x^2 - 4x$ for the following values of x: 0.4000, 0.4900, 0.4990, 0.4999 and for 0.6000, 0.5100, 0.5010, 0.5001. Find

$$\lim_{x \to 0.5} (2x^2 - 4x)$$

11. Evaluate $f(x) = \dfrac{x^3 - x}{x - 1}$ for the following values of x: 0.900, 0.990, 0.999 and for 1.100, 1.010, 1.001. Find

$$\lim_{x \to 1} \frac{x^3 - x}{x - 1}$$

12. Evaluate $f(x) = \dfrac{2 - \sqrt{x}}{4 - x}$ for the following values of x: 3.900, 3.990, 3.999 and for 4.100, 4.010, 4.001. Find

$$\lim_{x \to 4} \frac{2 - \sqrt{x}}{4 - x}$$

In Exercises 13 through 32 evaluate the given limits.

13. $\lim\limits_{x \to 3} (x + 4)$

14. $\lim\limits_{x \to -2} (2x - 1)$

15. $\lim\limits_{x \to 2} \dfrac{x^2 - 1}{x + 1}$

16. $\lim\limits_{x \to 5} \left(\dfrac{3}{x^2 + 2} \right)$

17. $\lim\limits_{x \to 0} \dfrac{x^2 + x}{x}$

18. $\lim\limits_{x \to 2} \dfrac{x^2 - 2x}{x - 2}$

19. $\lim\limits_{x \to -1} \dfrac{x^2 - 1}{x + 1}$

20. $\lim\limits_{x \to 5} \dfrac{x - 5}{x^2 - 25}$

21. $\lim\limits_{x \to 1} \dfrac{x^3 - x}{x - 1}$

22. $\lim\limits_{x \to 1/3} \dfrac{3x - 1}{3x^2 + 5x - 2}$

23. $\lim\limits_{x \to 3} \dfrac{x^2 - 2x - 3}{3 - x}$

24. $\lim\limits_{x \to 0} \dfrac{(2 + x)^2 - 4}{x}$

25. $\lim\limits_{x \to -1} \sqrt{x}\,(x + 1)$

26. $\lim\limits_{x \to 1} (x - 1)\sqrt{x^2 - 4}$

27. $\lim\limits_{x \to \infty} \dfrac{\frac{2}{x}}{1 - 2x}$

28. $\lim\limits_{x \to \infty} \dfrac{6}{1 + \dfrac{2}{x^2}}$

29. $\lim\limits_{x \to \infty} \dfrac{3x^2 + 5}{x^2 - 2}$

30. $\lim\limits_{x \to \infty} \dfrac{x - 1}{7x + 4}$

31. $\lim\limits_{x \to \infty} \dfrac{2x - 6}{x^2 - 9}$

32. $\lim\limits_{x \to \infty} \dfrac{2 - x^2}{3x^3 - 1}$

In Exercises 33 through 36 solve the given problems involving limits.

33. The magnetic field B at the center of a loop of wire due to a current i in the wire is given by $B = ki/r$, where r is the radius of the loop and k is a constant. If $k = 6.30 \times 10^{-7}$ Wb/A-m and $r = 0.100$ m, find $\lim_{i \to 0.3} B$ (in teslas) by evaluating the function for the following values of i: 0.200, 0.290, 0.299, 0.301, 0.310, 0.400.

34. Velocity can be determined by dividing the displacement of an object by the time elapsed in moving through the displacement. In a certain experiment the following values were determined for the displacements and elapsed times for the motion of an object. Determine the limit of the velocity.

displacement (centimeters)	0.480000	0.280000	0.029800	0.002998	0.00029998
time (seconds)	0.200000	0.100000	0.010000	0.001000	0.00010000

35. A certain object, after being heated, cools at such a rate that its temperature T (in degrees Celsius) decreases 10% each minute. If the object is originally heated to 100°C find $\lim_{t \to 10} T$ and $\lim_{t \to \infty} T$, where t is the time (in minutes).

36. Show that

$$\lim_{x \to 0^+} 2^{1/x} \neq \lim_{x \to 0^-} 2^{1/x}$$

where $\lim_{x \to 0^+}$ means to find the limit as x approaches zero through positive values only, and $\lim_{x \to 0^-}$ means to find the limit as x approaches zero through negative values only. These are called "one-sided" limits.

2–3 The Slope of a Tangent to a Curve

Having developed the basic operations with functions and the concept of a limit, we shall now turn our attention to a graphical interpretation of the rate of change of a function. This interpretation, basic to an understanding of the calculus, deals with the slope of a line tangent to the curve of a function.

Consider the points $P(x_1, y_1)$ and $Q(x_2, y_2)$ in Fig. 2–11. From Chapter 1 we know that the slope of the line through these points is given by

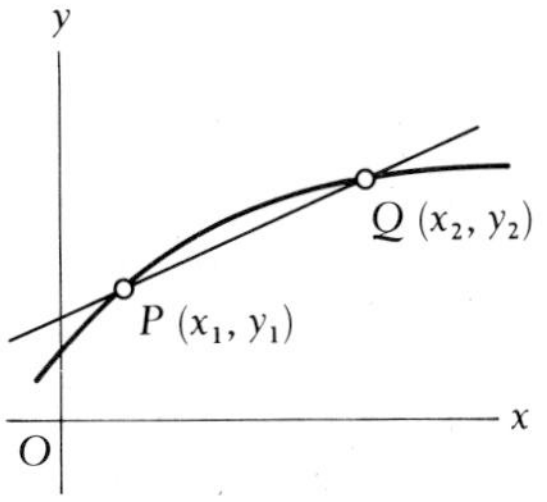

$$m = \frac{y_2 - y_1}{x_2 - x_1}$$

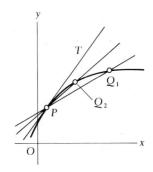

Figure 2—11

This, however, represents the slope of the line through P and Q and no other line. If we now allow Q to be a point closer to P, the slope of PQ will more closely approximate the slope of a line drawn tangent to the curve at P (see Fig. 2–12). In fact, the closer Q is to P, the better this approximation becomes. It is not possible to allow Q to coincide with P, for then it would not be possible to define the slope of PQ in terms of two points. The slope of the tangent line, often referred to as the slope of the curve, is the limiting value of the slope of PQ as Q approaches P.

Example A

Find the slope of a line tangent to the curve $y = x^2 + 3x$ at the point $P(2, 10)$ by determining the limit of slopes of lines PQ as Q approaches P.

We shall let point Q have the x-values of 3.0, 2.5, 2.1, 2.01, and 2.001, and tabulate the necessary values. Since P is the point (2, 10), $x_1 = 2$ and $y_1 = 10$. Thus, using the values of x_2 we tabulate the values of y_2, $y_2 - 10$, $x_2 - 2$, and thereby the values of the slope m.

Point	Q_1	Q_2	Q_3	Q_4	Q_5	P
x_2	3.0	2.5	2.1	2.01	2.001	2
y_2	18.0	13.75	10.71	10.0701	10.007001	10
$y_2 - 10$	8.0	3.75	0.71	0.0701	0.007001	
$x_2 - 2$	1.0	0.5	0.1	0.01	0.001	
$m = \dfrac{y_2 - 10}{x_2 - 2}$	8.0	7.5	7.1	7.01	7.001	

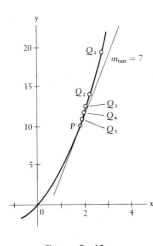

Figure 2—13

We can see from the above table that the slope of PQ approaches the value of 7 as Q approaches P. Thus, the slope of the tangent line at (2,10) is 7. See Fig. 2–13.

With the proper notation, it is possible to express the coordinates of Q in terms of the coordinates of P. If we define the quantity Δx ("delta" x) by the equation

$$x_2 = x_1 + \Delta x \qquad (2\text{--}16)$$

and Δy by the equation

$$y_2 = y_1 + \Delta y \qquad (2\text{--}17)$$

Figure 2—12

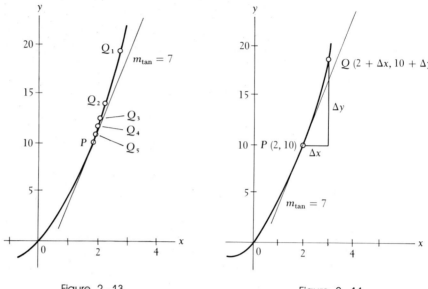

Figure 2–13   Figure 2–14

the coordinates of Q become $(x_1 + \Delta x, y_1 + \Delta y)$. The quantities Δx and Δy represent the difference in the coordinates of P and Q. They are not to be thought of as "Δ times x," for the symbol Δ used here has no meaning by itself. *The name* **increment** *is given to the difference of the coordinates of two points, and therefore Δx and Δy are the increments in x and y, respectively.*

Using Eqs. (2–16) and (2–17) along with the definition of slope, we can express the slope of PQ as

$$m_{PQ} = \frac{(y_1 + \Delta y) - y_1}{(x_1 + \Delta x) - x_1} = \frac{\Delta y}{\Delta x} \tag{2–18}$$

By the previous discussion, as Q approaches P, the slope of the tangent line is more *nearly* approximated by $\Delta y/\Delta x$.

Example B

Find the slope of a line tangent to the curve $y = x^2 + 3x$ at the point $(2, 10)$ by the increment method indicated in Eq. (2–18). (This is the same slope as calculated in Example A.)

As in Example A, point P has the coordinates $(2, 10)$. Thus the coordinates of any other point Q can be expressed as $(2 + \Delta x, 10 + \Delta y)$ (see Fig. 2–14). The slope of PQ then becomes

$$m_{PQ} = \frac{(10 + \Delta y) - 10}{(2 + \Delta x) - 2} = \frac{\Delta y}{\Delta x}$$

This expression itself does not enable us to find the slope, since values of Δx and Δy are not known. If, however, we can express Δy in terms of Δx, we might derive more information. Both P and Q are on the curve of the function, which means that the coordinates of each must satisfy the function. Using the coordinates of Q, we have

$$(10 + \Delta y) = (2 + \Delta x)^2 + 3(2 + \Delta x)$$
$$= 4 + 4\Delta x + (\Delta x)^2 + 6 + 3\Delta x$$

Subtracting 10 from each side, we have $\Delta y = 7\Delta x + (\Delta x)^2$. We then substitute this into the equation for m_{PQ}, which gives us

$$m_{PQ} = \frac{7\Delta x + (\Delta x)^2}{\Delta x} = 7 + \Delta x$$

As Q approaches P, Δx becomes smaller and smaller. The limiting value of Δx is zero. Using this fact, we can see that the slope of the tangent line is

$$m_{\tan} = \lim_{\Delta x \to 0} m_{PQ} = 7$$

We see that this result agrees with that found in Example A.

Example C
Find the slope of a line tangent to the curve $y = 4x - x^2$ at the point (x_1, y_1).

The points P and Q are $P(x_1, y_1)$ and $Q(x_1 + \Delta x, y_1 + \Delta y)$. Using the coordinates of Q in the function, we obtain

$$(y_1 + \Delta y) = 4(x_1 + \Delta x) - (x_1 + \Delta x)^2$$
$$= 4x_1 + 4\Delta x - x_1^2 - 2x_1\Delta x - (\Delta x)^2$$

Using the coordinates of P, we obtain

$$y_1 = 4x_1 - x_1^2$$

Subtracting the second of these expressions from the first to solve for Δy, we obtain

$$(y_1 + \Delta y) - y_1 = (4x_1 + 4\Delta x - x_1^2 - 2x_1\Delta x - (\Delta x)^2) - (4x_1 - x_1^2)$$

or

$$\Delta y = 4\Delta x - 2x_1\,\Delta x - (\Delta x)^2$$

Dividing each term by Δx to obtain an expression for $\Delta y/\Delta x$, we obtain

$$\frac{\Delta y}{\Delta x} = 4 - 2x_1 - \Delta x$$

In this last equation, the desired expression is on the left, but all we can determine from $\Delta y/\Delta x$ itself is that the ratio will become one very small number divided by another very small number as $\Delta x \to 0$. The right side, however, approaches $4 - 2x_1$ as $\Delta x \to 0$. This indicates that the slope of a tangent at the point (x_1, y_1) is given by

$$m_{\tan} = 4 - 2x_1$$

This method has an advantage over that of Example B—we now have a general expression for the slope of a tangent line for any value x_1. If $x_1 = -1$, $m_{\tan} = 6$, and if $x_1 = 3$, $m_{\tan} = -2$. These tangent lines are indicated in Fig. 2–15.

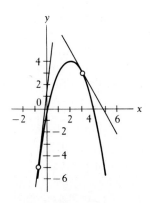

Figure 2–15

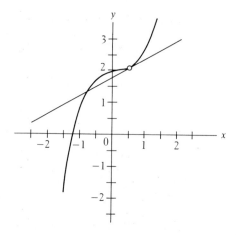

Figure 2–16

Example D

Find the expression for the slope of a line tangent to the curve $y = x^3 + 2$ at the general point (x_1, y_1), and use this expression to determine the slope when $x = \frac{1}{2}$.

Using the coordinates of the point $Q(x_1 + \Delta x, y_1 + \Delta y)$ in the function, we obtain

$$y_1 + \Delta y = (x_1 + \Delta x)^3 + 2 = x_1^3 + 3x_1^2 \, \Delta x + 3x_1 (\Delta x)^2 + (\Delta x)^3 + 2$$

and using the coordinates of the point $P(x_1, y_1)$, we obtain $y_1 = x_1^3 + 2$. Subtracting this second expression from the first, we have

$$\Delta y = 3x_1^2 \Delta x + 3x_1 (\Delta x)^2 + (\Delta x)^3$$

Dividing by Δx, we have

$$\frac{\Delta y}{\Delta x} = 3x_1^2 + 3x_1 \Delta x + (\Delta x)^2$$

As $\Delta x \to 0$, the right side approaches the value $3x_1^2$. This means that

$$m_{\text{tan}} = 3x_1^2$$

When $x_1 = \frac{1}{2}$, we find that the slope of the tangent is $3(\frac{1}{4}) = \frac{3}{4}$. The curve and this tangent line are indicated in Fig. 2–16.

In interpreting this analysis as the rate of change of a function, we see that if $\Delta x = 1$ and the corresponding value of Δy is found, it can be said that y changes by the amount Δy as x changes one unit. If x changed by some lesser amount, we could still calculate a ratio for the amount of change in y for the given change in x. Therefore, as long as x changes at all, there will be a corresponding change in y. In this way the ratio $\Delta y/\Delta x$ is looked on as the **average rate of change** of y, which is a function of x, with respect to x. As $\Delta x \to 0$ the limit of the ratio $\Delta y/\Delta x$ gives us the **instantaneous rate of change** of y with respect to x.

Exercises 2–3

In Exercises 1 through 4 use the method of Example A to calculate the slope of a line tangent to the curve of each of the given functions. Let Q_1, Q_2, Q_3, and Q_4 have the indicated x-values. Sketch the curves and tangent lines.

1. $y = x^2$; P is $(2, 4)$; let Q have x-values of 1.5, 1.9, 1.99, 1.999
2. $y = 1 - x^2$; P is $(2, -3)$; let Q have x-values of 1.5, 1.9, 1.99, 1.999
3. $y = x^2 + 2x$; P is $(-3, 3)$; let Q have x-values of -2.5, -2.9, -2.99, -2.999
4. $y = x^3 + 1$; P is $(-1, 0)$; let Q have x-values of -0.5, -0.9, -0.99, -0.999

In Exercises 5 through 8 use the method of Example B to calculate the slope of a line tangent to the curve of each of the given functions for the given points P. (These are the same functions and points P as for Exercises 1 through 4.)

5. $y = x^2$; P is $(2, 4)$
6. $y = 1 - x^2$; P is $(2, -3)$
7. $y = x^2 + 2x$; P is $(-3, 3)$
8. $y = x^3 + 1$; P is $(-1, 0)$

In Exercises 9 through 20 use the method of Example C to find a general expression for the slope of a tangent line to each of the indicated curves. Then find the slope for the given values of x. Sketch the curves and tangent lines.

9. $y = x^2$; $x = 2$, $x = -1$
10. $y = 1 - x^2$; $x = 2$, $x = -2$
11. $y = x^2 + 2x$; $x = -3$, $x = 1$
12. $y = 4 - 3x^2$; $x = 0$, $x = 2$
13. $y = x^2 + 4x + 5$; $x = -3$, $x = 2$
14. $y = 2x^2 - 4x$; $x = 1$, $x = 1.5$
15. $y = 6x - x^2$; $x = -2$, $x = 3$
16. $y = 2x - 3x^2$; $x = 0$, $x = 0.5$
17. $y = x^3 - 2x$; $x = -1$, $x = 0$, $x = 1$
18. $y = 3x - x^3$; $x = -2$, $x = 0$, $x = 2$
19. $y = x^4$; $x = 0$, $x = 0.5$, $x = 1$
20. $y = 1 - x^4$; $x = 0$, $x = 1$, $x = 2$

2–4 The Derivative

Using the terminology and techniques developed in earlier sections of this chapter, we are now ready to establish one of the fundamental definitions of calculus. Generalizing on the method of Example B of the preceding section, if we consider that the point $P(x, y)$ is held constant while the point $Q(x + \Delta x, y + \Delta y)$ approaches it, then both Δx and Δy approach zero. The points P and Q both lie on the curve of the function $f(x)$, which in turn means that the coordinates of each point must satisfy the function. For the point P this means that $y = f(x)$, and for the point Q this means that $y + \Delta y = f(x + \Delta x)$. Subtracting the first expression from the second, we have

$$(y + \Delta y) - y = f(x + \Delta x) - f(x)$$

$$\Delta y = f(x + \Delta x) - f(x) \tag{2-19}$$

From the discussion of the previous section, we saw that the slope of a line tangent to $f(x)$ at point $P(x, y)$ was found as the limiting value of the ratio $\Delta y / \Delta x$ as Δx approaches zero. Formally, *this limiting value of the ratio of $\Delta y / \Delta x$ is known as the* **derivative** *of the function.* Therefore, the derivative of a function $f(x)$ is defined by the relation

$$\lim_{\Delta x \to 0} \frac{\Delta y}{\Delta x} = \lim_{\Delta x \to 0} \frac{f(x + \Delta x) - f(x)}{\Delta x} \qquad (2\text{--}20)$$

The process of finding the derivative of a function from its definition is called the Δ-process. *The general process of finding a derivative is called* **differentiation.**

Example A

Find the derivative of the function $y = 2x^2 + 3x$ by the delta-process.

To find Δy, we must first derive the quantity $f(x + \Delta x) - f(x)$. Thus, we find $y + \Delta y = f(x + \Delta x)$ first:

$$y + \Delta y = 2(x + \Delta x)^2 + 3(x + \Delta x)$$

Then we subtract the original function:

$$y + \Delta y - y = 2(x + \Delta x)^2 + 3(x + \Delta x) - 2x^2 - 3x$$

Simplifying, we find that the result is

$$\Delta y = 4x\Delta x + 2(\Delta x)^2 + 3\Delta x$$

Next, dividing through by Δx, we obtain

$$\frac{\Delta y}{\Delta x} = 4x + 2\Delta x + 3$$

When we find the limit as Δx approaches zero, we see that the $2\Delta x$-term on the right approaches zero. Thus,

$$\lim_{\Delta x \to 0} \frac{\Delta y}{\Delta x} = 4x + 3$$

Therefore, we find that the derivative of the function $y = 2x^2 + 3x$ is the function $4x + 3$. This means that we can calculate the slope of a tangent line for any point on the graph of $y = 2x^2 + 3x$ by substituting the x-coordinate into the expression $4x + 3$. For example, the slope of a tangent line is 7 if $x = 1$.

Example B

Find the derivative of $y = 6x - 2x^3$ by the Δ-process.

First we find $y + \Delta y = f(x + \Delta x)$.

$$y + \Delta y = 6(x + \Delta x) - 2(x + \Delta x)^3$$
$$= 6x + 6(\Delta x) - 2[x^3 + 3x^2(\Delta x) + 3x(\Delta x)^2 + (\Delta x)^3]$$
$$= 6x + 6(\Delta x) - 2x^3 - 6x^2(\Delta x) - 6x(\Delta x)^2 - 2(\Delta x)^3$$

Next we subtract the original function.

$$y + \Delta y - y = [6x + 6(\Delta x) - 2x^3 - 6x^2(\Delta x) - 6x(\Delta x)^2 - 2(\Delta x)^3]$$
$$- (6x - 2x^3)$$
$$\Delta y = 6(\Delta x) - 6x^2(\Delta x) - 6x(\Delta x)^2 - 2(\Delta x)^3$$

Dividing through by Δx is the next step.

$$\frac{\Delta y}{\Delta x} = \frac{6(\Delta x) - 6x^2(\Delta x) - 6x(\Delta x)^2 - 2(\Delta x)^3}{\Delta x}$$
$$= 6 - 6x^2 - 6x(\Delta x) - 2(\Delta x)^2$$

Now we find the limit as Δx approaches zero.

$$\lim_{\Delta x \to 0} \frac{\Delta y}{\Delta x} = \lim_{\Delta x \to 0} [6 - 6x^2 - 6x(\Delta x) - 2(\Delta x)^2]$$
$$= 6 - 6x^2$$

Therefore, the derivative of $y = 6x - 2x^3$ is $6 - 6x^2$.

Example C
Find the derivative $y = \dfrac{1}{x}$ by the Δ-process.

We write

$$y + \Delta y = \frac{1}{x + \Delta x}$$

Then we find that

$$y + \Delta y - y = \frac{1}{x + \Delta x} - \frac{1}{x}$$
$$\Delta y = \frac{x - (x + \Delta x)}{x(x + \Delta x)} = \frac{-\Delta x}{x(x + \Delta x)}$$
$$\frac{\Delta y}{\Delta x} = \frac{-1}{x(x + \Delta x)}$$
$$\lim_{\Delta x \to 0} \frac{\Delta y}{\Delta x} = \lim_{\Delta x \to 0} \frac{-1}{x(x + \Delta x)}$$
$$\lim_{\Delta x \to 0} \frac{\Delta y}{\Delta x} = \frac{-1}{x^2}$$

Example D

Find the derivative of $y = x^2 + \dfrac{1}{x+1}$ by the Δ-process.

$$y + \Delta y = (x + \Delta x)^2 + \frac{1}{x + \Delta x + 1}$$

$$y + \Delta y - y = x^2 + 2x\,\Delta x + (\Delta x)^2 + \frac{1}{x + \Delta x + 1} - x^2 - \frac{1}{x + 1}$$

The problem is handled algebraically most easily if the fractions are combined and the other terms are handled separately:

$$\Delta y = 2x\,\Delta x + (\Delta x)^2 + \frac{1}{x + \Delta x + 1} - \frac{1}{x + 1}$$

$$= 2x\,\Delta x + (\Delta x)^2 + \frac{(x + 1) - (x + \Delta x + 1)}{(x + \Delta x + 1)(x + 1)}$$

$$= 2x\,\Delta x + (\Delta x)^2 - \frac{\Delta x}{(x + \Delta x + 1)(x + 1)}$$

$$\frac{\Delta y}{\Delta x} = 2x + \Delta x - \frac{1}{(x + \Delta x + 1)(x + 1)}$$

$$\lim_{\Delta x \to 0} \frac{\Delta y}{\Delta x} = 2x - \frac{1}{(x + 1)^2}$$

Several shorter notations are used for the derivative. The notation of the definition is clumsy for general use, so the notations

$$y', \quad D_x y, \quad f'(x), \quad \text{and} \quad \frac{dy}{dx}$$

are commonly used. Thus, the answer to Example A could have been written as $y' = 4x + 3$, or as $dy/dx = 4x + 3$.

One might ask why, when we are finding a derivative, we take a limit as Δx approaches zero and do not simply let Δx equal zero. If we did this, the ratio $\Delta y/\Delta x$, would be exactly $0/0$, which would then require division by zero. As we know, this is an undefined operation in mathematics, and therefore Δx cannot *equal* zero. However, it can equal any value as near zero as necessary. This idea is inherent in the meaning of the word limit.

Example E

Find dy/dx of the function $y = \sqrt{x}$ by the Δ-process.

We first square both sides of the equation, thus obtaining $y^2 = x$. (Here $y \geq 0$ as it is in $y = \sqrt{x}$.) Substituting the coordinates of the point $(x + \Delta x, y + \Delta y)$ into this equation, we have

$$(y + \Delta y)^2 = x + \Delta x$$

Expanding the left side and then subtracting the equation $y^2 = x$, we have

$$y^2 + 2y\,\Delta y + (\Delta y)^2 - y^2 = x + \Delta x - x$$

or

$$2y\,\Delta y + (\Delta y)^2 = \Delta x$$

Next, dividing through by Δx, we obtain

$$2y\frac{\Delta y}{\Delta x} + \Delta y\frac{\Delta y}{\Delta x} = 1$$

Solving for $\Delta y/\Delta x$, we have

$$\frac{\Delta y}{\Delta x} = \frac{1}{2y + \Delta y}$$

Finally, taking the limit as $\Delta x \to 0$, we obtain

$$\lim_{\Delta x \to 0}\frac{\Delta y}{\Delta x} = \lim_{\Delta x \to 0}\frac{1}{2y + \Delta y} = \frac{1}{2y}$$

The Δy is omitted, since $\Delta y \to 0$ as $\Delta x \to 0$. Now, substituting in the original function, we obtain the final result:

$$\frac{dy}{dx} = \frac{1}{2\sqrt{x}}$$

Exercises 2—4

In Exercises 1 through 24 find the derivative of each of the functions by using the Δ-process.

1. $y = 3x - 1$
2. $y = 6x + 3$
3. $y = 1 - 2x$
4. $y = 2 - 5x$
5. $y = x^2 - 1$
6. $y = 4 - x^2$
7. $y = 5x^2$
8. $y = -6x^2$
9. $y = x^2 - 7x$
10. $y = x^2 + 4x$
11. $y = 8x - 2x^2$
12. $y = 3x - 5x^2$
13. $y = x^3 + 4x - 6$
14. $y = 2x - 4x^3$
15. $y = \dfrac{1}{x + 2}$
16. $y = \dfrac{1}{x + 1}$
17. $y = x + \dfrac{1}{x}$
18. $y = \dfrac{x}{x - 1}$
19. $y = \dfrac{2}{x^2}$
20. $y = \dfrac{1}{x^2 + 1}$
21. $y = x^4 + x^3 + x^2 + x$
22. $y = \frac{1}{3}x^3 + \frac{1}{2}x^2 + x$
23. $y = x^4 - \dfrac{2}{x}$
24. $y = \dfrac{1}{x} + \dfrac{1}{x^2}$

In Exercises 25 through 28 find dy/dx for the given functions by the method of Example E.

25. $y = \sqrt{x + 1}$ 26. $y = \sqrt{x - 2}$ 27. $y = \sqrt{1 - 3x}$ 28. $y = \sqrt{x^2 + 3}$

2–5 The Meaning of the Derivative

In Section 2–3, we saw that the slope of a curve was the limiting value of the slope of the line through points P and Q as Q approaches P. At the end of the section, we discussed the idea that $\Delta y/\Delta x$ indicated the rate of change of y with respect to x. In Section 2–4, we defined the limit of the ratio of $\Delta y/\Delta x$ as $\Delta x \to 0$ as the derivative. Thus, the derivative is a measure of the rate of change of y with respect to x at point P. However, P may represent any point, which means that the value of the derivative changes from one point on a curve to another point. **We therefore interpret the derivative as the *instantaneous rate of change of y with respect to x.***

Example A

In Examples A and B of Section 2–3, y is changing at the rate of 7 units for every change of 1 unit in x, *when x is 2, and only when x is 2.* In Example C of Section 2–3, y is increasing 6 units for every increase of 1 unit of x *when x = -1.* When $x = 3$, y is decreasing 2 units for 1 unit of increase in x.

If a functional relationship exists between any two variables, then one can be thought of as varying with respect to the other. There are many applications of this principle, one of them being the velocity of an object. We shall consider here the case of rectilinear motion, that is, motion along a straight line.

As we have seen, the velocity of an object is found by dividing the change in displacement by the time required for this change. This, however, gives a value only for the **average velocity** for the specified time interval. If the time interval considered becomes smaller and smaller, then the average velocity which is calculated more nearly approximates the **instantaneous velocity** at some particular time. In the limit, the value of the average velocity gives the value of the instantaneous velocity. Using the symbols as defined for the derivative, the instantaneous velocity of an object moving in rectilinear motion at a particular time t is given by

$$v = \lim_{\Delta t \to 0} \frac{\Delta s}{\Delta t} \tag{2–21}$$

where s is the displacement.

Example B

Find the instantaneous velocity when $t = 4$ s of an object for which the displacement s (in feet) is given by $s = 16t^2$ by calculating values of $\Delta s/\Delta t$ and determining the limit as Δt approaches zero.

Here we shall let t take on values of 3.5, 3.9, 3.99, and 3.999 s. When $t = 4$ s, $s = 256$ ft. Therefore, we calculate Δt by subtracting values of t from 4, and Δs is calculated by subtracting values from 256. The values of velocity are then calculated by using $v = \Delta s/\Delta t$.

t (seconds)	3.5	3.9	3.99	3.999
s (feet)	196.0	243.36	254.7216	255.872016
Δs (feet)	60.0	12.64	1.2784	0.127984
Δt (seconds)	0.5	0.1	0.01	0.001
v (feet per second)	120.0	126.4	127.84	127.984

We can see that the limit is approaching 128 ft/s, which is therefore the instantaneous velocity when $t = 4$ s.

Example C

By use of the Δ-process, determine an expression for the instantaneous velocity of the object of Example B, for which the displacement s (in feet) is given by $s = 16t^2$ and t represents the time (in seconds). By use of the derived expression, determine the instantaneous velocity when $t = 2$ s and when $t = 4$ s.

Applying the Δ-process to this function, we find the derivative of s with respect to t:

$$s + \Delta s = 16(t + \Delta t)^2 = 16t^2 + 32t\,\Delta t + 16(\Delta t)^2$$
$$\Delta s = 32t\,\Delta t + 16(\Delta t)^2$$

$$\frac{\Delta s}{\Delta t} = 32t + 16\,\Delta t$$

$$v = \lim_{\Delta t \to 0} \frac{\Delta s}{\Delta t} = \frac{ds}{dt} = 32t$$

If $t = 2$ s, the velocity is 64 ft/s. When $t = 4$ s, the velocity is 128 ft/s. We see that this second result agrees with that obtained in Example B.

By finding $\lim_{\Delta x \to 0} \Delta y/\Delta x$, we can find the instantaneous rate of change of y with respect to x. The expression $\lim_{\Delta t \to 0} \Delta s/\Delta t$ gives the velocity, or instantaneous rate of change of displacement with respect to time. Generalizing, we can say that *the derivative can be interpreted as giving the instantaneous rate of change of the dependent variable with respect to the independent variable.* This is true no matter what the variables represent, so long as a function is defined between the variables.

Example D

Find an expression for the rate of change of the volume of a sphere with respect to its radius.

$$V = \frac{4}{3}\pi r^3$$

$$V + \Delta V = \frac{4\pi}{3}(r + \Delta r)^3$$

$$= \frac{4\pi}{3}[r^3 + 3r^2\,\Delta r + 3r\,(\Delta r)^2 + (\Delta r)^3]$$

$$\Delta V = \frac{4\pi}{3}[3r^2\,\Delta r + 3r\,(\Delta r)^2 + (\Delta r)^3]$$

$$\frac{\Delta V}{\Delta r} = \frac{4\pi}{3}[3r^2 + 3r\,\Delta r + (\Delta r)^2]$$

$$\lim_{\Delta r \to 0}\frac{\Delta V}{\Delta r} = 4\pi r^2$$

We can see that, as the radius increases, the rate of change of the volume with respect to the radius also increases. This should be expected, since a sphere changing from a radius of 1 to a radius of 2 has an increase in volume from $4\pi/3$ to $32\pi/3$, whereas a change in the radius from 2 to 3 involves a volume increase from $32\pi/3$ to $108\pi/3$.

Example E

The power P produced by an electric current i in a resistor varies directly as the square of the current. Given that 1.2 W of power are produced by a current of 0.5 A in a particular resistor, find an expression for the rate of change of power with respect to current.

We must first find the functional relationship between power and current. This we can do by solving the indicated problem in variation:

$$P = ki^2, \qquad 1.2 = k(0.5)^2, \qquad k = 4.8 \text{ W/A}^2, \qquad P = 4.8i^2$$

Now, knowing the function, we may determine the expression for the rate of change of P with respect to i by use of the Δ-process:

$$P + \Delta P = 4.8(i + \Delta i)^2 = 4.8[i^2 + 2i\,(\Delta i) + (\Delta i)^2]$$

$$\Delta P = 4.8[2i\,(\Delta i) + (\Delta i)^2]$$

$$\frac{\Delta P}{\Delta i} = 4.8(2i + \Delta i)$$

$$\lim_{\Delta i \to 0}\frac{\Delta P}{\Delta i} = 9.6i$$

This last expression tells us that as the current increases, the power changes at a rate of $9.6i$ W/A. For example, if the current is 2 A, the power is changing at the rate of 19.2 W/A. The larger the current, the greater the increase in power. This should be expected, since the power varies as the square of the current.

Exercises 2–5

In Exercises 1 through 4 calculate the instantaneous velocity for the indicated value of the time (in seconds) of an object for which the displacement (in feet) is given by the indicated function. Use the method of Example A, and calculate values of $\Delta s/\Delta t$ for the given values of t, and determine the limit as Δt approaches zero.

1. $s = 4t + 10$; when $t = 3$; use values of t of 2.0, 2.5, 2.9, 2.99, 2.999
2. $s = 20 - 8t$; when $t = 2$; use values of t of 1.0, 1.5, 1.9, 1.99, 1.999
3. $s = 1 - 2t^2$; when $t = 4$; use values of t of 3.0, 3.5, 3.9, 3.99, 3.999
4. $s = 120t - 16t^2$; when $t = 0.5$; use values of t of 0.4, 0.45, 0.49, 0.499, 0.4999

In Exercises 5 through 8 use the Δ-process to find an expression for the instantaneous velocity of an object moving with rectilinear motion according to the given functions (the same as those of Exercises 1 through 4) relating s (in feet) and t (in seconds). Then calculate the instantaneous velocity for the given value of t.

5. $s = 4t + 10$; $t = 3$ 6. $s = 20 - 8t$; $t = 2$
7. $s = 1 - 2t^2$; $t = 4$ 8. $s = 120t - 16t^2$; $t = 0.5$

In Exercises 9 through 12 use the Δ-process to find an expression for the instantaneous velocity of an object moving with rectilinear motion according to the given functions relating s and t.

9. $s = 3t - \dfrac{2}{t}$ 10. $s = \dfrac{1}{2t + 3}$

11. $s = t^3 - 6t + 2$ 12. $s = s_0 + v_0 t - \frac{1}{2}at^2$ (s_0, v_0, and a are constants)

In Exercises 13 through 16 use the Δ-process to find an expression for the instantaneous acceleration a of an object moving with rectilinear motion according to the given functions. The instantaneous acceleration of an object is defined as the instantaneous rate of change of velocity with respect to time.

13. $v = 6t^2 - 4t + 2$ 14. $v = v_0 - gt$ (v_0 and g are constants)
15. $s = 1 - 2t^2$ (find v, then find a)
16. $s = s_0 + v_0 t - \frac{1}{2}at^2$ (s_0, v_0, and a are constants) (find v, then find a)

In Exercises 17 through 28 find the indicated rates of change.

17. The perimeter p of a rectangular field is a function of its width w according to $p = 2(50 + w)$, where p and w are measured in meters. Find the expression for the instantaneous rate of change of p with respect to w.

18. The profit P in selling 100 items at price p, when the cost of each item is $10 is given by $P = 100(p - 10)$. Find the expression for the instantaneous rate of change of P with respect to p.

19. Find a general expression for the instantaneous rate of change of the area of a circle with respect to its radius. Then find the instantaneous rate of change of A with respect to r, when $r = 3$.

20. The centripetal acceleration a of an object moving in a circular path of radius r with a velocity v is given by $a = v^2/r$. Find the expression for the instantaneous rate of change of the acceleration with respect to the velocity for an object moving in a circular path of radius 2 ft. The acceleration is in feet per square second and the velocity is in feet per second.

21. The resistance of a certain resistor as a function of the temperature is given by the relation $R = 50(1 + 0.0053T + 0.00001T^2)$. Determine the instantaneous rate of change of resistance with respect to temperature (in ohms per degree Celsius) when the temperature is 20°C.

22. At very low temperatures, near absolute zero, the specific heat c of solids is given by the Debye equation $c = kT^3$, where T is the thermodynamic temperature and k is a constant dependent on the material. Determine the expression for the instantaneous rate of change for the specific heat with respect to temperature. The units of the specific heat are joules per kilogram kelvin and T is measured in kelvins.

23. A 5-Ω resistor and a variable resistor of resistance R are placed in parallel. The expression for the resulting resistance R_T is given by $R_T = 5R/(5 + R)$. Determine the instantaneous rate of change of R_T with respect to R, and then evaluate this expression for $R = 8\ \Omega$.

24. The value (in dollars) of a certain car is given by the function $V = \frac{6000}{t+1}$, where t is measured in years. Find a general expression for the rate of change of V with respect to t, and then evaluate this expression when $t = 3$ years.

25. The volume of a certain right circular cylinder is 100 in.^{3}. Find the expression for the instantaneous rate of change of the total surface area of the cylinder with respect to the radius of the base.

26. The total energy supplied to an inductor while the current increases from 0 to I is proportional to the square of the current I. If 8 J are supplied to a certain inductor as the current increases from 0 to 2 A, evaluate the expression for the instantaneous rate of change of energy with respect to I when $I = 1.5$ A.

27. The illuminance of a light source varies inversely as the square of the distance from the source. Given that the illuminance is 25 lx(lux) at a distance of 0.20 m, find the expression for the instantaneous rate of change of illuminance with respect to distance. Calculate this rate for distances of 0.01 m and 0.10 m.

28. The frequency of vibration of a wire varies directly as the square root of the tension on the wire. Find the expression for the instantaneous rate of change of frequency with respect to tension. Find the value of this rate of change for $T = 1$ N if the frequency is 400 Hz when $T = 1$ N.

2—6 Derivatives of Polynomials

The task of finding the derivative of a function can be considerably shortened from that involved in the direct use of the Δ-process. We can use the Δ-process to derive certain basic formulas for finding derivatives of particular types of functions. These formulas will then be used to find the derivatives. In this section, we shall derive the formulas for finding the derivatives of polynomial functions of the form $f(x) = a_0 x^n + a_1 x^{n-1} + \cdots + a_n$.

First we shall find the derivative of a constant. By letting $y = c$, and applying the Δ-process to this function, we can obtain the desired result:

$$y = c, \qquad y + \Delta y = c, \qquad \Delta y = 0, \qquad \frac{\Delta y}{\Delta x} = 0, \qquad \lim_{\Delta x \to 0} \frac{\Delta y}{\Delta x} = 0$$

From this we conclude that *the derivative of a constant is zero.* No assumption was made as to the value of the constant, and thus this result holds for all constants. Therefore, if $y = c$, $dy/dx = 0$. This is more commonly written as

$$\frac{dc}{dx} = 0 \tag{2–22}$$

Graphically this means that for any function of the type $y = c$ the slope is always zero. From our knowledge of straight lines, we know that $y = c$ represents a straight line parallel to the x-axis. From the definition of slope, we know that any line parallel to the x-axis has a slope of zero. We can see that the two results are consistent.

Next we shall find the derivative of any integral power of x. If $y = x^n$, where n is an integer, by use of the binomial theorem we have the following:

$$(y + \Delta y) = (x + \Delta x)^n$$

$$= x^n + nx^{n-1}\,\Delta x + \frac{n(n-1)}{2}x^{n-2}\,(\Delta x)^2 + \cdots + (\Delta x)^n$$

$$\Delta y = nx^{n-1}\,\Delta x + \frac{n(n-1)}{2}x^{n-2}\,(\Delta x)^2 + \cdots + (\Delta x)^n$$

$$\frac{\Delta y}{\Delta x} = nx^{n-1} + \frac{n(n-1)}{2}x^{n-2}\,\Delta x + \cdots + (\Delta x)^{n-1}$$

$$\lim_{\Delta x \to 0} \frac{\Delta y}{\Delta x} = nx^{n-1}$$

Thus, *the derivative of the nth power of x is found to be*

$$\frac{dx^n}{dx} = nx^{n-1} \tag{2–23}$$

Example A
Find the derivative of the function $y = -5$.
Applying the result of Eq. (2–22), since -5 is a constant, we have

$$\frac{dy}{dx} = \frac{d(-5)}{dx} = 0$$

or

$$\frac{dy}{dx} = 0$$

Example B
Find the derivative of the function $y = x$.
   Applying the result of Eq. (2–23) (here $n = 1$), we have

$$\frac{dy}{dx} = \frac{d(x)}{dx} = (1)x^{1-1} = 1$$

or simply

$$\frac{dy}{dx} = 1$$

Thus, the derivative of $y = x$ is 1, which means that the slope of the line $y = x$ is always 1. This is consistent with our previous discussion of the straight line.

Example C
Find the derivative of $y = x^3$.
   We find that

$$\frac{dy}{dx} = \frac{d(x^3)}{dx}$$

$$= 3x^{3-1}$$

$$= 3x^2$$

We can see that this result is consistent with the results found in the examples and exercises of the previous sections in this chapter.

Example D
Find the derivative of $y = x^{10}$.
   We find that

$$\frac{dy}{dx} = \frac{d(x^{10})}{dx}$$

$$= 10x^{10-1}$$

$$= 10x^9$$

Next we shall find the derivative of a constant times a function of x. If $y = cu$, where $u = f(x)$, we have the following result:

$$y + \Delta y = c(u + \Delta u)$$

(As x increases by Δx, u increases by Δu, since u is a function of x.) Then

$$\Delta y = c\,\Delta u, \qquad \frac{\Delta y}{\Delta x} = c\frac{\Delta u}{\Delta x}, \qquad \lim_{\Delta x \to 0} \frac{\Delta y}{\Delta x} = c \lim_{\Delta x \to 0} \frac{\Delta u}{\Delta x}$$

Therefore, *the derivative of the product of a constant and a function of x is the product of the constant and the derivative of the function of x.* This is written as

$$\frac{d(cu)}{dx} = c\frac{du}{dx} \tag{2−24}$$

Example E
Find the derivative of $y = 3x^2$.
In this case $c = 3$ and $u = x^2$. Thus, $du/dx = 2x$. Therefore,

$$\frac{dy}{dx} = 3(2x) = 6x$$

Occasionally the derivative of a constant times a function of x is confused with the derivative of a constant (alone). The distinction between a constant multiplying a function and an isolated constant must be made.

Finally, if the types of functions for which we have found derivatives are added, the result is a polynomial function with more than one term. The derivative of such a function is found by letting $y = u + v$, where u and v are functions of x. Applying the Δ-process, since u and v are functions of x, each has an increment corresponding to an increment in x. Thus, we have the following result:

$$y + \Delta y = (u + \Delta u) + (v + \Delta v)$$

$$\Delta y = \Delta u + \Delta v$$

$$\frac{\Delta y}{\Delta x} = \frac{\Delta u}{\Delta x} + \frac{\Delta v}{\Delta x}$$

$$\lim_{\Delta x \to 0} \frac{\Delta y}{\Delta x} = \lim_{\Delta x \to 0} \frac{\Delta u}{\Delta x} + \lim_{\Delta x \to 0} \frac{\Delta v}{\Delta x}$$

This tells us that the derivative of *the sum of functions of x is the sum of the derivatives of the functions.* This is written as

$$\frac{d(u + v)}{dx} = \frac{du}{dx} + \frac{dv}{dx} \tag{2−25}$$

Example F

Find the derivative of $y = 4x^2 + 5$.

Here $u = 4x^2$ and $v = 5$. Thus, $du/dx = 8x$ and $dv/dx = 0$. Hence we have

$$\frac{dy}{dx} = \frac{d(4x^2)}{dx} + \frac{d(5)}{dx}$$

$$= 8x + 0$$
$$= 8x$$

Example G

Find the derivative of $y = 5x^6 - 2x^3 - x + 7$.

$$\frac{dy}{dx} = \frac{d(5x^6)}{dx} - \frac{d(2x^3)}{dx} - \frac{dx}{dx} + \frac{d(7)}{dx}$$

$$= 30x^5 - 6x^2 - 1$$

Example H

Find the slope of a line which is tangent to the curve of the function $y = 4x^7 - x^4$ at the point $(1, 3)$.

We must find, and then evaluate, the derivative of the function for the value $x = 1$:

$$\frac{dy}{dx} = 28x^6 - 4x^3$$

For evaluating this, we write

$$\frac{dy}{dx}\bigg|_{x=1} = 28(1) - 4(1) = 24$$

Thus, the slope of the tangent line is 24. Again, we note that the substitution $x = 1$ must be made after the differentiation has been performed.

Example I

Find the instantaneous velocity when $t = 3$ s for an object moving with rectilinear motion according to the function $s = 8t - 2t^3$, where s is measured in centimeters.

In this example we must find the derivative of s with respect to t, ds/dt, and then evaluate it for $t = 3$. Thus,

$$s = 8t - 2t^3$$

$$\frac{ds}{dt} = 8 - 6t^2$$

$$\frac{ds}{dt}\bigg|_{t=3} = 8 - 6(3^2) = 8 - 54 = -46 \text{ cm/s}$$

Therefore, the object is moving at -46 cm/s (it is moving in a negative direction) when $t = 3$ s.

Exercises 2–6

In Exercises 1 through 16 find the derivative of each of the given functions.

1. $y = x^5$
2. $y = x^{12}$
3. $y = 4x^9$
4. $y = -7x^6$
5. $y = x^4 - 6$
6. $y = 3x^5 - 1$
7. $y = x^2 + 2x$
8. $y = x^3 - 2x^2$
9. $y = 5x^3 - x - 1$
10. $y = 6x^2 - 6x + 5$
11. $y = x^8 - 4x^7 - x$
12. $y = 4x^4 - 2x + 9$
13. $y = 6x^7 - 5x^3 + 2$
14. $y = 13x^4 - 6x^3 - x - 1$
15. $y = \frac{1}{3}x^3 + \frac{1}{2}x^2$
16. $y = \frac{1}{4}x^8 - \frac{1}{2}x^4 - \sqrt{3}$

In Exercises 17 through 20 find the slope of a tangent line to the curve of each of the given functions for the given values of x.

17. $y = 2x^6 - 6x^2 \quad (x = 2)$
18. $y = 3x^3 - 9x \quad (x = 1)$
19. $y = 6x - 2x^4 \quad (x = -1)$
20. $y = x^4 - \frac{1}{2}x^2 + 2 \quad (x = -2)$

In Exercises 21 through 24 find an expression for the instantaneous velocity of objects moving according to the functions given, if s represents displacement in terms of time t.

21. $s = 6t^5 - 5t + 2.$
22. $s = 4t^2 - 7t$
23. $s = 120t - 16t^2$
24. $s = s_0 + v_0 t + \frac{1}{2}at^2$

In Exercises 25 through 32 solve the given problems by finding the appropriate derivative.

25. For what value(s) of x is the tangent to the curve of $y = 3x^2 - 6x$ parallel to the x-axis? (That is, where is the slope zero?)

26. For what value(s) of x is the tangent to the curve of $y = 4x^3 - 12x^2$ parallel to the x-axis? (See Exercise 25.)

27. If the radius of a right circular cylinder always equals its height, find an expression for the instantaneous rate of change of volume with respect to the radius.

28. If the length of a pillar is constant, the crushing load varies as the fourth power of its radius. If 18 tons will crush a certain pillar 5 in. in radius, find an expression which gives the instantaneous rate of change of crushing load with respect to a change in radius.

29. The voltage V produced by a certain thermocouple as a function of the temperature T is given by $V = 4.0T + 0.005T^2$. If the temperature changes slowly, find the instantaneous rate of change of voltage with respect to temperature when the temperature is 200°C.

30. The cost C (in dollars) to a firm producing x units is given by the equation $C = 0.02x^3 - 2.40x^2 + 100x$. Determine the instantaneous rate of change of C with respect to x for $x = 50$ units.

31. The altitude h (in meters) of an airplane as a function of the horizontal distance x (in kilometers) it has traveled is given by $h = 30x^2 - x^3$. Find the instantaneous rate of change of h with respect to x for $x = 10$ km.

32. The ends of a 10-ft beam are supported at different levels. The deflection y of the beam is given by $y = kx^2(x^3 + 450x - 3500)$, where x is the horizontal distance from one end and k is a constant. Determine the expression for the instantaneous rate of change of deflection with respect to x.

2—7 Derivatives of Products and Quotients of Functions

The formulas developed in the previous section are valid for polynomial functions. However, many functions are not polynomial in form. Some functions can best be expressed as the product of two or more simpler functions, others are the quotient of two simpler functions, and some are expressed as powers of a function. In this section, we shall develop the formula for the derivative of a product of functions and the formula for the derivative of the quotient of two functions.

Example A

The functions $f(x) = x^2 + 2$ and $g(x) = 3 - 2x$ can be combined to form new functions which represent the types mentioned above. For example, the function $p(x) = f(x)g(x) = (x^2 + 2)(3 - 2x)$ is an example of a function expressed as the product of two simpler functions. The function

$$q(x) = \frac{g(x)}{f(x)} = \frac{3 - 2x}{x^2 + 2}$$

is an example of a rational function which is the quotient of two other functions. The function $F(x) = (g(x))^3 = (3 - 2x)^3$ is an example of a power of a function.

If u and v both represent functions of x, the derivative of the product of u and v is found by applying the Δ-process. This leads to the following result:

$$y = u \cdot v$$
$$y + \Delta y = (u + \Delta u)(v + \Delta v)$$

(Since u and v are functions of x, each has an increment corresponding to an increment in x.) Then we have

$$\Delta y = u\,\Delta v + v\,\Delta u + \Delta u\,\Delta v$$

$$\frac{\Delta y}{\Delta x} = u\frac{\Delta v}{\Delta x} + v\frac{\Delta u}{\Delta x} + \Delta u\frac{\Delta v}{\Delta x}$$

$$\lim_{\Delta x \to 0}\frac{\Delta y}{\Delta x} = u\lim_{\Delta x \to 0}\frac{\Delta v}{\Delta x} + v\lim_{\Delta x \to 0}\frac{\Delta u}{\Delta x} + \lim_{\Delta x \to 0}\left(\Delta u\frac{\Delta v}{\Delta x}\right)$$

(The functions u and v are not affected by Δx approaching zero, but Δu and Δv both approach zero as Δx approaches 0.) Thus,

$$\frac{dy}{dx} = u\frac{dv}{dx} + v\frac{du}{dx} + 0\frac{dv}{dx}$$

✓   We conclude that *the derivative of the product of two functions equals the first function times the derivative of the second function plus the second function times the derivative of the first function.* This is written as

$$\frac{d(uv)}{dx} = u\frac{dv}{dx} + v\frac{du}{dx} \qquad\qquad (2\text{–}26)$$

Example B

For the product function in Example A, the derivative is found as follows:

$$y = (x^2 + 2)(3 - 2x), \qquad u = x^2 + 2, \qquad v = 3 - 2x$$

$$\frac{dy}{dx} = (x^2 + 2)(-2) + (3 - 2x)(2x) = -2x^2 - 4 + 6x - 4x^2$$

$$= -6x^2 + 6x - 4$$

Example C

Find the derivative of the function $y = (3 - x - 2x^2)(x^4 - x)$.

In this problem, $u = 3 - x - 2x^2$ and $v = x^4 - x$. Hence

$$\frac{dy}{dx} = (3 - x - 2x^2)(4x^3 - 1) + (x^4 - x)(-1 - 4x)$$

$$= 12x^3 - 3 - 4x^4 + x - 8x^5 + 2x^2 - x^4 - 4x^5 + x + 4x^2$$
$$= -12x^5 - 5x^4 + 12x^3 + 6x^2 + 2x - 3$$

In both these examples, we could have multiplied the functions first and then taken the derivative as a polynomial. However, we shall soon meet functions for which this latter method would not be applicable.

We shall now find the derivative of the quotient of two functions. We apply the Δ-process to the function $y = u/v$, and obtain the result as follows:

$$y + \Delta y = \frac{u + \Delta u}{v + \Delta v}$$

$$\Delta y = \frac{u + \Delta u}{v + \Delta v} - \frac{u}{v} = \frac{vu + v\,\Delta u - uv - u\,\Delta v}{v(v + \Delta v)}$$

$$\frac{\Delta y}{\Delta x} = \frac{v(\Delta u/\Delta x) - u(\Delta v/\Delta x)}{v(v + \Delta v)}$$

$$\lim_{\Delta x \to 0}\frac{\Delta y}{\Delta x} = \frac{v\lim_{\Delta x \to 0}(\Delta u/\Delta x) - u\lim_{\Delta x \to 0}(\Delta v/\Delta x)}{\lim_{\Delta x \to 0}v(v + \Delta v)}$$

Thus,

$$\frac{dy}{dx} = \frac{v\,(du/dx) - u\,(dv/dx)}{v^2}$$

This last result says that *the derivative of the quotient of two functions equals the denominator times the derivative of the numerator minus the* numerator times the derivative of the denominator, all divided by the square of the denominator. This is written as

$$\checkmark \quad \frac{d\dfrac{u}{v}}{dx} = \frac{v\dfrac{du}{dx} - u\dfrac{dv}{dx}}{v^2} \qquad\qquad (2\text{--}27)$$

Example D

Find the derivative of the quotient indicated in Example A.

$$y = \frac{3 - 2x}{x^2 + 2}, \qquad u = 3 - 2x, \qquad v = x^2 + 2$$

$$\frac{dy}{dx} = \frac{(x^2 + 2)(-2) - (3 - 2x)(2x)}{(x^2 + 2)^2} = \frac{-2x^2 - 4 - 6x + 4x^2}{(x^2 + 2)^2}$$

$$= \frac{2(x^2 - 3x - 2)}{(x^2 + 2)^2}$$

Example E

Find the derivative of

$$y = \frac{x^3 - x - 1}{1 - 2x^2}$$

$$\frac{dy}{dx} = \frac{(1 - 2x^2)(3x^2 - 1) - (x^3 - x - 1)(-4x)}{(1 - 2x^2)^2}$$

$$= \frac{3x^2 - 1 - 6x^4 + 2x^2 + 4x^4 - 4x^2 - 4x}{(1 - 2x^2)^2}$$

$$= \frac{-2x^4 + x^2 - 4x - 1}{(1 - 2x^2)^2}$$

Exercises 2—7

In Exercises 1 through 8 find the derivative of each function by Eq. (2–26). Then check by multiplying out before finding dy/dx, treating the function as a polynomial.

1. $y = x^2(3x + 2)$ 2. $y = 3x(x^3 + 1)$
3. $y = 6x(3x^2 - 5x)$ 4. $y = 2x^3(3x^4 + x)$
5. $y = (x + 2)(2x - 5)$ 6. $y = (3x - 1)(x^2 + 1)$
7. $y = (x^4 - 3x^2 + 3)(1 - 2x^3)$ 8. $y = (x^3 - 6x)(2 - 4x^3)$

In Exercises 9 through 16 find the derivative of each function by Eq. (2–27).

9. $y = \dfrac{x}{2x + 3}$ 10. $y = \dfrac{2x}{x + 1}$ 11. $y = \dfrac{1}{x^2 + 1}$ 12. $y = \dfrac{x + 2}{2x + 3}$

13. $y = \dfrac{x^2}{3 - 2x}$ 14. $y = \dfrac{x + 8}{x^2 + x + 2}$ 15. $y = \dfrac{3x}{4x^5 - 3x - 4}$ 16. $y = \dfrac{2}{3x^2 - 5x}$

In Exercises 17 through 28 solve the given problems by finding the appropriate derivatives.

17. Find the derivative of $y = \dfrac{x^2(1 - 2x)}{3x - 7}$, but do not multiply out the numerator before taking the derivative.

18. Find the derivative of $y = 4x^2 - \dfrac{1}{x - 1}$, but do not combine the terms before finding the derivative.

19. Find the slope of a tangent line to the curve $y = (4x + 1)(x^4 - 1)$ at the point $(-1, 0)$. Do not multiply the factors together before taking the derivative.

20. Find the slope of a line tangent to the curve of the function $y = (3x + 4)(1 - 4x)$ at the point $(2, -70)$. Do not multiply before finding the derivative.

21. For what value(s) of x is the slope of a tangent to the curve $y = \dfrac{x}{x^2 + 1}$ equal to zero?

22. Determine the sign of the derivative of the function $y = \dfrac{2x - 1}{1 - x^2}$ for the following values of x: $-2, -1, 0, 1, 2$. Is the slope of a tangent to this curve ever negative?

23. If the distance (in feet) traveled by a particle in a given time is found from the expression $s = 2/t^2$, find the velocity after 4 s. (Use the quotient rule.)

24. During a chemical change the number n of grams of a compound being formed is given by $n = \dfrac{6t}{t + 2}$, where t is measured in seconds. How many grams per second are being formed after 3 s?

25. The voltage across a resistor in an electric circuit is the product of the resistance and the current. If the current I (in amperes) varies with time (in seconds) according to the relation $I = 5.00 + 0.01t^2$, and the resistance varies with time according to the relation $R = 15.00 - 0.10t$, find the time rate of change of the voltage when $t = 5$ s.

26. In thermodynamics, an equation relating pressure p, volume V, and temperature T is

$$\left(p + \frac{a}{V^2}\right)(V - b) = RT$$

where a, b, and R are constants. This equation is known as the van der Waals equation. Solve this equation for p, and then find the expression for the rate of change of p with respect to V, assuming a constant temperature.

27. The electric power produced by a certain source is given by

$$P = \frac{E^2 r}{R^2 + 2Rr + r^2}$$

where E is the voltage of the source, R is the resistance of the source, and r is the resistance in the circuit. Find the expression for the rate of change of power with respect to the resistance in the circuit, assuming that the other quantities remain constant.

28. In the study of simple harmonic motion, under certain conditions the tangent of the angle by which the displacement lags the impress force is given by

$$\tan \varphi = \frac{30\omega}{600 - \omega^2}$$

where ω, (in hertz) is the frequency of vibration. Find the rate of change of $\tan \varphi$ with respect to ω when $\omega = 3$ Hz.

2–8 The Derivative of a Power of a Function

In Example A of Section 2–7 we illustrated $y = (3 - 2x)^3$ as the power of a function of x. Here, $3 - 2x$ is the function of x, and it is being raised to the third power. If we let $u = 3 - 2x$, we can write

$$y = u^3 \quad \text{where } u = 3 - 2x$$

Writing it this way, y is a function of u and u is a function of x. This means that y is a function of a function of x, often referred to as a **composite function**. However, y is still a function of x, since u is a function of x.

Since we will frequently need to find the derivative of the power of a function, we now develop the necessary formula. Considering $y = f(u)$ and $u = g(x)$, we can express the derivative dy/dx in terms of dy/du and du/dx. If Δx is the increment in x, then Δy and Δu are the corresponding increments in y and u, respectively. We may then write

$$\frac{\Delta y}{\Delta x} = \frac{\Delta y}{\Delta u} \cdot \frac{\Delta u}{\Delta x}$$

When Δx approaches zero, Δu and Δy both approach zero, for u and y are functions of x. Thus,

$$\lim_{\Delta x \to 0} \frac{\Delta y}{\Delta x} = \lim_{\Delta u \to 0} \frac{\Delta y}{\Delta u} \cdot \lim_{\Delta x \to 0} \frac{\Delta u}{\Delta x}$$

or

$$\frac{dy}{dx} = \frac{dy}{du} \cdot \frac{du}{dx} \qquad (2-28)$$

(Here we have assumed $\Delta u \neq 0$, although it can be shown that this condition is not necessary.) *Equation (2–28) is known as the* **chain rule** *for derivatives.*

Using Eq. (2–28) for $y = u^n$, where u is a function of x, we have

$$\frac{dy}{dx} = \frac{d(u^n)}{du} \cdot \frac{du}{dx}$$

or

$$\frac{du^n}{dx} = nu^{n-1}\left(\frac{du}{dx}\right) \tag{2–29}$$

By use of Eq. (2–29) we may find the derivative of a power of a function of x.

Example A

Find the derivative of $y = (3 - 2x)^3$.

For this function, we identify $n = 3$ and $u = 3 - 2x$. This means that $du/dx = -2$, and therefore

$$\frac{dy}{dx} = 3(3 - 2x)^2(-2)$$

$$= -6(3 - 2x)^2$$

A common type of error in finding this type of derivative is to omit the du/dx factor; in this case it is the -2. The derivative is incomplete, and therefore incorrect without this factor.

Example B

Find the derivative of $y = (1 - 3x^2)^4$.

In this example, $n = 4$ and $u = 1 - 3x^2$. Hence

$$\frac{dy}{dx} = 4(1 - 3x^2)^3(-6x) = -24x(1 - 3x^2)^3$$

(We must not forget the $-6x$.)

Example C

Find the derivative of

$$y = \frac{(3x - 1)^3}{1 - x}$$

Here we must use the quotient rule in combination with the power rule. We find the derivative of the numerator by using the power rule.

$$\frac{dy}{dx} = \frac{(1 - x)3(3x - 1)^2(3) - (3x - 1)^3(-1)}{(1 - x)^2}$$

$$= \frac{(3x - 1)^2(9 - 9x + 3x - 1)}{(1 - x)^2} = \frac{(3x - 1)^2(8 - 6x)}{(1 - x)^2}$$

$$= \frac{2(3x - 1)^2(4 - 3x)}{(1 - x)^2}$$

It is normally better to have the derivative in a factored, simplified form, since this is the only form from which useful information may readily be obtained. In this form we can determine where the derivative is undefined (denominator equal to zero) or where the slope is zero (numerator equal to zero); we can also make other required analyses of the derivative. Thus, all derivatives should be simplified.

So far, we have derived the formulas for the derivatives of powers of x and for functions of x for integral powers. We shall now establish that these formulas are also valid for any rational number used as an exponent. If $y = u^{p/q}$, and if each side is raised to the qth power, we have $y^q = u^p$. Applying the power rule to each side of this equation, we have

$$qy^{q-1}\left(\frac{dy}{dx}\right) = pu^{p-1}\left(\frac{du}{dx}\right)$$

Solving for dy/dx, we have

$$\frac{dy}{dx} = \frac{pu^{p-1}(du/dx)}{qy^{q-1}} = \frac{p}{q}\frac{u^{p-1}}{(u^{p/q})^{q-1}}\frac{du}{dx} = \frac{p}{q}\frac{u^{p-1}}{u^{p-p/q}}\frac{du}{dx}$$

$$= \frac{p}{q}u^{p-1-p+(p/q)}\frac{du}{dx}$$

Thus,

$$\frac{du^{p/q}}{dx} = \frac{p}{q}u^{(p/q)-1}\frac{du}{dx} \qquad (2\text{--}30)$$

We see that in finding the derivative we multiply the function by the rational exponent and subtract 1 from it to find the exponent of the function in the derivative. This, of course, is the same rule derived for integral exponents. Since the only restriction on the values of p and q is that they are integers ($q \neq 0$), *the power rule can be used for all rational exponents, positive and negative.*

Example D

We can now find the derivative of $y = \sqrt{x^2 + 1}$.

By use of Eq. (2–30), or Eq. (2–29), and writing the square root as the fractional exponent $\frac{1}{2}$, we can easily derive the result:

$$y = (x^2 + 1)^{1/2}$$

$$\frac{dy}{dx} = \frac{1}{2}(x^2 + 1)^{-1/2}(2x) = \frac{x}{(x^2 + 1)^{1/2}}$$

To avoid introducing apparently significant factors into the numerator, we do not usually rationalize such fractions.

Having shown that we may use fractional exponents to find derivatives of roots of functions of x, we may also use them to find derivatives of roots of x itself.. Consider the following example.

Example E

Find the derivative of $y = 6\sqrt[3]{x}$.

We can write this function as $y = 6x^{1/3}$. In finding the derivative we may use Eq. (2–29), where we have $u = x$. This means that $du/dx = dx/dx = 1$. Therefore, using Eq. (2–29) is equivalent to using Eq. (2–23) for a power of x (not for a power of a function of x). Thus,

$$y = 6x^{1/3}$$

$$\frac{dy}{dx} = 6\left(\frac{1}{3}\right)x^{-2/3} = \frac{2}{x^{2/3}}$$

Example F

Find the derivative of

$$y = \frac{1}{(1 - 4x)^2}$$

We can handle this more easily by direct use of the power rule, rather than the quotient rule. By writing the function as $y = (1 - 4x)^{-2}$, we can easily find the derivative:

$$\frac{dy}{dx} = (-2)(1 - 4x)^{-3}(-4) = \frac{8}{(1 - 4x)^3}$$

(Remember, subtracting 1 from -2 gives -3.)

We can now see the value of fractional exponents in calculus. They are useful in algebraic operations, but are almost essential in calculus. Without fractional exponents, it would be necessary to develop more formulas to find derivatives of radicals. To find the derivative of an algebraic function, we need only those equations which we have already developed. Often it is necessary to combine these, as we saw in Example C. Actually, most derivatives are combinations. The problem in finding the derivative is recognizing the form of the function with which you are dealing. When you have recognized the form, completing the problem is only a matter of mechanics and algebra. You should now see the importance of being able to handle algebraic operations with facility.

Example G
Find the derivative of

$$y = \frac{x}{\sqrt{1 - 4x}}$$

Here we have a quotient, and in order to find the derivative of this quotient, we must also use the power rule (and a derivative of a polynomial form). With sufficient practice in taking derivatives, we can recognize the rule to use almost automatically. Thus, we find the derivative:

$$\frac{dy}{dx} = \frac{(1 - 4x)^{1/2}(1) - x(\frac{1}{2})(1 - 4x)^{-1/2}(-4)}{1 - 4x}$$

$$= \frac{(1 - 4x)^{1/2} + \dfrac{2x}{(1 - 4x)^{1/2}}}{1 - 4x}$$

$$= \frac{\dfrac{(1 - 4x)^{1/2}(1 - 4x)^{1/2} + 2x}{(1 - 4x)^{1/2}}}{1 - 4x} = \frac{(1 - 4x) + 2x}{(1 - 4x)^{1/2}(1 - 4x)}$$

$$= \frac{1 - 2x}{(1 - 4x)^{3/2}}$$

Exercises 2−8

In Exercises 1 through 24 find the derivative of each of the given functions.

1. $y = \sqrt{x}$

2. $y = \sqrt[4]{x}$

3. $y = \dfrac{1}{x^2}$

4. $y = \dfrac{2}{x^4}$

5. $y = \dfrac{3}{\sqrt[3]{x}}$

6. $y = \dfrac{1}{\sqrt{x}}$

7. $y = x\sqrt{x} - \dfrac{1}{x}$

8. $y = 2x^{-3} - x^{-2}$

9. $y = (x^2 + 1)^5$

10. $y = (1 - 2x)^4$

11. $y = 2(7 - 4x^3)^8$

12. $y = 3(8x^2 - 1)^6$

13. $y = (2x^3 - 3)^{1/3}$

14. $y = (1 - 6x)^{3/2}$

15. $y = \dfrac{1}{(1 - x^2)^4}$

16. $y = \dfrac{1}{\sqrt{1 - x}}$

17. $y = 4(2x^4 - 5)^{3/4}$

18. $y = 5(3x^7 - 4)^{2/3}$

19. $y = \sqrt[4]{1 - 8x^2}$

20. $y = \sqrt[3]{4x^6 + 2}$

21. $y = x\sqrt{x - 1}$

22. $y = x^2(1 - 3x)^5$

23. $y = \dfrac{\sqrt{4x + 3}}{8x + 1}$

24. $y = \dfrac{x\sqrt{x - 1}}{2x + 1}$

In Exercises 25 through 36 solve the given problems by finding the appropriate derivative.

25. Find any values of x for which the derivative of $y = \dfrac{x^2}{\sqrt{x^2 + 1}}$ is zero.

26. Find any values of x for which the derivative of $y = \dfrac{x}{\sqrt{4x - 1}}$ is zero.

27. Find the slope of a line tangent to the parabola $y^2 = 4x$ at the point $(1, 2)$.

28. Find the slope of a tangent to the circle $x^2 + y^2 = 25$ at the point $(4, 3)$.

29. Find the velocity at $t = 4$, given that the distance in terms of the time for a particular object is $s = (t^3 - t)^{4/3}$.

30. The period of a pendulum is directly proportional to the square root of its length. Find the derivative of the period with respect to the length for a pendulum 2 ft long which has a period of $\pi/2$ s.

31. When the volume of a gas changes very rapidly, an approximate relation is that the pressure varies inversely as the $\frac{3}{2}$ power of the volume. If P is 300 kPa when $V = 100$ cm^3, find the derivative of P with respect to V. Evaluate this derivative for $V = 100$ cm^3.

32. Under certain conditions, the velocity of an object moving with simple harmonic motion is given by $v = k\sqrt{A^2 - x^2}$, where x is the x-coordinate of the position of the object, A is the amplitude, and k is a constant. Find the derivative of v with respect to x.

33. Under certain conditions, due to the presence of a charge q, the electric potential V along a line is given by

$$V = \frac{kq}{\sqrt{x^2 + b^2}}$$

where k is a constant and b is the minimum distance from the charge to the line. Find the expression for the rate of change of V with respect to x.

34. The current in a circuit containing a resistance R and an inductance L is found from the expression

$$I = \frac{V}{\sqrt{R^2 + (\omega L)^2}}$$

Find the expression for the rate of change of current with respect to L, assuming that the other quantities remain constant.

35. The magnetic field H due to a magnet of length l at a distance r is given by

$$H = \frac{k}{[r^2 + (l/2)^2]^{3/2}}$$

where k is a constant for a given magnet. Find the expression for the derivative of H with respect to r.

36. Under certain circumstances the velocity of a water wave is given by

$$v = \sqrt{\frac{a}{\lambda} + b\lambda}$$

where λ is the wavelength and a and b are constants. Find the derivative of v with respect to λ.

2—9 Differentiation of Implicit Functions

To this point the functions which we have differentiated have been of the form $y = f(x)$. There are, however, occasions when we need to find the derivative of a function determined by an equation which does not express the dependent variable explicitly in terms of the independent variable.

An equation in which y is not expressed explicitly in terms of x may determine one or more functions. *Any such function, where y is defined implicitly as a function of x, is called an* **implicit function.** Some equations defining implicit functions may be solved to determine the explicit functions, and for others it is not possible to solve for the explicit functions. Also, not all such equations define y as a function of x for real values of x.

Example A

The equation $3x + 4y = 5$ is an equation which defines a function, although it is not in explicit form. In solving for y as $y = -\frac{3}{4}x + \frac{5}{4}$, we have the explicit form of the function.

The equation $y^2 + x = 3$ is an equation which defines two functions, although we do not have the explicit forms. When we solve for y, we obtain the explicit functions $y = \sqrt{3 - x}$ and $y = -\sqrt{3 - x}$.

The equation $y^5 + xy^2 + 3x^2 = 5$ defines y as a function of x, although we cannot actually solve for the algebraic form of the explicit form of the function.

The equation $x^2 + y^2 + 4 = 0$ is not satisfied by any pair of real values of x and y.

Even when it is possible to determine the explicit form of a function given in implicit form, it is not always desirable to do so. In some cases the implicit form is more convenient than the explicit form.

The derivative of an implicit function may be found directly without having to solve for the explicit function. Thus, *to find dy/dx when y is defined as an implicit function of x, we differentiate each term of the equation with respect to x, regarding y as a function of x.* We then solve for dy/dx, which will usually be in terms of x and y.

Example B

Find dy/dx if $y^2 + 2x^2 = 5$.

Here we find the derivative of each term, and then solve for dy/dx. Thus,

$$\frac{d(y^2)}{dx} + \frac{d(2x^2)}{dx} = \frac{d(5)}{dx}$$

$$2y\frac{dy}{dx} + 4x = 0$$

$$\frac{dy}{dx} = -\frac{2x}{y}$$

The factor dy/dx arises from the derivative of the first term as a result of using the derivative of a power of a function of x (Eq. 2–29). The factor dy/dx corresponds to the du/dx of the formula. In the second term, no factor of dy/dx appears, since there are no y factors in the term.

Example C

Find dy/dx if $3y^4 + xy^2 + 2x^3 - 6 = 0$.

In finding the derivative, we note that the second term is a product, and we must use the product rule for derivatives on it. Thus, we have

$$\frac{d(3y^4)}{dx} + \frac{d(xy^2)}{dx} + \frac{d(2x^3)}{dx} - \frac{d(6)}{dx} = \frac{d(0)}{dx}$$

$$12y^3 \frac{dy}{dx} + \left[x\left(2y\frac{dy}{dx}\right) + y^2(1) \right] + 6x^2 - 0 = 0$$

$$12y^3 \frac{dy}{dx} + 2xy\frac{dy}{dx} + y^2 + 6x^2 = 0$$

$$(12y^3 + 2xy)\frac{dy}{dx} = -y^2 - 6x^2$$

$$\frac{dy}{dx} = \frac{-y^2 - 6x^2}{12y^3 + 2xy}$$

Example D

Find the slope of a line tangent to the curve of $2y^3 + xy + 1 = 0$ at the point $(-3, 1)$.

Here we must find dy/dx and evaluate it for $x = -3$ and $y = 1$.

$$\frac{d(2y^3)}{dx} + \frac{d(xy)}{dx} + \frac{d(1)}{dx} = \frac{d(0)}{dx}$$

$$6y^2 \frac{dy}{dx} + x\frac{dy}{dx} + y + 0 = 0$$

$$\frac{dy}{dx} = \frac{-y}{6y^2 + x}$$

$$\left.\frac{dy}{dx}\right|_{(-3,1)} = \frac{-1}{6(1^2) - 3} = \frac{-1}{6 - 3}$$

$$= -\frac{1}{3}$$

Thus, the slope is $-\frac{1}{3}$.

Exercises 2–9

In Exercises 1 through 16 find dy/dx by differentiating implicitly. When applicable, express the result in terms of x and y.

1. $3x + 2y = 5$

2. $6x - 3y = 4$

3. $4y - 3x^2 = x$

4. $x^5 - 5y = 6 - x$

5. $x^2 - y^2 - 9 = 0$

6. $x^2 + 2y^2 - 11 = 0$

7. $y^5 = x^2 - 1$

8. $y^4 = 3x^3 - x$

9. $y^2 + y = x^2 - 4$

10. $2y^3 - y = 7 - x^4$

11. $y + 3xy - 4 = 0$

12. $8y - xy - 7 = 0$

13. $xy^3 + 3y + x^2 = 9$

14. $y^2x - x^2y + 3x = 4$

15. $2(x^2 + 1)^3 + (y^2 + 1)^2 = 17$

16. $(2x + 1)(1 - 3y) + y^2 = 13$

In Exercises 17 through 20 solve the given problems by use of implicit differentiation.

17. Find the slope of a line tangent to the curve of $xy + y^2 + 2 = 0$ at the point $(-3, 1)$.

18. Find the velocity of an object moving such that its displacement s and the time t are related by the equation $s^3 - t^2 = 7$ for $s = 2$ and $t = 1$.

19. For a right triangle in which the hypotenuse is always a units in length, the legs x and y are related by the equation $x^2 + y^2 = a^2$. Find the expression for dy/dx.

20. The polar moment of inertia I of a rectangular area is given by $I = \frac{1}{12}(b^3h + bh^3)$, where b and h are the base and height, respectively, of the rectangle. If I is constant, find the expression for db/dh.

2–10 Graphical Differentiation

In the previous sections of this chapter we developed certain important formulas by which we can differentiate the algebraic functions described in Section 2–1. Also, as we mentioned in that section, we will take up the derivatives of certain important transcendental functions in later chapters. However, these formulas can be used only when we know the equation relating the variables. Often empirical data may be known, but the exact equation is not known. Under these circumstances it is possible to plot the graph and then use techniques based on the definition of the derivative to obtain values of the rate of change of one variable with respect to the other. Such methods are referred to as **graphical differentiation**. In this section one of the principal methods of graphical differentiation will be developed.

First, by means of the following example, we shall show how the interpretation of the derivative as the slope of a tangent line can be used to obtain the graph of a derivative of a given curve of a function.

Example A

Determine the graph of the derivative of the function whose graph is shown in Fig. 2–17(a).

First draw tangent lines (this must be done carefully) at various points along the curve. Then draw a triangle for each with the tangent line along the hypotenuse and sides parallel to each of the axes. Then measure the length of these sides to calculate $\Delta y/\Delta x$ for each tangent line. In Fig. 2–17(b) tangent lines have been drawn at each of the points

x	Δy	$\dfrac{\Delta y}{\Delta x} = y'$
-4	$3.0 - (-3.4) = 6.4$	3.2
-3	$2.8 - (-1.2) = 4.0$	2.0
-2	$2.0 - 0.2 = 1.8$	0.9
-1	$1.2 - 0.8 = 0.4$	0.2
0	$0.1 - 0.7 = -0.6$	-0.3
1	$-0.1 - 0.5 = -0.6$	-0.3
2	$0.3 - 0.3 = 0$	0
3	$2.2 - 0.6 = 1.6$	0.8

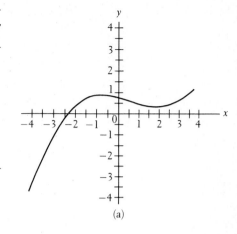

(a)

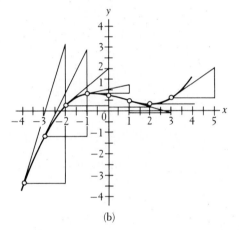

(b)

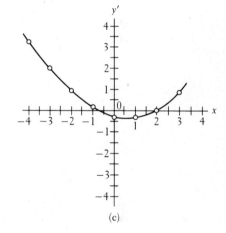

(c)

Figure 2–17

where x is an integer. Triangles have been constructed such that Δx is 2 in each case. This is done for convenience of calculation. In this way the following table is constructed, and then used to plot the graph of y' vs. x in Fig. 2–17(c), which is the desired curve.

Example A demonstrates how a derivative curve can be found from the definition of the derivative. It is possible to modify this method by eliminating the actual calculation of $\Delta y/\Delta x$ and make the method a completely graphical one. Example B demonstrates the method for a curve which exists only for positive x-values. This is done so that the method may be followed easily in the figure. Also, most physical problems involve only positive values for the variables involved, particularly the independent variable.

Example B

Determine the graph of the derivative of the function, curve F, given in Fig. 2–18.

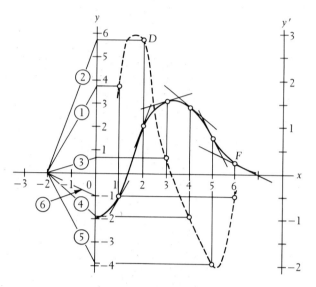

Figure 2–18

Tangent lines have been drawn to the given curve F at the points where x is integral. Here, the only thing which matters is that enough points are picked. Then a point on the negative x-axis is chosen and called the *pole*. In Fig. 2–18, the point where $x = -2$ is chosen. Next, lines parallel to each of the tangent lines are drawn through the pole until they cross the y-axis. A line is then drawn through this y-intercept, of each line from the pole, parallel to the x-axis until it crosses a line through the point on curve F drawn parallel to the y-axis. This point of intersection is a point on the curve D of the derivative.

Since the lines through the pole are parallel to the tangent lines, they have the same slopes. The slopes of the lines through the pole are the values of the y-intercept divided by the absolute value of the x-value of the pole. In this case, by dividing each y-intercept by 2, we have the proper values of the slopes. This scale for curve D is indicated at the right of Fig. 2–18. Thus, the derivative curve D is drawn through the derived points.

A final example will now be given to show the use of graphical differentiation of a curve found from empirical data.

Example C

An experiment which measured current i (in milliamperes) and voltage e (in volts) in a certain transistor under specified conditions gave the following results.

i	0	5	13	26	52	90
e	250	275	300	325	350	375

Plot current vs. voltage and then determine the derivative curve. From this curve find the rate of change of i with respect to e when e is 360 V.

Plotting the curve of i vs. e and then following the method of Example B, the solution is indicated in Fig. 2–19. The value of the derivative for $e = 360$ V is read directly from the derivative curve and gives the desired result, 1.45 mA/V, for the rate of change of i with respect to e. Note that the pole is 50 units to the left of the i-axis.

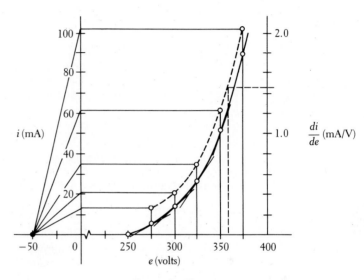

Figure 2–19

In Exercises 1 through 4 find the derivative curve for the given curves by the method of Example A.

1.

2.

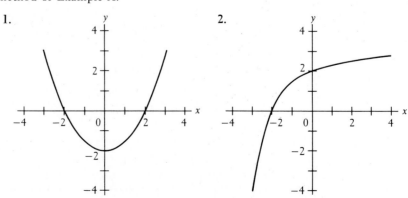

3.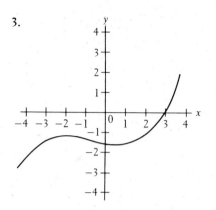

4.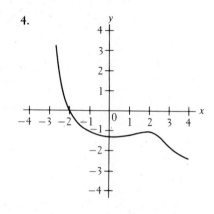

In Exercises 5 through 8 (a) sketch the graph of the given equation and then determine its derivative curve by means of the method of Example B, (b) find the derivative directly by formulas of the previous sections and then sketch the graph of the derivative, and (c) compare the results of (a) and (b).

5. $y = x^2$ 6. $y = x^3$ 7. $y = x^2 - 4x$ 8. $y = \dfrac{4}{x}$

In Exercises 9 and 10 the distance s is given as a function of the time t. Repeat the instructions for Exercises 5 through 8, noting that the derivative curve represents the velocity. The distance is in feet and the time is in seconds.

9. $s = t - \frac{1}{3}t^3$ 10. $s = 60t - 16t^2$

In Exercises 11 through 16 use the method of Example B to find the derivative curve of the given sets of empirical data. In Exercises 14 through 16 evaluate the derivative for the indicated value.

11. The power P (in watts) measured as a function of the load resistance R (in ohms) for a certain electric circuit gives the following results:

P	50.0	44.4	32.0	24.5	19.8	16.5	9.1
R	0.5	1.0	2.0	3.0	4.0	5.0	10.0

12. The vapor pressure (in kilopascals) of ethane gas as a function of temperature (in degrees Celsius) was measured as follows:

Pressure	260	380	570	790	1060	1430	1840	2390	3270
Temperature	−70	−60	−50	−40	−30	−20	−10	0	15

13. The density (in kilograms per cubic meter) of water from 0°C to 10°C is given in the following table:

Density	999.85	999.90	999.94	999.96	999.97	999.96
Temperature	0	1	2	3	4	5

Density	999.94	999.90	999.85	999.78	999.69
Temperature	6	7	8	9	10

14. The voltage (in volts) across a certain portion of an electric circuit as a function of time (in milliseconds) was measured as follows:

Voltage	0	110	160	180	195	200
Time	0	4	8	12	16	20

Evaluate dv/dt for $t = 15$ ms.

15. The force (in pounds) applied to an object in a certain experiment as a function of time (in seconds) was measured as follows:

Force	6.0	6.0	6.0	4.6	3.7	3.0	2.5	2.1	1.8	1.6	1.5
Time	0	1	2	3	4	5	6	7	8	9	10

Evaluate the time rate of change of force when $t = 8.5$ s.

16. When a certain photoelectric surface is illuminated with light of various wavelengths (in nanometers), the stopping potential (the voltage necessary to stop the emission of photoelectrons) is measured as follows:

Wavelength	360	400	430	490	540	580
Potential	1.45	1.10	0.90	0.60	0.35	0.25

Find the derivative of the wavelength with respect to the stopping potential, when the stopping potential is 0.75 V.

2–11 Review Exercises for Chapter 2

In Exercises 1 through 12 evaluate the given limits.

1. $\lim_{x \to 4} (8 - 3x)$

2. $\lim_{x \to 3} (2x^2 - 10)$

3. $\lim_{x \to -3} \dfrac{2x + 5}{x - 1}$

4. $\lim_{x \to 1} \dfrac{x^2 - 1}{x + 1}$

5. $\lim_{x \to 2} \dfrac{4x - 8}{x^2 - 4}$

6. $\lim_{x \to 5} \dfrac{x^2 - 25}{3x - 15}$

7. $\lim_{x \to 2} \dfrac{x^2 + 3x - 10}{x^2 - x - 2}$

8. $\lim_{x \to 3} \dfrac{x^2 - x - 6}{x - 3}$

9. $\lim_{x \to \infty} \dfrac{2 + \dfrac{1}{x + 4}}{3 - \dfrac{1}{x^2}}$

10. $\lim_{x \to \infty} \left(7 - \dfrac{1}{x + 1}\right)$

11. $\lim_{x \to \infty} \dfrac{x - 2x^3}{1 + x^3}$

12. $\lim_{x \to \infty} \dfrac{2x + 5}{3x^3 - 2x}$

In Exercises 13 through 20 use the Δ-process to find the derivative of each of the given functions.

13. $y = 7 + 5x$

14. $y = 6x - 2$

15. $y = 6 - 2x^2$

16. $y = 2x^2 - x^3$

17. $y = \dfrac{2}{x^2}$

18. $y = \dfrac{1}{1 - 4x}$

19. $y = \sqrt{x + 5}$

20. $y = \dfrac{1}{\sqrt{x}}$

In Exercises 21 through 36 find the derivative of each of the given functions.

21. $y = 2x^7 - 3x^2 + 5$

22. $y = 8x^5 - x - 1$

23. $y = 4\sqrt{x} - \dfrac{1}{x}$

24. $y = \dfrac{3}{x^2} - 8\sqrt[4]{x}$

25. $y = \dfrac{x}{1 - x}$

26. $y = \dfrac{2x - 1}{x^2 + 1}$

27. $y = (2 - 3x)^4$

28. $y = (2x^2 - 3)^6$

29. $y = \dfrac{3}{(5 - 2x^2)^{3/4}}$

30. $y = \dfrac{7}{(3x - 1)^3}$

31. $y = x^2\sqrt{1 - 6x}$

32. $y = (x - 1)^3(x^2 - 2)^2$

33. $y = \dfrac{\sqrt{4x + 3}}{2x}$

34. $y = \dfrac{x}{\sqrt{x^2 + 1}}$

35. $(2x - 3y)^3 = x^2 - y$

36. $x^2y^2 = x^2 + y^2$

In Exercises 37 through 54 solve the given problems by finding the appropriate derivative or limit.

37. The combined capacitance for two capacitors connected in series is given by

$$C_T = \dfrac{C_1 C_2}{C_1 + C_2}$$

Find $\lim\limits_{C_1 \to C_2} C_T$ and $\lim\limits_{C_1 \to 0} C_T$.

38. The illuminance I is inversely proportional to the square of the distance x from the source. Find $\lim\limits_{x \to 0} I$.

39. Find the slope of a line tangent to the curve of $y = 7x^4 - x^3$ at $(-1, 8)$.

40. Find the slope of a line tangent to the curve of $y = \sqrt[3]{3 - 8x}$ at $(-3, 3)$.

41. Find the velocity of an object after 3 s if its displacement s (in meters) is given by $s = \dfrac{t}{3t + 1}$.

42. The distance s (in feet) traveled by a subway train after the brakes are applied is given by $s = 40t - 5t^2$. How far does it travel, after the brakes are applied, in coming to a stop?

43. The voltage induced in an inductor L is given by

$$E = L\dfrac{dI}{dt}$$

where I is the current in the circuit and t is the time. Find the voltage induced in a 0.4-H inductor if the current I (in amperes) is related to the time (in seconds) by $I = 0.1t + 5.0$.

44. The specific heat c_p of a certain gas as a function of the temperature T is given by $c_p = 6.50 + 0.01T + 0.000002T^2$. Find the expression for the rate of change of the specific heat with respect to temperature.

45. The electric field E at a distance r from a point charge is $E = k/r^2$, where k is a constant. Find an expression for the rate of change of the electric field with respect to r.

46. The distance d between ion layers of a crystalline solid such as table salt is given by $d = \sqrt[3]{M/2N\rho}$ where M is the molecular weight, N is called Avogadro's number, and ρ is the density. Find the rate of change of d with respect to ρ.

47. The velocity of an object moving with constant acceleration can be found from the equation $v = \sqrt{v_0^2 + 2as}$, where v_0 is the initial velocity, a is the acceleration, and s is the distance traveled. Find dv/ds.

48. The time rate of change of angular velocity is angular acceleration. If a wire is moving through a magnetic field in a circular path such that its angular velocity is given by

$$\omega = \frac{2t^2}{1 + t}$$

where t is the time (in seconds), find the expression for the angular acceleration (in radians per second squared). What is the value of the angular acceleration when $t = 2$ s?

49. Under certain conditions, the efficiency of an internal combustion engine is given by

$$\text{eff (in percent)} = 100\left(1 - \frac{1}{(V_1/V_2)^{0.4}}\right)$$

where V_1 and V_2 are, respectively, the maximum and minimum volumes of air in a cylinder. Assuming that V_2 is kept constant, find the expression for the rate of change of efficiency with respect to V_1.

50. Water is being drained from a pond such that the volume of water in the pond (in cubic meters) after t hours is given by $V = 5000(60 - t)^2$. Find the rate at which the pond is being drained after 4 h.

51. A rectangle is inscribed in the first quadrant under the parabola $y = 4 - x^2$. Express the area of the rectangle in terms of x. Then find the expression for dA/dx.

52. An airplane flies over an observer with a velocity of 400 mi/h at an altitude of 2640 ft. If the plane flies horizontally in a straight line, find the rate at which the distance from the observer to the plane is changing 0.6 min after the plane passes over the observer.

53. In an experiment to study the pressure and volume of a gas under the condition of constant temperature, the following results were obtained:

Pressure (kilopascals)	142	157	164	176	182	193	203
Volume (cm³)	43.0	39.0	37.2	34.8	33.0	31.2	30.3

By graphical differentiation, determine the derivative curve of pressure with respect to volume. Evaluate this derivative when $V = 36.0$ cm³.

54. In the production of steel, the rate at which the temperature of the furnace changes is important. In studying this relationship, the following results were obtained:

Temperature (°C)	25	290	480	700	790	770	710	650	610
Time (min)	0	1	2	3	4	5	6	7	8

By graphical differentiation, determine the graph of the derivative curve. Find the rate of change of temperature when t is 3.5 min.

3

Applications of the Derivative

3—1 Tangents and Normals

In Chapter 2 we developed the meaning of the derivative of a function and then went on to find several formulas by which we can differentiate functions. In establishing the concept of the derivative as an instantaneous rate of change, numerous applications of the derivative were indicated in the examples and exercises. It was not necessary to develop any of these applications in any detail, since finding the derivative was the primary concern. There are, however, certain types of problems in geometry and technology in which the derivative plays a key role in the solution, although other concepts also are involved. The first of these problems involves finding the equations of lines tangent and lines normal (perpendicular) to a given curve.

To find the equation of a line tangent to a curve at a given point, we first find the derivative of the function. The derivative is then evaluated at the point, and this gives us the slope of a line tangent to the curve at the point. Then, by using the point-slope form of the equation of a straight line, we find the equation of the tangent line. The following examples illustrate the method.

Example A

Find the equation of the line tangent to the parabola $y = x^2 - 1$ at the point $(-2, 3)$.

The derivative of this function is

$$\frac{dy}{dx} = 2x$$

The value of this derivative for the value $x = -2$ (the y-value of 3 is not used since the derivative does not directly contain y) is

$$\frac{dy}{dx} = -4$$

which means that the slope of the tangent line at $(-2, 3)$ is -4. Thus, by using the point-slope form of the equation of the straight line we obtain the desired equation. Thus, we have

$$y - 3 = -4(x + 2)$$
$$y - 3 = -4x - 8$$

or

$$y = -4x - 5$$

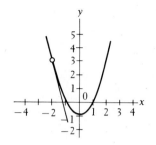

Figure 3–1

The parabola and the tangent line $y = -4x - 5$ are shown in Fig. 3–1.

Example B

Find the equation of the line tangent to the ellipse $4x^2 + 9y^2 = 40$ at the point $(1, 2)$.

The easiest method of finding the derivative of this equation is to treat it as an implicit function. In this way we have

$$8x + 18yy' = 0 \qquad \text{or} \qquad y' = -\frac{4x}{9y}$$

Evaluating this derivative at the point $(1, 2)$, we have

$$y' = -\frac{4}{18} = -\frac{2}{9}$$

Thus, the slope of the desired tangent line is $-\frac{2}{9}$. Using the point-slope form of the equation of the straight line we have

$$y - 2 = -\frac{2}{9}(x - 1)$$

$$9y - 18 = -2x + 2$$

$$2x + 9y - 20 = 0$$

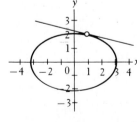

Figure 3–2

The ellipse and the tangent line $2x + 9y - 20 = 0$ are shown in Fig. 3–2.

It might be noted that the derivative could also have been found by solving for y. In this way we would have to differentiate $y = \frac{2}{3}\sqrt{10 - x^2}$.

If we wish to obtain the equation of a line normal (perpendicular to a tangent) to a curve, we recall that the slopes of perpendicular lines are negative reciprocals. Thus, the derivative is found and evaluated at the specified point. Since this gives the slope of a tangent line, we take the negative reciprocal of this number to find the slope of the normal line.

Then, by using the point-slope form of the equation of a straight line we find the equation of the normal. The following examples illustrate the method.

Example C

Find the equation of the line normal to the hyperbola $y = 2/x$ at the point (2, 1).

The derivative of this function is $dy/dx = -2/x^2$, which evaluated at $x = 2$ gives $dy/dx = -\frac{1}{2}$. Therefore, the slope of a line normal to the curve at the point (2, 1) is 2. The equation of the normal is then

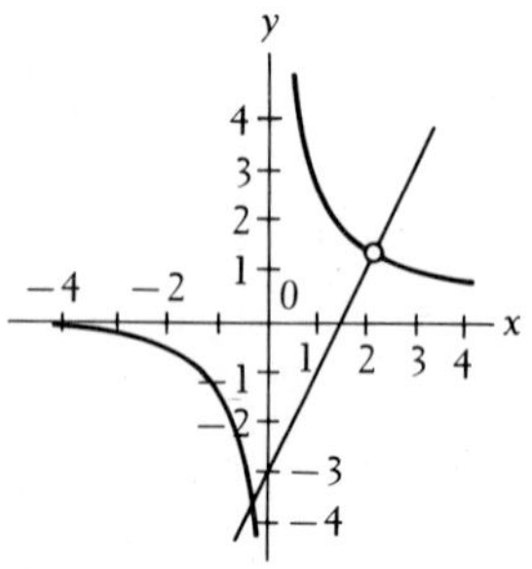

$$y - 1 = 2(x - 2)$$

or

$$y = 2x - 3$$

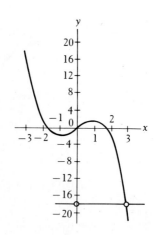

Figure 3-3

The hyperbola and the normal line are shown in Fig. 3-3.

Example D

Find the y-intercept of the line normal to the curve $y = 3x - x^3$ at (3, -18).

The derivative of this function is $dy/dx = 3 - 3x^2$, which evaluated at (3, -18) is

$$\left.\frac{dy}{dx}\right|_{x=3} = 3 - 3(9) = -24$$

Therefore, the slope of a normal line is $\frac{1}{24}$. The equation of the normal line is

$$y + 18 = \frac{1}{24}(x - 3)$$

or

$$24y - x + 435 = 0$$

To find the y-intercept of this line we set $x = 0$. In this way we find the y-intercept to be $-\frac{145}{8}$. The curve, normal line, and intercept are indicated in Fig. 3-4.

Figure 3-4

Many of the applications of tangents and normals are geometric. However, some of the areas of technology where they may be applied are indicated in the exercises.

Exercises 3-1

In Exercises 1 through 4 find the equations of the lines tangent to the indicated curves at the given points. In each, sketch the curve and the tangent line.

1. $y = x^2 + 2$ at (2, 6)

2. $y = x^3 - 2x$ at (2, 4)

3. $y = \dfrac{1}{x^2 + 1}$ at $(1, \frac{1}{2})$

4. $x^2 + y^2 = 25$ at (3, 4)

In Exercises 5 through 8 find the equations of the lines normal to the indicated curves at the given points. In each, sketch the curve and the normal line.

5. $y = 6x - x^2$ at $(1, 5)$

6. $y = 2x^3 + 10$ at $(-2, -6)$

7. $y = \dfrac{1}{(x^2 + 1)^2}$ at $(1, \frac{1}{4})$

8. $x^2 - y^2 = 8$ at $(3, 1)$

In Exercises 9 through 12 find the equations of the tangent lines and the normal lines to the indicated curves.

9. $y = \dfrac{1}{\sqrt{x^2 + 1}}$ where $x = \sqrt{3}$

10. $y = \dfrac{4}{(5 - 2x)^2}$ where $x = 2$

11. The parabola with vertex at $(0, 3)$ and focus at $(0, 0)$, where $x = -1$

12. The ellipse with focus at $(4, 0)$, vertex at $(5, 0)$, and center at $(0, 0)$ where $x = 2$

In Exercises 13 through 18 solve the given problems by finding the equation of the appropriate tangent or normal line.

13. Find the x-intercept of the line tangent to the parabola $y = 4x^2 - 8x$ at $(-1, 12)$.

14. Find the y-intercept of the line normal to the curve $y = x^{3/4}$ where $x = 16$.

15. A certain suspension cable with supports on the same level is closely approximated as being parabolic in shape. If the supports are 200 ft apart and the sag at the center is 30 ft, what is the equation of the line along which the tension acts (tangentially) at the right support. (Choose the origin of the coordinate system at the lowest point of the cable.)

16. In an electric field the lines of force are perpendicular to the curves of equal electric potential. A certain curve of equal electric potential is given by $y = 2x^{2/3}$. What is the equation of the line along which the force acts on an electron positioned on this curve where $x = 8$?

17. An object is moving in a horizontal circle of radius 2 ft at the rate of 2 rad/s. If the object is on the end of a string, and the string breaks after $\pi/3$ s, causing the object to travel along a line tangent to the circle, what is the equation of the path of the object after the string breaks? Sketch the path of the object. (Choose the origin of the coordinate system at the center of the circle and assume the object started on the positive x-axis moving counterclockwise.)

18. On a particular drawing, a pulley wheel can be described by the equation $x^2 + y^2 = 100$, where distances are measured in centimeters. The pulley belt is directed along the lines $y = -10$, and $4y - 3x - 50 = 0$ when first and last making contact with the wheel. What are the first and last points on the wheel where the belt makes contact?

3-2 Curvilinear Motion

A great many phenomena in technology involve the time rate of change of certain quantities. We encountered one of the most fundamental and most important of these when we discussed velocity, the time rate of change of displacement, in Section 2-5. In this section we further develop the concept of velocity and certain other concepts necessary for the discussion. Other time-rate-of-change problems are discussed in the next section.

When velocity was introduced in Section 2–5, the discussion was limited to rectilinear motion, or motion along a straight line. A more general discussion of velocity is necessary when we discuss the motion of an object in a plane. There are many important applications of motion in the plane, a principal one among these being the motion of a projectile.

An important concept necessary in developing this topic is that of a **vector.** A good elementary definition of *a vector is a quantity which has direction and magnitude.* For example, velocity, acceleration, force, and momentum are vector quantities. Each is specified by giving both its magnitude and direction. For instance, we cannot determine the effect of a force unless we know the magnitude of the force as well as the direction in which it is acting. *To represent a vector, we draw a line segment of length proportional to its magnitude, and in a direction which shows the direction of the vector.*

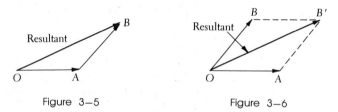

Figure 3–5   Figure 3–6

The addition of two vectors **OA** *and* **AB**, *directed from O to A and from A to B, respectively, gives the* **resultant** *vector* **OB**, *as shown in Fig. 3–5. The* **initial point** *is O and the* **terminal point** *is B.* This is equivalent to making the two vectors being added the sides of a parallelogram, with the diagonal being the resultant (see Fig. 3–6). (Note: A vector quantity is represented by a letter printed in boldface type. The same letter in italic (lightface) type represents the magnitude only.) In general, *a resultant is a single vector which can replace any number of other vectors and still produce the same physical effect.*

Vectors may be subtracted by reversing the direction of the vector being subtracted, and then proceeding as in adding vectors.

Not only do we need to add vectors, but it is also frequently useful to consider a given vector as the sum of two vectors. *These two vectors when added together give the original vector and are called* **components** *of the original vector.*

In practice there are certain components of a vector which are of particular importance. If a vector is so placed that its initial point is at the origin and its direction is indicated by an angle with respect to the positive x-axis (standard position), we may find its components which are directed along the x-axis and along the y-axis. These are known as its x-component and its y-component. *Finding these component vectors is called* **resolving** *the vector into its components.*

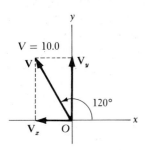

Figure 3–7

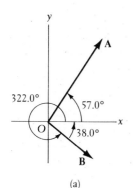

(a)

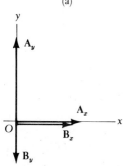

(b)

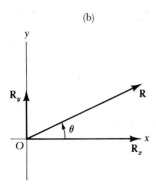

(c)

Figure 3–8

Example A

Resolve a vector 10.0 units long and directed at an angle of 120.0° into its x- and y-components (see Fig. 3–7).

Placing the initial point of the vector at the origin, and putting the angle in standard position, we see that the vector directed along the x-axis, V_x, is related to the vector V, of magnitude V by

$$V_x = V \cos 120.0° = -V \cos 60.0°$$

(The minus sign indicates that the x-component is directed in the negative direction, that is, to the left.) Since the vector directed along the y-axis, V_y, could also be placed along the dashed line, it is related to the vector V by

$$V_y = V \sin 120.0° = V \sin 60.0°$$

Thus, the vectors V_x and V_y have the magnitudes

$$V_x = -10.0(0.5000) = -5.00, \qquad V_y = 10.0(0.8660) = 8.66$$

Therefore, we have resolved the given vector into two components, one directed along the negative x-axis of magnitude 5.00, and the other directed along the positive y-axis of magnitude 8.66.

Adding vectors by diagrams gives only approximate results. It is possible to find accurate numerical results for the sum of two vectors by the use of components. The method is to resolve (break into components) all vectors being added into their x- and y-components. These can then be added algebraically, since they are all directed along either the x- or the y-axis. The result will give us the component of the resultant in the x-direction and the component of the resultant in the y-direction. By addition we may then find the value of the resultant vector.

Example B

Find the resultant R of the two vectors given in Fig. 3–8(a), A of magnitude 8.00 and direction 57.0° and B of magnitude 5.00 and direction 322.0°.

$$A_x = (8.00)(\cos 57.0°) = (8.00)(0.5446) = 4.36$$
$$A_y = (8.00)(\sin 57.0°) = (8.00)(0.8387) = 6.71$$
$$B_x = (5.00)(\cos 38.0°) = (5.00)(0.7880) = 3.94$$
$$B_y = -(5.00)(\sin 38.0°) = -(5.00)(0.6157) = -3.08$$
$$R_x = A_x + B_x = 4.36 + 3.94 = 8.30$$
$$R_y = A_y + B_y = 6.71 - 3.08 = 3.63$$
$$R = \sqrt{(8.30)^2 + (3.63)^2} = \sqrt{68.9 + 13.2} = \sqrt{82.1} = 9.06$$

$$\tan \theta = \frac{R_y}{R_x} = \frac{3.63}{8.30} = 0.4373, \qquad \theta = 23.6°$$

The resultant vector is 9.06 units long and is directed at an angle of 23.6°, as shown in Fig. 3–8(c).

Some general formulas can be derived from the previous examples. For a given vector $\mathbf{A}$, directed at an angle θ, and of magnitude A, with components $\mathbf{A}_x$ and $\mathbf{A}_y$, we have the following relations:

$$A_x = A \cos \theta, \qquad A_y = A \sin \theta \tag{3-1}$$

$$A = \sqrt{A_x^2 + A_y^2} \tag{3-2}$$

$$\tan \theta = \frac{A_y}{A_x} \tag{3-3}$$

Although vectors can be used to represent many physical quantities, we shall restrict our attention to their use in describing the velocity and acceleration of an object moving in a plane along a specified path. Such motion is called **curvilinear motion.**

In describing an object undergoing curvilinear motion, it is very common to express the x and y coordinates of its position separately as functions of time. Equations given in this form, that is, *x and y both given in terms of a third variable (in this case t), are said to be in* **parametric form.** The third variable, t, is called the **parameter.**

To find the velocity of an object whose coordinates are given in parametric form, we find its x-component of velocity v_x by determining dx/dt and its y-component of velocity v_y by determining dy/dt. These are then evaluated, and the resultant velocity is found from $v = \sqrt{v_x^2 + v_y^2}$. The direction in which the object is moving is found from $\tan \theta = v_y/v_x$. The following examples illustrate the method.

Example C

If the horizontal distance x that an object has moved is given by $x = 3t^2$, and the vertical distance y is given by $y = 1 - t^2$, find the resultant velocity when $t = 2$.

To find the resultant velocity, we must find v and θ, by first finding v_x and v_y as indicated above. Thus,

$$v_x = \frac{dx}{dt} = 6t \qquad v_x|_{t=2} = 12$$

$$v_y = \frac{dy}{dt} = -2t \qquad v_y|_{t=2} = -4$$

$$v = \sqrt{144 + 16} = \sqrt{160} = 12.6$$

$$\tan \theta = \frac{-4}{12} = -0.333 \qquad \theta = -18.4°$$

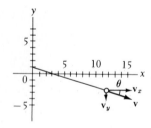

Figure 3–9

The path and velocity vectors are shown in Fig. 3–9.

Example D

Find the velocity and direction of motion when $t = 2$ of an object moving such that its x and y coordinates of position are given by $x = 1 + 2t$ and $y = t^2 - 3t$.

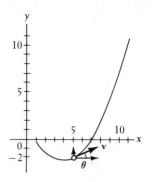

Figure 3-10

$$v_x = \frac{dx}{dt} = 2 \qquad v_x|_{t=2} = 2$$

$$v_y = \frac{dy}{dt} = 2t - 3 \qquad v_y|_{t=2} = 1$$

$$v|_{t=2} = \sqrt{4 + 1} = \sqrt{5} = 2.24$$

$$\tan \theta = \frac{1}{2} = 0.500 \qquad \theta = 26.6°$$

These quantities are shown in Fig. 3-10.

Acceleration *is the time rate of change of velocity.* Therefore, if the velocity, or its components, is known as a function of time, the acceleration of an object can be found by taking the derivative of the velocity with respect to time. For this reason it may be necessary to find the derivative of a derivative. This is called a **second derivative**. The notation for the second derivative of x with respect to t, for example, is d^2x/dt^2.

Example E

Find the magnitude and direction of the acceleration when $t = 2$ for an object moving such that its x and y coordinates of position are given by $x = t^3$ and $y = 1 - t^2$:

$$v_x = \frac{dx}{dt} = 3t^2 \qquad a_x = \frac{dv_x}{dt} = \frac{d^2x}{dt^2} = 6t \qquad a_x|_{t=2} = 12$$

$$v_y = \frac{dy}{dt} = -2t \qquad a_y = \frac{dv_y}{dt} = \frac{d^2y}{dt^2} = -2 \qquad a_y|_{t=2} = -2$$

$$a|_{t=2} = \sqrt{144 + 4} = \sqrt{148} = 12.2$$

$$\tan \theta = \frac{a_y}{a_x} = -\frac{2}{12} = -0.167 \qquad \theta = -9.5°$$

The quadrant in which θ lies is determined from the fact that a_y is negative and a_x is positive. Thus, θ must be a fourth-quadrant angle (see Fig. 3-11).

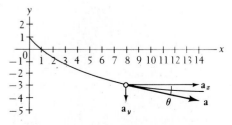

Figure 3-11

If the curvilinear path which an object follows is given as y as a function of x, the velocity (and acceleration) is found by taking derivatives of each term of the equation with respect to time. In finding the derivatives we must be very careful to use the power rule for finding derivatives, Eq. (2–29), so that the general factor du/dx is not neglected. The following example illustrates the method.

Example F

A particle moves along the parabola $y = \frac{1}{3}x^2$ so that its vertical velocity v_y is always 8. Find the magnitude of the resultant velocity when the particle is at the point $(2, \frac{4}{3})$.

Since both y and x change with time, both can be considered functions of time. Thus, we can take derivatives of $y = \frac{1}{3}x^2$ with respect to time. When we do this we obtain the following result:

$$\frac{dy}{dt} = \frac{2}{3}x\frac{dx}{dt} \qquad v_y = \frac{2}{3}xv_x$$

At the point $(2, \frac{4}{3})$ we have $8 = \frac{2}{3}(2)v_x$. Thus, $v_x = 6$ at this point (see Fig. 3–12). Therefore, the magnitude of the resultant velocity is

$$v = \sqrt{64 + 36} = 10$$

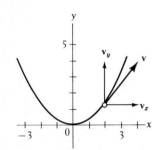

Figure 3–12

Exercises 3–2

In Exercises 1 through 4, given that the x- and y-coordinates of a moving particle are given by the indicated parametric equations, find the magnitude and direction of the velocity for the specific value of t. Plot the curves and show the appropriate components of the velocity.

1. $x = 3t$, $y = 1 - t$, $t = 4$

2. $x = 2t^3$, $y = 4t^2$, $t = 1$

3. $x = \dfrac{10}{2t + 3}$, $y = \dfrac{1}{t^2}$, $t = 2$

4. $x = \sqrt{1 + 2t}$, $y = t - t^2$, $t = 4$

In Exercises 5 through 8 use the parametric equations and values of t of Exercises 1 through 4 to find the magnitude and direction of the acceleration in each case.

In Exercises 9 through 20 find the indicated velocities and accelerations.

9. A particle moves along the curve $y = 5x^3$ with a constant x component of velocity equal to 20. Find the magnitude and direction of the velocity at the point $(1, 5)$.

10. A particle moves along the curve $y = 2\sqrt{x}$. What is the ratio of the x-component of velocity to the y-component at $(9, 6)$?

11. A particle moves along the upper half of the parabola $y^2 = 2x + 1$ such that $v_x = \sqrt{2x + 3}$. Find v_y when $x = 4$.

12. A particle moves counterclockwise along the ellipse $x^2 + 9y^2 = 9$ such that $v_y = 2y$. Find v_x at the point $(2, \frac{1}{3}\sqrt{5})$.

13. A projectile moves according to the parametric equations $x = 120t$ and $y = 160t - 16t^2$, where distances are in feet and time is in seconds. Find the magnitude and direction of the velocity when $t = 3.0$ s and when $t = 6.0$ s.

14. A projectile moves according to the parametric equations $x = 100t$ and $y = -16t^2$, where distances are in feet and time is in seconds. Find the magnitude and direction of each of the velocity and acceleration when $t = 3.0$ s.

15. Find the acceleration at the specified times for the projectile in Exercise 13.

16. Find the magnitude and direction of the acceleration of the particle in Exercise 9.

17. A spacecraft moves such that it follows the equation $x = 10(\sqrt{1 + t^4} - 1)$, $y = 40t^{3/2}$ for the first hundred seconds after launch. Here, x and y are measured in meters and t is measured in seconds. Find the magnitude and direction of the velocity of the spacecraft 10 s and 100 s after launch.

18. An electron moves in an electric field according to the parametric equations

$$x = \frac{20}{\sqrt{1 + t^2}} \quad \text{and} \quad y = \frac{20t}{\sqrt{1 + t^2}}$$

where distances are in meters and time in seconds. Find the magnitude and direction of the velocity when $t = 1$ s.

19. A rocket follows a path given by $y = x - \frac{1}{90}x^3$ (distances in miles). If the horizontal velocity is given by $v_x = x$, find the magnitude and direction of the velocity when the rocket hits the ground (assume level terrain) if time is in minutes.

20. A meteor traveling toward the earth has a velocity inversely proportional to the square root of the distance from the earth's center. Show that its acceleration is inversely proportional to the square of the distance from the center of the earth.

3–3 Related Rates

Any two variables which vary with respect to time and between which a relation is known to exist, can have the time rate of change of one expressed in terms of the time rate of change of the other. We do this by taking the derivative with respect to time of the expression which relates the variables as we did in Example F of Section 3–2. Since the rates of change are related, this type of problem is referred to as a **related-rate** problem. The following examples illustrate the basic method of solution.

Example A

The voltage of a certain thermocouple as a function of temperature is given by $E = 2.800\,T + 0.006\,T^2$. If the temperature is increasing at the rate $1.00°C/min$, how fast is the voltage increasing when $T = 100°C$?

Since we are asked to find the time rate of change of voltage, we first take derivatives with respect to time. This gives us

$$\frac{dE}{dt} = 2.800\,\frac{dT}{dt} + 0.012\,T\,\frac{dT}{dt}$$

again being careful to include the factor dT/dt. From the given information we know that $dT/dt = 1.00°C/min$ and that we wish to know dE/dt when $T = 100°C$. Thus,

$$\frac{dE}{dt}\bigg|_{T=100} = 2.800(1.00) + 0.012(100)(1.00) = 4.00 \text{ V/min}$$

The derivative must be taken before values are substituted. In this problem we are finding the time rate of change of the voltage for a specified value of T. For other values of T, dE/dt would have different values.

Example B

The distance q that an image is from a certain lens in terms of p, the distance the object is from the lens, is given by

$$q = \frac{10p}{p - 10}$$

If the object distance is increasing at the rate of 0.200 cm/s, how fast is the image distance changing when $p = 15.0$ cm?

Taking derivatives with respect to time, we have

$$\frac{dq}{dt} = \frac{(p - 10)\left(10\,\dfrac{dp}{dt}\right) - 10p\left(\dfrac{dp}{dt}\right)}{(p - 10)^2}$$

Now substituting $p = 15.0$ and $dp/dt = 0.200$, we have

$$\frac{dq}{dt}\bigg|_{p=15} = \frac{5(10)(0.200) - 10(15.0)(0.2)}{25.0} = -0.800 \text{ cm/s}$$

Thus, the image distance is decreasing (the significance of the minus sign) at the rate of 0.800 cm/s when $p = 15.0$ cm.

In many related-rate problems the function is not given, but must be set up according to the statement of the problem. The following examples illustrate this type of problem.

Example C

A spherical balloon is being blown up at the constant rate of 2.00 ft^3/min. Find the rate at which the radius is increasing when it is 3.00 ft.

We are asked to find the relation between the rate of change of the volume of a sphere with respect to time and the corresponding rate of change of the radius. Thus, we are to take derivatives of the expression for the volume of a sphere with respect to time.

$$V = \frac{4}{3}\pi r^3 \qquad \frac{dV}{dt} = 4\pi r^2 \left(\frac{dr}{dt}\right)$$

From the information given, we know the value of dV/dt to be 2.00. We want to find dr/dt when $r = 3.00$. Substituting these values, we have

$$2.00 = 4\pi(3.00)^2 \left(\frac{dr}{dt}\right) \qquad \text{or} \qquad \frac{dr}{dt}\bigg|_{r=3} = \frac{1}{18.0\pi} = 0.0177 \text{ ft/min}$$

Example D

Two ships, A and B, leave a port at noon. Ship A travels north at 6.00 mi/h, and ship B travels east at 8.00 mi/h. How fast are they separating at 2 PM?

We are asked to find how fast the length of a line drawn directly from A to B is increasing at a particular time. First a relation must be found between the distances traveled. The Pythagorean theorem provides the necessary relation (see Fig. 3–13). If y is the distance traveled by A and x is the distance traveled by B, we can find the distance between them, z, from $z^2 = x^2 + y^2$. Taking derivatives of this expression, we find that

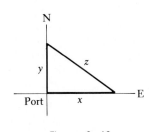

Figure 3—13

$$2z\frac{dz}{dt} = 2x\frac{dx}{dt} + 2y\frac{dy}{dt}$$

From the information given, we know that $dx/dt = 8.00$ mi/h, $dy/dt = 6.00$ mi/h, and for the specified time, $x = 16.0$ mi, $y = 12.0$ mi, and $z = 20.0$ mi. Thus we have

$$\frac{dz}{dt}\bigg|_{z=20} = \frac{(16.0)(8.00) + (12.0)(6.00)}{20.0} = 10.0 \text{ mi/h}$$

Exercises 3—3

In the following exercises solve the given problems in related rates.

1. The electric resistance of a certain resistor as a function of temperature is given by $R = 4.000 + 0.003\,T^2$, where R is measured in ohms and T in degrees Celsius. If the temperature is increasing at the rate of 0.100°C/s, find how fast the resistance changes when $T = 150$°C.

2. The kinetic energy of an object is given by $KE = \frac{1}{2}mv^2$, where m is the mass of the object and v is its velocity. If a 20.0-kg object is accelerating (remember that the time rate of change of velocity is acceleration) at 5.00 m/s², how fast is the kinetic energy (in joules) changing when the velocity is 30.0 m/s?

3. A firm found that its profit was p dollars for the production of x tons per week of a product according to the function $p = 30\sqrt{10x - x^2} - 50$. Determine the rate of change of profit if production is increasing such that $dx/dt = 0.200$ tons/week², when $x = 4.00$ tons/week.

4. A refinery is capable of producing two grades of oil. If x is the number of gallons of the lower-grade oil produced per day, and y is the number of gallons of the higher grade oil produced per day, where $y = \dfrac{100(20 - x)}{50 - x}$, find how fast y is changing if $dx/dt = 2.00$ gal/d², when $x = 8.00$ gal/d.

5. The length of a pendulum is slowly decreasing at the rate of 0.100 in./s. What is the rate of change of the period of the pendulum when the length is 16.0 in., if the relation between the period and length is $T = \pi\sqrt{L/96}$?

6. When air expands so that there is no change in heat (an adiabatic change) the relationship between the pressure and volume is $pv^{1.4} = k$, where k is a constant. At a certain instant, the pressure is 300 kPa and the volume is 10.0 cm³. The volume is increasing at the rate of 2.00 cm³/s. What is the time rate of change of pressure at this instant?

7. The relation between the voltage V which produces a current I in a wire of radius r is $V = 0.03I/r^2$. If the current increases at the rate of 0.020 A/s in a wire of 0.040 in. radius, find the rate at which the voltage is increasing.

8. Two resistances in parallel vary with time. At a certain instant R_1 is 8.00 Ω and is increasing at the rate of 0.500 Ω/s. At the same instant, R_2 is 6.00 Ω and is increasing at the rate of 0.600 Ω/s. What is the rate of change of the resistance R of the combination if

$$R = \frac{R_1 R_2}{R_1 + R_2}$$

9. The side of a square is increasing at the rate of 5.00 ft/min. How fast is the area changing when the side is 10.0 ft long?

10. A circular plate is contracting by being cooled. Find the rate of decrease in area if the radius decreases at the rate of 0.100 ft/h, when the radius is 4.00 ft.

11. The radius of a certain steam engine cylinder is 3.00 in. What is the speed of the piston if steam is entering the cylinder at 2.00 ft³/s? Assume no compression of the steam at this moment.

12. An ice cube is melting so that an edge decreases in length by 0.100 cm/min. How fast is the volume changing when an edge is 4.50 cm?

13. One statement of Boyle's law is that the pressure of a gas varies inversely as the volume for constant temperature. If a certain gas occupies 600 cm³ when the pressure is 200 kPa and the volume is increasing at the rate of 20.0 cm³/min, how fast is the pressure changing when the volume is 800 cm³?

14. The intensity of heat varies directly as the strength of the source and inversely as the square of the distance from the source. If an object approaches a heated object of strength of 8.00 units at the rate of 50.0 cm/s, how fast is the intensity changing when it is 100 cm from the source?

15. A spherical metal object is ejected from an Earth satellite and reenters the atmosphere. It heats up so that the radius increases at the rate of 5.00 mm/s. What is the rate of change of volume when the radius is 200 mm?

16. A pile of sand in the shape of a cone has a radius which always equals the altitude. If 100 ft³ of sand are poured onto the pile each minute, how fast is the radius increasing when the pile is 10.0 ft high?

17. Two cars leave the same point, traveling on straight roads which are at right angles. If the cars are traveling at 80.0 km/h and 100 km/h, how fast is the distance between them changing after 3.00 min (0.0500 h)?

18. A ladder is slipping down along a vertical wall. If the ladder is 10.0 ft long, and the top of it is slipping at the constant rate of 10.0 ft/s, how fast is the bottom of the ladder moving along the ground when the bottom is 6.00 ft from the wall?

19. A rope attached to a boat is being pulled in at a rate of 10.0 ft/s. If the water is 20.0 ft below the level at which the rope is being drawn in, how fast is the boat approaching the wharf when 36.0 ft of rope are yet to be pulled in?

20. A man 6.00 ft tall approaches a street light 15.0 ft above the ground at the rate of 5.00 ft/s. How fast is the end of the man's shadow moving when he is 10.0 ft from the base of the light?

3-4 Using Derivatives in Curve Sketching

Derivatives can be used effectively in the sketching of curves. An analysis of the first two derivatives can provide useful information as to the graph of a function. This information, possibly along with two or three key points on the curve, is often sufficient to obtain a good, although approximate, graph of the function. Graphs where this information must be supplemented with other analyses are the subject of the next section.

Considering the function $f(x)$ as shown in Fig. 3-14, we see that as x increases (from left to right) the y-values also increase until the point M is reached. From M to m the values of y decrease. To the right of m the values of y again increase. We also note that any tangent line to the left of M and to the right of m will have a positive slope. Any tangent line between M and m will have a negative slope. Since the derivative of a function determines the slope of a tangent line, we can conclude that, *as x increases, y increases if the derivative is positive, and decreases if the derivative is negative.* This can be stated as

$$f(x) \text{ increases if } f'(x) > 0$$

and

$$f(x) \text{ decreases if } f'(x) < 0$$

It is always assumed that x is increasing. Also, we assume in our present analysis that f(x) and its derivatives are continuous over the indicated interval.

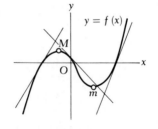

Figure 3-14

Example A

Determine those values of x for which the function $f(x) = x^3 - 3x^2$ is increasing and those values for which it is decreasing.

We find the solution to this problem by determining the values of x for which the derivative is positive and those for which it is negative. Finding the derivative, we have

$$f'(x) = 3x^2 - 6x = 3x(x - 2)$$

This now becomes a problem of solving an inequality. To find the values of x for which $f(x)$ is increasing we must solve the inequality

$$3x(x - 2) > 0$$

We now recall that the solution of an inequality consists of *all* values of x which may satisfy it. Normally, this consists of certain intervals of values of x. These intervals are found by first setting the left side of the inequality equal to zero, thus obtaining the **critical values** of x. The function on the left will have the same *sign* for all values of x less than the leftmost critical value. The sign of the function will also be the same within any given interval between critical values and to the right of the rightmost critical value. Those intervals which give the proper sign will satisfy the inequality. In this case the critical values are $x = 0$ and $x = 2$. Therefore, we have the following analysis:

$$\text{If } x < 0, \quad 3x(x - 2) > 0 \quad \text{or} \quad f'(x) > 0$$

$$\text{If } 0 < x < 2, \quad 3x(x - 2) < 0 \quad \text{or} \quad f'(x) < 0$$

$$\text{If } x > 2, \quad 3x(x - 2) > 0 \quad \text{or} \quad f'(x) > 0$$

Therefore, the solution of the above inequality is $x < 0$ or $x > 2$, which means that for these values $f(x)$ is increasing. We can also see that for $0 < x < 2$, $f'(x) < 0$, which means that $f(x)$ is decreasing for these values of x. The solution is now complete.

The points M and m in Fig. 3–14 are called a **relative maximum point** and a **relative minimum point**, respectively. *This means that M has a greater y-value than any other point near it, and that m has the least y-value of any point near it.* This does not necessarily mean that M has the greatest y-value of any point on the curve, or that m has the least y-value of any point on the curve. However, the points M and m are the greatest or least values of y for that part of the curve (that is why we use the word "relative"). Observation of Fig. 3–14 verifies this point. *The characteristic of both M and m is that the derivative is zero at each point.* (We see that this is so since a tangent line would have a slope of zero at each.) *This is how relative maximum and relative minimum points are located. The derivative is found and then set equal to zero. The solutions of the resulting equation give the x-coordinates of the maximum and minimum points.*

It remains now to determine whether a given value of x, for which the derivative is zero, is the coordinate of a maximum or a minimum point (or neither, which is also possible). From the discussion of increasing and decreasing values for y, we can see that *the derivative changes sign from plus to minus when passing through a relative maximum point, and from minus to plus when passing through a relative minimum point.* Thus, we find maximum and minimum points by determining those values of x for which the derivative is zero and by properly analyzing the sign change of the derivative. If the sign of the derivative does not change, it is neither a maximum nor a minimum point. This is known as the **first-derivative test for maxima and minima.**

Example B

Find any maximum and minimum points of the function

$$y = 3x^5 - 5x^3$$

Finding the derivative and setting it equal to zero, we have

$$y' = 15x^4 - 15x^2 = 15x^2(x^2 - 1)$$
$$15x^2(x - 1)(x + 1) = 0 \quad \text{for} \quad x = 0, \quad x = 1 \quad \text{and} \quad x = -1$$

Thus, the sign of the derivative is the same for all points to the left of $x = -1$. For these values $y' > 0$ (thus y is increasing). For values of x between -1 and 0, $y' < 0$. For values of x between 0 and 1, $y' < 0$. For values of x greater than 1, $y' > 0$. Thus, the curve has a maximum at $(-1, 2)$ and a minimum at $(1, -2)$. The point $(0, 0)$ is neither a maximum nor a minimum, since the sign of the derivative did not change for this value of x.

Looking again at the slope of tangents drawn to a curve (see Fig. 3–15a), we see that the slope goes from large to small positive values and then to negative values for this curve. At the point I the values of the slope again start to increase so that values of the slope again become positive after the point m. *The conclusion is that the slope decreases as a point approaches I, and that the slope increases as a point moves to the right of I.* The curve in Fig. 3–15(b) is that of the derivative, and therefore indicates the values of the slope of $f(x)$. If the slope changes, this means that we are dealing with the rate of change of slope, or the rate of change of the derivative. This function is the second derivative.

The curve in Fig. 3–15(c) is that of the second derivative. This means that we wish to be able to find an expression for the derivative of a derivative. We encountered this type of function before, in the discussion of acceleration. *Where the second derivative of a function is negative, the slope is decreasing, or the curve is* **concave down** *(opens down). Where the second derivative is positive, the slope is increasing, or the curve is* **concave up** *(opens up).* This may be summarized as follows:

If $f''(x) > 0$, the curve is concave up.
If $f''(x) < 0$, the curve is concave down.

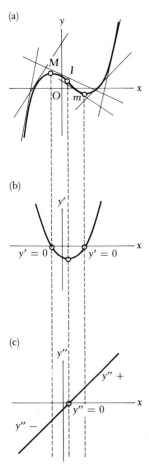

(a)

(b)

(c)

Figure 3–15

We can also now use this information in the determination of maximum and minimum points. By the nature of the definition of maximum and minimum points and of concavity, it is apparent that *a curve is concave down at a maximum point and concave up at a minimum point.* We can see these properties when we make a close analysis of the curves in Fig. 3–15. Therefore, at $x = a$,

$$\text{if } f'(a) = 0 \text{ and } f''(a) < 0,$$

then $f(x)$ has a relative maximum at $x = a$, or

$$\text{if } f'(a) = 0 \text{ and } f''(a) > 0,$$

then $f(x)$ has a relative minimum at $x = a$.

These statements comprise what is known as the **second-derivative test for maxima and minima.** This test is often easier to use than the first-derivative test. However, it can happen that $y'' = 0$ at a maximum or minimum point, and in such cases it is necessary that we use the first-derivative test.

The points at which the curve changes from concave up to concave down, or from concave down to concave up, are known as **points of inflection.** Thus point I in the figure is a point of inflection. Inflection points are found by determining those values of x for which the second derivative changes sign. This is analogous to finding maximum and minimum points by the first-derivative test.

Example C

Determine the concavity and find any points of inflection of the function $y = x^3 - 3x$.

This requires inspection and analysis of the second derivative. So we write

$$y' = 3x^2 - 3, \qquad y'' = 6x$$

The second derivative is positive where the function is concave up, and this occurs if $x > 0$. The curve is concave down for $x < 0$, since y'' is negative. Thus, $(0, 0)$ is a point of inflection, since the concavity changes there.

At this point we shall summarize the information regarding the derivatives of a function.

$f(x)$ increases if $f'(x) > 0$; $f(x)$ decreases if $f'(x) < 0$.

$f'(x) = 0$ at a maximum point if $f'(x)$ changes from $+$ to $-$, or if $f''(x) < 0$.

$f'(x) = 0$ at a minimum point if $f'(x)$ changes from $-$ to $+$, or if $f''(x) > 0$.

$f''(x) > 0$ where $f(x)$ is concave up; $f''(x) < 0$ where $f(x)$ is concave down.

$f''(x) = 0$ at a point of inflection if $f''(x)$ changes from $+$ to $-$ or $-$ to $+$.

The following examples illustrate how the above information is put together to obtain the graph of a function.

Example D

Sketch the graph of $y = 6x - x^2$.

Finding the first two derivatives, we have

$$y' = 6 - 2x = 2(3 - x)$$

and

$$y'' = -2$$

We now note that $y' = 0$ for $x = 3$. For $x < 3$ we see that $y' > 0$, which means that y is increasing over this interval. Also, for $x > 3$, we note that $y' < 0$, which means that y is decreasing over this interval.

Since y' changes from positive on the left of $x = 3$, to negative on the right of $x = 3$, the curve has a maximum point where $x = 3$. Since $y = 9$ for $x = 3$, this maximum point is $(3, 9)$.

Since $y'' = -2$, this means that its value remains constant for all values of x. Therefore, there are no points of inflection and the curve is concave down for all values of x. This also shows that the point $(3, 9)$ is a maximum point.

Summarizing, we know that y is increasing for $x < 3$, y is decreasing for $x > 3$, there is a maximum point at $(3, 9)$, and that the curve is always concave down. Using this information we sketch the curve shown in Fig. 3–16.

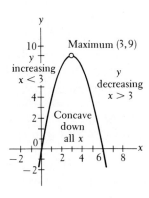

Figure 3–16

Example E

Sketch the graph of $y = 2x^3 + 3x^2 - 12x$.

Finding the first two derivatives, we have

$$y' = 6x^2 + 6x - 12 = 6(x + 2)(x - 1)$$

and

$$y'' = 12x + 6 = 6(2x + 1)$$

We note that $y' = 0$ when $x = -2$ and $x = 1$. Using these values in the second derivative we find that y'' is negative (-18) for $x = -2$ and y'' is positive $(+18)$ when $x = 1$. When $x = -2$, $y = 20$; and when $x = 1$, $y = -7$. Therefore, $(-2, 20)$ is a maximum point and $(1, -7)$ is a minimum point.

Next we see that $y' > 0$ if $x < -2$ or $x > 1$. Also, $y' < 0$ for the interval $-2 < x < 1$. Therefore, y is increasing if $x < -2$ or $x > 1$, and y is decreasing if $-2 < x < 1$.

Now we note that $y'' = 0$ when $x = -\frac{1}{2}$, and that $y'' < 0$ when $x < -\frac{1}{2}$, and $y'' > 0$ when $x > -\frac{1}{2}$. When $x = -\frac{1}{2}$, $y = \frac{13}{2}$. Therefore, there is a point of inflection at $(-\frac{1}{2}, \frac{13}{2})$, the curve is concave down if $x < -\frac{1}{2}$, and the curve is concave up if $x > -\frac{1}{2}$.

Finally, by locating the points $(-2, 20)$, $(-\frac{1}{2}, \frac{13}{2})$, and $(1, -7)$, we draw the curve *up* to $(-2, 20)$ and then *down* to $(-\frac{1}{2}, \frac{13}{2})$, with the curve *concave down*. Continuing *down*, but *concave up*, we draw the curve to $(1, -7)$ at which point we start *up* and continue up. This gives us an excellent graph indicating the shape of the curve. If more precision is required, additional points may be used (see Fig. 3–17).

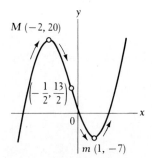

Figure 3–17

Example F

Sketch the graph of $y = x^5 - 5x^4$.

The first two derivatives are

$$y' = 5x^4 - 20x^3 = 5x^3(x - 4) \quad \text{and} \quad y'' = 20x^3 - 60x^2 = 20x^2(x - 3)$$

We now see that $y' = 0$ when $x = 0$ and $x = 4$. When $x = 0$, $y'' = 0$ also, which means we cannot use the second derivative test for maximum and minimum points for this value of x. When $x = 4$, y'' is positive $(+320)$ which means $(4, -256)$ is a minimum point. Next we note that $y' > 0$ if $x < 0$ or $x > 4$ and $y' < 0$ if $0 < x < 4$. Thus, by the first derivative test, there is a maximum at $(0, 0)$. Also, we see that y is increasing if $x < 0$ or $x > 4$ and decreasing if $0 < x < 4$. The second derivative indicates that there is point of inflection at $(3, -162)$. It also indicates that the function is concave down if $x < 3$ $(x \neq 0)$ and concave up if $x > 3$. There is no point of inflection at $(0, 0)$ since the second derivative does not change sign when $x = 0$. From this information we sketch the curve in Fig. 3–18.

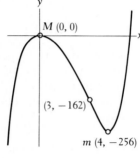

Figure 3–18

Exercises 3–4

In Exercises 1 through 4 determine those values of x for which the given functions are increasing and those values of x for which they are decreasing.

1. $y = x^2 - 2x$ 2. $y = 4 - x^2$
3. $y = 12x - x^3$ 4. $y = x^4 - 6x^2$

In Exercises 5 through 8 find any maximum or minimum points of the given functions. (These are the same functions as in Exercises 1 through 4.)

5. $y = x^2 - 2x$ 6. $y = 4 - x^2$
7. $y = 12x - x^3$ 8. $y = x^4 - 6x^2$

In Exercises 9 through 12 determine the values of x for which the curve of the given function is concave up, those values for which it is concave down, and any points of inflection. (These are the same functions as in Exercises 1 through 4.)

9. $y = x^2 - 2x$ 10. $y = 4 - x^2$
11. $y = 12x - x^3$ 12. $y = x^4 - 6x^2$

In Exercises 13 through 16 use the information from Exercises 1 through 12 to sketch the graphs of the given functions.

13. $y = x^2 - 2x$ 14. $y = 4 - x^2$
15. $y = 12x - x^3$ 16. $y = x^4 - 6x^2$

In Exercises 17 through 24 sketch the graphs of the given functions by determining the appropriate information and points from the first and second derivatives.

17. $y = 12x - 2x^2$ 18. $y = 3x^2 - 1$

19. $y = 2x^3 + 6x^2$ 20. $y = x^3 - 9x^2 + 15x + 1$

21. $y = x^5 - 5x$ 22. $y = x^4 + 8x + 2$

23. $y = 4x^3 - 3x^4$ 24. $y = x^5 - 20x^2$

In Exercises 25 through 28 sketch the indicated curves.

25. A certain projectile follows a path given by $y = x - 0.00025x^2$, where distances are measured in meters. By the methods of this section sketch the graph of the path of the projectile.

26. The deflection y of a beam at a horizontal distance x from one end is given by $y = k(x^3 - 60x^2)$. Sketch the curve representing the beam if it is 10 ft long and $k = 1/5000$.

27. A rectangular box is made from a piece of cardboard 8 in. by 12 in., by cutting a square from each corner and bending up the sides. Express the volume of the box as a function of the side of the square which is cut out, and then sketch the curve of the resulting equation.

28. A tool box whose end is square is to be made from 600 in.² of sheet metal. Express the volume of the box as a function of the side of the square of the end. Sketch the curve of the resulting function.

3–5 More on Curve Sketching

We are now in a position to combine the information from the derivative with information obtainable from the function itself to sketch the graph of a function. The information we can obtain from the function includes topics introduced previously. We can obtain intercepts, symmetry, the behavior of the curve as x becomes large, and the vertical asymptotes of the curve. Also, continuity is important in sketching certain functions. The examples which follow demonstrate how these concepts are used along with the information from the derivatives to sketch the graph of a function. We will find that some of these considerations are of much more value than others in graphing any particular curve.

Example A

Sketch the graph of

$$y = \frac{2}{x^2 + 1}$$

Intercepts: If $x = 0$, $y = 2$, which means that $(0, 2)$ is an intercept. We see that if $y = 0$, there is no corresponding value of x, since $2/(x^2 + 1)$ is a fraction greater than zero for all x. This also indicates that all points on the curve are above the x-axis.

Symmetry: The curve is symmetric to the y-axis since

$$y = \frac{2}{(-x)^2 + 1} \quad \text{is the same as} \quad y = \frac{2}{x^2 + 1}$$

The curve is not symmetric to the x-axis since

$$-y = \frac{2}{x^2 + 1} \quad \text{is not the same as} \quad y = \frac{2}{x^2 + 1}$$

The curve is not symmetric to the origin since

$$-y = \frac{2}{(-x)^2 + 1} \quad \text{is not the same as} \quad y = \frac{2}{x^2 + 1}$$

The value in knowing the symmetry is that we should find those portions of the curve on either side of the y-axis reflections of the other. It is possible to use this fact directly or to use it as a check.

Behavior as x becomes large: We note that as $x \to +\infty$, $y \to 0$ since $2/(x^2 + 1)$ is always a fraction greater than zero, but which becomes smaller as x becomes larger. Therefore, we see that $y = 0$ is an asymptote. Either from the symmetry or the function, we also see that $y \to 0$ as $x \to -\infty$.

Vertical asymptotes: From the previous discussions we recall that an asymptote is a line which a curve approaches. We have already noted that $y = 0$ is an asymptote for this curve. This asymptote, the x-axis, is a horizontal line. As noted previously, *vertical asymptotes, if any exist, are found by determining those values of x for which the denominator of any term is zero.* Such a value of x makes y undefined. Since $x^2 + 1$ cannot be zero, this curve has no vertical asymptotes. The next example illustrates a curve which has a vertical asymptote.

Derivatives: Since

$$y = \frac{2}{x^2 + 1} = 2(x^2 + 1)^{-1}$$

then

$$y' = -2(x^2 + 1)^{-2}(2x) = \frac{-4x}{(x^2 + 1)^2}$$

Since $(x^2 + 1)^2$ is positive for all values of x, the sign of y' is determined by the numerator. Thus, we note that $y' = 0$ for $x = 0$ and that $y' > 0$

for $x < 0$ and $y' < 0$ for $x > 0$. The curve, therefore, is increasing for $x < 0$, decreasing for $x > 0$, and has a maximum point at $(0, 2)$. Now,

$$y'' = \frac{(x^2 + 1)^2(-4) + 4x(2)(x^2 + 1)(2x)}{(x^2 + 1)^4} = \frac{-4(x^2 + 1) + 16x^2}{(x^2 + 1)^3}$$

$$= \frac{12x^2 - 4}{(x^2 + 1)^3} = \frac{4(3x^2 - 1)}{(x^2 + 1)^3}$$

We note that y'' is negative for $x = 0$, which confirms that $(0, 2)$ is a maximum point. Also, points of inflection are found for the values of x satisfying $3x^2 - 1 = 0$. Thus, $(-\frac{1}{3}\sqrt{3}, \frac{3}{2})$ and $(\frac{1}{3}\sqrt{3}, \frac{3}{2})$ are points of inflection. The curve is concave up if $x < -\frac{1}{3}\sqrt{3}$, or $x > \frac{1}{3}\sqrt{3}$, and concave down if $-\frac{1}{3}\sqrt{3} < x < \frac{1}{3}\sqrt{3}$.

Putting this information together we sketch the curve shown in Fig. 3–19. It might be noted that this curve could have been sketched primarily by use of the fact that $y \to 0$ as $x \to +\infty$ and as $x \to -\infty$, and the fact that a maximum point exists at $(0, 2)$. However, the other parts of the analysis, such as symmetry and concavity, serve as excellent checks and also make the curve more accurate.

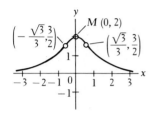

Figure 3–19

Example B

Sketch the graph of $y = x + \dfrac{4}{x}$.

Intercepts: If we set $x = 0$, y is undefined. This means that the curve is not continuous at $x = 0$ and there are no y-intercepts. If we set $y = 0$, $x + 4/x = (x^2 + 4)/x$ cannot be zero since $x^2 + 4$ cannot be zero. Therefore, there are no intercepts. This may seem to be of little value, but we must realize *this curve does not cross either axis.* This, in itself, will be of value when we sketch the graph in Fig. 3–20.

Symmetry: In testing for symmetry, we find that it is not symmetric to either axis. However, this curve does possess symmetry to the origin. This is determined by the fact that when $-x$ replaces x and at the same time $-y$ replaces y, the equation does not change.

Behavior as x becomes large: As $x \to +\infty$, and as $x \to -\infty$, $y \to x$ since $4/x \to 0$. Thus, $y = x$ is an asymptote of the curve.

Vertical asymptotes: As we noted in Example A, vertical asymptotes exist for values of x for which y is undefined. In this equation, $x = 0$ makes the second term on the right undefined and therefore y is undefined. In fact as $x \to 0$ from the positive side, $y \to +\infty$, and as $x \to 0$ from the negative side, $y \to -\infty$. This is derived from the sign of $4/x$ in each case.

Derivatives: Finding the first derivative we have

$$y' = 1 - \frac{4}{x^2} = \frac{x^2 - 4}{x^2}$$

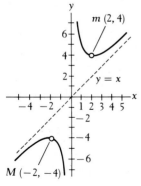

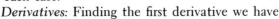

Figure 3–20

The x^2 in the denominator indicates that the sign of the first derivative is the same as its numerator. The numerator is zero if $x = -2$ or $x = 2$. If $x < -2$ or $x > 2$, then $y' > 0$; and if $-2 < x < 2$, $x \neq 0$, $y' < 0$.

Thus, y is increasing if $x < -2$ or $x > 2$, and also y is decreasing if $-2 < x < 2$, except at $x = 0$ (y is undefined). Also, $(-2, -4)$ is a maximum point and $(2, 4)$ is a minimum point. The second derivative is $y'' = 8/x^3$. This cannot be zero, but it is negative if $x < 0$ and positive if $x > 0$. Thus, the curve is concave down if $x < 0$ and concave up if $x > 0$. Using this information we have the graph shown in Fig. 3–20.

Example C
Sketch the graph of

$$y = \frac{1}{\sqrt{1 - x^2}}$$

Intercepts: If $x = 0$, $y = 1$. If $y = 0$, $1/\sqrt{1 - x^2}$ would have to be zero, but it cannot since it is a fraction with 1 as the numerator for all values of x. Thus, $(0, 1)$ is an intercept. We also note that if x is numerically greater than 1, we have a negative value under the radical. Thus, all points on the curve are between the lines $x = -1$ and $x = 1$.
Symmetry: The curve is symmetric to the y-axis.
Behavior as x becomes large: The values of x cannot be considered beyond 1 or -1, for any value of $x < -1$ or $x > 1$ gives imaginary values for y. Thus, the curve does not exist for values of $x < -1$ or $x > 1$.
Vertical asymptotes: If $x = 1$ or $x = -1$, y is undefined. In each case as $x \to 1$ and as $x \to -1$, $y \to +\infty$.
Derivatives:

$$y' = -\tfrac{1}{2}(1 - x^2)^{-3/2}(-2x) = \frac{x}{(1 - x^2)^{3/2}}$$

We see that $y' = 0$ if $x = 0$. If $-1 < x < 0$, $y' < 0$ and also if $0 < x < 1$, $y' > 0$. Thus the curve is decreasing if $-1 < x < 0$ and increasing if $0 < x < 1$. There is a minimum point at $(0, 1)$:

$$y'' = \frac{(1 - x^2)^{3/2} - x(\tfrac{3}{2})(1 - x^2)^{1/2}(-2x)}{(1 - x^2)^3} = \frac{(1 - x^2) + 3x^2}{(1 - x^2)^{5/2}}$$

$$= \frac{2x^2 + 1}{(1 - x^2)^{5/2}}$$

The second derivative cannot be zero since $2x^2 + 1$ is positive for all values of x. The second derivative is also positive for all permissible values of x which means the curve is concave up for these values.
Using this information we sketch the graph in Fig. 3–21.

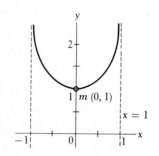

Figure 3–21

Exercises 3–5

In the following exercises use the method of the examples of this section to sketch the indicated curves.

1. $y = \dfrac{4}{x^2}$ 2. $y = \dfrac{x}{x - 2}$ 3. $y = x^2 + \dfrac{2}{x}$

4. $y = x + \dfrac{4}{x^2}$ **5.** $y = \dfrac{x^2}{x+1}$ **6.** $y = \dfrac{9x}{x^2+9}$

7. $y = \dfrac{1}{x^2 - 1}$ **8.** $y = x\sqrt{1 - x^2}$

9. Sketch R_T as a function of R for the function relating the combined resistance of the circuit given in Exercise 23 of Section 2–5.

10. A fence is to be constructed to enclose a rectangular area of 20,000 m². A previously constructed wall is to be used for one side. Sketch the length of fence to be built as a function of the side of the fence parallel to the wall.

11. A cylindrical oil drum is to be made such that it will contain 20 kL. Sketch the area of sheet metal required for construction as a function of the radius of the drum.

12. Sketch P as a function of r for the function relating power and resistance of the circuit given in Exercise 27 of Section 2–7. Assume values of $E = 6$ V and $R = 1\ \Omega$.

3–6 Applied Maximum and Minimum Problems

Problems from various applied situations frequently occur which require finding a maximum or minimum value of some function. If the function is known, the methods we have already discussed can be used directly. This is illustrated in the following example.

Example A

An automobile manufacturer, in testing a new engine on one of its models, found that the efficiency e of the engine as a function of the speed s of the car was given by $e = 0.768s - 0.00004s^3$. Here, e is given in percent and s is given in km/h. What is the maximum efficiency of the engine?

Since we wish to find a maximum value, we find the derivative of e with respect to s.

$$\frac{de}{ds} = 0.768 - 0.00012s^2$$

We then set the derivative equal to zero in order to find the value of s for which a maximum may occur.

$$0.768 - 0.00012s^2 = 0$$
$$0.00012s^2 = 0.768$$
$$s^2 = 6400$$
$$s = 80.0 \text{ km/h}$$

We know that s must be positive to have meaning in this problem. Therefore, the apparent solution of $s = -80$ is discarded. The second derivative is

$$\frac{d^2s}{de^2} = -0.00024s$$

which is negative for any positive value of s. Thus, a maximum occurs for $s = 80.0$. Substituting $s = 80.0$ in the function for e, we obtain

$$e = 0.768(80.0) - 0.00004(80.0^3)$$
$$= 61.44 - 20.48 = 40.96$$

Therefore, to the nearest tenth, the maximum efficiency is 41.0%, and this occurs for $s = 80.0$ km/h.

In many problems for which a maximum or minimum value is to be determined, the function is itself not given explicitly. It is necessary to find the function from the statement of the problem. The principal difficulty which arises in these problems is finding this function. Once we determine this function, we take a derivative with respect to a variable in which it is expressed, and then set the derivative equal to zero. Sufficient information must be given so that a unique solution is possible. Often this information is available, but we must analyze the wording of the problem carefully to find it. Also, many times it is more convenient to set up more than one function, based on the given information, and to find a derivative of each with respect to the same variable. We can then eliminate by substitution any undesired variables or derivatives which appear in each. The following examples illustrate these methods.

Example B
Find the number which exceeds its square by the greatest amount.
    We must set up a function which expresses the difference between a number and its square. Let the number be x, and let the difference be D. Then $D = x - x^2$. Taking a derivative, we have

$$\frac{dD}{dx} = 1 - 2x$$

Setting the derivative equal to zero, we have $0 = 1 - 2x$. Solving for x, we obtain the result $x = \frac{1}{2}$. The second derivative gives $d^2D/dx^2 = -2$, which states that the second derivative is always negative. This means that whenever the first derivative is zero, it represents a maximum. In many problems it is not necessary to test for maximum or minimum, since the nature of the problem will indicate which must be the case. For example, in this problem we know that numbers greater than 1 do not exceed their squares at all. The same is true for all negative numbers. Thus, the answer must lie between 0 and 1.

Example C

Find the maximum possible area of a rectangle whose perimeter is 16 units.

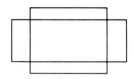

When we reflect on this problem, we see that there are limitless possibilities for rectangles of a perimeter of 16 units and differing areas (see Fig. 3–22). For example, if the sides are 7 and 1, the area is 7. If the sides are 6 and 2, the area is 12. Therefore, we set up a function for the area of a rectangle in terms of its sides x and y:

$$A = xy$$

We know, however, one other important fact: that the perimeter is 16. Thus, $2x + 2y = 16$. Solving for y, we have $y = 8 - x$. Substituting this equation into the expression for the area, we have

$$A = x(8 - x) = 8x - x^2$$

Figure 3–22

Taking a derivative of this function and setting it equal to zero, we have

$$\frac{dA}{dx} = 8 - 2x$$

$$8 - 2x = 0$$
$$x = 4$$

By noting values of the derivative near 4 or by finding the second derivative, we can show that we have a maximum for $x = 4$. Thus, $x = 4$ and $y = 4$ give the maximum area of 16 square units.

Example D

Find the point on the parabola $y = x^2$ which is nearest the point $(6, 3)$.

In this example we must set up a function for this distance between a general point (x, y) on the parabola and the point $(6, 3)$. This relation is

$$D = \sqrt{(x - 6)^2 + (y - 3)^2}$$

However, to make it easier for us to take derivatives, we shall square both sides of this expression. If a function is a minimum, then so is its square. We shall also use the fact that the point (x, y) is on $y = x^2$ by replacing y by x^2. Thus, we have

$$D^2 = (x - 6)^2 + (x^2 - 3)^2 = x^2 - 12x + 36 + x^4 - 6x^2 + 9$$
$$= x^4 - 5x^2 - 12x + 45$$

$$\frac{dD^2}{dx} = 4x^3 - 10x - 12$$

$$0 = 2x^3 - 5x - 6$$

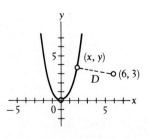

Figure 3–23

Using synthetic division or some similar method, we find that the solution to this equation is $x = 2$. (There are also two complex number solutions.) Thus, the required point on the parabola is $(2, 4)$ (see Fig. 3–23). We can show that we have a minimum by analyzing the first derivative, the second derivative, or by noting that points at much greater distances exist (therefore, it cannot be a maximum).

Example E
A company determines that it can sell 1000 units of a product per month if the price is \$5 for each unit. It also estimates that for each 1-cent reduction in unit price, 10 more units can be sold. Under these conditions, what is the maximum possible income and what price per unit gives this income?

If we let $x =$ the number of units over 1000 sold, the total number of units sold is $1000 + x$. The price for each unit is \$5 less 1 cent (0.01 dollars) for each block of ten units over 1000 which are sold. Thus, the price for each unit is

$$5 - 0.01\left(\frac{x}{10}\right) \quad \text{or} \quad 5 - 0.001x \text{ dollars}$$

The income I is the number of units sold times the price of each unit. Therefore, we have

$$I = (1000 + x)(5 - 0.001x)$$

Multiplying and finding the first derivative, we have

$$I = 5000 + 4x - 0.001x^2$$

$$\frac{dI}{dx} = 4 - 0.002x$$

Setting this derivative equal to zero, we have

$$0 = 4 - 0.002x$$
$$x = 2000$$

We note that if $x < 2000$, the derivative is positive, and if $x > 2000$, the derivative is negative. Therefore, if $x = 2000$, I is at a maximum. This means that the maximum income is derived if 2000 units over 1000 are sold, or 3000 units in all. This in turn means that the maximum income is \$9000 and the price per unit is \$3. These values are found by substituting $x = 2000$ into the expression for I and for the price.

Example F
Find the most economical dimensions for a cylindrical cup which holds a specified volume.

When we analyze the wording of the problem carefully, we see that we are to minimize the surface area of a right circular cylinder which has a bottom, but no top. Also the volume is to be considered as constant. Thus, we set up expressions for the surface area and volume:

$$A = \pi r^2 + 2\pi rh, \qquad V = \pi r^2 h$$

Since the volume is not specified numerically, it is easier to take the derivative of each of these formulas separately. The derivatives may be taken with respect to either r or h. Let us choose r, and write

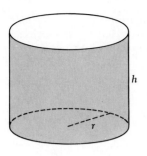

Figure 3-24

$$\frac{dA}{dr} = 2\pi r + 2\pi r \frac{dh}{dr} + 2\pi h, \qquad \frac{dV}{dr} = 0 = 2\pi rh + \pi r^2 \left(\frac{dh}{dr}\right)$$

Setting dA/dr equal to zero, and at the same time eliminating the derivative dh/dr by substitution, we have

$$0 = 2\pi r + 2\pi r \left(-\frac{2h}{r}\right) + 2\pi h$$

or

$$0 = r - 2h + h$$

This last equation tells us that $h = r$, which is the relation between dimensions for the most economical cup (see Fig. 3-24).

Example G

The illuminance of a light source at any point varies directly as the strength of the source and inversely as the square of the distance from the source. Given two sources, of strengths 8 units and 1 unit, respectively, which are 100 ft apart, determine at what point between them the illuminance is the least, assuming that the illuminance at any point is the sum of the illuminances of the two sources.

Let $I =$ the sum of the illuminances and
$x =$ the distance from the source of strength 8.

Then we find that

$$I = \frac{8}{x^2} + \frac{1}{(100 - x)^2}$$

is the function between the illuminance and the distance from the source of strength 8. We must now take a derivative of I with respect to x, set it equal to zero, and solve for x to find the point at which the illuminance is a minimum:

$$\frac{dI}{dx} = -\frac{16}{x^3} + \frac{2}{(100 - x)^3} = \frac{-16(100 - x)^3 + 2x^3}{x^3(100 - x)^3}$$

This function will be zero if the numerator is zero. Therefore, we have

$$2x^3 - 16(100 - x)^3 = 0 \qquad \text{or} \qquad x^3 = 8(100 - x)^3$$

Taking cube roots of each side, we have

$$x = 2(100 - x) \qquad \text{or} \qquad x = 66.7 \text{ ft}$$

The point where the illuminance is a minimum is 66.7 ft from the 8-unit source of illuminance.

Exercises 3-6

In the following exercises solve the given maximum and minimum problems.

1. The height s (in feet) of an object thrown vertically upward from the ground is given by $s = 112t - 16t^2$, where t is measured in seconds. What is the greatest height to which the object will go?

2. A company determines that the profit p (in dollars) from the sale of x units of a product, is given by $p = -2x^2 + 400x - 100$. What is the maximum profit which can be made from the sale of this product?

3. The cutting speed s (in feet per minute) of a saw in cutting a particular type of metal piece is given by $s = \sqrt{t - 4t^2}$ where t is the time in seconds. What is the maximum cutting speed in this operation?

4. The electric resistance (in ohms) of a particular type of resistor as a function of the temperature (in degrees Celsius) is given by

$$R = \sqrt{0.001 T^4 - 4T + 100}$$

What is the minimum resistance of this type of resistor?

5. The product of two positive numbers is 64. Find the numbers if their sum is a minimum.

6. The sum of two numbers is 40. Find the numbers such that their product is a maximum.

7. A rectangular field bounded on one side by a river is to be fenced in. If no fencing is required along the river, what is the maximum area which can be enclosed if 2000 ft of fencing are used?

8. An architect is designing a rectangular building in which the front wall costs twice as much per linear meter as the other three walls. The building is to cover 1350 m². What dimensions must it have such that the cost of the walls is a minimum?

9. Find the dimensions of the largest rectangle which can be inscribed in a quarter-circle of radius 4.

10. What is the area of the largest rectangle which can be inscribed in the first quadrant under the parabola $y = 4 - x^2$?

11. If the hypotenuse of a right triangle is 10, what must the sides be so that the area is a maximum?

12. A box with a square base is open at the top. If 108 ft² of material are used, what is the maximum volume possible for the box?

13. What is the maximum slope of the curve $y = 6x^2 - x^3$?

14. What is the minimum slope of the curve $y = x^5 - 10x^2$?

15. A person is in a boat 4 km from the nearest point P on a straight shoreline. The person wishes to go to point A which is on the shore, 5 km from P. If the person can row at 3 km/h and walk at 5 km/h, how far along the shore from P towards A should the boat land in order that the person can reach A in the least time?

16. A poster is to have margins of 4 in. at the top and bottom and margins of 2 in. at each side. The printed matter of the poster is to cover 50 in.². Find the dimensions of the poster if the total area is to be a minimum.

17. An open box is to be made from a square piece of cardboard whose sides are 8 in. long by cutting equal squares from the corners and bending up the sides. Determine the side of the square which is to be cut out so that the volume of the box may be a maximum.

18. Find the relation between the radius and the height of a right circular cylinder for which the total surface area is to be a minimum for a given volume.

19. A certain area is a rectangle with a semicircle at each end. Find the open side (not between the rectangle and the semicircle) of the rectangle if the perimeter is 8 units and the area of the rectangle is to be a maximum.

20. A company finds that there is a net profit of $10 for each of the first 1000 units produced each week. For each unit over 1000 produced, there is 2 cents less profit per unit. How many units should be produced each week to net the greatest profit?

21. The deflection y of a beam of length L at a horizontal distance x from one end is given by $y = k(2x^4 - 5Lx^3 + 3L^2x^2)$, where k is a constant. For what value of x does the maximum deflection occur?

22. The electric power (in watts) produced by a certain source is given by

$$P = \frac{144r}{(r + 0.6)^2}$$

where r is the resistance in the circuit. For what value of r is the power a maximum?

23. The strength of a rectangular beam is proportional to the product of its width and the square of its depth. Find the dimensions of the strongest beam that can be cut from a circular log 16 in. in diameter.

24. A specially made cylindrical container is made of stainless steel sides and bottom and a silver top. If silver is ten times as expensive as stainless steel, what are the most economical dimensions of the container if it is to hold 10π cm^3?

3−7 Review Exercises for Chapter 3

In Exercises 1 through 4 find the equations of the tangent and normal lines.

1. Find the equation of the line tangent to the parabola $y = 3x - x^2$ at the point $(-1, -4)$.

2. Find the equation of the line tangent to the curve $y = x^2 - \dfrac{2}{x}$ at the point $(2, 3)$.

3. Find the equation of the line normal to $y = \dfrac{x}{4x - 1}$ at the point $(1, \tfrac{1}{3})$.

4. Find the equation of the line normal to $y = \dfrac{1}{\sqrt{x - 2}}$ at the point $(6, \tfrac{1}{2})$.

In Exercises 5 through 8 determine the required velocities and accelerations.

5. Given that the x- and y-coordinates of a moving particle are given as a function of time by the parametric equations $x = t^4 - t$, $y = \dfrac{1}{t}$, find the magnitude and direction of the velocity when $t = 2$.

6. A particle moves along the curve of $y = \dfrac{1}{x + 2}$ with a constant velocity in the x-direction of 4 cm/s. Find v_y at $(2, \tfrac{1}{4})$.

7. Find the magnitude and direction of the acceleration for the particle in Exercise 5.

8. Find the magnitude and direction of the acceleration of the particle in Exercise 6.

In Exercises 9 through 16 sketch the graph of the given functions by information obtained from the function as well as information obtained from the derivatives.

9. $y = 4x^2 + 16x$

10. $y = 2x^2 + x - 1$

11. $y = 27x - x^3$

12. $y = x^3 + 2x^2 + x + 1$

13. $y = x^4 - 32x$

14. $y = x^4 + 4x^3 - 16x$

15. $y = \dfrac{x}{x + 1}$

16. $y = x^3 + \dfrac{3}{x}$

In Exercises 17 through 32 solve the given problems.

17. The parabolas $y = x^2 + 2$ and $y = 4x - x^2$ are tangent to each other. Find the equation of the line tangent to them at the point of tangency.

18. Find the equation of the line tangent to $y = x^4 - 8x$ and parallel to $y + 4x + 3 = 0$.

19. A projectile moves according to the equations $x = 1200t$ and $y = -490t^2$, where distances are measured in centimeters and time is measured in seconds. Find the magnitude and direction of velocity after 3.00 s.

20. An electron is moving along the path $y = 1/x$ at the constant velocity of 100 m/s. Find the velocity in the x-direction when the electron is at the point $(2, \frac{1}{2})$.

21. Sketch a continuous curve having the following characteristics.

$f(0) = 2$ $f'(x) < 0$ for $x < 0$

$f''(x) > 0$ for all x $f'(x) > 0$ for $x > 0$

22. Sketch a continuous curve having the following characteristics.

$f(0) = 1$ $f'(0) = 0$

$f''(x) < 0$ for $x < 0$ $f'(x) > 0$ for $|x| > 0$

$f''(x) > 0$ for $x > 0$

23. Side a of a triangle is increasing at the rate of 2.50 in./min, and side b is increasing at the rate of 3.70 in./min. If the angle between sides a and b is always $60°$, find the rate of change of side c when $a = 5.60$ in. and $b = 8.00$ in.

24. The current I through a circuit with a resistance R and a battery whose voltage is E and whose internal resistance is r is given by $I = E/(R + r)$. If R changes at the rate of 0.250 Ω/min, how fast is the current changing when $R = 6.25$ Ω, if $E = 3.10$ V and $r = 0.230$ Ω?

25. A machine part is to be in the shape of a circular sector of radius r and central angle θ. Find r and θ if the area is one unit and the perimeter is a minimum.

26. A beam of rectangular cross section is to be cut from a log 2 ft in diameter. The stiffness varies as the width and the cube of the depth. What dimensions will give the beam maximum stiffness?

27. Sketch a graph of van der Waals equation (see Exercise 26 of Section 2-7), assuming the following values: $R = T = a = 1$ and $b = 0$. For many gases the value of a is much greater than that for b. Even though the values of $R = T = 1$ are not realistic, the *shape* of the curve will be correct for the assumed value of $b = 0$.

28. The weekly profits of a corporation for a particular year are given by $p = x^4 - 40x^3$, where x is the week of the year. Sketch the graph of p vs. x, assuming it to be a continuous function over the appropriate interval.

29. Two ships leave the same port at the same time, one traveling south at the rate of 12 km/h and the other traveling east at the rate of 5 km/h. At what rate is the distance between them changing two hours after they leave the port?

30. The base of a conical machine part is being milled such that the height is decreasing at the rate of 0.05 cm/min. If the part originally had a radius of 1 cm and height of 3 cm, how fast is the volume changing when the height is 2.8 cm?

31. An alpha-particle moves through a magnetic field along the parabolic path $y = x^2 - 4$. Determine the closest that the particle comes to the origin.

32. A Norman window has the form of a rectangle surmounted by a semicircle. Find the dimensions (radius of circular part and height of rectangular part) of the window that will admit the most light if the perimeter of the window is 12 ft.

4

Integration

4–1 Differentials

Physical investigations of phenomena often lead to information regarding the rate of change of a variable. Then to arrive at the functional relationship between variables, we have to reverse the process of differentiation. Other applications of this reverse process involve finding areas under curves and finding volumes of solids. As we shall see, these basic problems, although apparently rather distinct, have a very similar mathematical interpretation.

Although the primary concern of this chapter is the development of the inverse process of differentiation, it is necessary to first introduce certain concepts and notation in this section before proceeding to the inverse process. The material presented here is necessary to properly relate differentiation and the methods we shall develop.

We shall now define the **differential** of a function $y = f(x)$ as

$$dy = f'(x) \ dx \tag{4-1}$$

In Eq. (4–1), the quantity *dy is the differential of y*, and *dx is the differential of x. The differential dx is defined as equal to Δx, the increment in x.* We define it purposely in this way, so that $f'(x) = dy/dx$. In this way we can interpret the derivative as the ratio of the differential of y to the differential of x. That is, now the derivative can be considered as a fraction. Although we had previously used the notation dy/dx for the derivative, we had not interpreted it as a fraction.

Example A

Find the differential of $y = 3x^5 - x$.

In this example, $f(x) = 3x^5 - x$, which in turn means that $f'(x) = 15x^4 - 1$. Thus,

$$dy = (15x^4 - 1)\,dx$$

Example B

Find the differential of $y = (2x^3 - 1)^4$.

$$dy = 4(2x^3 - 1)^3(6x^2)\,dx = 24x^2(2x^3 - 1)^3\,dx$$

Example C

Find the differential of $y = \dfrac{4x}{x^2 + 4}$.

$$dy = \frac{(x^2 + 4)(4) - (4x)(2x)}{(x^2 + 4)^2}\,dx = \frac{4x^2 + 16 - 8x^2}{(x^2 + 4)^2}\,dx$$

$$= \frac{-4x^2 + 16}{(x^2 + 4)^2}\,dx = \frac{-4(x^2 - 4)}{(x^2 + 4)^2}\,dx$$

We shall find that the definition of the differential will clarify the connection between the various interpretations of the inverse process of differentiation. Although the principal purpose in introducing it here is to be able to use the notation, there are useful direct applications of the differential.

The applications of the differential are based upon the fact that the differential of y, dy, closely approximates the increment in y, Δy, if the differential of x, dx, is small. To understand this statement, let us look at Fig. 4–1. Recalling the meaning of Δx and Δy, we see that the points $P(x, y)$ and $Q(x + \Delta x, y + \Delta y)$ lie on the curve of $f(x)$. However, $f'(x) = dy/dx$ at P, which means if we draw a tangent line at P, its slope may be indicated by dy/dx. By choosing $\Delta x = dx$, we can see the difference between Δy and dy. It can be seen that as dx becomes smaller, Δy more nearly equals dy.

For given changes in x, it is necessary to use the Δ-process to find the exact change, Δy, in y. However, *for small values of Δx, dy can be used to approximate Δy closely.* Generally dy is much more easily determined than is Δy.

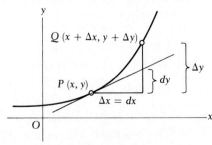

Figure 4–1

Example D

Calculate Δy and dy for $y = x^3 - 2x$ for $x = 3$ and $\Delta x = 0.1$.

By the Δ-process we find

$$\Delta y = 3x^2 \, \Delta x + 3x(\Delta x)^2 + (\Delta x)^3 - 2 \, \Delta x$$

Using the given values we find

$$\Delta y = 3(9)(0.1) + 3(3)(0.01) + (0.001) - 2(0.1) = 2.591$$

The differential of y is

$$dy = (3x^2 - 2) \, dx$$

Since $dx = \Delta x$, we have

$$dy = [3(9) - 2](0.1) = 2.5$$

Thus, $\Delta y = 2.591$ and $dy = 2.5$. We see that in this case dy is very nearly equal to Δy.

The fact that dy can be used to approximate Δy is useful in determining the error in a result, if the data are in error, or the equivalent problem of finding the change in a result if a change is made in the data. Even though such changes could be found by use of a calculator, the differential can be used to set up a general expression for the change in a particular function. The following example illustrates the method.

Example E

The edge of a cube was determined to be 9.00 in. From this value the volume was found. Later it was discovered that the value of the edge was 0.20 in. too small. Approximately by how much was the volume in error?

The volume V of a cube, in terms of an edge, e, is $V = e^3$. Since we wish to find the change in V for a given change in e, this means we want the value of dV for $de = 0.20$ in. Thus,

$$dV = 3e^2 \, de$$
$$dV = 3(9.00)^2(0.20) = 48.6 \text{ in.}^3$$

We see that in this case the volume was in error by about 49 in.^{3}. As long as de is small compared to e, we can calculate an error or change in the volume of a cube by calculating the value of $3e^2 \, de$.

Often when considering the error of a given value or result, the actual numerical value of the error, the **absolute error,** is not as important as its size is in relation to the size of the quantity itself. *The ratio of the absolute error to the size of the quantity itself is known as the* **relative error.** The relative error is commonly expressed as a percentage.

Example F

Referring to Example E, we see that the absolute error in the edge was 0.20 in. The relative error in the edge was $0.20/9.00 = 0.022 = 2.2\%$. The absolute error in the volume is 49 in.^{3}, whereas the relative error in the volume is $49/729 = 0.067 = 6.7\%$.

Exercises 4–1

In Exercises 1 through 8 find the differential of each of the given functions.

1. $y = x^5 + x$

2. $y = 3x^2 + 6$

3. $y = (x^2 - 1)^4$

4. $y = (4 + 3x)^{1/3}$

5. $y = x^2(1 - x)^3$

6. $y = x\sqrt{1 - 4x}$

7. $y = \dfrac{x}{x + 1}$

8. $y = \dfrac{1}{\sqrt{1 - x^3}}$

In Exercises 9 through 12 find the values of Δy and dy for the given values of x and Δx.

9. $y = 7x^2 + 4x, x = 4, \Delta x = 0.2$

10. $y = 2x^2 - 3x + 1, x = 5, \Delta x = 0.15$

11. $y = 2x^3 - 4x, x = 2.5, \Delta x = 0.05$

12. $y = x^4, x = 3.2, \Delta x = 0.08$

In Exercises 13 through 16 determine the value of dy for the given values of x and Δx. Compare with values of $f(x + \Delta x) - f(x)$ found by use of a calculator.

13. $y = (1 - 3x)^5, x = 1, \Delta x = 0.01$

14. $y = (x^2 + 2x)^3, x = 7, \Delta x = 0.02$

15. $y = x\sqrt{1 + 4x}, x = 12, \Delta x = 0.06$

16. $y = \dfrac{x}{\sqrt{6x - 1}}, x = 3.5, \Delta x = 0.025$

In Exercises 17 through 24 solve the given problems by finding the appropriate differential.

17. If the side of a square is measured to be 16.0 in., and an error of 0.5 in. is found in the measurement, approximately by how much is the original calculation of area in error?

18. If the radius of a circle is measured to be 8.00 cm, and the measurement has a maximum possible error of 0.03 cm, what is the maximum possible relative error in the area of the circle?

19. The voltage of a certain thermocouple as a function of temperature is given by $E = 6.2T + 0.0002T^3$. What is the approximate change in voltage if the temperature changes from 100°C to 101°C?

20. The velocity of an object rolling down a certain inclined plane is given by $v = \sqrt{100 + 16h}$, where h is the distance traveled along the plane by the object. What is the increase in velocity (in feet per second) of an object in moving from 20.0 to 20.5 ft along the plane? What is the relative change in the velocity?

21. What is the volume of metal used to make a right circular cylindrical container of radius 3.00 in. and height 4.00 in., if the thickness of sheet metal it was made from is 0.02 in.?

22. A precisely measured 10.0-Ω resistor (any error in its value will be considered negligible) is put in parallel with a variable resistor of resistance R. The combined resistance of the two resistors is $R_T = 10R/(10 + R)$. What is the relative error of the combined resistance, if R is measured at 40.0 Ω, with a possible error of 1.5 Ω?

23. Show that an error of 2% in the measurement of the side of a square results in an error of approximately 4% in the calculation of the area.

24. Show that the relative error in the calculation of the volume of a sphere is approximately three times the relative error in the measurement of the radius.

4–2 Antiderivatives

Since many kinds of problems in many areas, including science and technology, can be solved by reversing the process of finding a derivative or a differential, we shall introduce the basic technique of this procedure in this section. *This reverse process is known as* **antidifferentiation.** In the next section we shall formalize the process, but it is only the basic idea that is the topic of this section. Many of the applications are found in Chapter 5. The following example illustrates the method.

Example A

Find a function for which the derivative is $8x^3$. That is, find an antiderivative of $8x^3$.

We know that, when we set out to find the derivative of a polynomial, we reduce the power of x by 1. Also we multiply the coefficient by the power of x. Thus, the power of the function must have been 4, since the power in the derivative is 3. A factor of 4 of the derivative must also be divided out in the process of finding the function. If we write the derivative as $2(4x^3)$, we recognize $4x^3$ as the derivative of x^4. Therefore, the desired function is $2x^4$, which means that an antiderivative of $8x^3$ is $2x^4$. We can verify our result by finding the derivative.

Example B

Find an antiderivative of $x^2 + 2x$.

As for the x^2, we know that the power of x required in an antiderivative is 3. Also, to make the coefficient correct, we must multiply by $\frac{1}{3}$. The $2x$ should be recognized as the derivative of x^2. Therefore, we have the antiderivative as $\frac{1}{3}x^3 + x^2$.

In Examples A and B, we note that we could add any constant to the antiderivative given as the result, and still have a correct antiderivative. This is due to the fact that the derivative of a constant is zero. This is considered further in the following section. For the examples and exercises in this section, we will not include any constants in the results.

A great many functions to which we must apply the process of antidifferentiation are not polynomials. It is these functions which may cause more difficulty in the general process of antidifferentiation. The reader is advised to pay special attention to the following examples, for they illustrate a type of problem which *will* be found to be very important.

Example C

Find an antiderivative of $3(x^3 - 1)^2(3x^2)$.

Noting that we have a power of $x^3 - 1$ in the derivative, it is reasonable that the antiderivative may include a power of $x^3 - 1$. Since, in the derivative, $x^3 - 1$ is raised to the power 2, the antiderivative would then have $x^3 - 1$ raised to the power 3. Noting that the derivative of $(x^3 - 1)^3$ is $3(x^3 - 1)^2(3x^2)$, the desired antiderivative is $(x^3 - 1)^3$. We note that the factor of $3x^2$ does not appear in the antiderivative, for it was included from the process of finding a derivative. Therefore, it must be present for $(x^3 - 1)^3$ to be the proper antiderivative, but it must also be excluded in the process of antidifferentiation.

Example D

Find an antiderivative of $(2x + 1)^{1/2}$.

Here we note a power of $2x + 1$ in the derivative, which infers that the antiderivative has a power of $2x + 1$. Since in finding a derivative 1 is subtracted from the power of $2x + 1$, we should add 1 in finding the antiderivative. Thus, we should have $(2x + 1)^{3/2}$ as part of the antiderivative. Finding a derivative of $(2x + 1)^{3/2}$ we obtain $\frac{3}{2}(2x + 1)^{1/2}(2)$ $= 3(2x + 1)^{1/2}$. This differs from the given derivative by the factor of 3. Thus, if we write $(2x + 1)^{1/2} = \frac{1}{3}[3(2x + 1)^{1/2}]$, we have the required antiderivative as $\frac{1}{3}(2x + 1)^{3/2}$. Checking, the derivative of $\frac{1}{3}(2x + 1)^{3/2}$ is $\frac{1}{3}(\frac{3}{2})(2x + 1)^{1/2}(2) = (2x + 1)^{1/2}$.

Exercises 4–2

In the following exercises find antiderivatives of the given derivatives. That is, find functions for which the derivatives are given.

1. $3x^2$

2. $5x^4$

3. $6x^5$

4. $10x^9$

5. $6x^3 + 1$

6. $12x^5 + 2x$

7. $2x^2 - x$

8. $x^2 - 5$

9. $-\dfrac{1}{x^2}$

10. $-\dfrac{2}{x^3}$

11. $-\dfrac{6}{x^4}$

12. $-\dfrac{8}{x^5}$

13. $2x^4 + 1$

14. $3x^3 - 5x^2 + 3$

15. $6(2x + 1)^5(2)$

16. $3(x^2 + 1)^2(2x)$

17. $4(x^2 - 1)^3(2x)$

18. $5(2x^4 + 1)^4(8x^3)$

19. $x^3(2x^4 + 1)^4$

20. $x(1 - x^2)^7$

21. $\frac{3}{2}(6x + 1)^{1/2}(6)$

22. $\frac{5}{4}(1 - x)^{1/4}(-1)$

23. $(3x + 1)^{1/3}$

24. $(4x + 3)^{1/2}$

4–3 The Indefinite Integral

In the previous section, in developing the basic technique of finding an antiderivative, we noted that the results given are not unique. That is, we could have added any constant to the answers and the result would still have been correct. Again, this is the case since the derivative of a constant is zero.

Example A

The derivatives of x^3, $x^3 + 4$, $x^3 - 7$, and $x^3 + 4\pi$ are all $3x^2$. This means that any of the functions listed, as well as others, would be a proper answer to the problem of finding an antiderivative of $3x^2$.

From Section 4–1, we know that the differential of a function $F(x)$ can be written as $d[F(x)] = F'(x)dx$. Therefore, since finding a differential of a function is very closely related to finding the derivative, so is the antiderivative very closely related to the process of finding the function for which the differential is known.

The notation used for finding the general form of the antiderivative, the **indefinite integral,** *is written in terms of the differential.* Thus, the indefinite integral of a function $f(x)$, for which $dF(x)/dx = f(x)$, or $dF(x) = f(x)dx$, is defined as

$$\int f(x)\, dx = F(x) + C \qquad (4\text{–}2)$$

where C is an arbitrary constant, called the **constant of integration.** It represents any of the constants which may be attached to an antiderivative to have a proper result. We must have additional information beyond a knowledge of the differential to assign a specific value to C. *The symbol $\int$ is the* **integral sign,** *and indicates that the inverse of the differential is to be found. Determining the indefinite integral is called* **integration,** which we can see is essentially the same as finding an antiderivative.

Example B

$$\int 5x^4\, dx = x^5 + C.$$

We might think that the inclusion of this constant C would affect the derivative of the function x^5. However, the only effect of the C is to raise or lower the curve. The slope of $x^5 + 2$, $x^5 - 2$, or any function of the form $x^5 + C$ is the same for any given value of x. As Fig. 4–2 shows, tangents drawn to the curves are all parallel for the same value of x.

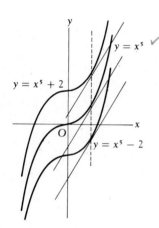

$y = x^5$

$y = x^5 + 2$

$y = x^5 - 2$

Figure 4–2

At this point we shall derive some basic formulas for integration. Since

$$\frac{d(cu)}{dx} = c\frac{du}{dx}$$

where u is a function of x, we can write

$$\int c\, du = c\int du = cu + C \qquad (4\text{–}3)$$

Also, since the derivative of a sum of functions equals the sum of the derivatives, we write

$$\int (du + dv) = u + v + C \qquad (4\text{–}4)$$

To find the derivative of a power of a function, we multiply by the power and subtract 1 from it. *To find the integral, we reverse this by adding 1 to the power of $f(x)$ in $f(x)\ dx$, and dividing by this new power.* The power formula for integration is therefore

$$\int u^n\ du = \frac{u^{n+1}}{n+1} + C \qquad n \neq -1 \tag{4–5}$$

When we use the formulas above, we must be able to recognize not only the proper form but also the component parts of the given formula. Unless you can do this and unless you have a good knowledge of differentiation, you will have trouble in applying Eq. (4–5). Most of the difficulty, if it exists, arises from improper identification of du. The following examples illustrate the use of the above formulas.

Example C
Integrate: $\int 6x\ dx$.
   First we must identify u, n, du, and any multiplying constants. By noting that 6 is a multiplying constant, we identify x as u, which in turn means that dx must be du and $n = 1$. Hence,

$$\int 6x\ dx = 6\int x\ dx = 6\left(\frac{x^2}{2}\right) + C = 3x^2 + C$$

We see that our result checks since the differential of $3x^2 + C$ is $6x\ dx$.

Example D
Integrate: $\int (5x^3 - 6x^2 + 1)\ dx$.
   In solving this problem, we have to use a combination of all three of the above formulas. Thus, we have

$$\int (5x^3 - 6x^2 + 1)\ dx = \int 5x^3\ dx + \int (-6x^2)\ dx + \int dx$$

$$= 5\int x^3\ dx - 6\int x^2\ dx + \int dx$$

In the first of these integral, $u = x$, $n = 3$, and $du = dx$. In the second, $u = x$, $n = 2$, and $du = dx$. The third is a direct application of Eq. (4–3), with $c = 1$ and $du = dx$. Thus, we have

$$5\int x^3\ dx - 6\int x^2\ dx + \int dx = 5\left(\frac{x^4}{4}\right) - 6\left(\frac{x^3}{3}\right) + x + C$$

$$= \frac{5}{4}x^4 - 2x^3 + x + C$$

Example E

Integrate: $\int \left(\sqrt{x} - \dfrac{1}{x^3} \right) dx.$

We write

$$\int \left(\sqrt{x} - \frac{1}{x^3} \right) dx = \int x^{1/2} \, dx - \int x^{-3} \, dx$$

$$= \frac{1}{3/2} x^{3/2} - \frac{1}{-2} x^{-2} + C$$

$$= \frac{2}{3} x^{3/2} + \frac{1}{2} x^{-2} + C = \frac{2}{3} x^{3/2} + \frac{1}{2x^2} + C$$

Example F

Integrate: $\int (x^2 + 1)^3 (2x \, dx).$

We first note that $n = 3$, for this is the power involved in the function being integrated. If $n = 3$, then $x^2 + 1$ must be u. If $u = x^2 + 1$, then $du = 2x \, dx$. Thus the integral is in proper form for integration *as it stands*. Using the power formula, we have

$$\int (x^2 + 1)^3 (2x \, dx) = \frac{(x^2 + 1)^4}{4} + C$$

It is in this type of problem that recognition of the proper form of a differential is very important. Only a good knowledge of differential forms, along with sufficient experience, makes ready recognition of this type possible. *It cannot be overemphasized that the entire quantity* $(2x \, dx)$ *must be equated to du if we are to integrate properly.* Normally u and n are recognized first, and then the proper form of du is derived from u.

Example G

Integrate: $\int x^2 \sqrt{x^3 + 2} \, dx.$

We first note that $n = \frac{1}{2}$ and u is then $x^3 + 2$. Since $u = x^3 + 2$, $du = 3x^2 \, dx$. In order to integrate properly, we must group the quantity $3x^2 \, dx$ as du, with other factors isolated from this quantity. Since there is no 3 under the integral sign, we introduce one. In order not to change the numerical value, we also introduce a $\frac{1}{3}$, normally before the integral sign. In this way we take full advantage of the fact that a constant (and only a constant) factor may be moved across the integral sign. The next form we should write is

$$\int x^2 \sqrt{x^3 + 2} \, dx = \frac{1}{3} \int \sqrt{x^3 + 2} \, (3x^2 \, dx)$$

In this way we indicate the proper grouping to result in the correct form of Eq. (24–5), Thus,

$$\int x^2 \sqrt{x^3 + 2} \, dx = \frac{1}{3} \cdot \frac{2}{3} (x^3 + 2)^{3/2} + C = \frac{2}{9} (x^3 + 2)^{3/2} + C$$

The $1/\frac{3}{2}$ was written as $\frac{2}{3}$, since this form is generally more convenient when we are working with fractions.

We mentioned earlier that, in addition to knowing the differential, we need more information in order to find the constant of integration. Such information usually consists of a set of values which the function is known to satisfy. That is, a point through which the curve of the function passes would provide the necessary information.

Example H

Find y in terms of x, given that $dy/dx = 3x - 1$, and the curve passes through $(1, 4)$.

When we write the equation as $dy = (3x - 1) \, dx$, then indicate the integration $\int dy = \int (3x - 1) \, dx$, the result of this integration is $y = \frac{3}{2}x^2 - x + C$. We know that the required curve passes through $(1, 4)$, which means that the coordinates of this point must satisfy the equation. Thus, $4 = \frac{3}{2} - 1 + C$, or $C = \frac{7}{2}$. The complete solution is

$$y = \frac{3}{2}x^2 - x + \frac{7}{2}$$

or

$$2y = 3x^2 - 2x + 7$$

Example I

At each point of a certain curve, the slope is given by $x\sqrt{2 - x^2}$. Given that the curve passes through $(1, 2)$, find its equation.

We write

$$\frac{dy}{dx} = x\sqrt{2 - x^2}, \qquad y = \int x\sqrt{2 - x^2} \, dx$$

To integrate this last expression, we recognize that $n = \frac{1}{2}$, $u = 2 - x^2$, and that therefore $du = -2x \, dx$. Hence,

$$y = -\frac{1}{2}\int (2 - x^2)^{1/2}(-2x \, dx) = -\frac{1}{2} \cdot \frac{2}{3}(2 - x^2)^{3/2} + C$$

$$= -\frac{1}{3}(2 - x^2)^{3/2} + C$$

Using the coordinates of the point $(1, 2)$, we have

$$2 = -\frac{1}{3}(2 - 1)^{3/2} + C \qquad \text{or} \qquad C = \frac{7}{3}$$

and

$$3y = 7 - (2 - x^2)^{3/2}$$

In this section we have discussed the integration of certain basic types of functions. There are many other methods used to integrate other functions, and some of these methods are discussed in Chapter 8. Also, there are many functions which cannot be integrated.

Exercises 4–3

In Exercises 1 through 28 integrate each of the given expressions.

1. $\int 2x \; dx$ 2. $\int 5x^4 \; dx$ 3. $\int x^7 \; dx$

4. $\int x^5 \; dx$ 5. $\int x^{3/2} \; dx$ 6. $\int \sqrt[8]{x} \; dx$

7. $\int x^{-4} \; dx$ 8. $\int \dfrac{1}{\sqrt{x}} \; dx$ 9. $\int (x^2 - x^5) \; dx$

10. $\int (1 - 3x) \; dx$ 11. $\int (1 + 2x)^2 \; dx$ 12. $\int (1 - x)^2 \; dx$

13. $\int (x\sqrt{x} - 5x^2) \; dx$ 14. $\int \left(4x - \dfrac{2}{x^3}\right) dx$ 15. $\int \sqrt{x}(x^2 - x) \; dx$

16. $\int (x^{1/3} + x^{1/5} + x^{-1/7}) \; dx$ 17. $\int (x^2 - 1)^5 (2x \; dx)$

18. $\int (x^3 - 2)^6 (3x^2 \; dx)$ 19. $\int (x^4 + 3)^4 (4x^3 \; dx)$ 20. $\int (1 - 2x)^{1/3} (-2 \; dx)$

21. $\int (x^5 + 4)^7 x^4 \; dx$ 22. $\int x^2 (1 - x^3)^{4/3} \; dx$ 23. $\int \sqrt{8x + 1} \; dx$

24. $\int \sqrt[3]{4 - 3x} \; dx$ 25. $\int \dfrac{x \; dx}{\sqrt{6x^2 + 1}}$ 26. $\int \dfrac{x^2 \; dx}{\sqrt{2x^3 + 1}}$

27. $\int \dfrac{x - 1}{\sqrt{x^2 - 2x}} \; dx$ 28. $\int (x^2 - x)(x^3 - \tfrac{3}{2}x^2)^8 \; dx$

In Exercises 29 through 32 find y in terms of x.

29. $\dfrac{dy}{dx} = 6x^2$, curve passes through $(0, 2)$

30. $\dfrac{dy}{dx} = x + 1$, curve passes through $(-1, 4)$

31. $\dfrac{dy}{dx} = x^2 (1 - x^3)^5$, curve passes through $(1, 5)$

32. $\dfrac{dy}{dx} = 2x^3 (x^4 - 6)^4$, curve passes through $(2, 10)$

In Exercises 33 through 36 find the required equations.

33. Find the equation of the curve whose slope is $-x\sqrt{1 - 4x^2}$ and which passes through $(0, 7)$.

34. Find the equation of the curve whose slope is $\sqrt{6x - 3}$ and which passes through $(2, -1)$.

35. Find the equation of the curve for which the second derivative is 6. The curve passes through $(1, 2)$ with a slope of 8.

36. Find the equation of the curve for which the second derivative is $12x^2$. The curve passes through $(1, 6)$ with a slope of 4.

4–4 The Area Under a Curve

Another basic problem which can be solved by integration is that of finding the area under a curve. In geometry, there are methods and formulas for finding the area of regular figures. By means of the calculus we will find it possible to find the area between curves for which we know the equations. First let us look at an example which illustrates the basic idea behind the method.

Example A

Approximate the area in the first quadrant and to the left of the line $x = 4$ under the parabola $y = x^2 + 1$. First make this approximation by inscribing two rectangles of equal width in the area and finding the sum of the areas of these rectangles. Then improve the approximation by repeating the process using eight rectangles.

The area to be approximated is shown in Fig. 4–3(a). The area with two rectangles inscribed under the curve is shown in Fig. 4–3(b). The first approximation, admittedly small, of the area can be found by adding the areas of the two rectangles. Both rectangles have a width of 2. The left rectangle is 1 unit high and the right rectangle is 5 units high. Thus, the area of the two rectangles is

$$A = 2(1 + 5) = 12$$

A much better approximation is found by inscribing the eight rectangles as shown in Fig. 4–3(c). Each of these rectangles has a width of $\frac{1}{2}$. The leftmost rectangle has a height of 1. The next has a height of $\frac{5}{4}$, which is determined by finding y for $x = \frac{1}{2}$. The next rectangle has a height of 2, which is found by evaluating y for $x = 1$. Finding the

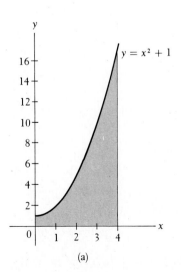

(a)

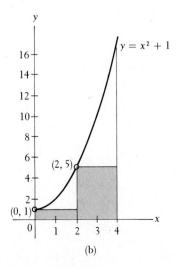

(b)

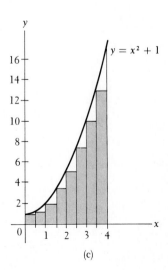

(c)

Figure 4–3

heights of all the rectangles and multiplying their sum by $\frac{1}{2}$ gives the area of the eight rectangles as

$$A = \frac{1}{2}\left(1 + \frac{5}{4} + 2 + \frac{13}{4} + 5 + \frac{29}{4} + 10 + \frac{53}{4}\right)$$

$$= \frac{43}{2} = 21.5$$

An even better approximation could be obtained by inscribing more rectangles under the curve. The greater the number of rectangles, the more nearly the sum of their areas equals the area under the curve. By a method involving integration developed later in this section we determine the *exact* area to be $\frac{76}{3} = 25\frac{1}{3}$.

We can now develop the basic method involved in finding an area under a curve. We consider the sum of the areas of rectangles inscribed under the curve as the number of rectangles is assumed to increase without bound. The reason for this last condition is that, as we saw in Example A, as the number of rectangles increases, the approximation of the area is better. Following is a specific example of this technique.

Example B
Find the area under the straight line $y = 2x$, above the x-axis, and to the left of the line $x = 4$.

We might mention at the beginning that this area can easily be found, since the figure is a triangle. However, the method we shall use here is the important concept at the moment. We first subdivide the interval from $x = 0$ to $x = 4$ into n inscribed rectangles of Δx in width. Then the extremities of the intervals are labeled $a, x_1, x_2, \ldots, b(=x_n)$, as shown in Fig. 4-4, where

$$x_1 = \Delta x$$
$$x_2 = 2\,\Delta x$$
$$\vdots$$
$$x_{n-1} = (n-1)\,\Delta x$$
$$b = n\,\Delta x$$

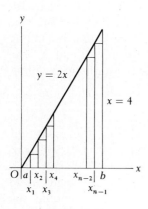

Figure 4–4

We then find the area for each of the rectangles. These areas are (for the respective rectangles):

First $f(a)\,\Delta x$, where $f(a) = f(0) = 2(0) = 0$ is the height

Second $f(x_1)\,\Delta x$, where $f(x_1) = 2(\Delta x) = 2\,\Delta x$ is the height

Third $f(x_2)\,\Delta x$, where $f(x_2) = 2(2\,\Delta x) = 4\,\Delta x$ is the height

Fourth $f(x_3)\,\Delta x$, where $f(x_3) = 2(3\,\Delta x) = 6\,\Delta x$ is the height

$$\vdots$$

Last $f(x_{n-1})\,\Delta x$, where $f[(n-1)\,\Delta x] = 2(n-1)\,\Delta x$ is the height

These areas are summed up as follows:

$$A_n = f(a)\,\Delta x + f(x_1)\,\Delta x + f(x_2)\,\Delta x + \cdots + f(x_{n-1})\,\Delta x$$
$$= 0 + 2\,\Delta x(\Delta x) + 4\,\Delta x(\Delta x) + \cdots + 2[(n-1)\,\Delta x]\,\Delta x$$
$$= \{2\,\Delta x[(1+2+3+\cdots+(n-1)]\}\,\Delta x$$
$$= 2(\Delta x)^2[1+2+3+\cdots+(n-1)]$$

Now $b = n\,\Delta x$, or $4 = n\,\Delta x$, or $\Delta x = 4/n$. Thus,

$$A_n = 2\left(\frac{4}{n}\right)^2[1+2+3+\cdots+(n-1)]$$

The sum $1 + 2 + 3 + \cdots + n - 1$ is the indicated sum of an arithmetic progression with the first term equal to 1 and the nth term equal to $n - 1$. This sum is

$$s = \frac{n-1}{2}(1+n-1) = \frac{n(n-1)}{2} = \frac{n^2-n}{2}$$

Now the expression for the sum of the areas can be set forth as

$$A_n = \frac{32}{n^2}\left(\frac{n^2-n}{2}\right) = 16\left(1 - \frac{1}{n}\right)$$

This expression is an approximation to the actual area under consideration. The larger n becomes, the better the approximation. If we let $n \to \infty$ (which is equivalent to letting $\Delta x \to 0$), the limit of this sum will equal the area in question. Thus,

$$A = \lim_{n \to \infty} 16\left(1 - \frac{1}{n}\right) = 16$$

(This checks with the geometric result.) The area under the curve can be considered as the limit of the sum of the inscribed rectangles as the number of rectangles approaches infinity.

The method indicated in Example B illustrates the interpretation of finding an area as a summation process, although it should not be considered as a proof. However, we shall find that integration proves to be a much more useful method for finding an area. Let us now see how integration can be used directly.

Let ΔA represent the area $BCEG$ under the curve, as indicated in Fig. 4–5. We see that the following inequality is true for the indicated areas:

$$A_{BCDG} < \Delta A < A_{BCEF}$$

If the point G is now designated as (x, y) and E as $(x + \Delta x, y + \Delta y)$, we have $y\,\Delta x < \Delta A < (y + \Delta y)\,\Delta x$. Dividing through by Δx, we have

$$y < \frac{\Delta A}{\Delta x} < y + \Delta y$$

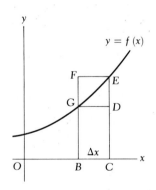

Figure 4–5

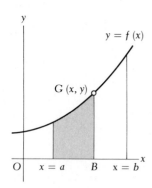

Figure 4–6

Now we take the limit as $\Delta x \to 0$ (Δy then also approaches 0). This results in the relation

$$\frac{dA}{dx} = y \tag{4-6}$$

This is true since the left member of the inequality is y and the right member approaches y. Also remember that

$$\lim_{\Delta x \to 0} \frac{\Delta A}{\Delta x} = \frac{dA}{dx}$$

We shall now use Eq. (4–6) to show the method of finding the complete area under a curve. We now let $x = a$ be the left boundary of the desired area, and $x = b$ be the right boundary (Fig. 4–6). The area under the curve to the right of $x = a$ and bounded on the right by the line GB is now designated as A_{ax}. From Eq. (4–6), we have

$$dA_{ax} = [y \, dx]_a^x \qquad \text{or} \qquad A_{ax} = \left[\int y \, dx \right]_a^x$$

where $[\ ]_a^x$ is the notation used to indicate the boundaries of the area. Thus,

$$A_{ax} = \left[\int f(x) \, dx \right]_a^x = [F(x) + C]_a^x \tag{4-7}$$

But we know that if $x = a$, then $A_{aa} = 0$. Thus, $0 = F(a) + C$, or $C = -F(a)$. Therefore,

$$A_{ax} = \left[\int f(x) \, dx \right]_a^x = F(x) - F(a) \tag{4-8}$$

Now, to find the area under the curve that reaches from a to b, we write

$$A_{ab} = F(b) - F(a) \tag{4-9}$$

Thus the area under the curve that reaches from a to b is given by

$$A_{ab} = \left[\int f(x) \, dx \right]_a^b = F(b) - F(a) \tag{4-10}$$

This shows that *the area under the curve may be found by integrating the function $f(x)$ to find the function $F(x)$, which is then evaluated at each boundary value. The area is the difference between these values of $F(x)$.*

In Example B, we found an area under a curve by finding the limit of the sum of the inscribed rectangles as the number of rectangles approaches infinity. Equation (4–10) expresses the area under a curve in terms of integration. We can now see that we have obtained the area by summation and also expressed it in terms of integration. *Thus, we conclude that summations can be evaluated by integration.* Also, we have grasped the connection between the problem of finding the slope of a tangent to a curve (differentiation) and the problem of finding an area (integration). We would not normally suspect that these two problems would have solutions which lead to reverse processes. We have also seen that the definition of integration has much more application than originally anticipated.

Example C

Find the area under the curve $y = x^2 + 1$ between the y-axis and the line $x = 4$. This is the area of Example A, which is shown in Fig. 4–3(a) on page 151.

In Eq. (4–10), we note that $f(x) = x^2 + 1$. This means that

$$F(x) = \int (x^2 + 1) \, dx = \frac{1}{3}x^3 + x + C$$

Therefore, the area is given by

$$A_{0,4} = F(4) - F(0)$$

$$= \left[\frac{1}{3}(4^3) + 4 + C\right] - \left[\frac{1}{3}(0^3) + 0 + C\right]$$

$$= \frac{1}{3}(64) + 4 = \frac{76}{3}$$

We note that this is about 4 sq units greater than the value obtained by the approximation in Example A. This result means that the exact area is $25\frac{1}{3}$, as stated at the end of Example A.

Example D

Find the area under the curve $y = x^3$ between the lines $x = 1$ and $x = 2$.

In Eq. (4–10), $f(x) = x^3$. Therefore,

$$F(x) = \int x^3 \, dx = \frac{1}{4}x^4 + C$$

$$A_{1,2} = F(2) - F(1) = \left[\frac{1}{4}(2^4) + C\right] - \left[\frac{1}{4}(1^4) + C\right]$$

$$= 4 - \frac{1}{4} = \frac{15}{4}$$

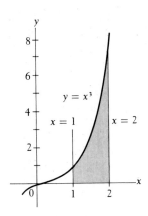

Figure 4–7

Therefore, the required area is $\frac{15}{4}$. Again, this is the exact area, not an approximation (see Fig. 4–7).

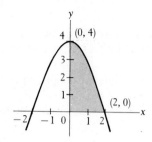

Figure 4–8

Example E

Find the area under the curve $y = 4 - x^2$ which lies in the first quadrant.

By solving the equation $4 - x^2 = 0$, we determine that the area to be found extends from $x = 0$ to $x = 2$ (see Fig. 4–8). Thus,

$$\int (4 - x^2)\, dx = 4x - \frac{x^3}{3} + C$$

$$A_{0,2} = \left(8 - \frac{8}{3} + C\right) - (0 - 0 + C) = \frac{16}{3}$$

We can see from these examples that we do not have to include the constant of integration when we are finding areas. It cancels out when the values are subtracted.

Exercises 4–4

In each of Exercises 1 through 8 find the approximate area under the given curves by dividing the indicated intervals into n subintervals, and add up the areas of the inscribed rectangles. There are two values of n for each, and therefore two approximations for each area. The height of each rectangle may be found by evaluating the function for the proper value of x. See Example A.

1. $y = 3x$, between $x = 0$ and $x = 3$, for (a) $n = 3$ ($\Delta x = 1$),
   (b) $n = 10$ ($\Delta x = 0.3$)

2. $y = 2x$, between $x = 0$ and $x = 2$, for (a) $n = 4$ ($\Delta x = 0.5$),
   (b) $n = 10$ ($\Delta x = 0.2$)

3. $y = x^2$, between $x = 0$ and $x = 2$, for (a) $n = 5$ ($\Delta x = 0.4$),
   (b) $n = 10$ ($\Delta x = 0.2$)

4. $y = x^2 + 2$, between $x = 0$ and $x = 3$, for (a) $n = 3$ ($\Delta x = 1$),
   (b) $n = 10$ ($\Delta x = 0.3$)

5. $y = 4x - x^2$, between $x = 1$ and $x = 4$, for (a) $n = 6$, (b) $n = 10$

6. $y = 1 - x^2$, between $x = 0.5$ and $x = 1$, for (a) $n = 5$, (b) $n = 10$

7. $y = \dfrac{1}{x^2}$, between $x = 1$ and $x = 5$, for (a) $n = 4$, (b) $n = 8$

8. $y = \sqrt{x}$, between $x = 1$ and $x = 4$, for (a) $n = 3$, (b) $n = 12$

In Exercises 9 through 16 find the exact area under the given curves between the indicated values of x. The functions are the same as those for which approximate areas were found in Exercises 1 through 8.

9. $y = 3x$, between $x = 0$ and $x = 3$

10. $y = 2x$, between $x = 0$ and $x = 2$

11. $y = x^2$, between $x = 0$ and $x = 2$

12. $y = x^2 + 2$, between $x = 0$ and $x = 3$

13. $y = 4x - x^2$, between $x = 1$ and $x = 4$

14. $y = 1 - x^2$, between $x = 0.5$ and $x = 1$

15. $y = \dfrac{1}{x^2}$, between $x = 1$ and $x = 5$

16. $y = \sqrt{x}$, between $x = 1$ and $x = 4$

4–5 The Definite Integral

Using reasoning similar to that in the preceding section, *we define the definite integral* of *a function $f(x)$ as*

$$\int_a^b f(x)\ dx = F(b) - F(a) \tag{4-11}$$

where $F'(x) = f(x)$. We call this a *definite integral* because the final result of integrating and evaluating is a number. (The *indefinite* integral had an arbitrary constant in the result.) *The numbers a and b are called the* **lower limit** *and the* **upper limit**, *respectively. We can see that the value of a definite integral is found by evaluting the function (found by integration) at the upper limit and subtracting the value of this function at the lower limit.*

We know, from the analysis in the preceding section, that this definite integral can be interpreted as a summation process, where the size of the subdivision approaches a limit of zero. This fact explains the choice of the $\int$ symbol for integration: It is an elongated S, representing the sum. It is this interpretation of integration which we shall apply to many kinds of problems.

Example A

Evaluate the integral $\int_0^2 x^4\ dx$.

We write

$$\int_0^2 x^4\ dx = \frac{x^5}{5}\bigg|_0^2 = \frac{2^5}{5} - 0 = \frac{32}{5}$$

Note that a vertical line—with the limits written at the top and the bottom—is the way the value is indicated after integration, but before evaluation.

Example B

Evaluate $\int_1^3 (x^{-2} - 1)\ dx$.

We set this forth as

$$\int_1^3 (x^{-2} - 1)\ dx = -\frac{1}{x} - x\bigg|_1^3 = \left(-\frac{1}{3} - 3\right) - (-1 - 1)$$

$$= 2 - \frac{10}{3} = -\frac{4}{3}$$

Example C

Evaluate $\int_0^1 5x(x^2 + 1)^5\ dx$.

For purposes of integration, $n = 5$, $u = x^2 + 1$, and $du = 2x\ dx$. Hence

$$\int_0^1 5x(x^2 + 1)^5\ dx = \frac{5}{2}\int_0^1 (x^2 + 1)^5 (2x\ dx) = \frac{5}{2} \cdot \frac{1}{6}(x^2 + 1)^6\bigg|_0^1$$

$$= \frac{5}{12}(2^6 - 1^6) = \frac{5(63)}{12} = \frac{105}{4}$$

Example D

Evaluate $\int_{-1}^{3} x(1 - 3x^2)^{1/3}\, dx$.

For purposes of integration, $n = \frac{1}{3}$, $u = 1 - 3x^2$, and $du = -6x\, dx$. Thus,

$$\int_{-1}^{3} x(1 - 3x^2)^{1/3}\, dx = -\frac{1}{6} \int_{-1}^{3} (1 - 3x^2)^{1/3}(-6x\, dx)$$

$$= -\frac{1}{6} \cdot \frac{3}{4}(1 - 3x^2)^{4/3}\Big|_{-1}^{3}$$

$$= -\frac{1}{8}(-26)^{4/3} + \frac{1}{8}(-2)^{4/3}$$

$$= \frac{1}{8}(2\sqrt[3]{2} - 26\sqrt[3]{26}) = \frac{1}{4}(\sqrt[3]{2} - 13\sqrt[3]{26})$$

Example E

Evaluate

$$\int_{0}^{4} \frac{x + 1}{(x^2 + 2x + 2)^3}\, dx$$

For purposes of integration,

$$n = -3, \quad u = x^2 + 2x + 2 \quad \text{and} \quad du = (2x + 2)\, dx$$

Therefore,

$$\int_{0}^{4} (x^2 + 2x + 2)^{-3}(x + 1)\, dx = \frac{1}{2} \int_{0}^{4} (x^2 + 2x + 2)^{-3}[2(x + 1)\, dx]$$

$$= \frac{1}{2} \cdot \frac{1}{-2}(x^2 + 2x + 2)^{-2}\Big|_{0}^{4}$$

$$= -\frac{1}{4}(16 + 8 + 2)^{-2} + \frac{1}{4}(0 + 0 + 2)^{-2}$$

$$= \frac{1}{4}\left(-\frac{1}{26^2} + \frac{1}{2^2}\right) = \frac{1}{4}\left(\frac{1}{4} - \frac{1}{676}\right)$$

$$= \frac{1}{4}\left(\frac{168}{676}\right) = \frac{21}{338}$$

It should be pointed out that the definition of the definite integral is valid regardless of the source of $f(x)$. That is, we may apply the definite integral whenever we want to sum a function in a manner similar to that which we use to find an area.

Exercises 4–5

In the following exercises evaluate the given definite integrals.

1. $\int_0^1 2x \, dx$

2. $\int_0^2 3x^2 \, dx$

3. $\int_1^4 x^{5/2} \, dx$

4. $\int_4^9 (x^{3/2} - 1) \, dx$

5. $\int_{-1}^1 (1 - x)^{1/3} \, dx$

6. $\int_1^5 \sqrt{2x - 1} \, dx$

7. $\int_0^3 (x^4 - x^3 + x^2) \, dx$

8. $\int_1^2 (3x^5 - 2x^3) \, dx$

9. $\int_0^4 (1 - \sqrt{x})^2 \, dx$

10. $\int_1^4 \frac{x^2 + 1}{\sqrt{x}} \, dx$

11. $\int_1^2 2x(4 - x^2)^3 \, dx$

12. $\int_0^1 x(3x^2 - 1)^3 \, dx$

13. $\int_0^4 \frac{x \, dx}{\sqrt{x^2 + 9}}$

14. $\int_0^3 x^2 (x^3 + 2)^{3/2} \, dx$

15. $\int_3^7 3\sqrt{4x - 3} \, dx$

16. $\int_{-5}^1 \sqrt{6 - 2x} \, dx$

17. $\int_0^2 2x(9 - 2x^2)^2 \, dx$

18. $\int_{-1}^0 x^3 (1 - 2x^4)^3 \, dx$

19. $\int_0^1 (x^2 + 3)(x^3 + 9x + 6) \, dx$

20. $\int_2^3 \frac{x^2 + 1}{(x^3 + 3x)^2} \, dx$

4–6 Graphical Integration

To this point we are limited to integrating algebraic functions, primarily by use of the power rule. There are many other kinds of functions and methods of integration. These methods, many of which are taken up later in Chapter 8, are based on the analytic use of the equation relating the variables. If empirical data are known, or we have a function which cannot be integrated by present methods, it is possible to plot the graph and develop methods of performing the integration graphically. In this section we shall develop a graphical method based upon the area under a curve. This method, which has similarities to the method of graphical differentiation presented in Chapter 2, is developed in the following example.

Example A

For the curve given in Fig. 4–9(a), find the curve of the integral by graphical means.

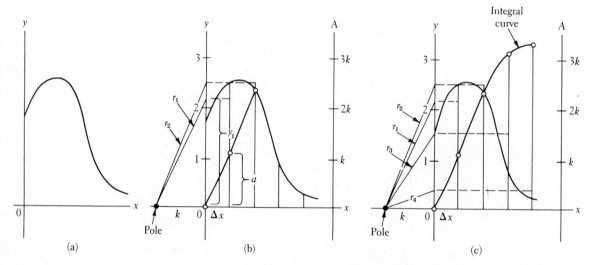

Figure 4–9

Let us first divide the area under the curve into equal (this is not essential, but is usually convenient) intervals along the x-axis as shown in Fig. 4–9(b). A horizontal line is then drawn through the leftmost vertical strip such that the area of the rectangle created is estimated as best possible as equal to the area under that portion of the curve. This requires that the area of the "triangle" above the curve equals the area of the "triangle" below the curve. A **pole**, k units from the origin, is chosen on the negative x-axis, just as in the case of graphical differentiation. A line, called a **ray**, is drawn from the pole to the point of the y-axis where the horizontal line cuts the y-axis. This is the line r_1. Finally a line is drawn from the origin to the point at which it crosses the right boundary of the rectangle such that it is parallel to the ray r_1. The two end points of this last line are points on the integral curve. In practice, the scale of these drawings should be larger than that of Fig. 4–9 so that the ray and the parallel line from the origin can be more easily and accurately drawn.

Since the ray and the line from the origin are parallel, two similar triangles are created. Referring to Fig. 4–9(b), we see that

$$\frac{y_1}{k} = \frac{a}{\Delta x}$$

or

$$y_1 \, \Delta x = ak$$

However, $y_1 \, \Delta x$ is the area of the rectangle and therefore ak equals this area. Since the area under the curve equals the area of the rectangle, ak equals the area under this portion of the curve. Thus, a scale k times the y-scale is indicated on the right. This is the proper scale for the area A, and therefore of the integral curve.

A horizontal line is now drawn through the second vertical strip such that the area of the rectangle equals the area under this portion of the curve. This line is then extended to the y-axis. A ray r_2 is drawn from the pole, and a line parallel to this ray is then drawn from the end of the similar line of the first area on the left boundary of the second rectangle to the right boundary of the second rectangle. This gives another point on the integral curve. This process is kept up for the desired number of intervals. The result gives points on the integral curve which can then be joined by a smooth curve representing the area A under the given curve, which therefore represents the integral (see Fig. 4-9c).

This method of integration is valuable, as was mentioned previously, for finding the integral curve for a set of empirical data or a function which cannot be integrated by methods developed thus far. The following examples illustrate the method of graphical integration in these cases.

Example B
By graphical integration, find the integral curve of the function defined by the following table. (Such values could easily be arrived at empirically.)

x	0.0	1.0	2.0	3.0	4.0	5.0	6.0	7.0
y	1.0	1.4	2.1	4.0	5.2	4.8	4.0	3.2

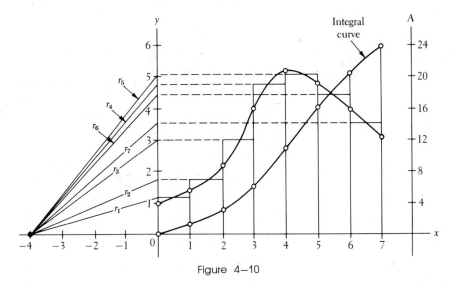

Figure 4-10

Observation of Fig. 4-10 indicates the curve found by plotting the data and the resulting curve. Since $k = 4$, the A-scale is four times that of the y-scale.

We might note that the integral curve can be used to evaluate the definite integral $\int_0^c y \, dx$, where c is a particular value of x. The A-coordinate of the integral curve is the value of this integral. For example, in this case $\int_0^4 y \, dx = 11$, and $\int_0^6 y \, dx = 21$ approximately.

Example C

By graphical integration, find the value of $\int_0^5 \sqrt{x^2 + 1}\, dx$.

By letting $y = \sqrt{x^2 + 1}$, we plot the graph of y vs. x from the following table (see Fig. 4–11).

x	0	1	2	3	4	5
y	1.0	1.4	2.2	3.2	4.1	5.1

Thus, $\int_0^5 \sqrt{x^2 + 1}\, dx = 14.0$, since this is the estimated value of the A-coordinate for $x = 5$. Methods of direct integration of this function give the value 13.9.

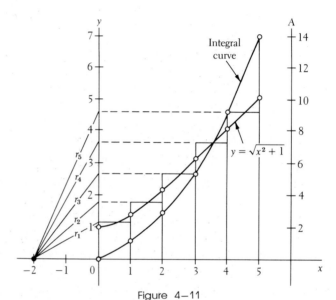

Figure 4–11

Two other points deserve mention. First, if a definite integral with a lower limit of a value other than zero is to be evaluated, subtraction of the appropriate A-values gives the result. For example, in Example B, $\int_4^6 y\, dx = 10$ approximately. Second, the integral curve found is the graph of the function of the indefinite integral, where the constant of integration is evaluated such that the curve passes through the origin. For other constants of integration, the entire curve can be raised or lowered by the appropriate number of units. Applications of this method to problems in technology are found in Chapter 5.

Exercises 4–6

In Exercises 1 through 4 copy the given curve to a larger scale, and then find the integral curve.

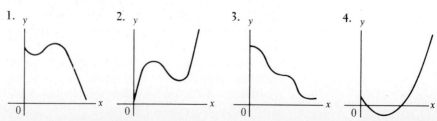

In Exercises 5 through 8 plot the given sets of data, and then find the integral curve.

5.

x	0	2	4	6	8	10
y	5.0	3.8	1.4	0.6	1.5	2.8

6.

x	0	0.1	0.2	0.3	0.4	0.5
y	0.1	0.9	1.1	0.7	0.6	1.2

7.

x	0	15	30	45	60	75
y	5	40	75	65	45	20

8.

x	0	5	10	15	20	25
y	0	-1	-3	-2	2	10

In Exercises 9 through 12 use graphical integration to evaluate the given integrals. In Exercises 9 and 10 check the results by direct integration.

9. $\displaystyle\int_0^4 \sqrt{1 + x}\, dx$ 10. $\displaystyle\int_2^5 \frac{2x\, dx}{\sqrt{x^2 + 5}}$ 11. $\displaystyle\int_1^3 \frac{dx}{x + 1}$ 12. $\displaystyle\int_3^{15} \frac{x\, dx}{\sqrt{x + 1}}$

In Exercises 13 through 16 solve the given problems by graphical integration.

13. Find the integral curve for

$$y = \frac{1}{x^2 + 1}$$

Then draw a curve "parallel" to the given curve such that it passes through $(0, 2)$. In this way the constant of integration is 2 greater than that of the original integral curve.

14. Find the integral curve for

$$y = \frac{x^3 + 1}{x + 2}$$

Then draw a curve "parallel" to the integral curve such that it passes through $(2, -4)$. Find the difference between the constants of integration.

15. Find the area in the first quadrant under the curve $y = \dfrac{10}{x + 1}$, and to the left of the line $x = 9$.

16. Find the area in the first quadrant under the curve $y = (6x - x^2 - 5)^{3/2}$.

4–7 Numerical Integration: The Trapezoidal Rule

For data and functions which cannot be directly integrated by available methods, it is possible to develop numerical methods of integration. These numerical methods are of greater importance today since they are readily adaptable for use on a calculator or computer. There are a great many such numerical techniques for approximating the value of an integral. In this section we shall develop one of these, the trapezoidal rule. In the following section another numerical method is discussed.

We know from Sections 4–4 and 4–5 that we can interpret a definite integral as the area under a curve. We shall therefore show how we can approximate the value of the integral by approximating the appropriate area by a set of inscribed trapezoids. The basic idea here is very similar to that used when rectangles were inscribed under a curve. However, the use of trapezoids reduces the error and provides a better approximation.

The area to be found is subdivided into n intervals of equal width. Perpendicular lines are then dropped from the curve (or points, if only a given set of numbers is available). If the points on the curve are joined by straight-line segments, the area of successive parts under the curve is then approximated by finding the area of each of the trapezoids formed. However, if these points are not too far apart, the approximation will be very good (see Fig. 4–12). From geometry we recall that the area of a trapezoid equals one-half the product of the sum of the bases times the altitude. For these trapezoids the bases are the y-coordinates and the altitudes are Δx. When we thus indicate the sum of these trapezoidal areas, we have

$$A_T = \frac{1}{2}(y_0 + y_1)\,\Delta x + \frac{1}{2}(y_1 + y_2)\,\Delta x + \frac{1}{2}(y_2 + y_3)\,\Delta x + \cdots$$

$$+ \frac{1}{2}(y_{n-2} + y_{n-1})\,\Delta x + \frac{1}{2}(y_{n-1} + y_n)\,\Delta x$$

We note, when this addition is performed, that the result is

Trapezoidal Rule

$$A_T = \left(\frac{1}{2}y_0 + y_1 + y_2 + \cdots + y_{n-1} + \frac{1}{2}y_n\right)\Delta x \qquad (4\text{–}12)$$

The y-values to be used are either derived from the function as $y_1 = f(x_1)$, or are the y-coordinates of a set of data.

Since A_T approximates the area under the curve, it also approximates the value of the definite integral, or

$$\frac{upper - lower}{n}$$

$$\int_a^b f(x)\,dx \approx \left(\frac{1}{2}y_0 + y_1 + y_2 + \cdots + y_{n-1} + \frac{1}{2}y_n\right)\Delta x \qquad (4\text{–}13)$$

Equation (4–13) is known as the **trapezoidal rule**.

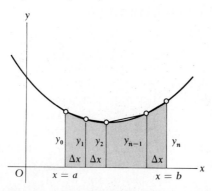

Figure 4–12

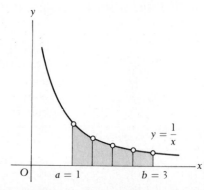

Figure 4–13

Example A

Approximate the value of $\int_1^3 \frac{1}{x}\, dx$ by the trapezoidal rule. Let $n = 4$.

We are to approximate the area under $y = 1/x$ from $x = 1$ to $x = 3$ by dividing the area into 4 trapezoids. This area is found by applying Eq. (4–12), which is the approximate value of the integral, as shown in Eq. (4–13). Figure 4–13 shows the graph. In this example, $f(x) = 1/x$, and

$$\Delta x = \frac{3 - 1}{4} = \frac{1}{2}, \qquad y_0 = f(a) = f(1) = 1$$

$$y_1 = f\left(\frac{3}{2}\right) = \frac{2}{3}, \qquad y_2 = f(2) = \frac{1}{2}$$

$$y_3 = f\left(\frac{5}{2}\right) = \frac{2}{5} \quad \text{and} \quad y_n = y_4 = f(b) = f(3) = \frac{1}{3}$$

And so we write

$$A_T = \left[\frac{1}{2}(1) + \frac{2}{3} + \frac{1}{2} + \frac{2}{5} + \frac{1}{2}\left(\frac{1}{3}\right)\right]\frac{1}{2}$$

$$= \left(\frac{15 + 20 + 15 + 12 + 5}{30}\right)\frac{1}{2} = \frac{67}{30}\left(\frac{1}{2}\right) = \frac{67}{60}$$

Therefore,

$$\int_1^3 \frac{1}{x}\, dx \approx \frac{67}{60}$$

Note that we cannot perform this integration directly by methods developed up to this point.

Example B

Approximate the value of $\int_0^1 \sqrt{x^2 + 1}\, dx$ by the trapezoidal rule. Let $n = 5$.

Figure 4–14 shows the graph. In this example,

$$\Delta x = \frac{1 - 0}{5} = 0.2$$

Therefore,

$$y_0 = f(0) = 1 \qquad\qquad y_1 = f(0.2) = \sqrt{1.04} = 1.02$$
$$y_2 = f(0.4) = \sqrt{1.16} = 1.08 \qquad y_3 = f(0.6) = \sqrt{1.36} = 1.17$$
$$y_4 = f(0.8) = \sqrt{1.64} = 1.28 \qquad y_5 = f(1) = \sqrt{2.00} = 1.41$$

Hence we have

$$A_T = \left[\frac{1}{2}(1) + 1.02 + 1.08 + 1.17 + 1.28 + \frac{1}{2}(1.41)\right](0.2) = 1.15$$

This means that

$$\int_0^1 \sqrt{x^2 + 1}\, dx \approx 1.15$$

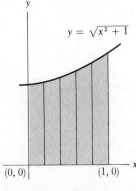

$y = \sqrt{x^2 + 1}$

$(0, 0)$ $(1, 0)$

Figure 4–14

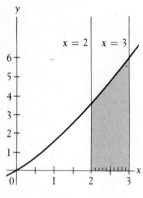

Figure 4–15

Example C

Approximate the value of $\int_2^3 x\sqrt{x+1}\,dx$ by use of the trapezoidal rule. Use $n = 10$.

In Fig. 4–15 the graph of the function and the area used in the trapezoidal rule are shown. From the given values we have $\Delta x = \frac{3-2}{10} = 0.1$. Therefore,

$$y_0 = f(2) = 2\sqrt{3} = 3.464 \qquad y_1 = f(2.1) = 2.1\sqrt{3.1} = 3.697$$
$$y_2 = f(2.2) = 2.2\sqrt{3.2} = 3.935 \qquad y_3 = f(2.3) = 2.3\sqrt{3.3} = 4.178$$
$$y_4 = f(2.4) = 2.4\sqrt{3.4} = 4.425 \qquad y_5 = f(2.5) = 2.5\sqrt{3.5} = 4.677$$
$$y_6 = f(2.6) = 2.6\sqrt{3.6} = 4.933 \qquad y_7 = f(2.7) = 2.7\sqrt{3.7} = 5.194$$
$$y_8 = f(2.8) = 2.8\sqrt{3.8} = 5.458 \qquad y_9 = f(2.9) = 2.9\sqrt{3.9} = 5.727$$
$$y_{10} = f(3) = 3\sqrt{4} = 6.000$$

$$A_T = \left[\frac{1}{2}(3.464) + 3.697 + 3.935 + \cdots + 5.727 + \frac{1}{2}(6.000)\right](0.1)$$

$$= 4.6956$$

Therefore, $\int_2^3 x\sqrt{x+1}\,dx \approx 4.6956$. (The actual value, to five significant digits, is 4.6954.)

Example D

The following points were found empirically.

x	0	1	2	3	4	5
y	5.68	6.75	7.32	7.35	6.88	6.24

Approximate the value of the integral of the function defined by these points between $x = 0$ and $x = 5$ by the trapezoidal rule.

In order to find A_T, we use the values of y_0, y_1, etc. directly from the table. We also note that $\Delta x = 1$. The graph is shown in Fig. 4–16. Therefore, we have

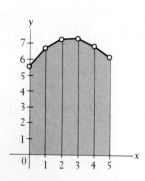

Figure 4–16

$$A_T = \left[\frac{1}{2}(5.68) + 6.75 + 7.32 + 7.35 + 6.88 + \frac{1}{2}(6.24)\right] = 34.26$$

Although we do not know the algebraic form of the function, we can state that

$$\int_0^5 f(x)\,dx \approx 34.26$$

Exercises 4–7

In Exercises 1 through 4, (a) approximate the value of each of the given integrals by use of the trapezoidal rule, using the given value of n, and (b) check by direct integration.

1. $\int_0^2 2x^2\,dx, \; n = 4$ \qquad\qquad 2. $\int_0^1 (1 - x^2)\,dx, \; n = 3$

3. $\int_{1}^{4} (1 + \sqrt{x}) \, dx$, $n = 6$ \qquad 4. $\int_{3}^{8} \sqrt{1 + x} \, dx$, $n = 5$

In Exercises 5 through 12 approximate each of the given integrals by use of the trapezoidal rule, using the given value of n.

5. $\int_{2}^{3} \frac{1}{2x} \, dx$, $n = 2$ \qquad 6. $\int_{2}^{4} \frac{dx}{x + 3}$, $n = 4$

7. $\int_{0}^{2} \sqrt{4 - x^2} \, dx$, $n = 4$ \qquad 8. $\int_{0}^{2} \sqrt{x^3 + 1} \, dx$, $n = 4$

9. $\int_{1}^{5} \frac{1}{x^2 + x} \, dx$, $n = 10$ \qquad 10. $\int_{2}^{4} \frac{1}{x^2 + 1} \, dx$, $n = 10$

11. $\int_{0}^{4} 2^x \, dx$, $n = 12$ \qquad 12. $\int_{0}^{1.5} 10^x \, dx$, $n = 15$

In Exercises 13 through 16 approximate the value of the integrals defined by the given sets of points.

13.

x	3	4	5	6	7
y	1.08	1.65	3.23	3.67	2.97

14.

x	2	4	6	8	10	12	14
y	0.67	2.34	4.56	3.67	3.56	4.78	6.87

15.

x	1.4	1.7	2.0	2.3	2.6	2.9	3.2
y	0.18	7.87	18.23	23.53	24.62	20.93	20.76

16.

x	0.4	0.8	1.2	1.6	2.0	2.4
y	1.24	3.54	5.52	4.40	3.63	3.88

4–8 Simpson's Rule

The numerical method of integration developed in this section is obtained by interpreting the definite integral as the area under a curve, as we did in developing the trapezoidal rule, and by approximating the curve by a set of parabolic arcs. The use of parabolic arcs, rather than chords as with the trapezoidal rule, usually gives a better approximation.

Since we will be using parabolic arcs, we first derive a formula for the area under a parabolic arc. The curve in Fig. 4–17 represents the parabola $y = ax^2 + bx + c$. The indicated points on the curve are $(-h, y_0)$, $(0, y_1)$ and (h, y_2). The area under the parabola is given by

$$A = \int_{-h}^{h} y \, dx = \int_{-h}^{h} (ax^2 + bx + c) \, dx = \frac{ax^3}{3} + \frac{bx^2}{2} + cx \Big|_{-h}^{h}$$

$$= \frac{2}{3} ah^3 + 2ch$$

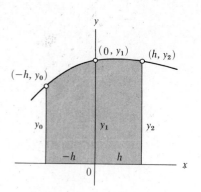

Figure 4–17

or

$$A = \frac{h}{3}(2ah^2 + 6c) \qquad (4\text{--}14)$$

The coordinates of the three points also satisfy the equation $y = ax^2 + bx + c$. This means that

$$y_0 = ah^2 - bh + c$$
$$y_1 = c$$
$$y_2 = ah^2 + bh + c$$

By finding the sum of $y_0 + 4y_1 + y_2$, we have

$$y_0 + 4y_1 + y_2 = 2ah^2 + 6c \qquad (4\text{--}15)$$

Substitution of Eq. (4–15) into Eq. (4–14), we have

$$A = \frac{h}{3}(y_0 + 4y_1 + y_2) \qquad (4\text{--}16)$$

We note that the area depends only on the distance h and the three y-coordinates.

Now let us consider the area under the curve in Fig. 4–18. If a parabolic arc is passed through the points (x_0, y_0), (x_1, y_1), and (x_2, y_2) we may use Eq. (4–16) to approximate the area under the curve between x_0 and x_2. We also note that the distance h used in finding Eq. (4–16) is the difference in the x-coordinates, or $h = \Delta x$. Therefore, the area under the curve between x_0 and x_2 is

$$A_1 = \frac{\Delta x}{3}(y_0 + 4y_1 + y_2)$$

Similarly, if a parabolic arc is passed through the three points starting with (x_2, y_2), the area between x_2 and x_4 is

$$A_2 = \frac{\Delta x}{3}(y_2 + 4y_3 + y_4)$$

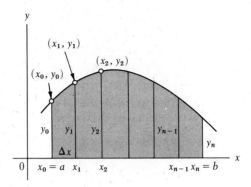

Figure 4-18

The sum of these areas is

$$A_1 + A_2 = \frac{\Delta x}{3}(y_0 + 4y_1 + 2y_2 + 4y_3 + y_4) \qquad (4-17)$$

We can continue this procedure until the approximate value of the entire area has been found. We must note, however, that *the number of intervals n of width Δx must be even.* Therefore, generalizing on Eq. (4-17), and recalling again that the value of the definite integral is the area under the curve, we have

$$\int_a^b f(x)\,dx = \frac{\Delta x}{3}(y_0 + 4y_1 + 2y_2 + 4y_3 + 2y_4 + \cdots + 4y_{n-1} + y_n) \qquad (4-18)$$

Equation (4-18) is known as **Simpson's rule.**

Example A

Approximate the value of $\int_0^1 \frac{dx}{x+1}$ by Simpson's rule. Let $n = 2$.

We are to approximate the integral by use of Eq. (4-18). We therefore note that $f(x) = 1/(x+1)$. Also, $x_0 = a = 0$, $x_1 = 0.5$, and $x_2 = b = 1$. This is due to the fact that $n = 2$ and $\Delta x = 0.5$ since the total interval is one unit (from $x = 0$ to $x = 1$). Therefore,

$$y_0 = \frac{1}{0+1} = 1.0000, \quad y_1 = \frac{1}{0.5+1} = 0.6667, \quad y_2 = \frac{1}{1+1} = 0.5000$$

Substituting, we have

$$\int_0^1 \frac{dx}{x+1} = \frac{0.5}{3}[1.0000 + 4(0.6667) + 0.5000] = 0.694$$

To three decimal places, the actual value of the integral is 0.693. The method of integrating this function will be considered in a later chapter.

Example B

Approximate the value of $\int_2^3 x\sqrt{x+1}\,dx$ by Simpson's rule. Let $n = 10$.

Since the necessary values for this function are shown in Example C of Section 4–7, we shall simply tabulate them here.

$$\Delta x = 0.1 \qquad y_0 = 3.464 \qquad y_1 = 3.697 \qquad y_2 = 3.935$$
$$y_3 = 4.178 \qquad y_4 = 4.425 \qquad y_5 = 4.677$$
$$y_6 = 4.933 \qquad y_7 = 5.194 \qquad y_8 = 5.458$$
$$y_9 = 5.727 \qquad y_{10} = 6.000$$

Therefore, we evaluate the integral as follows.

$$\int_2^3 x\sqrt{x+1}\,dx = \frac{0.1}{3}[3.464 + 4(3.697) + 2(3.935) + 4(4.178)$$
$$+ 2(4.425) + 4(4.677) + 2(4.933) + 4(5.194)$$
$$+ 2(5.458) + 4(5.727) + 6.000]$$

$$= \frac{0.1}{3}(140.86) = 4.6953$$

This result compares with 4.6956 from the trapezoidal rule, and with 4.6954, the actual value to five significant digits.

Exercises 4—8

In Exercises 1 through 4 (a) approximate the value of each of the given integrals by use of Simpson's rule, using the given value of n, and (b) check by direct integration.

1. $\int_0^2 (1 + x^3)\,dx, \ n = 2$

2. $\int_0^8 x^{1/3}\,dx, \ n = 2$

3. $\int_1^4 (2x + \sqrt{x})\,dx, \ n = 6$

4. $\int_0^2 x\sqrt{x^2+1}\,dx, \ n = 4$

In Exercises 5 through 10 approximate each of the given integrals by use of Simpson's rule, using the given values of n. These are the same as Exercises 5 through 10 of Section 4–7.

5. $\int_2^3 \frac{1}{2x}\,dx, \ n = 2$

6. $\int_2^4 \frac{dx}{x+3}, \ n = 4$

7. $\int_0^2 \sqrt{4 - x^2}\,dx, \ n = 4$

8. $\int_0^2 \sqrt{x^3 + 1}\,dx, \ n = 4$

9. $\int_1^5 \frac{dx}{x^2 + x}, \ n = 10$

10. $\int_2^4 \frac{1}{x^2 + 1}\,dx, \ n = 10$

In Exercises 11 and 12 approximate the value of the integrals defined by the given set of points. These are the same as Exercises 13 and 15 of Section 4–7.

11.

x	3	4	5	6	7
y	1.08	1.65	3.23	3.67	2.97

12.

x	1.4	1.7	2.0	2.3	2.6	2.9	3.2
y	0.18	7.87	18.23	23.53	24.62	20.93	20.76

4–9 Review Exercises for Chapter 4

In Exercises 1 through 20 evaluate the given integrals.

1. $\displaystyle\int (4x^3 - x)\, dx$

2. $\displaystyle\int (5 - 3x^2)\, dx$

3. $\displaystyle\int 2(x - x^{3/2})\, dx$

4. $\displaystyle\int x(x - 3x^4)\, dx$

5. $\displaystyle\int_1^4 \left(\sqrt{x} + \frac{1}{\sqrt{x}}\right) dx$

6. $\displaystyle\int_1^2 \left(x + \frac{1}{x^2}\right) dx$

7. $\displaystyle\int_0^2 x(4 - x)\, dx$

8. $\displaystyle\int_0^1 x(x^3 + 1)\, dx$

9. $\displaystyle\int \frac{dx}{(2 - 5x)^2}$

10. $\displaystyle\int x(1 - 2x^2)^5\, dx$

11. $\displaystyle\int (7 - 2x)^{3/4}\, dx$

12. $\displaystyle\int (3x - 1)^{1/3}\, dx$

13. $\displaystyle\int_0^2 \frac{3x\, dx}{\sqrt[3]{1 + 2x^2}}$

14. $\displaystyle\int_1^6 \frac{2\, dx}{(3x - 2)^{3/4}}$

15. $\displaystyle\int x^2(1 - 2x^3)^4\, dx$

16. $\displaystyle\int 3x^3(1 - 5x^4)^{1/3}\, dx$

17. $\displaystyle\int \frac{(2 - 3x^2)\, dx}{(2x - x^3)^2}$

18. $\displaystyle\int \frac{x^2 - 3}{\sqrt{6 + 9x - x^3}}\, dx$

19. $\displaystyle\int_1^3 (x^2 + x + 2)(2x^3 + 3x^2 + 12x)\, dx$

20. $\displaystyle\int_0^2 (4x + 18x^2)(x^2 + 3x^3)^2\, dx$

In Exercises 21 through 24 find the differential of each of the given functions.

21. $y = \dfrac{1}{(x^2 - 1)^3}$

22. $y = \dfrac{1}{(2x - 1)^2}$

23. $y = x\sqrt[3]{1 - 3x}$

24. $y = \dfrac{3 + x}{4 - x^2}$

In Exercises 25 and 26 evaluate $\Delta y - dy$ for the given functions and values.

25. $y = x^3$, $x = 2$, $\Delta x = 0.1$

26. $y = 6x^2 - x$, $x = 3$, $\Delta x = 0.2$

In Exercises 27 and 28 find the required equations.

27. Find the equation of the curve which passes through $(-1, 3)$ for which the slope is given by $3 - x^2$.

28. Find the equation of the curve which passes through $(1, -2)$ for which the slope is $x(x^2 + 1)^2$.

In Exercises 29 and 30 use Eq. (4–10) to find the indicated areas.

29. The area under $y = 6x - 1$ between $x = 1$ and $x = 3$

30. The first quadrant area under $y = 8x - x^4$

In Exercises 31 and 32 solve the given problems by the trapezoidal rule.

31. Approximate $\displaystyle\int_1^3 \frac{dx}{2x - 1}$ with $n = 4$.

32. Approximate the value of the integral defined by the following set of points.

x	6.0	9.0	12	15	18	21
y	2.0	1.2	0.2	1.0	6.0	12

In Exercises 33 and 34 solve the given problems by Simpson's rule.

33. Approximate $\displaystyle\int_1^3 \frac{dx}{2x - 1}$ with $n = 4$ (see Exercise 31).

34. Approximate the value of the integral defined by the following set of points.

x	1.0	1.4	1.8	2.2	2.6	3.0	3.4
y	1.45	1.89	2.66	3.50	3.22	3.04	2.44

In Exercises 35 and 36 use the function $y = \dfrac{x}{x^2 + 2}$ and approximate the area under the curve in the first quadrant to the left of the line $x = 5$ by the indicated method.

35. Inscribe five rectangles and find the sum of the areas of the rectangles.

36. Use the trapezoidal rule with $n = 5$.

In Exercises 37 and 38 use the function $y = x\sqrt{x^3 + 1}$ and approximate the area under the curve between $x = 1$ and $x = 3$ by the indicated method.

37. Use Simpson's rule with $n = 10$.

38. Inscribe ten rectangles, and find the sum of the areas of the rectangles.

In Exercises 39 through 42 solve the given problems by finding the appropriate differentials.

39. A ball bearing 4.00 mm in radius is coated with a special metal of thickness 0.02 mm. Approximately what volume of the special metal is used?

40. The electric resistance of a certain resistor as a function of temperature is given by $R = 12.00 + 0.06T + 0.0001T^2$. Approximately by how much does the resistance (in ohms) change as T changes from $100°C$ to $102°C$?

41. The impedance Z of an electric circuit as a function of the resistance R and the reactance X is given by $Z = \sqrt{R^2 + X^2}$. Derive an expression for the relative error in impedance for an error in R and given value of X.

42. Show that the relative error of the nth root of a given measurement equals approximately $1/n$ of the relative error of the measurement.

5

Applications of Integration

5-1 Applications of the Indefinite Integral

The applications of integration in engineering and technology are numerous. In this section we shall present two basic applications of the indefinite integral, with other applications being indicated in the exercises. The sections which follow deal with many of the basic applications of the definite integral.

The first of these applications deals with velocity and acceleration. The concepts of velocity as a first derivative and acceleration as a second derivative were introduced in Chapters 2 and 3. Here we shall apply integration to the problem of finding the distance as a function of time, when we know the relationship between acceleration and time, as well as certain specific values of distance and velocity. These latter values are necessary for determining the values of the constants of integration which are introduced. Recalling now that the acceleration a of an object is given by $a = dv/dt$, we can find the expression for the velocity in terms of a, t, and the constant of integration. We write

$$dv = a\ dt \qquad \text{or}$$

$$v = \int a\ dt \tag{5-1}$$

If the acceleration is constant, we have

$$v = at + C_1 \tag{5-2}$$

Of course, Eq. (5–1) can be used in general to find the velocity as a function of time so long as we know the acceleration as a function of time. However, since the case of constant acceleration is often encountered, Eq. (5–2) is often encountered. If the velocity is known for some specified time, the constant C_1 may be evaluated.

Example A

Find the expression for the velocity if $a = 12t$, given that $v = 8$ when $t = 1$.

Using Eq. (5–1), we have

$$v = \int (12t)\ dt = 6t^2 + C_1$$

Substituting the known values, we obtain

$$8 = 6 + C_1 \quad \text{or} \quad C_1 = 2$$

Thus, $v = 6t^2 + 2$.

Example B

For an object falling under the influence of gravity, the acceleration due to gravity is essentially constant. Its value is -32 ft/s². (The minus sign is chosen so that all quantities directed up are positive, and all quantities directed down are negative.) Find the expression for the velocity of an object under the influence of gravity if $v = v_0$ when $t = 0$.

We write

$$v = \int (-32)\ dt = -32t + C_1 \qquad v_0 = 0 + C_1 \qquad v = v_0 - 32t$$

The velocity v_0 is called the initial velocity. If the object is given an initial upward velocity of 100 ft/s, $v_0 = 100$ ft/s. If the object is dropped, $v_0 = 0$. If the object is given an initial downward velocity of 100 ft/s, $v_0 = -100$ ft/s.

Once we obtain the expression for velocity, we can integrate to find the expression for displacement in terms of the time. Since $v = ds/dt$, we can write $ds = v\ dt$, or

$$s = \int v\ dt \tag{5–3}$$

Example C

Find the expression for displacement in terms of time, if $a = 6t^2$, $v = 0$ when $t = 2$, and $s = 4$ when $t = 0$.

$$v = \int 6t^2\ dt = 2t^3 + C_1; \qquad 0 = 2(2^3) + C_1, \qquad C_1 = -16$$

$$v = 2t^3 - 16$$

$$s = \int (2t^3 - 16)\, dt = \frac{1}{2}t^4 - 16t + C_2; \qquad 4 = 0 - 0 + C_2, \qquad C_2 = 4$$

$$s = \frac{1}{2}t^4 - 16t + 4$$

Example D

Find the expression for the distance above the ground of an object, given a vertical velocity of v_0 from the ground.

From Example B, we know that $v = v_0 - 32t$. In this problem we know that $s = 0$ when $t = 0$ (given velocity v_0 *from the ground*) if distances are measured from ground level. Therefore,

$$s = \int (v_0 - 32t)\, dt = v_0 t - 16t^2 + C_2; \qquad 0 = 0 - 0 + C_2, \qquad C_2 = 0$$

$$s = v_0 t - 16t^2$$

Example E

An object is thrown vertically from the top of a building 200 ft high, and hits the ground 5 s later. What initial velocity was the object given?

Measuring vertical distances from the ground, we know that $s = 200$ ft when $t = 0$. Also, we know that $v = v_0 - 32t$. Thus,

$$s = \int (v_0 - 32t)\, dt = v_0 t - 16t^2 + C$$

$$200 = v_0(0) - 16(0) + C, \qquad C = 200$$
$$s = v_0 t - 16t^2 + 200$$

We also know that $s = 0$ when $t = 5$ s. Thus,

$$0 = v_0(5) - 16(5^2) + 200$$
$$5v_0 = 200$$
$$v_0 = 40 \text{ ft/s}$$

This means that the initial velocity was 40 ft/s upward.

The second basic application of the indefinite integral which we shall discuss comes from the field of electricity. By definition, *the current i in an electric circuit equals the time rate of change of the charge q (in coulombs) which passes a given point in the circuit, or*

$$i = \frac{dq}{dt} \tag{5–4}$$

Rewriting this expression in differential notation as $dq = i\, dt$ and integrating both sides of the equation we have

$$q = \int i\, dt \tag{5–5}$$

Now, the voltage V_C across a capacitor C is given by $V_C = q/C$. By combining equations, the voltage V_C is given by

$$V_C = \frac{1}{C} \int i \, dt \tag{5-6}$$

Here, V_C is measured in volts, C in farads, i in amperes, and t in seconds.

Example F

The current in a certain electric circuit as a function of time is given by $i = 6t^2 + 4$. Find an expression for the amount of charge which passes a point in the circuit as a function of time. Assuming that $q = 0$ when $t = 0$, determine the total charge which passes the point in two seconds.

Since $q = \int i \, dt$, we have

$$q = \int (6t^2 + 4) \, dt$$

$$= 2t^3 + 4t + C$$

This last expression is the desired expression giving charge as a function of time. We note that when $t = 0$, then $q = C$, which means that the constant of integration represents the initial charge, or the change which passed a given point before we started timing. Using q_0 to represent this charge, we have

$$q = 2t^3 + 4t + q_0$$

Now, returning to the second part of the problem, we see that $q_0 = 0$. Therefore, evaluating q for $t = 2$ s, we have

$$q = 2(8) + 4(2) = 24 \text{ C}$$

(Here, the symbol C represents coulombs, and is not the C for capacitance of Eq. (5-6) or the constant of integration.) This is the charge which passes any specified point in the circuit in two seconds.

Example G

The voltage across a 5.0 μF capacitor is zero. What is the voltage after 20 ms if a current of 75 mA charges the capacitor?

From Eq. (5-6) we have

$$V_C = \frac{1}{C} \int i \, dt$$

In substituting we must use the proper power of 10 which corresponds to each prefix which is used. Since 5.0 μF $= 5.0 \times 10^{-6}$ F and 75 mA $= 7.5 \times 10^{-2}$ A, we have

$$V_C = \frac{1}{5.0 \times 10^{-6}} \int 7.5 \times 10^{-2} dt$$

$$= 1.5 \times 10^4 \int dt$$

$$= 1.5 \times 10^4 t + C_1$$

From the given information we know that $V_C = 0$ when $t = 0$. Thus,

$$0 = 1.5 \times 10^4 (0) + C_1 \quad \text{or} \quad C_1 = 0$$

This means that

$$V_C = 1.5 \times 10^4 t$$

Evaluating this expression for $t = 20 \times 10^{-3}$ s, we have

$$V_C = 1.5 \times 10^4 (20 \times 10^{-3})$$
$$= 30 \times 10 = 300 \text{ V}$$

Example H

A certain capacitor is measured to have a voltage of 100 V across it. At this instant a current as a function of time given by $i = 0.06 \sqrt{t}$ is sent through the circuit. After 0.25 s, the voltage across the capacitor is measured to be 140 V. What is the capacitance of the capacitor?

From Eq. (5–6) we have

$$V_C = \frac{1}{C} \int i \, dt$$

Substituting $i = 0.06 \sqrt{t}$, we find that

$$V_C = \frac{1}{C} \int (0.06 \sqrt{t} \, dt) = \frac{0.06}{C} \int t^{1/2} \, dt$$

$$= \frac{0.04}{C} t^{3/2} + C_1$$

From the given information we know that $V_C = 100$ V when $t = 0$. Thus,

$$100 = \frac{0.04}{C} (0) + C_1$$

or

$$C_1 = 100 \text{ V}$$

This means that

$$V_C = \frac{0.04}{C} t^{3/2} + 100$$

We also know that $V_C = 140$ V when $t = 0.25$ s. Therefore,

$$140 = \frac{0.04}{C} (0.25)^{3/2} + 100$$

$$40 = \frac{0.04}{C} (0.125)$$

or

$$C = 1.25 \times 10^{-4} \text{ F} = 125 \ \mu\text{F}$$

Exercises 5—1

1. What is the velocity after 3 s of an object which is dropped and falls under the influence of gravity?

2. The acceleration of an object rolling down an inclined plane is 6.0 ft/s². If it is given an upward velocity along the plane of 20 ft/s, what is its velocity after 4 s?

3. The acceleration of an object is given by $6t$. What is the relation between the velocity and time for this object if $v = 8$ when $t = 1$?

4. An object starts from rest and has an acceleration given by $a = \sqrt{t + 1}$. Find the velocity (in meters per second) after 8 s.

5. An object moves in a straight line with a constant velocity of 80 cm/s. How far is it from its starting point after 7 s?

6. The velocity of an object is given by $v = 2t + 1$. Find the relation between s and t if $s = 0$ when $t = 0$.

7. If an object is given an initial upward velocity from the ground of 120 ft/s, what is its distance above the ground after 3 s?

8. An object is given an upward initial velocity of 60 ft/s from the ground. What is its distance above the ground after 2.5 s?

9. Standing on a cliff, a woman throws a ball with a vertical velocity of 64 ft/s upward. If the ball hits the ground at the base of the cliff 7 s after it is released, how far is it from the top of the cliff to the ground below?

10. A man wishes to throw an object to a height 100 ft above him. What initial vertical velocity must he give the object?

11. The acceleration of an object is given by $a = 24t$. Find the displacement in meters after 1.5 s if $s = 0$ and $v = 0$ when $t = 0$.

12. The acceleration of a certain particle is given by $a = t^4$. Find the displacement in centimeters of the particle after 2 s if $v = 6$ cm/s and $s = 0$ when $t = 0$.

13. The current in a certain electric circuit is 0.5 A. How many coulombs of charge pass a given point in two seconds?

14. The current in a certain circuit is a function of the time and is given by $i = 2t + 3$. How many coulombs pass a given point in the circuit in the first four seconds?

15. The current in a certain wire changes with time according to the relation $i = \sqrt[3]{1 + 3t}$. How many coulombs of charge pass a specified point in the first 3 seconds?

16. The current in a certain circuit as a function of time is given by $i = 8 - t$. If $q_0 = 0$, for what value of t, greater than zero, is $q = 0$? What interpretation can be given to this result?

17. The voltage across a 3.0 μF capacitor is zero. What is the voltage after 10 ms if a current of 0.20 A charges the capacitor?

18. The voltage across a 7.5 μF capacitor is zero. What is the voltage after 5.0 ms if a current of $i = 0.1t$ charges the capacitor?

19. The voltage across a 150 μF capacitor is 50 V. What is the voltage after 1 s if a current $i = 0.012t^{1/5}$ further charges the capacitor?

20. A current $i = t/\sqrt{t^2 + 1}$ is sent into a circuit containing a previously uncharged 4.0 μF capacitor. How long does it take for the capacitor voltage to be 100 V?

21. The angular velocity ω is the time rate of change of the angular displacement θ of a rotating object. If the angular velocity of an object is given by $\omega = 4t$, find an expression for the angular displacement θ if $\theta = 0$ when $t = 0$.

22. The angular acceleration α is the time rate of change of the angular velocity. If the angular acceleration of a rotating object is given by $\alpha = \sqrt{2t + 1}$, find the expression for the angular displacement if $\omega = 0$ and $\theta = 0$ when $t = 0$.

23. An inductor in an electric circuit is essentially a coil of wire in which the voltage is affected by a changing current. By definition, the voltage caused by the changing current is given by

$$V_L = L\frac{di}{dt}$$

where L is the inductance and is measured in henrys. If $V_L = 12.0 - 0.2t$ for a 3-H inductor, find the current in the circuit after 20 s if the initial current was zero.

24. If the inner and outer walls of a container are at different temperatures, the rate of change of temperature with respect to the distance from one wall is a function of the distance from the wall. Symbolically this is stated as $dT/dx = f(x)$, where T is the temperature. If x is measured from the outer wall, at 20°C, and $f(x) = 72x^2$, find the temperature at the inner wall if the container walls are 0.5 cm thick.

25. Surrounding an electrically charged particle is an electric field. The rate of change of electric potential with respect to the distance from the particle creating the field equals the negative of the value of the electric field. That is,

$$\frac{dV}{dx} = -E$$

where E is the electric field. If $E = k/x^2$, where k is a constant, find the electric potential at a distance x_1 from the particle, if $V \to 0$ as $x \to \infty$.

26. The rate of change of the vertical deflection y with respect to the horizontal distance x from one end of a beam is a function of x. For a particular beam, this function is $k(x^5 + 1350x^3 - 7000x^2)$, where k is a constant. Find y as a function of x.

27. Fresh water is flowing into a brine solution, with an equal volume of mixed solution flowing out. The amount of salt in the solution decreases, but more slowly as time increases. Under certain conditions the time rate of change of mass of salt (in grams per minute) is given by $-1/\sqrt{t + 1}$. Find the mass of salt as a function of time if 1000 g were originally present. Under these conditions, how long would it take for all the salt to be removed?

28. The rate of change of resistance of a certain electric resistor with respect to the temperature is given by

$$\frac{0.002T}{\sqrt[3]{3T^2 + 1}}$$

Find the resistance as a function of temperature if $R = 0.5$ mΩ when $T = 0$°C.

5—2 Areas by Integration

In Section 4—4 we introduced the method of finding the area under a curve by means of integration. In the same section we also showed that the area under a curve can be found by a summation process performed on the rectangles inscribed under the curve. In this way it was shown that integration can be interpreted as a summation process. The basic applications of the definite integral use this summation interpretation of the integral. In this section we shall formulate a general procedure for finding the area for which the bounding curves are known. The method is based on the summing of the areas of inscribed rectangles and using integration for the summation.

The first step in finding any area is to make a sketch of the area to be found. Next a representative **element of area** dA (a typical rectangle) should be drawn. In Fig. 5—1 the width of the element is dx. The length of the element is determined by the y-coordinate (of the vertex of the element) of the point on the curve. Thus, we call the length y. The area of this element is $y\,dx$, which in turn means that $dA = y\,dx$, or

$$A = \int_a^b y\,dx = \int_a^b f(x)\,dx \qquad (5-7)$$

This equation states that the elements are to be summed (this is the meaning of the integral sign) from a (the leftmost element) to b (the rightmost element).

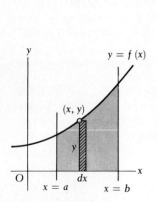

Figure 5—1

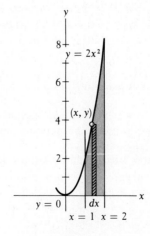

Figure 5—2

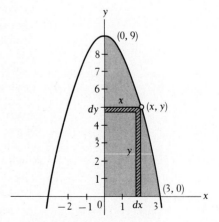

Figure 5—3

Example A

Find the area bounded by the curves $y = 2x^2$, $y = 0$, $x = 1$, and $x = 2$.

The desired area is shown in Fig. 5-2. The rectangle shown is the representative element. Its area is $y\ dx$. The elements are to be summed from $x = 1$ to $x = 2$. Thus,

$$A = \int_1^2 y\ dx = \int_1^2 2x^2\ dx = \frac{2}{3}x^3 \Big|_1^2 = \frac{2}{3}(8) - \frac{2}{3}(1) = \frac{14}{3}$$

In Figs. 5-1 and 5-2 the elements are vertical. It is also possible to use horizontal elements, and many problems are simplified by using them. The difference is that the length (longest dimension) is now measured in terms of the x-coordinate of the point on the curve, and the width becomes dy. In the following example, the area is found by both types of elements.

Example B

Find the area in the first quadrant bounded by $y = 9 - x^2$.

The area to be found is shown in Fig. 5-3. Using first the vertical element of length y and width dx, we have

$$A = \int_0^3 y\ dx = \int_0^3 (9 - x^2)\ dx = \left(9x - \frac{x^3}{3}\right)\Big|_0^3 = (27 - 9) - 0 = 18$$

Now using the horizontal element of length x and width dy, we have

$$A = \int_0^9 x\ dy = \int_0^9 \sqrt{9 - y}\ dy = -\int_0^9 (9 - y)^{1/2}(-dy)$$

$$= -\frac{2}{3}(9 - y)^{3/2}\Big|_0^9 = -\frac{2}{3}(9 - 9)^{3/2} + \frac{2}{3}(9 - 0)^{3/2} = \frac{2}{3}(27) = 18$$

Note that the limits for the vertical elements were 0 and 3, while those for the horizontal elements were 0 and 9. These limits are determined by the direction in which the elements are summed. *By definition, vertical elements are summed from left to right, and horizontal elements are summed from bottom to top.* Doing it this way means that the summation will be in a positive direction.

The choice of vertical or horizontal elements is determined by (1) which one leads to the simplest solution, or (2) the form of the resulting integral. In some problems it makes little difference which is chosen. However, our present methods of integration do not include many types of integrals.

It is also possible to find the area between two curves if one is not an axis. In such a case, the length of the element becomes the difference in the y- or x-coordinates, depending on which element is used. The following examples show how we find this type of area.

Example C

Find the area bounded by $y = x^2$ and $y = x + 2$.

This area is indicated in Fig. 5–4. Here we choose vertical elements, since they are all bounded on the top by the line and on the bottom by the parabola. If we choose horizontal elements, the bounding curves are different above the point $(-1, 1)$ (points of intersection are found by solving equations simultaneously) than below this point. Choosing horizontal elements would thus require two separate integrals for solution. Therefore, using vertical elements, we have

$$A = \int_{-1}^{2} (y \text{ of line} - y \text{ of parabola}) \, dx$$

(The difference in y's is taken so that it is positive.) Continuing, we have

$$A = \int_{-1}^{2} (x + 2 - x^2) \, dx = \left(\frac{x^2}{2} + 2x - \frac{x^3}{3} \right) \Bigg|_{-1}^{2}$$

$$= \left(2 + 4 - \frac{8}{3} \right) - \left(\frac{1}{2} - 2 + \frac{1}{3} \right)$$

$$= \frac{10}{3} + \frac{7}{6} = \frac{27}{6} = \frac{9}{2}$$

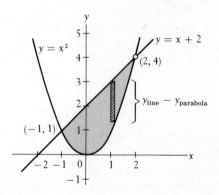

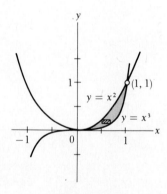

Figure 5–4   Figure 5–5

Example D

Find the area between $y = x^2$ and $y = x^3$.

Either horizontal or vertical elements may be used equally well (see Fig. 5–5). If we use horizontal elements, we have

$$A = \int_{0}^{1} (y^{1/3} - y^{1/2}) \, dy = \frac{3}{4} y^{4/3} - \frac{2}{3} y^{3/2} \Bigg|_{0}^{1} = \frac{3}{4} - \frac{2}{3} = \frac{1}{12}$$

It is important that the length of the element be positive. If the difference is taken incorrectly, the result will be negative. *Getting positive lengths can be assured for vertical elements if we subtract y of the lower curve from y of the upper curve. For horizontal elements we should sub-tract x of the left curve from x of the right curve.* This becomes of definite importance if part of the area considered is above, and the remainder below, the x-axis. In such a case, the area is found by two integrals. The following example illustrates the necessity of this procedure.

Example E
Find the area between $y = x^3 - x$ and the x-axis.

We note from Fig. 5–6 that the area to the left of the origin is above the axis and the area to the right is below. If we find the area from

$$A = \int_{-1}^{1} (x^3 - x) \, dx = \frac{x^4}{4} - \frac{x^2}{2} \Big|_{-1}^{1} = \left(\frac{1}{4} - \frac{1}{2}\right) - \left(\frac{1}{4} - \frac{1}{2}\right) = 0$$

we see that the apparent area is zero. From the figure we know this is not correct. Noting that the y-values (of the area) are negative to the right of the origin, we set up the integrals

$$A = \int_{-1}^{0} (x^3 - x) \, dx + \int_{0}^{1} - (x^3 - x) \, dx$$

$$= \left(\frac{x^4}{4} - \frac{x^2}{2}\right) \Big|_{-1}^{0} - \left(\frac{x^4}{4} - \frac{x^2}{2}\right) \Big|_{0}^{1}$$

$$= 0 - \left(\frac{1}{4} - \frac{1}{2}\right) - \left(\frac{1}{4} - \frac{1}{2}\right) + 0 = \frac{1}{2}$$

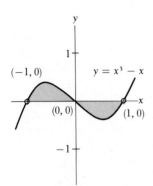

Figure 5–6

Exercises 5–2

In Exercises 1 through 28 find the areas bounded by the indicated curves.

1. $y = 4x$, $y = 0$, $x = 1$
2. $y = 2x$, $y = 0$, $x = 2$
3. $y = x^2$, $y = 0$, $x = 2$
4. $y = 3x^2$, $y = 0$, $x = 3$
5. $y = 6x$, $x = 0$, $y = 3$
6. $y = 8x$, $x = 0$, $y = 4$
7. $y = x^2$, $x = 0$, $y = 2$
8. $y = x^3$, $x = 0$, $y = 3$
9. $y = 2 - x$, $y = 0$, $x = 1$
10. $y = x^2 - 2x$, $y = 0$
11. $y = x^{-2}$, $y = 0$, $x = 2$, $x = 3$
12. $y = 16 - x^2$, $y = 0$, $x = 1$, $x = 2$
13. $y = \frac{1}{2}x$, $x = 0$, $y = 2$, $y = 4$
14. $y = \frac{1}{3}x$, $x = 0$, $y = 1$, $y = 5$
15. $y = \sqrt{x}$, $x = 0$, $y = 2$, $y = 3$
16. $x = y^2 - y$, $x = 0$
17. $y = 4 - 2x$, $x = 0$, $y = 0$, $y = 3$
18. $y = x$, $y = 2 - x$, $x = 0$
19. $y = x^2$, $y = 2 - x$, $x = 0$ (smaller)
20. $y = x^2$, $y = 2 - x$, $y = 2$ (smallest)
21. $y = x^4$, $y = 16$
22. $y = x^4 - 8x^2 + 16$, $y = 0$
23. $y = \sqrt{x - 1}$, $y = 0$, $x = 2$
24. $y = \sqrt{2x + 1}$, $y = 0$, $x = 4$
25. $y = x^2$, $y = \sqrt{x}$
26. $y = 8 - x^3$, $y = 7x$, $x = 0$
27. $y = x$, $x = -1$, $x = 1$, $y = 0$
28. $y = x^2 + 2x - 8$, $y = x + 4$

In Exercises 29 through 34 some applications of areas are shown.

29. Certain physical quantities are often represented as an area under a curve. By definition, power is defined as the time rate of change of performing work. Thus, $p = dw/dt$, or $dw = p\ dt$. Therefore, if $p = 12t - 4t^2$, find the work (in joules) performed in 3 s by finding the area under the curve of p versus t.

30. When considering the expansion of a gas, the expression for the work done in the expansion is $w = \int_{V1}^{V2} p\ dV$, where p is the pressure of the gas and V is its volume. For an adiabatic (no *heat* gained or lost) change, an approximate relation is $p = kV^{-1.4}$. Given that $k = 1$, find a numerical value for the work done if V changes from 1 to 4 units.

31. It is also possible to represent the change in displacement as an area. Since $s = \int v\ dt$, if $v = t^{2/3} + 1$, determine the distance (in feet) that an object moves from $t = 1$ s to $t = 8$ s, by finding the area under the curve of v versus t.

32. In designing a lawn area, an architect planned one such area of an estate to be between parabolas $y = 20 - 0.01x^2$ and $y = 0.005x^2 - 15$ and the lines $x = -27$ and $x = 27$ where dimensions are in meters. Find the area of this part of the lawn.

33. An experiment was performed by expanding a gas under the condition of constant temperature. The data found are tabulated below. How much work (in joules) was done in the expansion? Use the trapezoidal rule, with $n = 4$.

p (kilopascals)	102	77.0	60.0	47.0	43.0
V (cubic centimeters)	6.0	8.0	10.0	12.0	14.0

To obtain units of joules we must multiply the numerical value obtained by 10^{-3} to include the powers of 10 indicated by the prefix. 1 kPa $= 10^3$ Pa and 1 cm$^3 = (10^{-2}$ m$)^3 = 10^{-6}$ m^3.

34. A certain tract of land is bounded by a bend in a river and a straight fence. If the distances y from the fence to the river are given for various distances x along the fence from one end to the other as in the following table, find the approximate area of the piece of land by use of Simpson's rule, with $n = 8$.

y (feet)	0	15	50	70	95	80	35	20	0
x (feet)	0	10	20	30	40	50	60	70	80

5–3 Volumes by Integration

Consider an area and its representative element (see Fig. 5–7) to be rotated about the x-axis. When an area is rotated in this manner, it is said to generate a volume, which is also indicated in the figure. We shall now show methods of finding volumes which are generated in this manner.

As the area rotates about the x-axis, so does its representative element. The element generates a solid for which the volume is known. This is a thin right circular cylinder. We know that the volume of a right circular cylinder is π times the square of the radius times the height of the cylinder. We must now determine the radius and height of this cylinder.

Since the element is rotated about the x-axis, the y coordinate of the point on the curve which touches the element must represent the radius. Also, it can be seen that the height is dx (the cylinder is on its side). The representative **element of volume**, a circular **disk**, is $dV = \pi y^2\, dx$. Summing these elements of volume from left to right, we have the total volume V as

$$V = \pi \int_a^b y^2\, dx = \pi \int_a^b [f(x)]^2\, dx \tag{5–8}$$

Thus, by use of Eq. (5–8), we can find the volume generated by an area bounded by the x-axis, which is rotated about the x-axis.

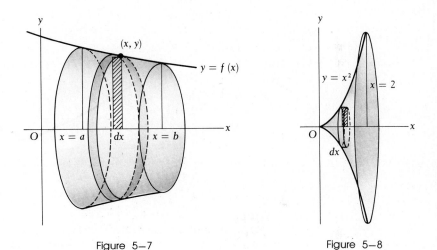

Figure 5–7   Figure 5–8

Example A
Find the volume generated by rotating the area bounded by $y = x^2$, $x = 2$, and $y = 0$ about the x-axis.
  See Fig. 5–8. By Eq. (5–8), we have

$$V = \pi \int_0^2 (x^2)^2\, dx = \pi \int_0^2 x^4\, dx = \left. \frac{\pi}{5} x^5 \right|_0^2 = \frac{32\pi}{5}$$

If an area bounded by the y-axis is rotated about the y-axis, the volume generated is given by

$$V = \pi \int_c^d x^2\, dy \tag{5–9}$$

In this case the radius of the element of volume is the x-coordinate of the point on the curve, and the height of the disk is dy. One should always be careful to identify the radius and height properly.

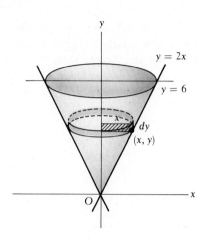

Figure 5—9   Figure 5—10

Example B

Find the volume generated by rotating the area bounded by $y = 2x$, $y = 6$, and $x = 0$ about the y-axis.

Figure 5—9 shows the volume to be found. We write

$$V = \pi \int_0^6 x^2 \, dy = \pi \int_0^6 \left(\frac{y}{2}\right)^2 \, dy = \frac{\pi}{4} \int_0^6 y^2 \, dy = \frac{\pi}{12} y^3 \Big|_0^6 = 18\pi$$

Since this volume is a right circular cone, it is possible to check the result:

$$V = \frac{1}{3}\pi r^2 h = \frac{1}{3}\pi(3^2)(6) = 18\pi$$

There is another method of finding a volume of a solid revolution. If the area in Fig. 5—8 is rotated about the y-axis, the element of area $y \, dx$ generates a different element of volume from that generated when it is rotated about the x-axis. We can see in Fig. 5—10 that this element of volume is a **cylindrical shell**. *The total volume is made up of an infinite number of concentric shells.* When the volumes of these shells are summed, we have the total volume generated. Thus we must now find the approximate volume dV of the representative shell. By finding the circumference of the base and multiplying this by the height, we can obtain an expression for the surface area of the shell. Then, by multiplying this by the thickness of the shell, we obtain its volume. The volume of the representative shell is

$$dV = 2\pi \text{ (radius)} \cdot \text{(height)} \cdot \text{(thickness)} \qquad (5\text{--}10)$$

Similarly, the volume of a disk is given by

$$dV = \pi \text{ (radius)}^2 \cdot \text{(height)} \qquad (5\text{--}11)$$

The elements of volume should be remembered in the general forms given in Eqs. (5–10) and (5–11), and not in specific forms such as Eqs. (5–8) and (5–9). If we remember them in this manner, we can readily apply these methods to finding any such volume of a solid of revolution.

Example C

Use the method of cylindrical shells to find the volume generated by rotating the area bounded by $y = 4 - x^2$, $x = 0$, and $y = 0$ about the y-axis.

From Fig. 5–11, we identify the radius, height, and thickness: $r = x$, $h = y$, $t = dx$ (this determines the limits as $x = 0$ and $x = 2$). And so

$$V = 2\pi \int_0^2 xy \, dx = 2\pi \int_0^2 x(4 - x^2) \, dx = 2\pi \int_0^2 (4x - x^3) \, dx = 8\pi$$

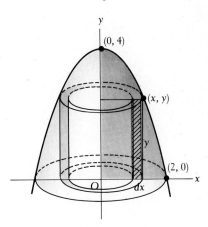

Figure 5–11

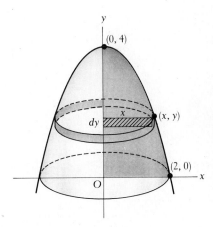

Figure 5–12

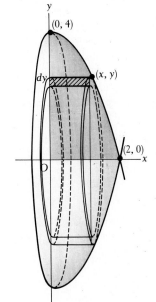

Figure 5–13

Example D

Use the disk method to find the indicated volume of Example C.

From Fig. 5–12, we identify the radius and height: $r = x$, $h = dy$ (which tells us that the limits are from $y = 0$ to $y = 4$). Thus,

$$V = \pi \int_0^4 x^2 \, dy = \pi \int_0^4 (4 - y) \, dy = 8\pi$$

Example E

Use shells to find the volume if the area of Example C is rotated about the x-axis.

From Fig. 5–13, we see that $r = y$, $h = x$, $t = dy$, (thus, the limits are $y = 0$ to $y = 4$). Hence

$$V = 2\pi \int_0^4 xy \, dy = 2\pi \int_0^4 \sqrt{4 - y}(y \, dy) = \frac{256\pi}{15}$$

(The method of integrating this function has not yet been discussed. We present the answer here for the reader's information at this time.)

Example F

By the use of disks, find the volume indicated in Example E.

From Fig. 5–14, we see that $r = y$, $h = dx$ (the limits are $x = 0$ to $x = 2$). Therefore

$$V = \pi \int_0^2 y^2 \, dx = \pi \int_0^2 (4 - x^2)^2 \, dx$$

$$= \pi \int_0^2 (16 - 8x^2 + x^4) \, dx = \frac{256\pi}{15}$$

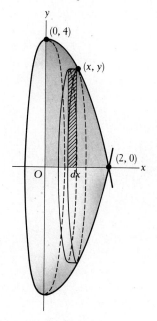

Figure 5–14

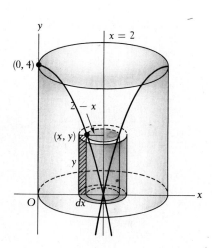

Figure 5–15

Example G

Find the volume generated if the area in Example C is rotated about the line $x = 2$.

Shells are convenient, since the volume of a shell can be expressed as a single integral. From Fig. 5–15, we see that

$$r = 2 - x \quad h = y \quad \text{and} \quad t = dx$$

Thus, we have

$$V = 2\pi \int_0^2 (2 - x) y \, dx = 2\pi \int_0^2 (2 - x)(4 - x^2) \, dx$$

$$= 2\pi \int_0^2 (8 - 2x^2 - 4x + x^3) \, dx = \frac{40\pi}{3}$$

(If the area had been rotated about the line $x = 3$, the only difference in the integral would have been that $r = 3 - x$. Everything else, including the limits, would have remained the same.)

Exercises 5–3

In Exercises 1 through 12 find the volume generated by the areas bounded by the given curves if they are rotated about the x-axis. Use the indicated method in each case.

1. $y = 1 - x$, $x = 0$, $y = 0$ (disks)
2. $y = x$, $y = 0$, $x = 2$ (disks)
3. Area of Exercise 1 (shells)
4. $y = x$, $x = 0$, $y = 2$ (shells)
5. $y = x^2 + 1$, $x = 0$, $x = 3$, $y = 0$ (disks)
6. $y = 2x - x^2$, $y = 0$ (disks)
7. $y = x^3$, $y = 8$, $x = 0$ (shells)
8. $y = x^2$, $y = x$ (shells)
9. $y = 3\sqrt{x}$, $y = 0$, $x = 4$ (disks)
10. $y = 6 - x - x^2$, $x = 0$, $y = 0$ (quad. I) (disks)
11. $x = 4y - y^2 - 3$, $x = 0$ (shells)
12. $y = x^4$, $x = 0$, $y = 1$, $y = 2$ (shells)

In Exercises 13 through 24 find the volume generated by the areas bounded by the given curves if they are rotated about the y-axis. Use the indicated method in each case.

13. Area of Exercise 1 (disks)
14. Area of Exercise 4 (disks)
15. Area of Exercise 1 (shells)
16. Area of Exercise 2 (shells)
17. $y = 2\sqrt{x}$, $x = 0$, $y = 2$ (disks)
18. $y^2 = x$, $y = 4$, $x = 0$ (disks)
19. Area of Exercise 6 (shells)
20. Area of Exercise 10 (shells)
21. Area of Exercise 11 (disks)
22. $x^2 + 4y^2 = 4$ (quad. I) (disks)
23. $y = \sqrt{4 - x^2}$ (quad. I) (shells)
24. $y = 8 - x^3$, $x = 0$, $y = 0$ (shells)

In Exercises 25 through 30 find the indicated volumes by integration.

25. Find the volume generated if the area of Exercise 6 is rotated about the line $x = 2$.
26. Find the volume generated if the area of Exercise 8 is rotated about the line $y = 4$.
27. Derive the expression for the volume of a right circular cone obtained by rotating the area bounded by $y = (r/h)x$, $y = 0$ and $x = h$ about the x-axis.
28. A conical funnel has an opening at the bottom of radius 0.5 cm. The diameter at the top of the funnel is 20 cm and the depth of the funnel is 19 cm. What is the maximum capacity of the funnel?
29. The water in a spherical tank 20 m in radius is 15 m deep at the deepest point. How much water is in the tank?
30. All horizontal cross-sections of a keg 4 ft tall are circular, and the sides of the keg are parabolic. The diameter at the top and the bottom is 2 ft, and the diameter in the middle is 3 ft. Find the volume that the keg holds.

5–4 Centroids

In the study of mechanics, a very important property of an object is its center of mass. In this section we shall show the meaning of center of mass and then show how integration is used to determine the center of mass for areas and solids of rotation.

If a mass m is at a distance d from a specified point O, the **moment**
of the mass about O is defined as md. If several masses $m_1, m_2, \ldots, m_n$
are at distances $d_1, d_2, \ldots, d_n$, respectively, from point O, their moment
(as a group) about O is defined as

$$m_1 d_1 + m_2 d_2 + \cdots + m_n d_n$$

If all the masses could be concentrated at one point $\overline{d}$ units from O, the
moment would be $(m_1 + m_2 + \cdots + m_n)\overline{d}$: this is, however, what is
meant by the above expression. Therefore, we may write

$$m_1 d_1 + m_2 d_2 + \cdots + m_n d_n = (m_1 + m_2 + \cdots + m_n)\overline{d} \qquad (5\text{--}12)$$

In Eq. (5–12), $\overline{d}$ is the distance from O to the **center of mass.** The
moment of a mass is a measure of its tendency to rotate about a point. A
weight far from the point of balance of a long rod is more likely to make
the rod turn than if the same weight were placed near the point of balance. It is easier to open a door if you push near the door knob than if
you push near the hinges. This is the type of physical property which the
moment of mass measures.

Example A
On the x-axis a mass of 3 units is placed at (2, 0), another of 6 units at
(5, 0) and a third, 7 units, at (6, 0). Find the center of mass of the three
objects.

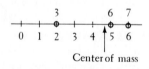

Taking the reference point as the origin, we find $d_1 = 2$, $d_2 = 5$, and
$d_3 = 6$. Thus, $m_1 d_1 + m_2 d_2 + m_3 d_3 = (m_1 + m_2 + m_3)\overline{d}$ becomes

$$3(2) + 6(5) + 7(6) = (3 + 6 + 7)\overline{d} \qquad \text{or} \qquad \overline{d} = 4.88$$

Figure 5–16

This means that the center of mass of the three objects is at (4.88, 0).
Therefore a mass of 16 units placed at this point has the same moment
as the three masses as a unit (see Fig. 5–16).

Example B
Find the center of mass of the area indicated in Fig. 5–17.

We first note that the center of mass is not *on* either axis. This can be
seen from the fact that the major portion of the area is in the first quadrant. *We shall therefore measure the moments with respect to each axis
to find the point which is the center of mass. This point is also called
the* **centroid** *of the area.*

The easiest method of finding the centroid is to divide the area into
rectangles, as indicated by the dashed line in Fig. 5–17, and assume that
we may consider the mass of each rectangle to be concentrated at its
center. In this way the left rectangle has its center $(-1, 1)$ and the right
rectangle has its center at $(\frac{5}{2}, 2)$. The mass of each rectangle, assumed

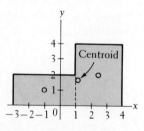

Figure 5–17

uniform, is porportional to the area. The area of the left rectangle is 8 units and that of the right rectangle is 12 units. Thus, taking moments with respect to the y-axis, we have

$$8(-1) + 12\left(\frac{5}{2}\right) = (8 + 12)\bar{x}$$

where $\bar{x}$ is the x-coordinate of the centroid. Solving for $\bar{x}$ we have $\bar{x} = \frac{11}{10}$.
Now taking moments with respect to the x-axis we have

$$8(1) + 12(2) = (8 + 12)\bar{y}$$

where $\bar{y}$ is the y-coordinate of the centroid. Thus, $\bar{y} = \frac{8}{5}$. This means that the coordinates of the centroid, the center of mass, are $(\frac{11}{10}, \frac{8}{5})$. This may be interpreted as meaning that an area of this shape would balance on a single support under this point. As an approximate check, we note from the figure that this point appears to be a reasonable balance point for the area.

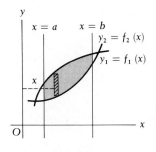

Figure 5–18

If an area is bounded on the curves of the functions $y_1 = f_1(x)$, $y_2 = f_2(x)$, $x = a$, and $x = b$, as shown in Fig. 5–18, the moment of the element of area about the y-axis is $(k\,dA)x$, where k is the mass per unit area. In this expression, $k\,dA$ is the mass of the element, and x is its distance (moment arm) from the y-axis. The element dA may be written as $(y_1 - y_2)\,dx$, which means that the moment may be written as $kx(y_1 - y_2)\,dx$. If we then sum up the moments of all the elements and express this as an integral (which, of course, means sum), we have $k \int_a^b x(y_1 - y_2)\,dx$. If we consider all the mass of the area to be concentrated at one point $\bar{x}$ units from the y-axis, the moment would be $(kA)\bar{x}$, where kA is the mass of the entire area, and $\bar{x}$ is the distance the center of mass is from the y-axis. By the previous discussion these two expressions should be equal. This means $k \int_a^b x(y_1 - y_2)\,dx = kA\bar{x}$. Since k appears on each side of the equation, we divide it out (we are assuming that the mass per unit area is constant). The area A is found by the integral $\int_a^b (y_1 - y_2)\,dx$. Therefore,

$$\bar{x} = \frac{\displaystyle\int_a^b x(y_1 - y_2)\,dx}{\displaystyle\int_a^b (y_1 - y_2)\,dx} \tag{5–13}$$

Equation (5–13) gives us the x-coordinate of the centroid of an area if vertical elements are used. It should be noted that the two integrals in Eq. (5–13) must be evaluated separately. We cannot cancel the apparent common factor $y_1 - y_2$, and we cannot combine quantities and perform one integration. The two integrals must be evaluated before cancellations are made.

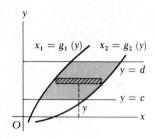

Figure 5–19

Following the same reasoning, if an area is bounded by $x_1 = g_1(y)$, $x_2 = g_2(y)$, $y = c$, and $y = d$ (Fig. 5–19), the y-coordinate of the centroid is

$$\bar{y} = \frac{\displaystyle\int_c^d y(x_2 - x_1)\,dy}{\displaystyle\int_c^d (x_2 - x_1)\,dy} \tag{5–14}$$

In this equation, horizontal elements are used.

In applying Eqs. (5–13) and (5–14), we should keep in mind that the denominators of the right-hand sides give the area, and once we have found this, we may use it for both $\bar{x}$ and $\bar{y}$. Also, we should utilize any symmetry a curve may have.

Example C

Find the coordinates of the centroid of the area bounded by $y = x^2$ and the line $y = 4$.

We sketch a graph indicating the area and an element of area (see Fig. 5–20). The curve is a parabola whose axis is the y-axis. Since the area is symmetrical to the y-axis, the centroid must be on this axis. This means that the x-coordinate of the centroid is zero, or $\bar{x} = 0$. To find the y-coordinate of the centroid, we use Eq. (5–14). For this area we have

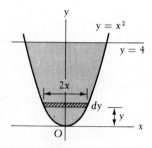

Figure 5–20

$$\bar{y} = \frac{\displaystyle\int_0^4 y(2x)\,dy}{\displaystyle\int_0^4 2x\,dy} = \frac{\displaystyle\int_0^4 y(2\sqrt{y})\,dy}{\displaystyle\int_0^4 2\sqrt{y}\,dy} = \frac{2\displaystyle\int_0^4 y^{3/2}\,dy}{2\displaystyle\int_0^4 y^{1/2}\,dy} = \frac{2\left(\dfrac{2}{5}\right)y^{5/2}\Big|_0^4}{2\left(\dfrac{2}{3}\right)y^{3/2}\Big|_0^4}$$

$$= \frac{\dfrac{4}{5}(32)}{\dfrac{4}{3}(8)} = \frac{128}{5}\cdot\frac{3}{32} = \frac{12}{5}$$

The coordinates of the centroid are $(0, \frac{12}{5})$. This area would balance if a single pointed support were to be put under this point.

Example D

Find the coordinates of the centroid of an isosceles right triangle with side a.

We must first set up this area in the xy-plane. One choice is to place the triangle with one vertex at the origin and the right angle on the x-axis (see Fig. 5–21). Since each side is a, the hypotenuse passes through the point (a, a). The equation of the hypotenuse is $y = x$. The x-coordinate of the centroid is

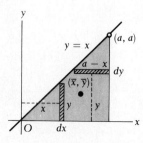

Figure 5–21

$$\bar{x} = \frac{\displaystyle\int_0^a xy\,dx}{\displaystyle\int_0^a y\,dx} = \frac{\displaystyle\int_0^a x(x)\,dx}{\displaystyle\int_0^a x\,dx} = \frac{\displaystyle\int_0^a x^2\,dx}{\dfrac{1}{2}x^2\Big|_0^a} = \frac{\dfrac{1}{3}x^3\Big|_0^a}{\dfrac{a^2}{2}} = \frac{\dfrac{a^3}{3}}{\dfrac{a^2}{2}} = \frac{2a}{3}$$

The *y*-coordinate of the centroid is

$$\overline{y} = \frac{\displaystyle\int_0^a y(a - x)\, dy}{\dfrac{a^2}{2}} = \frac{\displaystyle\int_0^a y(a - y)\, dy}{\dfrac{a^2}{2}} = \frac{\displaystyle\int_0^a (ay - y^2)\, dy}{\dfrac{a^2}{2}}$$

$$= \frac{\left.\dfrac{ay^2}{2} - \dfrac{y^3}{3}\right|_0^a}{\dfrac{a^2}{2}} = \frac{\dfrac{a^3}{6}}{\dfrac{a^2}{2}} = \frac{a}{3}$$

Thus, the coordinates of the centroid are $(\tfrac{2}{3}a, \tfrac{1}{3}a)$. The results indicate that the center of mass is $\tfrac{1}{3}a$ units from each of the equal sides.

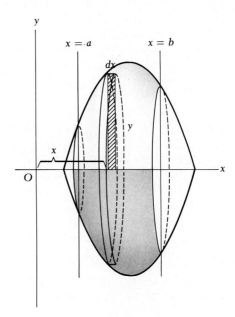

Figure 5–22

Another important figure for which we wish to find the centroid is the solid of revolution. If the density of the solid is constant, the coordinates of the centroid will be on the axis of rotation. The problem which remains is to find just where on the axis the centroid is located.

If a given area bounded by the *x*-axis, as shown in Fig. 5–22, is revolved about the *x*-axis, a vertical element of area generates a disk element of volume. The center of mass of the disk is at its center, and therefore, for purposes of finding moments, we may consider its mass concentrated there. The moment about the *y*-axis of a typical element is $x(k)(\pi y^2\, dx)$, where x is the moment arm, k is the density, and $\pi y^2\, dx$ is the volume. The sum of the moments of the elements can be expressed as an integral; it equals the volume times the density times the x-coordinate of the centroid of the volume. Since the density k and the factor of π would appear on each side of the equation, they need not be written. Therefore,

$$\bar{x} = \frac{\displaystyle\int_a^b xy^2 \, dx}{\displaystyle\int_a^b y^2 \, dx} \tag{5–15}$$

is the equation giving the x-coordinate of the centroid of a volume of a solid of revolution about the x-axis.

In the same manner we may find the y-coordinate of the centroid of a volume of a solid of revolution about the y-axis. It is

$$\bar{y} = \frac{\displaystyle\int_c^d yx^2 \, dy}{\displaystyle\int_c^d x^2 \, dy} \tag{5–16}$$

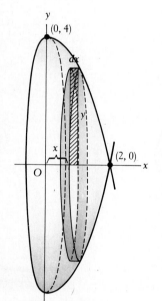

Figure 5–23

Example E

Find the coordinates of the centroid of the volume generated by rotating the first-quadrant area under the curve $y = 4 - x^2$ about the x-axis.

Since the curve (see Fig. 5–23) is rotated about the x-axis, the centroid is on the x-axis, which means that $\bar{y} = 0$. We find the x-coordinate by using Eq. (5–15):

$$\bar{x} = \frac{\displaystyle\int_0^2 xy^2 \, dx}{\displaystyle\int_0^2 y^2 \, dx} = \frac{\displaystyle\int_0^2 x(4 - x^2)^2 \, dx}{\displaystyle\int_0^2 (4 - x^2)^2 \, dx} = \frac{\displaystyle\int_0^2 (16x - 8x^3 + x^5) \, dx}{\displaystyle\int_0^2 (16 - 8x^2 + x^4) \, dx}$$

$$= \frac{8x^2 - 2x^4 + \dfrac{1}{6}x^6 \Big|_0^2}{16x - \dfrac{8}{3}x^3 + \dfrac{1}{5}x^5 \Big|_0^2} = \frac{32 - 32 + \dfrac{64}{6}}{32 - \dfrac{64}{3} + \dfrac{32}{5}} = \frac{5}{8}$$

The coordinates of the centroid are $(\frac{5}{8}, 0)$.

Example F

Find the coordinates of the centroid of the volume generated by rotating the area of Example E about the y-axis.

Since the curve is rotated about the y-axis, $\bar{x} = 0$ (see Fig. 5–24). The y-coordinate is found by Eq. (5–16):

$$\bar{y} = \frac{\int_0^4 yx^2\,dy}{\int_0^4 x^2\,dy} = \frac{\int_0^4 y(4-y)\,dy}{\int_0^4 (4-y)\,dy} = \frac{\int_0^4 (4y-y^2)\,dy}{\int_0^4 (4-y)\,dy} = \frac{2y^2 - \dfrac{1}{3}y^3\Big|_0^4}{4y - \dfrac{1}{2}y^2\Big|_0^4}$$

$$= \frac{32 - \dfrac{64}{3}}{16 - 8} = \frac{4}{3}$$

The coordinates of the centroid are $(0, \frac{4}{3})$.

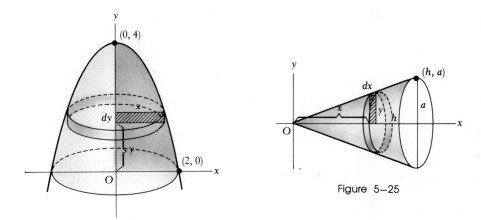

Figure 5–24

Figure 5–25

Example G

Find the centroid of a right circular cone of radius a and altitude h.

To generate a right circular cone, we may rotate a right triangle about one of its legs (Fig. 5–25). Placing the leg of length h along the x-axis, we rotate the right triangle whose hypotenuse is given by $y = (a/h)x$ about the x-axis. The x-coordinate of the centroid is

$$\bar{x} = \frac{\int_0^h xy^2\,dx}{\int_0^h y^2\,dx} = \frac{\int_0^h x\left[\left(\dfrac{a}{h}\right)x\right]^2\,dx}{\int_0^h \left[\left(\dfrac{a}{h}\right)x\right]^2\,dx} = \frac{\left(\dfrac{a^2}{h^2}\right)\left(\dfrac{1}{4}x^4\right)\Big|_0^h}{\left(\dfrac{a^2}{h^2}\right)\left(\dfrac{1}{3}x^3\right)\Big|_0^h} = \frac{3}{4}h$$

Therefore, the centroid is located along the altitude $\frac{3}{4}$ of the way from the vertex, or $\frac{1}{4}$ of the way from the base.

Exercises 5—4

In Exercises 1 through 4 find the center of mass of the particles of the given masses located at the given points.

1. 5 units at $(1, 0)$, 10 units at $(4, 0)$, 3 units at $(5, 0)$

2. 2 units at $(2, 0)$, 9 units at $(3, 0)$, 3 units at $(8, 0)$, 1 unit at $(12, 0)$

3. 4 units at $(-3, 0)$, 2 units at $(0, 0)$, 1 unit at $(2, 0)$, 8 units at $(3, 0)$

4. 2 units at $(-4, 0)$, 1 unit at $(-3, 0)$, 5 units at $(1, 0)$, 4 units at $(4, 0)$

In Exercises 5 through 8 find the coordinates of the centroid of the area shown.

5. Fig. 5—26(a)   6. Fig. 5—26(b)

7. Fig. 5—26(c)   8. Fig. 5—26(d)

In Exercises 9 through 24 find the coordinates of the centroids of the indicated figures.

9. Find the coordinates of the centroid of the area bounded by $y = x^2$ and $y = 2$.

10. Find the coordinates of the centroid of a semicircular area.

11. Find the coordinates of the centroid of the area bounded by $y = 4 - x$ and the axes.

12. Find the coordinates of the centroid of the area bounded by $y = x^3$, $x = 2$, and the x-axis.

13. Find the coordinates of the centroid of the area bounded by $y = x^2$ and $y = x^3$.

14. Find the coordinates of the centroid of the area bounded by $y^2 = x$, $y = 2$, $x = 0$.

15. Find the coordinates of the centroid of the volume generated by rotating the area in the first quadrant bounded by $y^2 = 4x$, $y = 0$, and $x = 1$ about the y-axis.

16. Find the coordinates of the centroid of the volume generated by rotating the area bounded by $y = x^2$, $x = 2$, and the x-axis about the x-axis.

17. Find the coordinates of the centroid of the volume generated by rotating the area bounded by $y^2 = 4x$ and $x = 1$ about the x-axis.

18. Find the coordinates of the centroid of the volume generated by rotating the area bounded by $x^2 - y^2 = 9$, $y = 4$, and the x-axis about the y-axis.

19. Find the coordinates of the centroid of the volume generated by rotating the area bounded by $y = 4/x^2$, $x = 1$, $x = 2$, and the x-axis about the x-axis.

20. Find the coordinates of the centroid of the volume generated by rotating the first-quadrant area bounded by the ellipse $4x^2 + 9y^2 = 36$ around the y-axis.

21. Find the location of the centroid of a right triangle with legs a and b.

22. Find the coordinates of the centroid of an isosceles trapezoid with bases b_1 and b_2 and altitude a.

23. Find the location of the centroid of a hemisphere of radius a.

24. Find the location of the centroid of the frustum of a right circular cone if the radii are 2 and 6 in., and the altitude is 5 in.

(a)
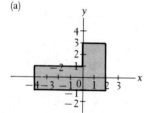

(b)
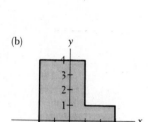

(c)
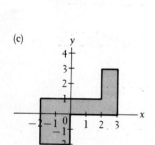

(d)
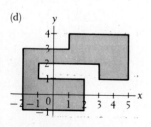

Figure 5—26

5–5 Moments of Inertia

In applying the laws of physics, we encounter an important quantity when we are discussing the rotation of an object: its **moment of inertia.** The moment of inertia of an object rotating about an axis is analogous to the mass of an object moving through space in reference to some point. *In each case the quantity (moment of inertia or mass) is the measure of the tendency of an object to resist a change in motion.*

Suppose that a particle of mass m is rotating about some point; we define its moment of inertia as md^2, where d is the distance from the particle to the point. If a group of particles of masses $m_1, m_2, \ldots, m_n$ are rotating about some axis, the moment of inertia of the group with respect to that axis is

$$I = m_1 \, d_1^2 + m_2 \, d_2^2 + \cdots + m_n \, d_n^2$$

where the d's are the respective distances of the particles from the axis. If all of the masses were at the same distance R from the axis of rotation, so that the total moment of inertia were the same, we would have

$$m_1 \, d_1^2 + m_2 \, d_2^2 + \cdots + m_n \, d_n^2 = (m_1 + m_2 + \cdots + m_n)R^2 \qquad (5\text{–}17)$$

where R is called the **radius of gyration.**

Example A

Find the moment of inertia and radius of gyration of three masses, one of 3 units at $(-2, 0)$, the second of 5 units at $(1, 0)$, and the third of 4 units at $(4, 0)$ with respect to the origin (see Fig. 5–27).

The moment of inertia of the group is

$$I = 3(-2)^2 + 5(1)^2 + 4(4)^2 = 81$$

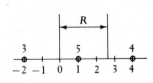

Figure 5–27

The radius of gyration is found from $I = (m_1 + m_2 + m_3)R^2$. Thus,

$$81 = (3 + 5 + 4)R^2 \qquad R^2 = \frac{81}{12} \qquad \text{or} \qquad R = 2.60$$

This means that a mass of 12 units placed at $(2.60, 0)$ [or at $(-2.60, 0)$] would have the same rotational inertia about the origin as the three objects as a unit.

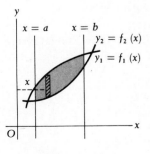

Figure 5–28

If an area is bounded by the curves of the functions $y_1 = f_1(x)$, $y_2 = f_2(x)$, and $x = a$ and $x = b$, as shown in Fig. 5–28, the moment of inertia of this area with respect to the y-axis, I_y, is given by the sum of the moments of inertia of the individual elements. The mass of each element is $k(y_1 - y_2) \, dx$ where k is the mass per unit area, and $(y_1 - y_2) \, dx$ is the area of the element. The distance of the element from the y-axis is x. Representing this sum as an integral, we have

$$I_y = k \int_a^b x^2(y_1 - y_2) \, dx \qquad (5\text{–}18)$$

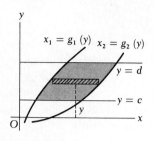

Figure 5–29

To find the *radius of gyration of the area with respect to the y-axis*, R_y, we would first find the moment of inertia, divide this by the mass of the area, and take the square root of this result.

In the same manner, the moment of inertia of an area, with respect to the x-axis, bounded by $x_1 = g_1(y)$ and $x_2 = g_2(y)$ is given by

$$I_x = k \int_c^d y^2 (x_2 - x_1) \, dy \qquad (5\text{–}19)$$

We find the radius of gyration of the area with respect to the x-axis, R_x, in the same manner as we find it with respect to the y-axis (see Fig. 5–29).

Example B

Find the moment of inertia and the radius of gyration of the area bounded by $y = 4x^2$, $x = 1$, and the x-axis with respect to the y-axis.

We find the moment of inertia of the area (see Fig. 5–30) by using Eq. (5–18):

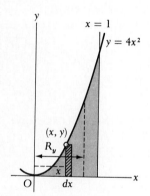

Figure 5–30

$$I_y = k \int_0^1 x^2 y \, dx = k \int_0^1 x^2 (4x^2) \, dx = 4k \int_0^1 x^4 \, dx$$

$$= 4k \left(\frac{1}{5} x^5 \right) \Big|_0^1 = \frac{4k}{5}$$

To find the radius of gyration, we first determine the mass of the area:

$$m = k \int_0^1 y \, dx = k \int_0^1 (4x^2) \, dx = 4k \left(\frac{1}{3} x^3 \right) \Big|_0^1 = \frac{4k}{3}$$

$$R_y^2 = \frac{I_y}{m} = \frac{4k}{5} \cdot \frac{3}{4k} = \frac{3}{5} \qquad R_y = \sqrt{\frac{3}{5}} = \frac{\sqrt{15}}{5}$$

Example C

Find the moment of inertia of a right triangle with sides a and b with respect to the side b. Assume that $k = 1$.

Placing the triangle as shown in Fig. 5–31, we see that the equation of the hypotenuse is $y = (a/b)x$. The moment of inertia is

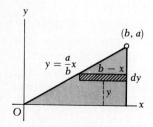

Figure 5–31

$$I_x = \int_0^a y^2 (b - x) \, dy = \int_0^a y^2 \left(b - \frac{b}{a} y \right) dy = b \int_0^a \left(y^2 - \frac{1}{a} y^3 \right) dy$$

$$= b \left(\frac{1}{3} y^3 - \frac{1}{4a} y^4 \right) \Big|_0^a = b \left(\frac{a^3}{3} - \frac{a^3}{4} \right) = \frac{ba^3}{12}$$

Among the most important moments of inertia are those of solids of revolution. Since all parts of an element of mass should be at the same distance from the axis, the most convenient element of volume to choose is the cylindrical shell (see Fig. 5–32). If the area bounded by the curves $y = f(x)$, $x = a$, $x = b$, and the x-axis is rotated about the y-axis, the moment of inertia of the element of volume is $k(2\pi xy\ dx)(x^2)$, where k is the density, $2\pi xy\ dx$ is the volume of the element, and x^2 is the square of its distance from the y-axis. Expressing the sum of the moments of the elements as an integral, the moment of inertia of the volume with respect to the y-axis, I_y, is

$$I_y = 2\pi k \int_a^b yx^3\ dx \qquad\qquad (5\text{–}20)$$

The radius of gyration of the volume with respect to the y-axis R_y is found by determining (1) the moment of inertia, (2) the mass of the volume, and (3) the square root of the quotient of the moment of inertia divided by the mass.

The moment of inertia of the volume (see Fig. 5–33) generated by rotating the area bounded by $x = g(y)$, $y = c$, $y = d$, and the y-axis about the x-axis, I_x, is given by

$$I_x = 2\pi k \int_c^d xy^3\ dy \qquad\qquad (5\text{–}21)$$

The radius of gyration of the volume with respect to the x-axis, R_x, is found in the same manner as R_y.

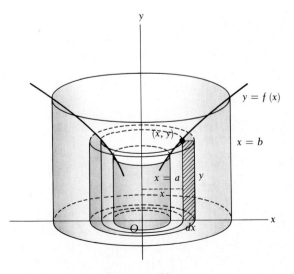

Figure 5–32

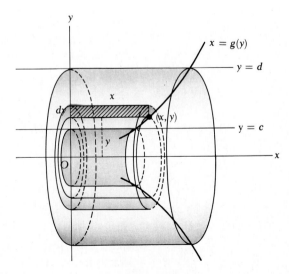

Figure 5–33

Example D

Find the moment of inertia and the radius of gyration of the solid (see Fig. 5–34) generated by rotating the area bounded by $y^3 = x$, $y = 2$, and the y-axis about the x-axis.

$$I_x = 2\pi k \int_0^2 xy^3 \, dy = 2\pi k \int_0^2 (y^3)y^3 \, dy = 2\pi k \left(\frac{1}{7}y^7\right)\Big|_0^2 = \frac{256\pi k}{7}$$

$$m = 2\pi k \int_0^2 xy \, dy = 2\pi k \int_0^2 y^3 y \, dy = 2\pi k \left(\frac{1}{5}y^5\right)\Big|_0^2 = \frac{64\pi k}{5}$$

$$R_x^2 = \frac{256\pi k}{7} \cdot \frac{5}{64\pi k} = \frac{20}{7} \qquad R_x = \sqrt{\frac{20}{7}} = \frac{2}{7}\sqrt{35}$$

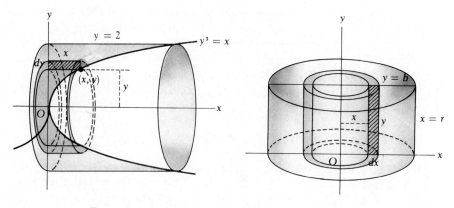

Figure 5–34   Figure 5–35

Example E

Find the moment of inertia of a disk of radius r with respect to its axis.

To generate a disk (see Fig. 5–35) we shall rotate the area bounded by the axes, $x = r$ and $y = b$, about the y-axis. We then have

$$I_y = 2\pi k \int_0^r x^3 y \, dx = 2\pi k \int_0^r x^3 (b) \, dx$$

$$= 2\pi k b \int_0^r x^3 \, dx = 2\pi k b \left(\frac{1}{4}x^4\right)\Big|_0^r = \frac{\pi k b r^4}{2}$$

The mass of the disk is $k(\pi r^2)b$. Rewriting the expression for I_y, we have

$$I_y = \frac{(\pi k b r^2)r^2}{2} = \frac{mr^2}{2}$$

This is the usual way of expressing the moment of inertia of a particular solid. That is, the mass is usually included in the formula for the moment of inertia.

Due to the limited methods of integration available at this point, we cannot integrate the expressions for the moments of inertia of circular areas or of a sphere. These will be introduced in Section 8–8 in the exercises, by which point the proper method of integration will have been developed.

Exercises 5–5

In Exercises 1 through 4 find the moment of inertia and the radius of gyration with respect to the origin of the given masses at the given points.

1. 5 units at $(2, 0)$ and 3 units at $(6, 0)$
2. 3 units at $(-1, 0)$, 6 units at $(3, 0)$, 2 units at $(4, 0)$
3. 4 units at $(-4, 0)$, 10 units at $(0, 0)$, 6 units at $(5, 0)$
4. 6 units at $(-3, 0)$, 5 units at $(-2, 0)$, 9 units at $(1, 0)$, 2 units at $(8, 0)$

In Exercises 5 through 20 find the indicated moment of inertia or radius of gyration.

5. Find the moment of inertia of the area bounded by $y^2 = x$, $x = 4$, and the x-axis with respect to the x-axis.
6. Find the moment of inertia of the area bounded by $y = 2x$, $x = 1$, $x = 2$, and the x-axis with respect to the y-axis.
7. Find the radius of gyration of the area bounded by $y = x^3$, $x = 2$, and the x-axis with respect to the y-axis.
8. Find the radius of gyration of the first-quadrant area bounded by the curve $y = 1 - x^2$ with respect to the y-axis.
9. Express the moment of inertia of the triangular area in Example C in terms of the mass of the area.
10. Find the moment of inertia of a rectangular area of sides a and b with respect to side a. Express the resul in terms of the mass.
11. Find the radius of gyration of the area bounded by $y = x^2$, $x = 2$, and the x-axis with respect to the x-axis.
12. Find the radius of gyration of the area bounded by $y^2 = x^3$, $y = 8$, and the y-axis with respect to the y-axis.
13. Find the radius of gyration of the area of Exercise 12 with respect to the x-axis.
14. Find the radius of gyration of the first-quadrant area bounded by $x = 1$, $y = 2 - x$, and the y-axis with respect to the y-axis.
15. Find the moment of inertia with respect to its axis of the solid generated by rotating the area bounded by $y^2 = x$, $y = 2$, and the y-axis about the x-axis.
16. Find the radius of gyration with respect to its axis of the volume generated by rotating the first-quadrant area under the curve $y = 4 - x^2$ about the y-axis.
17. Find the radius of gyration with respect to its axis of the volume generated by rotating the area bounded by $y = 2x - x^2$ and the x-axis about the y-axis.
18. Find the radius of gyration with respect to its axis of the volume generated by rotating the area bounded by $y = 2x$ and $y = x^2$ about the y-axis.
19. Find the moment of inertia in terms of its mass of a right circular cone of radius r and height h with respect to its axis.
20. Find the moment of inertia in terms of its mass of a circular disk with a hole in the center with respect to its axis. Let the inner radius be r_1 and the outer radius be r_2.

5–6 Work by a Variable Force

The next application of integration we shall demonstrate is to the physical problem of work done by a variable force. The concept of work is useful for comparing the necessary amounts of mechanical energy required to perform various physical tasks.

Work, *in its physical sense, is the product of a constant force times the distance through which it acts.* When we consider the work done in stretching a spring, the first thing we recognize is that the more the spring is stretched, the greater is the force necessary to stretch it. Thus, the force varies. However, if we are stretching the spring a distance Δx, where we are considering the limit as $\Delta x \to 0$, the force can be considered as approaching a constant. Adding the products of force$_1$ times Δx_1, force$_2$ times Δx_2, and so forth, we see that the total work is the sum of these products. Thus, the work can be expressed as a definite integral in the form

$$W = \int_a^b f(x) \; dx \qquad\qquad (5\text{–}22)$$

where $f(x)$ is the force as a function of the distance the spring is stretched. The limits a and b refer to the initial and the final distances the spring is stretched *from its normal length.*

One problem remains: We must find the function $f(x)$. From physics we learn that the force required to stretch a spring is proportional to the amount it is stretched (Hooke's law). If a spring is stretched x units from its normal length, then $f(x) = kx$. From conditions stated for a particular spring, the value of k may be determined. Thus, $W = \int_a^b kx \; dx$ is the formula for finding the total work done in stretching a spring.

Example A

A spring of natural length 12 in. requires a force of 6 lb to stretch it 2 in. (see Fig. 5–36). Find the work done in stretching it 6 in.

From Hooke's law, we have $F = kx$, or $6 = k2$, $k = 3$ lb/in. Thus,

$$W = \int_0^6 3x \; dx = \frac{3}{2}x^2 \Big|_0^6 = 54 \text{ lb-in.}$$

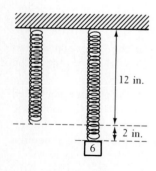

12 in.

2 in.

Figure 5–36

Problems involving work by a variable force arise in many fields of technology. An illustration from electricity is found in the motion of an electric charge through an electric field created by another electric charge.

Electric charges are of two types, designated as positive and negative. A basic law is that charges of the same sign repel each other and charges of opposite signs attract each other. *The force between charges is proportional to the product of their charges, and inversely proportional to the square of the distance between them.* Thus for this case

$$f(x) = \frac{kq_1 q_2}{x^2} \tag{5–23}$$

where q_1 and q_2 are the charges (in coulombs), x is the distance (in meters), the force is in newtons, and $k = 9 \times 10^9$ N·m²/C². For other systems of units, the numerical value of k is different. We can find the work done when electric charges move toward each other or when they separate by use of Eq. (5–23) in Eq. (5–22).

Example B
Find the work done when two α-particles, $q = 0.32$ aC each, move until they are 10 nm apart, if they were originally separated by one meter.

Applying Eqs. (5–22) and (5–23) we have, with $q = 0.32$ aC $= 0.32 \times 10^{-18}$ C $= 3.2 \times 10^{-19}$ C, and an upper limit of 10 nm $= 10 \times 10^{-9}$ m $= 10^{-8}$ m,

$$W = \int_1^{10^{-8}} \frac{9 \times 10^9 (3.2 \times 10^{-19})^2}{x^2}\, dx$$

$$= 9.2 \times 10^{-28} \int_1^{10^{-8}} \frac{dx}{x^2}$$

$$= 9.2 \times 10^{-28} \left(-\frac{1}{x}\right)\Bigg|_1^{10^{-8}} = -9.2 \times 10^{-28}(10^8 - 1)$$

$$= -9.2 \times 10^{-20} \text{ J}$$

Since $10^8 \gg 1$, where $\gg$ means "much greater than," the 1 may be neglected in the calculation. The meaning of the minus sign in the result is that work must be done *on* the system to move the particles together. If free to move, they would tend to separate.

The following two examples illustrate other types of problems in which work by a variable force is involved.

Example C
Find the work done in winding up 60 ft of a 100-ft cable which weighs 4 lb/ft.

First we let x denote the length of cable which has been wound up at any time. Then the force which is required to raise the remaining cable equals the weight of the cable which has not yet been wound up. This weight is the product of the unwound cable length $100 - x$ and its weight per unit length 4 lb/ft, or $4(100 - x)$. Therefore, since 60 ft are to be wound up, the work done is

$$W = \int_0^{60} 4(100 - x)\, dx = \int_0^{60} (400 - 4x)\, dx = 400x - 2x^2 \Big|_0^{60}$$

$$= 16{,}800 \text{ ft·lb}$$

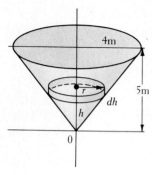

Figure 5–37

Example D

A conical tank with a circular cross-section has its axis vertical and its vertex at the bottom. The radius of the top of the tank is 4.00 m and the altitude is 5.00 m. If the tank is filled with water, find the work required to pump all of the water out over the top. The force of gravity on each cubic meter of water is 9800 N (see Fig. 5–37).

Consider the water to be divided into horizontal layers, each of which is dh meters thick. Each of these layers is circular, and therefore its volume is that of a circular disk (just as in finding volumes by use of disk elements). This means that the weight of each disk of water is $9800\pi(\text{radius}^2)(\text{thickness})$. From similar triangles in the figure we see that

$$\frac{r}{h} = \frac{4}{5}, \text{ or } r = 0.8h.$$

Also, the distance each disk of water must be raised is $5.00 - h$ meters. Therefore, summing the products of weight and distance, we find the work done.

$$W = \int_0^5 (9800\pi r^2 dh)(5 - h) = 9800\pi \int_0^5 (0.8h)^2(5 - h)\, dh$$

$$= 6270\pi \int_0^5 (5h^2 - h^3)\, dh = 6270\pi \left(\frac{5}{3}h^3 - \frac{1}{4}h^4\right)\Bigg|_0^5$$

$$= 6270\pi(52.1) = 1.03 \times 10^6 \text{ N} \cdot \text{m} = 1.03 \text{ MJ}.$$

Exercises 5–6

1. A spring of natural length 10 in. measures 14 in. when a 12-lb weight is attached to it. How much work is required to stretch it 5 in. from its natural length?

2. How much work is required to stretch the spring in Exercise 1 from a length of 14 in. to a length of 19 in.?

3. It requires 500 lb to stretch a certain spring 1 ft. How much work is done on the spring if we attach a 300-lb weight to it?

4. If another 300-lb weight is attached to the spring in Exercise 3, how much work is done by the additional weight?

5. Find the work done in separating a 2 μC positive charge from a 4 μC negative charge to a distance of 6 cm from a distance of 2 cm.

6. Find the work done in separating an 0.8 nC negative charge from an 0.6 nC charge to a distance of 200 mm from a distance of 100 mm.

7. The gravitational force of attraction between two objects which are x units apart is given by $F = k/x^2$. If two objects are 1 cm apart, find the work done in separating them to a distance of 1 m apart. Express the answer in terms of k.

8. Given two objects which are 10 ft apart, find the work required to separate them further, until they are 100 ft apart. Express the result in terms of k. (see Exercise 7).

9. Find the work done in winding up a 200-ft cable which weighs 100 lb.

10. In Exercise 9, if an object weighing 50 lb is attached at the end of the cable, find the work done in winding it up.

11. Find the work done in winding up 20 m of a 25-m rope on which the force of gravity is 6.00 N/m.

12. A chain on which the force of gravity is 12.0 N/m is being unwound from a winch. If 20 m have already been unwound, how much work is done by the force of gravity when an additional 30 m are unwound?

13. Find the work done in pumping the water out of the top of a cylindrical tank 3 ft in radius and 10 ft high, if the tank is initially full and water weighs 62.4 lb/ft^3.

14. Find the work done in pumping the water from the tank in Exercise 13, if the tank is initially half full.

15. Find the work done in emptying the tank in Example D, if the water is pumped to a level of 1 m above the top of the tank, and the water is initially 4 m deep in the tank.

16. A hemispherical tank of radius 10 ft is full of water. Find the work done in pumping the water to the top of the tank. The water weighs 62.4 lb/ft^3.

5–7 Force Due to Liquid Pressure

In physics it is shown that at any point within a liquid the pressure is the same in all directions. Also, it is shown that *the pressure is directly proportional to the depth below the surface of the liquid.* If the weight per unit volume of the liquid is w, the pressure at a depth h is

$$p = wh \tag{5–24}$$

For water, $w = 62.4$ lb/ft^3 or $w = 9800$ N/m^3. *Since pressure is the force per unit area, the force on an area at a given depth is*

$$F = pA \tag{5–25}$$

Substituting the expression from Eq. (5–24) into Eq. (5–25), we have

$$F = whA \tag{5–26}$$

as the force on an area A at a depth h.

Let us now assume that the plate shown in Fig. 5–38 is submerged vertically in water. At points on the plate at different depths the pressure will be different. Therefore, in using Eq. (5–26) to find the force on the plate, we must choose an area A such that all of its points are at the same depth. We approximate this by choosing the horizontal strip of length l and width dh. The force on this strip is wh (ldh). Using integration to sum the forces on all such strips from the top of the plate

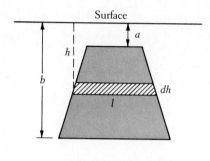

Figure 5–38

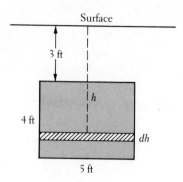

Figure 5–39

to its bottom, we have the total force on the plate as

$$F = w \int_a^b lh\, dh \qquad\qquad (5-27)$$

Here l is the length of the element of area, h is the depth of the element of area, w is the weight per unit volume of the liquid, a is the depth of the top and b is the depth of the bottom of the area on which the force is exerted.

Example A
A vertical rectangular floodgate is 5 ft long and 4 ft deep. Find the force on the gate if its upper edge is 3 ft below the surface (see Fig. 5–39).

Each element of area of the floodgate has a length of 5 ft, which means that $l = 5$ ft. Since the top of the gate is 3 ft below the surface, $a = 3$ ft, and since the gate is 4 ft deep, $b = 7$ ft. Using $w = 62.4$ lb/ft³ we have the force on the gate as

$$F = 62.4 \int_3^7 5h\, dh = 312 \int_3^7 h\, dh = 156h^2 \Big|_3^7 = 156(49 - 9)$$

$$= 6240 \text{ lb}$$

Example B
The vertical end of a tank of water is in the shape of a right triangle as shown in Fig. 5–40. What is the force on the end of the tank?

In setting up the figure it is convenient to use coordinate axes. It is also convenient to have the y-axis directed downward, since we integrate from the top to the bottom of the area. The equation of the line OA is $y = \frac{1}{2}x$. Thus, we see that the length of an element of area of the end of the tank is $4 - x$, the depth of the element of area is y, the top of the tank is 0 and the bottom is $y = 2$ m. Therefore, the force on the end of the tank is ($w = 9800$ N/m³)

$$F = 9800 \int_0^2 (4 - x)(y)(dy) = 9800 \int_0^2 (4 - 2y)(y\, dy)$$

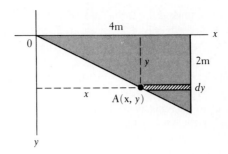

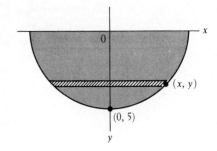

Figure 5–40   Figure 5–41

$$= 19600 \int_0^2 (2y - y^2)dy = 19600\left(y^2 - \frac{1}{3}y^3\right)\Big|_0^2 = 26,100 \text{ N}$$

Example C
The vertical end of a trough is a semicircle with a radius of 5 ft as shown in Fig. 5–41. Find the force on the end if the trough is full of water.

Again we select our axes such that the y-axis is directed downward. The equation of the circle is $x^2 + y^2 = 25$. The length of an element of area is $2x$, the depth of an element is y, the top of the water is at $y = 0$, and the bottom is at $y = 5$ ft. Therefore, the force on the end of the trough is

$$F = 62.4 \int_0^5 (2x)(y)dy = 62.4 \int_0^5 2y\sqrt{25 - y^2}\, dy$$

$$= -\frac{2}{3}(62.4)(25 - y^2)^{3/2}\Big|_0^5 = \frac{2}{3}(62.4)(125) = 5200 \text{ lb}$$

Exercises 5–7

1. Find the force on one end of a rectangular tank 4 ft long and 2 ft deep if the tank is full of water.

2. Find the force on one end of a square tank 2 m on an edge if the tank is full of water.

3. Find the force on a square floodgate 2 m on an edge if the upper edge is 4 m below the surface of the water.

4. Find the force on a rectangular floodgate 15 ft wide and 6 ft deep if the upper edge is 2 ft below the surface of the water.

5. The end of a trough in the shape of an isosceles right triangle has a horizontal upper edge of 8 ft and is 8 ft deep. Find the force on the end if the trough is full of water.

6. A right triangular plate of base 2 m and height 1 m is submerged vertically with its base parallel to the surface and its vertex at the surface. Find the force on one side of the plate.

7. Find the force on the plate in Exercise 6 if the vertex is 3 m below the surface.

8. The end of a tank is in the shape of a trapezoid. The upper edge is 6 ft long, the lower edge is 4 ft long, and the tank is 4 ft deep. Find the force on the end of the tank if it is full of water.

9. A horizontal circular tank has a radius of 4 ft. It is filled to a depth of 4 ft with oil of density 60 lb/ft³. Find the force on one end of the tank.

10. A horizontal cylindrical tank has elliptical ends with the minor axis horizontal. The major axis is 6 ft and the minor axis is 4 ft. Find the force on one end if the tank is half filled with fuel oil of density 50 lb/ft³.

11. A small dam is in the shape of the area bounded by $y = x^2$ and $y = 20$. Find the force on the area below $y = 4$ if the surface of the water is at the top of the dam. Assume all distances are in feet.

12. Find the force on the area bounded by $x = 2y - y^2$ and the y-axis if the upper point of the area is at the surface of the water. Assume all distances are in feet.

5–8 Other Applications

To illustrate the great variety of applications of integration, we shall demonstrate three more types of applied problems in this section. The formulas will be presented without proof, and then examples of their use will be given. The first application is that of finding the length of an arc of a curve.

If $y = f(x)$ is the equation of a given curve, *the length of arc s of the curve from $x = a$ to $x = b$ is given by*

$$s = \int_a^b \sqrt{1 + \left(\frac{dy}{dx}\right)^2}\, dx \qquad (5\text{--}28)$$

Example A

Find the length of arc of the curve $y = \frac{2}{3}x^{3/2}$ from $x = 0$ to $x = 3$ (see Fig. 5–42).

First we identify $a = 0$ and $b = 3$. Then we note that we must find the derivative of y with respect to x. Therefore, $dy/dx = x^{1/2}$. Substituting in Eq. (5–28) we find the length of arc.

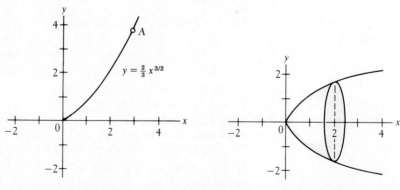

Figure 5–42   Figure 5–43

$$s = \int_0^3 \sqrt{1 + (x^{1/2})^2}\, dx = \int_0^3 \sqrt{1 + x}\, dx = \frac{2}{3}(1 + x)^{3/2}\Big|_0^3$$

$$= \frac{2}{3}[4^{3/2} - 1^{3/2}] = \frac{2}{3}(8 - 1) = \frac{14}{3}$$

This means that the distance between points O and A on the curve, moving along the curve, is $\frac{14}{3}$ units.

The second application of integration in this section is that of finding the area of a surface of revolution. *If the arc of a curve is rotated about the x-axis, it generates a surface of revolution from $x = a$ to $x = b$ for which the area is given by*

$$S = 2\pi \int_a^b y \sqrt{1 + \left(\frac{dy}{dx}\right)^2}\, dx \tag{5-29}$$

Example B

Find the area of the surface generated by rotating about the x-axis the arc of the parabola $y = \sqrt{x}$ from $x = 0$ to $x = 2$ (see Fig. 5-43).

First we note that $a = 0$ and $b = 2$. The derivative of $y = \sqrt{x}$ is $dy/dx = \dfrac{1}{2x^{1/2}}$. Substituting into Eq. (5-29) we have the area of the surface.

$$S = 2\pi \int_0^2 x^{1/2} \sqrt{1 + \left(\frac{1}{2x^{1/2}}\right)^2}\, dx = 2\pi \int_0^2 x^{1/2} \sqrt{1 + \frac{1}{4x}}\, dx$$

$$= \pi \int_0^2 \sqrt{4x + 1}\, dx = \frac{\pi}{4}\left(\frac{2}{3}\right)(4x + 1)^{3/2}\Big|_0^2 = \frac{\pi}{6}(9^{3/2} - 1^{3/2})$$

$$= \frac{\pi}{6}(27 - 1) = \frac{13\pi}{3}$$

This means that the area of the surface is $\frac{13}{3}\pi$ square units.

The third application of integration in this section is that of the **average value** of a function. In general *an average is found by summing up the quantities to be averaged, and then dividing by the total number of them.* Generalizing on this and using integration for the summation, the average value of a function y with respect to x from $x = a$ to $x = b$ is given by

$$y_{av} = \frac{\displaystyle\int_a^b y\, dx}{b - a} \tag{5-30}$$

The following examples illustrate the applications of the average value of a function.

Example C

The velocity of an object falling under the influence of gravity as a function of time is given by $v = 32t$. What is the average velocity (in feet per second) with respect to time for the first three seconds?

In this case, the average velocity with respect to t is

$$v_{av} = \frac{\int_0^3 v\,dt}{3-0} = \frac{\int_0^3 32t\,dt}{3} = \frac{16t^2}{3}\Big|_0^3$$

$$= 48 \text{ ft/s}$$

This result can be interpreted as meaning that an average velocity of 48 ft/s for 3 s would result in the same distance, 144 ft, being traveled as that with the variable velocity. Since $s = \int v\,dt$, the numerator represents the distance traveled.

Example D

The power developed in a certain resistor as a function of current is $P = 6i^2$. What is the average power (in watts) with respect to the current as the current changes from 2 to 5 A?

The average value of P with respect to i is

$$P_{av} = \frac{\int_2^5 P\,di}{5-2} = \frac{6\int_2^5 i^2\,di}{3} = \frac{2}{3}i^3\Big|_2^5 = \frac{2}{3}(125 - 8) = 78 \text{ W}$$

In general, it might be noted that the average value of y with respect to x is that value of y, which when multiplied by the length of the interval for x, gives the same area as that under the curve of y as a function of x.

Exercises 5–8

In Exercises 1 through 4 find the length of arc of each of the given curves.

1. $y = \frac{2}{3}x^{3/2}$ from $x = 3$ to $x = 8$
2. $y = \frac{2}{3}(x - 1)^{3/2}$ from $x = 1$ to $x = 4$
3. $y = \frac{2}{3}(x^2 + 1)^{3/2}$ from $x = 0$ to $x = 3$
4. $y = \frac{1}{6}x^3 + \dfrac{1}{2x}$ from $x = 1$ to $x = 3$

In Exercises 5 through 8 find the area of the area generated by rotating the given arc about the x-axis.

5. $y = x$ from $x = 0$ to $x = 2$
6. $y = \frac{4}{3}x$ from $x = 0$ to $x = 3$
7. $y = 2\sqrt{x}$ from $x = 0$ to $x = 3$
8. $y = x^3$ from $x = 1$ to $x = 3$

In Exercises 9 through 12 find the indicated average values.

9. The electric current as a function of time for a certain circuit is given by $i = 4t - t^2$. Find the average value of the current (in amperes), with respect to time, for the first 4 s.

10. Find the average value of the function $y = \sqrt{x}$, with respect to x, from $x = 1$ to $x = 9$.

11. An object moves along the x-axis according to the relation $x = t^3 - 3t$. Find the average value of the velocity with respect to t, as t changes from 1 to 3.

12. Find the average value of the volume of a sphere with respect to the radius.

5-9 Review Exercises for Chapter 5

1. What is the velocity after 5 s of an object which is dropped and falls due to gravity?

2. Find an expression for the distance traveled by an object for which the velocity is given by $v = 3t/\sqrt{t^2 + 1}$, if $s = 0$ when $t = 0$.

3. A weather balloon is rising at the rate of 20 ft/s when a small metal part drops off. If the balloon is 200 ft high at this instant, when will the part hit the ground?

4. The acceleration of an object rolling down an inclined plane is 12 ft/s². If it starts from rest at the top of the plane, find an expression for the distance the object travels from the top as a function of the time.

5. The current in a certain electric resistor is given by $i = 5t^{2/3}$. Find the charge which passes a given point in the circuit during the third second.

6. The current in a certain electric circuit is given by $i = \sqrt{1 + 4t}$, where i is in amperes and t in seconds. Find the charge which passes a given point during the first two seconds.

7. The voltage across a 55.0 nF capacitor is zero. What is the voltage after 20 ms if a current of 10 mA charges the capacitor?

8. What is the capacitance of a capacitor if a current $i = \sqrt[6]{3t + 1}$ charges it such that the voltage across it is 200 V at $t = 0$ and 280 V when $t = 0.7$ s?

9. The distribution of weight on a cable is not uniform. If the slope of the cable at any point is given by $dy/dx = 20 + \frac{1}{40}x^2$, and if the origin of the coordinate system is at the lowest point, find the equation which gives the curve described by the cable.

10. The slope of a certain cable is given by $dy/dx = -0.1/\sqrt{x}$. If the point $(0, 0)$ is on the cable, find the equation which gives the curve described by the cable.

11. Find the area between $y = \sqrt{1 - x}$ and the coordinate axes.

12. Find the area bounded by $y = 3x^2 - x^3$ and the x-axis.

13. Find the area bounded by $y^2 = 2x$ and $y = x - 4$.

14. Find the area bounded by $y = 1/(2x + 1)^2$, $y = 0$, $x = 1$, and $x = 2$.

15. Find the volume generated by rotating the area bounded by $y = 3 + x^2$ and the line $y = 4$ about the x-axis.

16. Find the volume generated by rotating the area of Exercise 12 about the x-axis.

17. Find the volume generated by rotating the area of Exercise 12 about the y-axis.

18. Find the volume generated by rotating the area bounded by $y = x$ and $y = 3x - x^2$ about the y-axis.

19. Find the centroid of the area bounded by $y^2 = x^3$ and $y = 2x$.

20. Find the centroid of the area bounded by $y = 2x - 4$, $x = 1$, and $y = 0$.

21. Find the centroid of the volume generated by rotating the area bounded by $y = \sqrt{x}$, $x = 1$, $x = 4$, and $y = 0$ about the x-axis.

22. Find the centroid of the volume generated by rotating the area bounded by $yx^4 = 1$, $y = 1$, and $y = 4$ about the y-axis.

23. Find the moment of inertia of the area bounded by $y = 3x - x^2$ and $y = x$ with respect to the y-axis.

24. Find the radius of gyration of the first-quadrant area which is bounded by $y = 8 - x^3$ with respect to the y-axis.

25. Find the moment of inertia with respect to its axis of the solid generated by rotating the area bounded by $y = x^{1/2}$, $y = 0$, and $x = 8$ about the x-axis.

26. Find the radius of gyration with respect to its axis of the solid formed by rotating the area bounded by $xy = 1$, $x = 1$, $x = 3$, and $y = \frac{1}{3}$ about the x-axis.

27. A certain spring is 16 in. long when 12 lb are hung from it. It measures 12 in. when 4 lb are hung from it. How much work is done when the spring is stretched from its natural length by hanging 6 lb on it?

28. If the pressure of a gas is 400 kPa when the volume is 243 cm³, how much work is done in compressing the gas to 32 cm³, if it undergoes an adiabatic change such that $pV^{1.4} = c$?

29. Find the work done in pumping water out at the top of a cylindrical tank 4 m in diameter and 3 m high, if the tank is initially full. ($w = 9800$ N/m³)

30. A pail and its contents weigh 80 lb. The pail is attached to the end of a 100 ft rope which weighs 10 lb and is hanging vertically. How much work is done in winding up the rope with the pail attached?

31. A square plate is submerged vertically in a tank of water with its upper edge 2 ft below the surface. If the edge of the plate is 4 ft, find the force on one side of the plate.

32. A section of a dam is in the shape of a right triangle. The base of the triangle is 6 ft and is in the surface of the water. If the triangular section goes to a depth of 4 ft, find the force on it.

33. Find the length of arc of the curve $y = \frac{2}{3}(x^2 - 1)^{3/2}$ from $x = 1$ to $x = 3$.

34. Find the area of the surface generated by rotating $y = \frac{3}{4}x$ from $x = 0$ to $x = 2$ about the x-axis.

35. The electric resistance of a wire is inversely proportional to the square of its radius. If a certain wire has a resistance of 0.3 Ω when its radius is 2.0 mm, find the average value of the resistance with respect to the radius if the radius changes from 2.0 mm to 2.1 mm.

36. The velocity of an object as a function of the distance fallen after being dropped is $v = 8\sqrt{s}$. Find the average value of v with respect to s during the third second, if $s = 16t^2$, and s is measured in feet.

37. The production rate (in pounds per hour) of a certain chemical plant is given by the following table. Using the trapezoidal rule, determine the approximate daily production.

Time	12 M	4 AM	8 AM	12 N	4 PM	8 PM	12 M
Production rate	2000	2000	7000	12,000	12,000	8000	2000

38. The current as a function of time was measured in a particular circuit, with the results tabulated below. What charge (in coulombs) was transmitted across a given point in the first 6 s? Use Simpson's rule.

i (amperes)	0.0	1.5	4.0	7.2	7.5	4.3	1.6
t (seconds)	0.0	1.0	2.0	3.0	4.0	5.0	6.0

6

Derivatives of the Trigonometric and Inverse Trigonometric Functions

6–1 The Trigonometric Functions

Many important applications in technology are periodic in nature. That is, these phenomena have properties that repeat after definite periods of time. Applications of this type arise in alternating-current theory and harmonic motion analysis, as well as in many other fields. Such phenomena may be represented by the use of the trigonometric functions, especially the sine and cosine functions. It is assumed that the reader has a background which includes basic trigonometry. However, it is the purpose in this section and the next to present a summary of the trigonometric functions and their properties for purposes of review and reference. This material is essential to the development of later topics in this chapter and later chapters.

An angle in **standard position** *is one with its vertex at the origin and its initial side the positive x-axis. If the angle is measured counterclockwise it is positive, and if it is measured clockwise it is negative* (see Fig. 6–1).

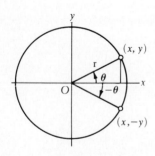

Figure 6–1

The six **trigonometric functions** *defined in terms of the coordinates (x, y) of a point on its terminal side, at a distance r (considered positive) from the origin are as follows:*

214

$$\sin \theta = \frac{y}{r} \qquad \cos \theta = \frac{x}{r} \qquad \tan \theta = \frac{y}{x}$$

$$\cot \theta = \frac{x}{y} \qquad \sec \theta = \frac{r}{x} \qquad \csc \theta = \frac{r}{y}$$

(6–1)

There are two widely used measurements of angles. These are the **degree** and the **radian**. *The degree, defined as* $\frac{1}{360}$ *of a complete revolution, is used primarily in triangle solution. The degree is divided into 60 equal parts called* **minutes,** *and each minute is divided into 60 equal parts called* **seconds.** Degrees and decimal parts are also widely used.

It is often useful to be able to change an angle expressed in degrees and minutes to an angle expressed in degrees and decimal parts of a degree, and conversely. One reason for this is that it has been very common to use degrees and minutes in tables, and most electronic calculators use degrees and decimals parts. The method of making such changes is illustrated in the following example.

Example A
To change the angle $43°24'$ to decimal form, we use the fact that $1' = \left(\frac{1}{60}\right)°$. Therefore, $24' = \left(\frac{24}{60}\right)° = 0.4°$. This means that

$$43°24' = 43.4°$$

Also, $17°53'$ is changed to decimal form as follows.

$$53' = \left(\frac{53}{60}\right)° = 0.88°$$

Thus, $17°53' = 17.88°$ (to the nearest hundredth), or $17°53' = 17.9°$ (to the nearest tenth).

To change $154.36°$ to an angle expressed to the nearest minute, we use the fact that $1° = 60'$. Therefore,

$$0.36° = 0.36(60') = 21.6'$$

To the nearest minute, we have $154.36° = 154°22'$.

The radian is a more convenient measure of an angle for theoretical discussions. Since it is our purpose to develop mathematics which has wide application in technology, we shall restrict our attention largely to radian measure. A **radian** *is the measure of an angle, with vertex at the center of a circle, whose intercepted arc is equal in length to the radius of the circle.*

Since the circumference of a circle is given by $c = 2\pi r$, the ratio of the circumference to the radius is 2π. This means that the radius may be laid off 2π times along the circumference, or that an angle of $360°$ equals an angle of 2π radians. Therefore,

$$\pi \text{ rad} = 180°$$

(6–2)

Example B

$$1° = \frac{\pi}{180}\text{rad} = 0.01745 \text{ rad}, \qquad 1 \text{ rad} = \frac{180°}{\pi} = 57.3°,$$

$$\frac{\pi}{4}\text{rad} = \frac{180°}{4} = 45°, \qquad \frac{\pi}{3}\text{rad} = \frac{180°}{3} = 60°,$$

$$\frac{\pi}{10}\text{rad} = \frac{180°}{10} = 18°$$

We see from Eq. (6–2) that (1) *to convert an angle measured in degrees to the same angle measured in radians, we multiply the number of degrees by $\pi/180$*, and (2) *to convert an angle measured in radians to the same angle measured in degrees, we multiply the number of radians by $180/\pi$.*

Example C

$$18.0° = \left(\frac{\pi}{180}\right)(18.0) = \frac{\pi}{10.0} = \frac{3.14}{10.0} = 0.314 \text{ rad}$$

$$120° = \left(\frac{\pi}{180}\right)(120) = \frac{6.28}{3.00} = 2.09 \text{ rad}$$

$$0.400 \text{ rad} = \left(\frac{180°}{\pi}\right)(0.400) = \frac{72.0°}{3.14} = 22.9°$$

$$2.00 \text{ rad} = \left(\frac{180°}{\pi}\right)(2.00) = \frac{360°}{3.14} = 114.6°$$

Due to the nature of the definition of the radian, it is very common to express radians in terms of π. Expressing angles in terms of π is illustrated in the following example.

Example D

$$30° = \left(\frac{\pi}{180}\right)(30) = \frac{\pi}{6}\text{rad}$$

$$45° = \left(\frac{\pi}{180}\right)(45) = \frac{\pi}{4}\text{rad}$$

$$\frac{\pi}{2}\text{rad} = \left(\frac{180°}{\pi}\right)\left(\frac{\pi}{2}\right) = 90°$$

$$\frac{3\pi}{4}\text{rad} = \left(\frac{180°}{\pi}\right)\left(\frac{3\pi}{4}\right) = 135°$$

We wish now to make a very important point. Since π is a special way of writing the number (slightly greater than 3) that is the ratio of the circumference of a circle to its diameter, it is the ratio of one distance to another. Thus radians really have no units and *radian measure amounts to measuring angles in terms of numbers.* It is this property of radians that makes them useful in many situations. Therefore, when radians are being used, it is customary that no units are indicated for the angle. *When no units are indicated, the radian is understood to be the unit of angle measurement.*

Example E

$$60.0° = \left(\frac{\pi}{180}\right)(60.0) = \frac{\pi}{3.00} = 1.05$$

$$2.50 = \left(\frac{180°}{\pi}\right)(2.50) = \frac{450°}{3.14} = 143°$$

Since no units are indicated for 1.05 and 2.50 in this example, they are known to be radian measure.

Tables of values of the trigonometric functions are found in the Appendix. In Table 6–1 we present a short list of commonly used angles and the exact values of trigonometric functions of these angles.

Table 6–1

Degrees	0°	30°	45°	60°	90°	180°	270°	360°
Radians	0	$\frac{\pi}{6}$	$\frac{\pi}{4}$	$\frac{\pi}{3}$	$\frac{\pi}{2}$	π	$\frac{3\pi}{2}$	2π
Sine	0	$\frac{1}{2}$	$\frac{\sqrt{2}}{2}$	$\frac{\sqrt{3}}{2}$	1	0	-1	0
Cosine	1	$\frac{\sqrt{3}}{2}$	$\frac{\sqrt{2}}{2}$	$\frac{1}{2}$	0	-1	0	1
Tangent	0	$\frac{\sqrt{3}}{3}$	1	$\sqrt{3}$	Undef.	0	Undef.	0

Since the tables include the values of functions only for angles between 0° and 90°, it is necessary to be able to determine the values of the functions of angles greater than 90°.

An angle in standard position has the same terminal side as some positive angle less than 360°. Since the trigonometric functions of angles with the same terminal side are equal, we need consider only the problem of finding values of the trigonometric functions of positive angles less than 360°.

We note that the *signs* of the functions differ, depending on the quadrant in which the terminal side of the angle lies. Noting the definitions, Eqs. (6–1), we see that *all the functions are positive for first quadrant angles, sin θ and csc θ are positive for second quadrant angles, tan θ and cot θ are positive for third quadrant angles, cos θ and sec θ are positive for fourth quadrant angles, and all other functions are negative.* An analysis of the signs of *x* and *y* in the various quadrants will verify this.

From Fig. 6–2 we see that the point (x_1, y_1) is on the terminal side of θ_1 and the point $(-x_1, y_1)$ is on the terminal side of θ_2. This means that if the terminal side of the angle θ_2 lies in the second quadrant, the functions of θ_2 are numerically equal to the corresponding functions of the angle θ_1, in the first quadrant. Thus, r is the same for the two angles, and $\alpha = \theta_1$, where α is called the **reference angle.** However, $\alpha = 180° - \theta_2$. Letting F represent any of the trigonometric functions we conclude that

$$F(\theta_2) = \pm F(180° - \theta_2) = \pm F(\pi - \theta_2) \qquad (6\text{–}3)$$

The sign to be used depends on whether the function is positive or negative in the second quadrant.

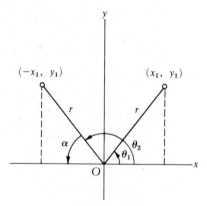

Figure 6–2

In the same way, we may derive the formulas for finding the trigonometric functions of a third- or fourth-quadrant angle (see Figs. 6–3 and 6–4):

$$F(\theta_3) = \pm F(\theta_3 - 180°) = \pm F(\theta_3 - \pi)$$
$$F(\theta_4) = \pm F(360° - \theta_4) = \pm F(2\pi - \theta_4) \qquad (6\text{–}4)$$

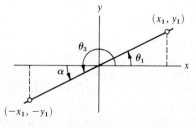

Figure 6–3

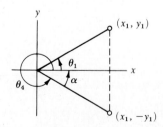

Figure 6–4

Example F

Let $\theta_2 = 135°$

$$\sin 135° = \sin(180° - 135°) = \sin 45° = \frac{\sqrt{2}}{2},$$

$$\tan 135° = -\tan(180° - 135°) = -\tan 45° = -1$$

Let $\theta_3 = 210°$

$$\tan 210° = \tan(210° - 180°) = \tan 30° = \frac{\sqrt{3}}{3},$$

$$\cos 210° = -\cos(210° - 180°) = -\cos 30° = -\frac{\sqrt{3}}{2}$$

Let $\theta_4 = 330°$

$$\sin 330° = -\sin(360° - 330°) = -\sin 30° = -\frac{1}{2},$$

$$\cos 330° = \cos(360° - 330°) = \cos 30° = \frac{\sqrt{3}}{2}$$

One of the clearest ways to demonstrate the variation in the trigonometric functions is by means of their graphs. Since the graphs of the sine and cosine functions are of primary importance, we shall restrict our attention here to them.

In plotting the trigonometric functions, it is normal to express the angle in radians. In this way both x and y are expressed as numbers.

In the equations $y = a \sin(bx + c)$ and $y = a \cos(bx + c)$ we recall that a *represents the* **amplitude** *of the curve. This is the maximum number of units the curve may be from the* x-axis. The **period** *of each of these curves is* $2\pi/b$. *For any function F, we say that it has a period P if, for all x, $F(x) = F(x + P)$. The effect of the c in the equation is to shift the curve of $y = a \sin bx$ or $y = a \cos bx$ to the left if $c > 0$, and to shift it to the right if $c < 0$. The amount of shift is given by c/b. The quantity c/b is called the* **displacement.**

Thus, by knowing the basic shape of the sine and cosine curves, which are shown in Figs. 6–5 and 6–6, and the information obtained from the equation, we may sketch the graphs of sine and cosine functions. The following examples illustrate the method.

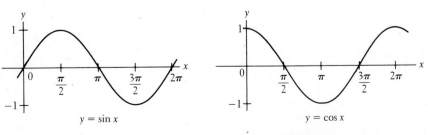

$y = \sin x$

Figure 6–5

$y = \cos x$

Figure 6–6

Example G
Sketch the graph of $y = 3 \sin(2x + \frac{1}{4}\pi)$.

First we note that $a = 3$, which means that the amplitude of the curve is 3. Next we see that $b = 2$. This means that the period of the curve is $2\pi/2 = \pi$. Therefore points on the curve will "repeat" every π units along the x-axis. Finally $c = \frac{1}{4}\pi$, which means the displacement is $(\frac{1}{4}\pi)/2 = \frac{1}{8}\pi$. This means the curve is displaced $\frac{1}{8}\pi$ units to the left of the curve of $y = 3 \sin 2x$. Therefore, "starting" the curve at $x = -\frac{1}{8}\pi$, we know it starts repeating at $x = \frac{7}{8}\pi$ (π units to the right of $x = -\frac{1}{8}\pi$), again at $15\pi/8$, etc. With this information we can construct the following table, from which the curve can quickly be sketched (see Fig. 6–7).

x	$-\dfrac{\pi}{8}$	$\dfrac{\pi}{8}$	$\dfrac{3\pi}{8}$	$\dfrac{5\pi}{8}$	$\dfrac{7\pi}{8}$	$\dfrac{9\pi}{8}$
y	0	3	0	-3	0	3

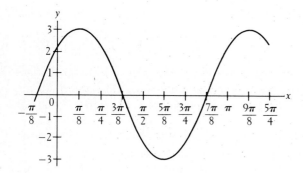

Figure 6–7

Figure 6–8

Example H
Sketch the graph of the function $y = 2 \cos(\frac{1}{2}x - \frac{1}{6}\pi)$; amplitude $= 2$, period $= 2\pi/\frac{1}{2} = 4\pi$, displacement $= \frac{1}{6}\pi/\frac{1}{2} = \frac{1}{3}\pi$ (to the right).

We construct the table of values:

x	$\dfrac{\pi}{3}$	$\dfrac{4\pi}{3}$	$\dfrac{7\pi}{3}$	$\dfrac{10\pi}{3}$	$\dfrac{13\pi}{3}$
y	2	0	-2	0	2

And then we sketch the graph in Fig. 6–8. We note that when the coefficient of x is less than 1, the period is extended.

Exercises 6–1

In Exercises 1 and 2 change the given angles to equal angles expressed in decimal form. In Exercises 3 and 4 change the given angles to equal angles expressed to the nearest minute.

1. 15°12′ 2. 301°16′ 3. 315.8° 4. 24.92°

In Exercises 5 through 8 express the given angle measurements in terms of π.

5. 15°, 150°   **6.** 12°, 225°   **7.** 75°, 330°   **8.** 36°, 315°

In Exercises 9 through 12 the given numbers express angle measurement. Express the measure of each angle in terms of degrees.

9. $\dfrac{2\pi}{5}, \dfrac{3\pi}{2}$ **10.** $\dfrac{3\pi}{10}, \dfrac{5\pi}{6}$ **11.** $\dfrac{\pi}{18}, \dfrac{7\pi}{4}$ **12.** $\dfrac{7\pi}{15}, \dfrac{4\pi}{3}$

In Exercises 13 and 14 express the given angles in radian measure. In Exercises 15 and 16 express the given numbers in terms of angle measure in degrees to the nearest 0.1°.

13. 23.0°   **14.** 178.5°   **15.** 0.750   **16.** 16.4

In Exercises 17 through 24 find the indicated trigonometric functions.

17. $\sin\dfrac{\pi}{4}$ **18.** $\cos\dfrac{\pi}{6}$ **19.** $\tan\dfrac{5\pi}{12}$ **20.** $\sin\dfrac{7\pi}{18}$

21. $\cot\dfrac{5\pi}{6}$ **22.** $\tan\dfrac{4\pi}{3}$ **23.** $\cos 4.59$ **24.** $\cot 3.27$

In Exercises 25 through 28 find θ for $0 \le \theta < 2\pi$.

25. $\sin\theta = 0.3420$ **26.** $\cos\theta = 0.4384$
27. $\tan\theta = -1.192$ **28.** $\cot\theta = -0.5095$

In Exercises 29 through 36 sketch the graph of each of the given functions.

29. $y = 3\sin x$ **30.** $y = 2\cos 2x$ **31.** $y = -4\sin 3x$

32. $y = 6\cos \pi x$ **33.** $y = 2\sin\left(3x - \dfrac{\pi}{2}\right)$ **34.** $y = 4\cos\left(\dfrac{x}{2} + \dfrac{\pi}{8}\right)$

35. $y = \cos\left(\pi x - \dfrac{\pi}{2}\right)$ **36.** $y = \sin(\pi x + 1)$

In Exercises 37 through 42 sketch the indicated graphs.

37. The electric current in a certain 60 Hz alternating-current circuit is given by $i = 10\sin 120\pi t$, where i is the current (in amperes) and t is the time (in seconds). Sketch the graph of i vs. t for $0 \le t \le 0.1$ s.

38. A generator produces a voltage given by $v = 200\cos 50\pi t$, where t is the time (in seconds). Sketch the graph of v vs. t for $0 \le t \le 0.1$ s.

39. A wave traveling in a string may be represented by the equation

$$y = A\sin\left(\dfrac{t}{T} - \dfrac{x}{\lambda}\right)$$

Here A is the amplitude, t is the time the wave has traveled, x is the distance from the origin, T is the time required for the wave to travel one wavelength λ (the Greek lambda). Sketch three cycles of the wave for which $A = 2.00$ cm, $T = 0.100$ s, $\lambda = 20.0$ cm, and $x = 5.00$ cm.

40. The cross-section of a particular water wave is

$$y = 0.5 \sin\left(\frac{\pi}{2}x + \frac{\pi}{4}\right)$$

where x and y are measured in feet. Sketch two cycles of y vs. x.

41. If the upper end of a spring is not fixed and is being moved with a sinusoidal motion, the motion of the bob at the end of the spring is affected. Plot the curve of the motion of the upper end of a spring which is being moved by an external force according to the equation $y = 4 \sin 2t - 2 \cos 2t$. This may be done graphically by plotting $y_1 = 4 \sin 2t$ and $y_2 = 2 \cos 2t$ and then finding $y = y_1 - y_2$. This method is referred to as *addition of ordinates.*

42. The current in a certain electric circuit is given by $i = 4 \sin 60\pi t + 2 \cos 120\pi t$. Sketch the graph of i vs. t.

6–2 Basic Trigonometric Relations

Referring to the definitions of the trigonometric functions, Eqs. (6–1), we see that there are a number of relations which exist among the functions. Certain other important relationships may also be established. We shall find that these have very definite application, since certain problems rely on a change of form for solution. The ability to effectively use these basic relations depends to a large extent on being familiar with them, and recognizing them in different forms. Therefore, in this section we shall review the important trigonometric relations.

Directly from the definitions we observe that

$$\sin \theta = \frac{1}{\csc \theta} \qquad \cos \theta = \frac{1}{\sec \theta} \qquad \tan \theta = \frac{1}{\cot \theta} \qquad (6\text{–}5)$$

Also, since

$$\tan \theta = \frac{y}{x} = \frac{y/r}{x/r}$$

we have

$$\tan \theta = \frac{\sin \theta}{\cos \theta} \qquad \cot \theta = \frac{\cos \theta}{\sin \theta} \qquad (6\text{–}6)$$

Certain important relations can be found by use of the definitions and the Pythagorean relation $x^2 + y^2 = r^2$. By dividing each term of the Pythagorean relation by r^2, we have

$$\sin^2\theta + \cos^2\theta = 1 \qquad\qquad (6\text{--}7)$$

By dividing each term of the Pythagorean relation by x^2, we obtain

$$1 + \tan^2\theta = \sec^2\theta \qquad\qquad (6\text{--}8)$$

By dividing each term of the Pythagorean relation by y^2, we obtain

$$\cot^2\theta + 1 = \csc^2\theta \qquad\qquad (6\text{--}9)$$

Many other relationships which exist among the trigonometric functions may be proven valid by the basic identities (6–5) to (6–9). The following example illustrates the method of proving such trigonometric identities.

Example A

Prove the identity $\sec^2 x + \csc^2 x = \sec^2 x \, \csc^2 x$.

Here we note the presence of $\sec^2 x$ and $\csc^2 x$ on each side. This suggests the possible use of the square relationships. By replacing the $\sec^2 x$ on the right-hand side by $1 + \tan^2 x$, we can create $\csc^2 x$ plus another term. The left-hand side is the $\csc^2 x$ plus another term, so this procedure should help. Thus

$$\sec^2 x + \csc^2 x = \sec^2 x \, \csc^2 x$$
$$= (1 + \tan^2 x)(\csc^2 x)$$
$$= \csc^2 x + \tan^2 x \, \csc^2 x$$

Now we note that $\tan x = \sin x / \cos x$ and $\csc x = 1/\sin x$. Thus

$$\sec^2 x + \csc^2 x = \csc^2 x + \left(\frac{\sin^2 x}{\cos^2 x}\right)\left(\frac{1}{\sin^2 x}\right)$$

$$= \csc^2 x + \frac{1}{\cos^2 x}$$

But $\sec x = 1/\cos x$, and therefore $\sec^2 x + \csc^2 x = \csc^2 x + \sec^2 x$, which proves the identity. We could have used many other variations of this procedure, and they would have been perfectly valid.

Referring to Fig. 6–1 and the definitions of the functions we see, for example, that $\sin\theta = y/r$ and $\sin(-\theta) = -y/r$. Thus, in this way we arrive at the following important relations concerning negative angles:

$$\sin(-\theta) = -\sin\theta \qquad \cos(-\theta) = \cos\theta \qquad \tan(-\theta) = -\tan\theta \qquad (6\text{--}10)$$

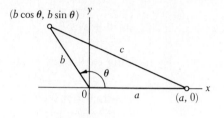

Figure 6–9

By use of the distance formula, Eq. (1–1), and Fig. 6–9, we may establish the **law of cosines**. Thus,

$$c^2 = (b \cos \theta - a)^2 + (b \sin \theta)^2$$
$$= b^2 \cos^2 \theta - 2ab \cos \theta + a^2 + b^2 \sin^2 \theta$$
$$= b^2 (\cos^2 \theta + \sin^2 \theta) + a^2 - 2ab \cos \theta$$

$$c^2 = a^2 + b^2 - 2ab \cos \theta \qquad\qquad (6\text{–}11)$$

We shall now develop a number of important equations. We will be using the distance formula, the law of cosines and some of the other relations already established. Also, some of those we derive will be used in turn to derive others. Since there are numerous uses of the various equations of the section in the following material, the reader will gain facility in the use of the relationships by following the development.

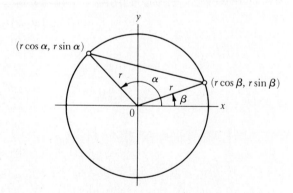

Figure 6–10

From Fig. 6–10 we have

$$(r \cos \alpha - r \cos \beta)^2 + (r \sin \alpha - r \sin \beta)^2$$
$$= r^2 + r^2 - 2(r)(r) \cos (\alpha - \beta)$$
$$r^2 \cos^2 \alpha - 2r^2 \cos \alpha \cos \beta + r^2 \cos^2 \beta$$
$$+ r^2 \sin^2 \alpha - 2r^2 \sin \alpha \sin \beta + r^2 \sin^2 \beta = 2r^2 - 2r^2 \cos(\alpha - \beta)$$

Dividing through by r^2 and collecting terms we have

$$\cos^2\alpha + \sin^2\alpha - 2(\cos\alpha\cos\beta + \sin\alpha\sin\beta)$$
$$+ \cos^2\beta + \sin^2\beta = 2 - 2\cos(\alpha - \beta)$$

Rearranging terms after using Eq. (6–7) we obtain

$$\cos(\alpha - \beta) = \cos\alpha\cos\beta + \sin\alpha\sin\beta \qquad (6\text{–}12)$$

Equation (6–12) is one of the important **addition formulas**, the remainder of which we shall now develop.

Writing $\cos(\alpha + \beta) = \cos[\alpha - (-\beta)]$, and using Eqs. (6–10) and (6–12), we have

$$\cos(\alpha + \beta) = \cos\alpha\cos\beta - \sin\alpha\sin\beta \qquad (6\text{–}13)$$

In Eq. (6–12), if we let $\alpha = 90°$, the result is

$$\cos(90° - \beta) = \cos 90° \cos\beta + \sin 90° \sin\beta$$

Since $\sin 90° = 1$, $\cos 90° = 0$, we obtain

$$\cos(90° - \beta) = \sin\beta \qquad (6\text{–}14)$$

Since β is a general representation of an angle, and $\cos\beta$ is a function of β, we may substitute other expressions for angles for β. Therefore, replacing β by $90° - \beta$ in Eq. (6–14) we have

$$\cos[90° - (90° - \beta)] = \sin(90° - \beta)$$

or

$$\sin(90° - \beta) = \cos\beta \qquad (6\text{–}15)$$

Now, replacing β by $\alpha + \beta$ in Eq. (6–14) we obtain

$$\sin(\alpha + \beta) = \cos[(90° - \alpha) - \beta]$$
$$= \cos(90° - \alpha)\cos\beta + \sin(90° - \alpha)\sin\beta$$

and by using Eqs. (6–14) and (6–15) we arrive at the equation

$$\sin(\alpha + \beta) = \sin\alpha\cos\beta + \cos\alpha\sin\beta \qquad (6\text{–}16)$$

If we replace β by $-\beta$ in Eq. (6–16) we have

$$\sin(\alpha - \beta) = \sin\alpha\cos\beta - \cos\alpha\sin\beta \qquad (6\text{–}17)$$

Example B

Show that

$$\sin\left(\frac{\pi}{4} + x\right)\cos\left(\frac{\pi}{4} + x\right) = \frac{1}{2}(\cos^2 x - \sin^2 x)$$

$$\sin\left(\frac{\pi}{4} + x\right)\cos\left(\frac{\pi}{4} + x\right)$$

$$= \left(\sin\frac{\pi}{4}\cos x + \cos\frac{\pi}{4}\sin x\right)\left(\cos\frac{\pi}{4}\cos x - \sin\frac{\pi}{4}\sin x\right)$$

$$= \sin\frac{\pi}{4}\cos\frac{\pi}{4}\cos^2 x - \sin^2\frac{\pi}{4}\sin x \cos x$$

$$+ \cos^2\frac{\pi}{4}\sin x \cos x - \sin^2 x \sin\frac{\pi}{4}\cos\frac{\pi}{4}$$

$$= \frac{\sqrt{2}}{2}\frac{\sqrt{2}}{2}\cos^2 x - \left(\frac{\sqrt{2}}{2}\right)^2 \sin x \cos x$$

$$+ \left(\frac{\sqrt{2}}{2}\right)^2 \sin x \cos x - \frac{\sqrt{2}}{2}\frac{\sqrt{2}}{2}\sin^2 x$$

$$= \frac{1}{2}\cos^2 x - \frac{1}{2}\sin^2 x = \frac{1}{2}(\cos^2 x - \sin^2 x)$$

Letting $A = \alpha + \beta$ and $B = \alpha - \beta$ leads to $\alpha = \frac{1}{2}(A + B)$ and $\beta = \frac{1}{2}(A - B)$. Using these in Eqs. (6–16) and (6–17), and subtracting, we obtain

$$\sin A - \sin B = 2\cos\frac{A + B}{2}\sin\frac{A - B}{2} \qquad (6\text{–}18)$$

We shall find this equation of use in the following section.

If $\alpha = \beta$ in Eq. (6–16) we have

$$\sin 2\alpha = 2\sin\alpha\cos\alpha \qquad (6\text{–}19)$$

Also, if $\alpha = \beta$ in Eq. (6–13), we obtain

$$\cos 2\alpha = \cos^2\alpha - \sin^2\alpha \qquad (6\text{–}20)$$

Using Eq. (6–7), we may rewrite Eq. (6–20) in the form

$$\cos 2\alpha = 2\cos^2\alpha - 1 = 1 - 2\sin^2\alpha \qquad (6\text{–}21)$$

Equations (6–19), (6–20), and (6–21) are the **double-angle formulas**.

By replacing 2α by α in Eq. (6–21) we obtain

$$\sin\frac{\alpha}{2} = \pm\sqrt{\frac{1 - \cos\alpha}{2}} \qquad (6\text{–}22)$$

and

$$\cos\frac{\alpha}{2} = \pm\sqrt{\frac{1 + \cos\alpha}{2}} \qquad (6\text{–}23)$$

Equations (6–22) and (6–23) are the **half-angle formulas.**

Example C

Given that $\cos\alpha = \frac{3}{5}$, where $3\pi/2 < \alpha < 2\pi$, find $\sin 2\alpha$ and $\cos\alpha/2$.

Knowing that $\cos\alpha = \frac{3}{5}$ for an angle in the fourth quadrant, we then determine that $\sin\alpha = -\frac{4}{5}$. Thus

$$\sin 2\alpha = 2\left(-\frac{4}{5}\right)\left(\frac{3}{5}\right) = -\frac{24}{25}$$

Also, since α is between $3\pi/2$ and 2π, $\alpha/2$ must be between $3\pi/4$ and π, which means that $\alpha/2$ is in the second quadrant and therefore $\cos\alpha/2$ is negative. Thus,

$$\cos\frac{\alpha}{2} = -\sqrt{\frac{1 + 3/5}{2}} = -0.8944$$

Two other useful relations are found from the definition of the radian. From geometry we know that *the length of an arc on a circle is proportional to the central angle,* and that the length of the arc of a complete circle is the circumference. Letting s represent the length of arc, we may state that $s = 2\pi r$ for a complete circle, or

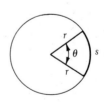

$$s = \theta r \qquad (6\text{–}24)$$

for a circular arc with central angle θ, measured in radians (see Fig. 6–11).

Figure 6–11

Another geometric fact is that *areas of sectors of circles are proportional to their central angles.* The area of a complete circle is $A = \pi r^2$. This can be written as $A = \frac{1}{2}(2\pi)r^2$. We now note that the angle for a complete circle is 2π, and therefore the area of a sector of a circle in terms of the radius and the central angle, in radians, is

$$A = \frac{1}{2}\theta r^2 \qquad (6\text{–}25)$$

Figure 6–12   (see Fig. 6–12).

Example D

Find the length of a circular arc if the central angle is 18° and the radius of the circle is 5 in. Also find the area of the corresponding circular sector.

First we must express θ in radians:

$$18° = \frac{\pi}{10}$$

Now, solving for s, we have

$$s = \frac{\pi}{10}(5) = \frac{\pi}{2} = 1.57 \text{ in.}$$

The area is

$$A = \frac{1}{2}\left(\frac{\pi}{10}\right)(25) = \frac{5\pi}{4} = 3.93 \text{ in.}^2$$

A great many formulas have been developed in this section. As stated previously, however, these will be essential to the development of the material to come. The development of these formulas has been included, primarily to help the reader review the use of the equations.

Exercises 6–2

In Exercises 1 through 20 prove the given trigonometric identities.

1. $\dfrac{\cot x}{\cos x} = \csc x$

2. $\dfrac{\sin x}{\tan x} = \cos x$

3. $\tan y \csc y = \sec y$

4. $\sin x \cot x = \cos x$

5. $\cos^2 x - \sin^2 x = 1 - 2\sin^2 x$

6. $\cos^2\theta(1 + \tan^2\theta) = 1$

7. $\sin x \tan x + \cos x = \sec x$

8. $\tan x + \cot x = \sec x \csc x$

9. $\tan x(\tan x + \cot x) = \sec^2 x$

10. $\tan^2 y \sec^2 y - \tan^4 y = \tan^2 y$

11. $\sin(x + y)\sin(x - y) = \sin^2 x - \sin^2 y$

12. $\cos(x + y)\cos(x - y) = \cos^2 x - \sin^2 y$

13. $\cos(x - y) + \sin(x + y) = (\cos x + \sin x)(\cos y + \sin y)$

14. $\cos^4 x - \sin^4 x = \cos 2x$

15. $2\sin x + \sin 2x = \dfrac{2\sin^3 x}{1 - \cos x}$

16. $2\csc 2x \tan x = \sec^2 x$

17. $\dfrac{\sin 3x}{\sin x} + \dfrac{\cos 3x}{\cos x} = 4\cos 2x$

18. $\dfrac{\sin 3x}{\sin x} - \dfrac{\cos 3x}{\cos x} = 2$

19. $\dfrac{1 - \cos\alpha}{2\sin(\alpha/2)} = \sin\dfrac{\alpha}{2}$

20. $\tan\dfrac{\alpha}{2} = \pm\sqrt{\dfrac{1 - \cos\alpha}{1 + \cos\alpha}}$

In Exercises 21 through 28 solve the given problems.

21. The displacements y_1 and y_2 of two waves traveling through the same medium are given by

$$y_1 = A\sin 2\pi\left(\frac{t}{T} - \frac{x}{\lambda}\right) \quad \text{and} \quad y_2 = A\sin 2\pi\left(\frac{t}{T} + \frac{x}{\lambda}\right)$$

Find an expression for the displacement $y_1 + y_2$ of the combination of the waves.

22. In the analysis of the angles of incidence i and reflection r of a light ray subject to certain conditions, the following expression is found:

$$E_2\left(\frac{\tan r}{\tan i} + 1\right) = E_1\left(\frac{\tan r}{\tan i} - 1\right)$$

Show that an equivalent expression is

$$E_2 = E_1\frac{\sin(r - i)}{\sin(r + i)}$$

23. The equation for the displacement of a certain object at the end of a spring is $y = A \sin 2t + B \cos 2t$. Show that this equation may be written as

$$y = C \sin(2t + \alpha)$$

where $C = \sqrt{A^2 + B^2}$ and $\tan \alpha = B/A$. [*Hint:* Let $A/C = \cos \alpha$ and $B/C = \sin \alpha$.]

24. In developing an expression for the power in an alternating current circuit, the expression $\sin \omega t \sin(\omega t + \phi)$ is found. Show that this expression can be written as $\frac{1}{2}[\cos \phi - \cos(2\omega t + \phi)]$.

25. A pendulum 3.00 ft long oscillates through an angle of 5.0°. Find the distance through which the end of the pendulum swings in going from one extreme position to the other.

26. An ammeter needle is deflected 52.0° by a current of 0.200 A. The needle is 3.00 cm long and a circular scale is used. How long is the scale for a maximum current of 1.00 A?

27. A horizontal water pipe has a radius of 6.00 cm. If the depth of water in the pipe is 3.00 cm, what percent of the volume of the pipe is filled?

28. A conical tent is made from a circular piece of canvas 15.0 ft in diameter, with a sector of central angle 120° removed. What is the surface area of the tent?

6–3 Derivatives of the Sine and Cosine Functions

In Chapter 2 we introduced the operation of differentiation. The functions we used, however, were algebraic functions in all cases. Now that we have developed the process of finding derivatives, and reviewed the properties of the trigonometric functions, we are in a position to develop the formulas necessary for finding the derivatives of the trigonometric functions.

If we find the derivative of the sine function, it is possible to use this derivative as a basis for finding the derivatives of the other trigonometric and inverse trigonometric functions. Therefore, we shall now find the derivative of the sine function, by using the Δ-process.

Let $y = \sin u$, where u is a function of x and is expressed in radians. If u changes by Δu, y then changes by Δy. Thus,

$$y + \Delta y = \sin(u + \Delta u)$$

$$\Delta y = \sin(u + \Delta u) - \sin u$$

$$\frac{\Delta y}{\Delta u} = \frac{\sin(u + \Delta u) - \sin u}{\Delta u}$$

Referring now to Eq. (6–18) we have

$$\frac{\Delta y}{\Delta u} = \frac{2 \sin \frac{1}{2}(u + \Delta u - u)\cos \frac{1}{2}(u + \Delta u + u)}{\Delta u}$$

$$= \frac{\sin(\Delta u/2)\cos[u + (\Delta u/2)]}{\Delta u/2}$$

Looking ahead to the next step of letting $\Delta u \to 0$, we see that the numerator and denominator both approach zero. This situation is precisely the same as that in which we were finding the derivatives of the algebraic functions. To find the limit, we must find

$$\lim_{\Delta u \to 0} \frac{\sin(\Delta u/2)}{\Delta u/2}$$

since these are the factors which cause the numerator and the denominator to approach zero.

In finding this limit, we shall let $\theta = \Delta u/2$ for convenience of notation. This means that we are to determine $\lim_{\theta \to 0} \frac{\sin\theta}{\theta}$. Of course, it would be convenient to know before proceeding if this limit does actually exist. Therefore, we now show a table of values of $\frac{\sin\theta}{\theta}$ as θ becomes very small.

θ (radians)	0.5	0.1	0.05	0.01
$\dfrac{\sin\theta}{\theta}$	0.958851	0.9983342	0.9995834	0.9999833

We see from this table that the limit of $\frac{\sin\theta}{\theta}$, as θ approaches zero, appears to be 1.

In order to prove that $\lim_{\theta \to 0} \frac{\sin\theta}{\theta} = 1$, we use a geometric approach. Considering Fig. 6–13, we can see that the following inequality is true for the indicated areas:

Area triangle OBD < area sector OBD < area triangle OBC

$$\frac{1}{2}r(r \sin\theta) < \frac{1}{2}r^2\theta < \frac{1}{2}r(r \tan\theta) \qquad \text{or} \qquad \sin\theta < \theta < \tan\theta$$

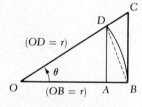

Figure 6–13

Remembering that we want to find the limit of $(\sin\theta)/\theta$, we next divide through by $\sin\theta$, and then take reciprocals:

$$1 < \frac{\theta}{\sin\theta} < \frac{1}{\cos\theta} \qquad \text{or} \qquad 1 > \frac{\sin\theta}{\theta} > \cos\theta$$

When we consider the limit as $\theta \to 0$, we see that the left member remains 1 and the right member approaches 1. Thus, $(\sin \theta)/\theta$ must approach 1. This means

$$\lim_{\theta \to 0} \frac{\sin \theta}{\theta} = \lim_{\Delta u \to 0} \frac{\sin(\Delta u/2)}{\Delta u/2} = 1 \tag{6–26}$$

Using the result expressed in Eq. (6–26) in the expression for $\Delta y/\Delta u$, we have

$$\lim_{\Delta u \to 0} \frac{\Delta y}{\Delta u} = \lim_{\Delta u \to 0} \left[\cos\left(u + \frac{\Delta u}{2}\right) \frac{\sin(\Delta u/2)}{\Delta u/2} \right] = \cos u$$

or

$$\frac{dy}{du} = \cos u \tag{6–27}$$

However, we want the derivative of y with respect to x. This requires the use of the chain rule, Eq. 2–28, which we repeat here for reference.

$$\frac{dy}{dx} = \frac{dy}{du} \frac{du}{dx} \tag{6–28}$$

Combining Eqs. (6–27) and (6–28), we have ($y = \sin u$)

$$\frac{d (\sin u)}{dx} = \cos u \frac{du}{dx} \tag{6–29}$$

Example A

Find the derivative of $y = \sin 2x$.

In this example $u = 2x$. Thus,

$$\frac{dy}{dx} = \cos 2x \left(\frac{d (2x)}{dx} \right)$$

$$= 2 \cos 2x$$

Example B

Find the derivative of $y = 2 \sin(x^2)$.

In this example $u = x^2$, which means that $du/dx = 2x$. Hence,

$$\frac{dy}{dx} = 2[\cos(x^2)](2x)$$

$$= 4x \cos(x^2)$$

It is important here, just as it is in finding the derivatives of powers of functions, to remember to include the factor du/dx.

Example C

Find the derivative of $y = \sin^2 x$.

This example is a combination of the use of the power rule, Eq. (2-29), and the derivative of the sine function, Eq. (6-29). Since $\sin^2 x$ means $(\sin x)^2$, in using the power rule we have $u = \sin x$. Thus,

$$\frac{dy}{dx} = 2(\sin x)\frac{d \sin x}{dx}$$

$$= 2 \sin x \cos x = \sin 2x$$

Example D

Find the derivative of $y = 2 \sin^3(2x^4)$.

In the general power rule, $u = \sin(2x^4)$. In the derivative of the sine function, $u = 2x^4$. Thus, we have

$$\frac{dy}{dx} = 2(3)\sin^2(2x^4)\frac{d(\sin 2x^4)}{dx}$$

$$= 6 \sin^2(2x^4)\cos(2x^4)\frac{d(2x^4)}{dx} = 48x^3\sin^2(2x^4)\cos(2x^4)$$

To find the derivative of the cosine function, we write it in the form $\cos u = \sin[(\pi/2) - u]$. Thus, if $y = \sin[(\pi/2) - u]$, we have

$$\frac{dy}{dx} = \cos\left(\frac{\pi}{2} - u\right)\frac{d[(\pi/2) - u]}{dx} = \cos\left(\frac{\pi}{2} - u\right)\left(-\frac{du}{dx}\right)$$

$$= -\cos\left(\frac{\pi}{2} - u\right)\frac{du}{dx}$$

Since $\cos[(\pi/2) - u] = \sin u$, we have

$$\frac{d(\cos u)}{dx} = -\sin u\, \frac{du}{dx} \qquad\qquad (6\text{-}30)$$

Example E

Find the derivative of $y = \cos^2(2x)$.

Using the power rule and Eq. (6-30), we have

$$\frac{dy}{dx} = 2 \cos(2x)(-\sin 2x)(2) = -4 \sin 2x \cos 2x$$

$$= -2 \sin 4x$$

Example F

Find the derivative of $y = \sin 2x \cos x^2$.

Using the product rule and the derivatives of the sine and cosine functions, we arrive at the following result:

$$\frac{dy}{dx} = \sin 2x(-\sin x^2)(2x) + \cos x^2(\cos 2x)(2)$$

$$= -2x \sin 2x \sin x^2 + 2 \cos 2x \cos x^2$$

Example G

Find the slope of a tangent line to the curve of $y = 5 \sin 3x$ where $x = 0.2$.

Here we are to find the derivative of $y = 5 \sin 3x$, and then evaluate the derivative for $x = 0.2$. Therefore, we have the following:

$$y = 5 \sin 3x$$

$$\frac{dy}{dx} = 5(\cos 3x)(3) = 15 \cos 3x$$

$$\left.\frac{dy}{dx}\right|_{x=0.2} = 15 \cos 3(0.2) = 15 \cos 0.6$$

$$= 15(0.8253) = 12.38$$

In evaluating the slope we must remember that $x = 0.2$ means the values are in radians. Therefore, the slope is 12.38.

Exercises 6–3

In Exercises 1 through 28 find the derivatives of the given functions.

1. $y = \sin(x + 2)$ **2.** $y = 3 \sin 4x$ **3.** $y = 2 \sin(2x - 1)$

4. $y = 5 \sin(3 - x)$ **5.** $y = 5 \cos 2x$ **6.** $y = \cos(1 - x)$

7. $y = 2 \cos(3x - 1)$ **8.** $y = 4 \cos(6x^2 + 5)$ **9.** $y = \sin^2 4x$

10. $y = 3 \sin^3(2x + 1)$ **11.** $y = 3 \cos^3(5x + 2)$ **12.** $y = 4 \cos^2 \sqrt{x}$

13. $y = x \sin 3x$ **14.** $y = x^2 \sin 2x$ **15.** $y = 3x^3 \cos 5x$

16. $y = x \cos 2x^3$ **17.** $y = \sin x^2 \cos 2x$ **18.** $y = \sin x \cos 4x$

19. $y = 2 \sin 2x \cos x$ **20.** $y = 6 \sin 4x \cos 2x^2$

21. $y = \dfrac{\sin 3x}{x}$ **22.** $y = \dfrac{2x}{\sin 4x}$ **23.** $y = \dfrac{2 \cos x^2}{3x}$

24. $y = \dfrac{5x}{\cos x}$ **25.** $y = \sin^3 x - \cos 2x$ **26.** $y = x \sin x + \cos x$

27. $y = x - \frac{1}{3}\sin^3 4x$ **28.** $y = 2x \sin x + 2 \cos x - x^2 \cos x$

In Exercises 29 through 36 solve the given problems.

29. Verify the values of $(\sin \theta)/\theta$ in the table where $(\sin \theta)/\theta$ is calculated for $\theta = 0.5, 0.1, 0.05,$ and 0.01.

30. Evaluate $\lim_{\theta \to 0}(\tan \theta)/\theta$. (Use the fact that $\lim_{\theta \to 0}(\sin \theta)/\theta = 1$.)

31. Show that $\dfrac{d^4 \sin x}{dx^4} = \sin x$

32. If $y = \cos 2x$, show that $\dfrac{d^2 y}{dx^2} = -4y$

33. Find the slope of a line tangent to the curve of $y = 3 \sin 2x$ where $x = \pi/8$.

34. Find the slope of a line tangent to the curve of $y = 6 \cos x^2$ where $x = 0.5$.

35. Find the velocity after 2 s, if the displacement from a reference point of a particle is given by $s = 2 \sin 3t + t^2$, where s is in feet.

36. The current (in amperes) in a certain electric circuit, as a function of time (in seconds), is given by $i = 10 \cos(120\pi t + \pi/6)$. Find the expression for the voltage across a 2-H inductor (see Exercise 23 of Section 5–1).

6–4 Derivatives of the Other Trigonometric Functions

We can find the derivatives of the other trigonometric functions by expressing the functions in terms of the sine and cosine. After we perform the differentiation, we use trigonometric relations to put the derivative in a convenient form.

We obtain the derivative of tan u by expressing tan u as sin u/cos u. Therefore, letting $y = \sin u/\cos u$, by employing the quotient rule we have

$$\frac{dy}{dx} = \frac{\cos u[\cos u\,(du/dx)] - \sin u[-\sin u\,(du/dx)]}{\cos^2 u}$$

$$= \frac{\cos^2 u + \sin^2 u}{\cos^2 u}\frac{du}{dx} = \frac{1}{\cos^2 u}\frac{du}{dx} = \sec^2 u\,\frac{du}{dx}$$

$$\frac{d(\tan u)}{dx} = \sec^2 u\,\frac{du}{dx} \tag{6–31}$$

We find the derivative of cot u by letting $y = \cos u/\sin u$, and again using the quotient rule:

$$\frac{dy}{dx} = \frac{\sin u[-\sin u\,(du/dx)] - \cos u[\cos u\,(du/dx)]}{\sin^2 u}$$

$$= \frac{-\sin^2 u - \cos^2 u}{\sin^2 u}\frac{du}{dx}$$

$$\frac{d(\cot u)}{dx} = -\csc^2 u\,\frac{du}{dx} \tag{6–32}$$

To obtain the derivative of sec u, we let $y = 1/\cos u$. Then

$$\frac{dy}{dx} = -(\cos u)^{-2}\left[(-\sin u)\left(\frac{du}{dx}\right)\right] = \frac{1}{\cos u}\frac{\sin u}{\cos u}\frac{du}{dx}$$

$$\frac{d(\sec u)}{dx} = \sec u \tan u\,\frac{du}{dx} \tag{6–33}$$

We obtain the derivative of csc u by letting $y = 1/\sin u$. And so

$$\frac{dy}{dx} = -(\sin u)^{-2}\left(\cos u\,\frac{du}{dx}\right) = -\frac{1}{\sin u}\frac{\cos u}{\sin u}\frac{du}{dx}$$

$$\frac{d(\csc u)}{dx} = -\csc u \cot u\,\frac{du}{dx} \tag{6–34}$$

Example A

Find the derivative of $y = 2 \tan 8x$.

Using Eq. (6–31), we have

$$\frac{dy}{dx} = 2(\sec^2 8x)(8)$$

$$= 16 \sec^2 8x$$

Example B

Find the derivative of $y = \cot^4 2x$.

Using the power rule and Eq. (6–32), we have

$$\frac{dy}{dx} = 4(\cot^3 2x)(-\csc^2 2x)(2)$$

$$= -8 \cot^3 2x \csc^2 2x$$

Example C

Find the derivative of $y = 3 \sec^2 4x$.

Using the power rule and Eq. (6–33), we have

$$\frac{dy}{dx} = 3(2)(\sec 4x)(\sec 4x \tan 4x)(4)$$

$$= 24 \sec^2 4x \tan 4x$$

Example D

Find the derivative of $y = x \csc^3 2x$.

Using the power rule, the product rule, and Eq. (6–34), we have

$$\frac{dy}{dx} = x(3 \csc^2 2x)(-\csc 2x \cot 2x)(2) + \csc^3 2x$$

$$= \csc^3 2x(-6x \cot 2x + 1)$$

Example E

Find the derivative of $y = (\tan 2x + \sec 2x)^3$.

Using the power rule, Eqs. (6–31) and (6–33), we have

$$\frac{dy}{dx} = 3(\tan 2x + \sec 2x)^2[\sec^2 2x(2) + \sec 2x \tan 2x(2)]$$

$$= 6 \sec 2x(\tan 2x + \sec 2x)^3$$

Example F

Find the differential of $y = \sin 2x \tan x^2$.

Here we are to find the derivative of the given function and multiply by dx. Therefore, using the product rule along with Eqs. (6–29) and (6–31), we have the following:

$$dy = [(\sin 2x)(\sec^2 x^2)(2x) + (\tan x^2)(\cos 2x)(2)] \, dx$$

$$= (2x \sin 2x \sec^2 x^2 + 2 \cos 2x \tan x^2) \, dx$$

Exercises 6–4

In Exercises 1 through 28 find the derivatives of the given functions.

1. $y = \tan 5x$

2. $y = 3 \tan(3x + 2)$

3. $y = \cot(1 - x)^2$

4. $y = 3 \cot 6x$

5. $y = 3 \sec 2x$

6. $y = \sec\sqrt{1 - x}$

7. $y = -3 \csc\sqrt{x}$

8. $y = \csc(1 - 2x)$

9. $y = 5 \tan^2 3x$

10. $y = 2 \tan^2(x^2)$

11. $y = 2 \cot^4 3x$

12. $y = \cot^2(1 - x)$

13. $y = \sqrt{\sec 4x}$

14. $y = \sec^3 x$

15. $y = 3 \csc^4 7x$

16. $y = \csc^2(2x^2)$

17. $y = x^2 \tan x$

18. $y = 3x \sec 4x$

19. $y = 4 \cos x \csc x^2$

20. $y = \frac{1}{2}\sin 2x \sec x$

21. $y = \dfrac{\csc x}{x}$

22. $y = \dfrac{\cot 4x}{2x}$

23. $y = \dfrac{\cos 4x}{1 + \cot 3x}$

24. $y = \dfrac{\tan^2 3x}{2 + \sin x^2}$

25. $y = \frac{1}{3}\tan^3 x - \tan x$

26. $y = \csc 2x - 2 \cot 2x$

27. $y = \tan 2x - \sec 2x$

28. $y = x - \tan x + \sec^2 2x$

In Exercises 29 through 32 find the differentials of the given functions.

29. $y = 4 \tan^2 3x$

30. $y = 5 \sec^3 2x$

31. $y = \tan 4x \sec 4x$

32. $y = 2x \cot 3x$

In Exercises 33 through 36 solve the given problems.

33. Find the slope of a line tangent to the curve of $y = 2 \cot 3x$ where $x = \frac{\pi}{12}$.

34. If the displacement of an object is given by $s = 2t \sec\sqrt{t}$, find the velocity v when $t = 0.04$.

35. An observer to a rocket launch was 1000 ft from the take-off position. He found the angle of elevation of the rocket as a function of time to be $\theta = 3t/(2t + 10)$. Therefore, the height h of the rocket was

$$h = 1000 \tan\frac{3t}{2t + 10}$$

Find the time-rate of change of height after 5 s.

36. Show that $y = \cos^3 x \tan x$ satisfies the equation

$$\cos x \frac{dy}{dx} + 3y \sin x - \cos^2 x = 0$$

6—5 The Inverse Trigonometric Functions

There are times when we wish to solve for the angle in trigonometric relations. Such occasions occur when we know the function of an angle and must find the angle. Also, if we wish to find the rate of change of an angle under certain conditions, it might be convenient to have an expression representing the angle, and not simply a function of it. Functions have been defined for this purpose, and as we shall see, they have significant application in calculus.

Therefore, we define the **inverse trigonometric relations.** *The* **inverse sine** *of x is defined by*

$$y = \arcsin x \quad \text{(the notation } y = \sin^{-1} x \text{ is also used)} \qquad (6\text{–}35)$$

Similar relations exist for the other inverse trigonometric relations. In Eq. (6–35), x is the value of the sine of the angle y, and therefore the most meaningful way of reading it is "y is the angle whose sine is x."

Example A
$y = \arccos x$ would be read as "y is the angle whose cosine is x."

The equation $y = \arctan 2x$ would be read as "y is the angle whose tangent is $2x$."

It is important to emphasize that $y = \arcsin x$ and $x = \sin y$ express the same relationship between x and y. The advantage of having both forms is that a trigonometric relation may be expressed in terms of a function of an angle or in terms of the angle itself.

If we consider closely the equation $y = \arcsin x$ and possible values of x, we note that there are an unlimited number of possible values of y for a given value of x. Consider the following example.

Example B
For $y = \arcsin x$, if $x = \frac{1}{2}$, we have $y = \arcsin \frac{1}{2}$. This means that we are to find an angle whose sine is $\frac{1}{2}$. We know that $\sin (\pi/6) = \frac{1}{2}$. Therefore, $y = \pi/6$.

However, we also know that $\sin (5\pi/6) = \frac{1}{2}$. Therefore, $y = 5\pi/6$ is also a proper value. If we consider negative angles, such as $-7\pi/6$, or angles generated by additional rotations, such as $13\pi/6$, we conclude that there are an unlimited number of possible values for y.

To have a properly defined function *in mathematics, there must be only one value of the dependent variable for a given value of the independent variable. A* relation, *on the other hand, may have more than one such value.* Therefore, we see that $y = \arcsin x$ is not really a function, although it is properly a relation. It is necessary to restrict the values of y in order to define the **inverse trigonometric functions** properly. Therefore, the following values are defined for the given functions.

$$-\frac{\pi}{2} \leq \text{Arcsin } x \leq \frac{\pi}{2}, \quad 0 \leq \text{Arccos } x \leq \pi, \quad -\frac{\pi}{2} < \text{Arctan } x < \frac{\pi}{2}$$

$$(6\text{–}36)$$

$$0 < \text{Arccot } x < \pi, \quad 0 \leq \text{Arcsec } x \leq \pi, \quad -\frac{\pi}{2} \leq \text{Arccsc } x \leq \frac{\pi}{2}$$

This means that when we are looking for a value of y to correspond to a given value for x, we must use a value of y as defined in Eqs. (6–36). The capital letter designates the use of the inverse trigonometric *function*.

Example C

$$\text{Arcsin}\left(\frac{1}{2}\right) = \frac{\pi}{6}$$

This is the only value of the function which lies within the defined range. The value $5\pi/6$ is not correct, since it lies outside the defined range of values.

Example D

$$\text{Arccos}\left(-\frac{1}{2}\right) = \frac{2\pi}{3}$$

Other values such as $4\pi/3$ and $-2\pi/3$ are not correct, since they are not within the defined range of values for the function Arccos x.

Example E

$$\text{Arctan}(-1) = -\frac{\pi}{4}$$

This is the only value within the defined range for the function Arctan x. We must remember that when x is negative for Arcsin x and Arctan x, the value of y is a fourth-quadrant angle, expressed as a *negative angle*. This is a direct result of the definition.

Example F

$$\text{Arcsin}\left(-\frac{\sqrt{3}}{2}\right) = -\frac{\pi}{3} \qquad \text{Arccos}(-1) = \pi$$

$$\text{Arctan } 0 = 0 \qquad \text{Arcsin}(-0.1564) = -\frac{\pi}{20}$$

$$\text{Arccos}(-0.8090) = \frac{4\pi}{5} \qquad \text{Arctan}(\sqrt{3}) = \frac{\pi}{3}$$

One might logically ask why these values are chosen when there are so many different possibilities. The values are so chosen that, if x is positive, the resulting answer gives an angle in the first quadrant. We must, however, account for the possibility that x might be negative. We could not choose second-quadrant angles for Arcsin x. Since the sine of a second-quadrant angle is also positive, this then would lead to ambiguity. The sine is negative for fourth-quadrant angles, and to have a continuous range of values we must express the fourth-quadrant angles in the form of negative angles. This range is also chosen for Arctan x, for similar reasons. However, Arccos x cannot be chosen in this way, since the cosine of fourth-quadrant angles is also positive. Thus, again to keep a continuous range of values for Arccos x, the second-quadrant angles are chosen for negative values of x.

As for the values for the other functions, we chose values such that if x is positive, the result is also an angle in the first quadrant. As for negative values of x, it rarely makes any difference, since either positive values of x arise, or we can use one of the other functions. Our definitions, however, are those which are generally used.

The graphs of the inverse trigonometric relations can be used to show the fact that many values of y correspond to a given value of x. We can also show the ranges used in defining the inverse trigonometric functions, and that these ranges are specific sections of the curves.

Since $y = \arcsin x$ and $x = \sin y$ are equivalent equations, we can obtain the graph of the inverse sine by sketching the sine curve *along the y-axis*. In Figs. (6–14), (6–15), and (6–16), the graphs of three inverse trigonometric relations are shown, with the heavier portions indicating the graphs of the inverse trigonometric functions. The graphs of the other inverse relations are found in the same way.

If we know the value of x for one of the inverse functions, we can find the trigonometric functions of the angle. If general relations are desired, a representative triangle is very useful. The following examples illustrate these methods.

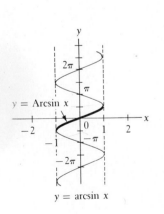

Figure 6–14

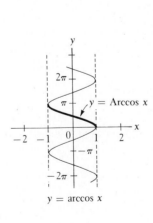

Figure 6–15

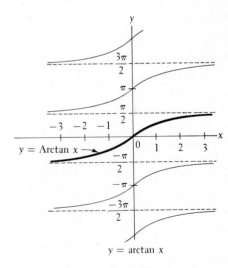

Figure 6–16

Example G

Find $\cos(\text{Arcsin } 0.5)$. [Remember again: the inverse functions yield *angles.*]

We know Arcsin 0.5 is a first-quadrant angle, since 0.5 is positive. Thus we find Arcsin $0.5 = \pi/6$. The problem now becomes one of finding $\cos(\pi/6)$. This is, of course, $\sqrt{3}/2$ or 0.8660.

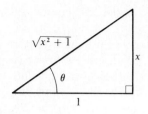

Figure 6—17

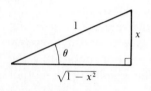

Figure 6—18

Example H

$$\sin(\text{Arccot } 1) = \sin \frac{\pi}{4} = \frac{\sqrt{2}}{2} = 0.7071$$

$$\tan[\text{Arccos}(-1)] = \tan \pi = 0$$

Example I

Find $\sin(\text{Arctan } x)$.

We know that Arctan x is another way of stating "the angle whose tangent is x." Thus, let us draw a right triangle (as in Fig. 6–17) and label one of the acute angles θ, the side opposite θ as x, and the side adjacent to θ as 1. In this way we see that, by definition, $\tan \theta = x/1$, or $\theta = \text{Arctan } x$, which means θ is the desired angle. By the Pythagorean theorem, the hypotenuse of this triangle is $\sqrt{x^2 + 1}$. Now we find that the $\sin \theta$, which is the same as $\sin(\text{Arctan } x)$, is $x/\sqrt{x^2 + 1}$, from the definition of the sine. Thus,

$$\sin(\text{Arctan } x) = \frac{x}{\sqrt{x^2 + 1}}$$

Example J

Find $\cos(2 \text{ Arcsin } x)$.

From Fig. 6–18, we see that $\theta = \text{Arcsin } x$. From the double-angle formulas, we have $\cos 2\theta = 1 - 2 \sin^2\theta$. Thus, since $\sin \theta = x$, we have

$$\cos(2 \text{ Arcsin } x) = 1 - 2x^2$$

Exercises 6—5

In Exercises 1 through 24 evaluate the given expressions.

1. $\text{Arccos}(\frac{1}{2})$

2. $\text{Arcsin}(1)$

3. $\text{Arcsin } 0$

4. $\text{Arccos } 0$

5. $\text{Arctan}(-\sqrt{3})$

6. $\text{Arcsin}(-\frac{1}{2})$

7. $\text{Arcsec } 2$

8. $\text{Arccot } \sqrt{3}$

9. $\text{Arctan}\left(\frac{\sqrt{3}}{3}\right)$

10. $\text{Arctan } 1$

11. $\text{Arcsin}\left(-\frac{\sqrt{2}}{2}\right)$

12. $\text{Arccos}\left(-\frac{\sqrt{3}}{2}\right)$

13. $\text{Arccsc } \sqrt{2}$

14. $\text{Arccot } 1$

15. $\text{Arctan}(-3.732)$

16. $\text{Arccos}(-0.5878)$

17. $\sin(\text{Arctan } \sqrt{3})$

18. $\tan\left(\text{Arcsin } \frac{\sqrt{2}}{2}\right)$

19. $\cos[\text{Arctan}(-1)]$

20. $\sec[\text{Arccos}(-\frac{1}{2})]$

21. $\tan[\text{Arccos}(-0.6561)]$

22. $\cot[\text{Arcsin}(-0.3827)]$

23. $\cos(2 \text{ Arcsin } 1)$

24. $\sin(2 \text{ Arctan } 2)$

In Exercises 25 through 32 find an algebraic expression for each of the expressions given.

25. tan(Arcsin x)
26. sin(Arccos x)
27. cos(Arcsec x)
28. cot(Arccot x)
29. sec(Arccsc $3x$)
30. tan(Arcsin $2x$)
31. sin(2 Arcsin x)
32. cos(2 Arctan x)

In Exercises 33 and 34, prove that the given expressions are equal. This can be done by use of the relation for sin $(\alpha + \beta)$ and by showing that the sine of the sum of angles on the left equals the sine of the angle on the right.

33. Arcsin $\frac{3}{5}$ + Arcsin $\frac{5}{13}$ = Arcsin $\frac{56}{65}$ **34.** Arctan $\frac{1}{3}$ + Arctan $\frac{1}{2}$ = $\frac{\pi}{4}$

In Exercises 35 and 36 derive the given expressions.

35. For the triangle in Fig. 6–19, show that $\alpha =$ Arctan($a +$ tan β).

36. Show that the length of the pulley belt indicated in Fig. 6–20 is given by the expression

$$L = 24 + 11\pi + 10 \text{ Arcsin } (\tfrac{5}{13}).$$

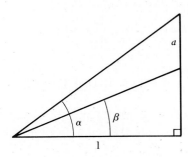

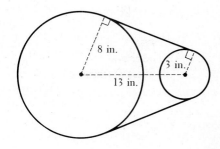

Figure 6–19   Figure 6–20

In Exercises 37 and 38 derive the required expressions.

37. For the figure shown in Fig. 6–21, express θ as a function of x, using inverse trigonometric functions ($x < 6$).

38. For the building and lines of sight shown in Fig. 6–22, express θ as a function of the distance x from the building, using inverse trigonometric functions.

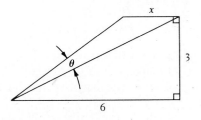

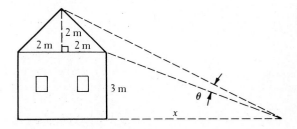

Figure 6–21   Figure 6–22

6–6 Derivatives of the Inverse Trigonometric Functions

To obtain the derivatives of $y = \text{Arcsin } u$, we first solve for u in the form $u = \sin y$ and then take derivatives with respect to x:

$$\frac{du}{dx} = \cos y \, \frac{dy}{dx}$$

Solving this equation for dy/dx, we obtain

$$\frac{dy}{dx} = \frac{1}{\cos y} \frac{du}{dx} = \frac{1}{\sqrt{1 - \sin^2 y}} \frac{du}{dx} = \frac{1}{\sqrt{1 - u^2}} \frac{du}{dx}$$

We choose the positive square root since $\cos y > 0$ for $-\pi/2 < y < \pi/2$, which is the range of the principal values of $\text{Arcsin } u$. Therefore, we obtain the following result:

$$\frac{d(\text{Arcsin } u)}{dx} = \frac{1}{\sqrt{1 - u^2}} \frac{du}{dx} \qquad (6\text{–}37)$$

We note here that the derivative of the inverse sine function is an algebraic function.

We find the derivative of the inverse cosine function by letting $y = \text{Arccos } u$, and by following the same procedure as that used in finding the derivative of $\text{Arcsin } u$:

$$u = \cos y, \qquad \frac{du}{dx} = -\sin y \, \frac{dy}{dx}$$

$$\frac{dy}{dx} = -\frac{1}{\sin y} \frac{du}{dx} = -\frac{1}{\sqrt{1 - \cos^2 y}} \frac{du}{dx}$$

Therefore,

$$\frac{d(\text{Arccos } u)}{dx} = -\frac{1}{\sqrt{1 - u^2}} \frac{du}{dx} \qquad (6\text{–}38)$$

The positive square root is chosen here since $\sin y > 0$ for $0 < y < \pi$, which is the range of the principal values of $\text{Arccos } u$. We note that the derivative of the inverse cosine is the negative of the derivative of the inverse sine.

By letting $y = \text{Arctan } u$, solving for u, and taking derivatives, we find the derivative of the inverse tangent function:

$$u = \tan y, \qquad \frac{du}{dx} = \sec^2 y \, \frac{dy}{dx}, \qquad \frac{dy}{dx} = \frac{1}{\sec^2 y} \frac{du}{dx} = \frac{1}{1 + \tan^2 y} \frac{du}{dx}$$

Therefore,

$$\frac{d(\text{Arctan } u)}{dx} = \frac{1}{1 + u^2}\frac{du}{dx} \qquad (6\text{--}39)$$

We can see that the derivative of the inverse tangent is an algebraic function also.

The inverse sine, inverse cosine, and inverse tangent prove to be of the greatest importance in applications and in further development of mathematics. Therefore, the formulas for the derivatives of the other inverse functions are not presented here, although they are included in the exercises.

Example A

Find the derivative of $y = \text{Arcsin } 4x$.

Using Eq. (6–37), we have

$$\frac{dy}{dx} = \frac{1}{\sqrt{1 - (4x)^2}}(4)$$

$$= \frac{4}{\sqrt{1 - 16x^2}}$$

Example B

Find the derivative of $y = \text{Arccos}(1 - 2x)$.

Using Eq. (6–38), we have

$$\frac{dy}{dx} = \frac{-1}{\sqrt{1 - (1 - 2x)^2}}(-2) = \frac{2}{\sqrt{1 - 1 + 4x - 4x^2}}$$

$$= \frac{2}{\sqrt{4x - 4x^2}} = \frac{1}{\sqrt{x - x^2}}$$

Example C

Find the derivative of $y = (x^2 + 1)\text{Arctan } x - x$.

Using the product rule along with Eq. (6–39) on the first term, we have

$$\frac{dy}{dx} = (x^2 + 1)\frac{1}{1 + x^2} + \text{Arctan } x(2x) - 1$$

$$= 2x \text{ Arctan } x$$

Example D

Find the derivative of $y = x \text{ Arctan}^2 2x$.

We have

$$\frac{dy}{dx} = x(2)(\text{Arctan } 2x)\left(\frac{1}{1 + 4x^2}\right)(2) + \text{Arctan}^2 2x$$

$$= \text{Arctan } 2x\left(\frac{4x}{1 + 4x^2} + \text{Arctan } 2x\right)$$

Example E

Find the derivative of $y = x \text{ Arcsin } 2x + \frac{1}{2}\sqrt{1 - 4x^2}$.

We write

$$\frac{dy}{dx} = x\left(\frac{2}{\sqrt{1 - 4x^2}}\right) + \text{Arcsin } 2x + \frac{1}{2} \cdot \frac{1}{2}(1 - 4x^2)^{-1/2}(-8x)$$

$$= \frac{2x}{\sqrt{1 - 4x^2}} + \text{Arcsin } 2x - \frac{2x}{\sqrt{1 - 4x^2}}$$

$$= \text{Arcsin } 2x$$

Exercises 6—6

In Exercises 1 through 24 find the derivatives of the given functions.

1. $y = \text{Arcsin}(x^2)$

2. $y = \text{Arcsin}(1 - x^2)$

3. $y = 2 \text{ Arcsin } 3x^3$

4. $y = \text{Arcsin } \sqrt{1 - x}$

5. $y = \text{Arccos } \frac{1}{2}x$

6. $y = 3 \text{ Arccos } 5x$

7. $y = 2 \text{ Arccos } \sqrt{2 - x}$

8. $y = 3 \text{ Arccos}(x^2 + 1)$

9. $y = \text{Arctan } \sqrt{x}$

10. $y = \text{Arctan}(1 + x)$

11. $y = \text{Arctan}\left(\frac{1}{x}\right)$

12. $y = 4 \text{ Arctan } 3x^4$

13. $y = x \text{ Arcsin } x$

14. $y = x^2 \text{Arccos } x$

15. $y = 2x \text{ Arctan } 2x$

16. $y = (x^2 + 1)\text{Arcsin } 4x$

17. $y = \dfrac{3x}{\text{Arcsin } 2x}$

18. $y = \dfrac{\text{Arctan } 2x}{x}$

19. $y = \text{Arcsin}^2 4x$

20. $y = \sqrt{\text{Arcsin }(x - 1)}$

21. $y = 3 \text{ Arctan}^3 x$

22. $y = \dfrac{1}{\text{Arccos } 2x}$

23. $y = \dfrac{1}{1 + 4x^2} - \text{Arctan } 2x$

24. $y = \text{Arcsin } x - \sqrt{1 - x^2}$

In Exercises 25 through 32 solve the given problems.

25. Find the differential of the function $y = \text{Arcsin}^3 x$.

26. Find the differential of the function $y = x^3 \text{Arccos } x^2$.

27. Find the second derivative of the function $y = \text{Arctan } 2x$.

28. Show that $\dfrac{d(\text{Arccot } u)}{dx} = -\dfrac{1}{1 + u^2}\dfrac{du}{dx}$

29. Show that $\dfrac{d(\text{Arcsec } u)}{dx} = \dfrac{1}{\sqrt{u^2(u^2 - 1)}}\dfrac{du}{dx}$

30. Show that $\dfrac{d(\text{Arccsc } u)}{dx} = -\dfrac{1}{\sqrt{u^2(u^2 - 1)}}\dfrac{du}{dx}$

31. The angle with the x-axis of the x-component of a vector $\mathbf{R}$ is given by $\theta = \text{Arccos } R_x/R$. If R is constant and R_x changes with time, find an expression for $d\theta/dt$.

32. As a person approaches a building of height h, the angle of elevation of the top of the building is a function of his distance from the building. Express the angle of elevation θ in terms of h and the distance x from the building and then find $d\theta/dx$. Assume the person's height is negligible to that of the building.

6—7 Applications

With our development of the formulas for the derivatives of the trigonometric and inverse trigonometric functions, it is now possible for us to apply these derivatives in the same manner as we applied the derivatives of algebraic functions. We can now use trigonometric and inverse trigonometric functions to solve time-rate of change, tangent and normal, curve tracing, maximum and minimum, and differential application problems. The following examples illustrate the use of these functions in these types of problems.

Example A

Sketch the curve $y = \sin^2 x - \dfrac{x}{2}$ $(0 \le x \le 2\pi)$.

First we write

$$\frac{dy}{dx} = 2 \sin x \cos x - \frac{1}{2} = \sin 2x - \frac{1}{2}, \qquad \frac{d^2y}{dx^2} = 2 \cos 2x$$

The only easily obtained intercept is $(0, 0)$ (see Fig. 6–23). Maximum and minimum points will occur for $\sin 2x = \frac{1}{2}$. Thus, maximum and minimum points occur if

$$2x = \frac{\pi}{6}, \frac{5\pi}{6}, \frac{13\pi}{6}, \frac{17\pi}{6}$$

Thus, the x-values are

$$x = \frac{\pi}{12}, \frac{5\pi}{12}, \frac{13\pi}{12}, \frac{17\pi}{12}$$

Using the second derivative, we find that d^2y/dx^2 is positive for $x = \pi/12$ and $x = 13\pi/12$, and negative for $x = 5\pi/12$ and $x = 17\pi/12$. Thus, the maximum points are $(5\pi/12, 0.279)$ and $(17\pi/12, -1.29)$. The minimum points are $(\pi/12, -0.064)$ and $(13\pi/12, -1.64)$. Inflection points occur for $\cos 2x = 0$, or

$$2x = \frac{\pi}{2}, \frac{3\pi}{2}, \frac{5\pi}{2}, \frac{7\pi}{2}, \qquad \text{or} \qquad x = \frac{\pi}{4}, \frac{3\pi}{4}, \frac{5\pi}{4}, \frac{7\pi}{4}$$

Therefore, the points of inflection are $(\pi/4, 0.11)$, $(3\pi/4, -0.68)$, $(5\pi/4, -1.47)$, and $(7\pi/4, -2.25)$. Using this information, we sketch the curve.

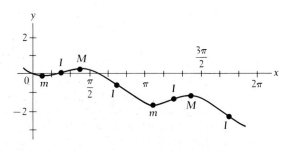

Figure 6—23

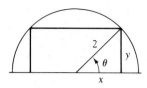

Figure 6–24

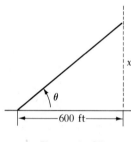

Figure 6–25

Example B

Find the area of the largest rectangle which can be inscribed in a semicircle of radius 2.

From Fig. 6–24, we see that $A = 2xy$, where $x = 2 \cos \theta$ and $y = 2 \sin \theta$. Thus, $A = 8 \sin \theta \cos \theta = 4 \sin 2\theta$. We shall find the value of θ corresponding to the maximum area. Therefore,

$$\frac{dA}{d\theta} = 8 \cos 2\theta, \qquad 8 \cos 2\theta = 0, \qquad 2\theta = \frac{\pi}{2}, \qquad \theta = \frac{\pi}{4}$$

Thus, the maximum area is $A = 4 \sin 2(\pi/4) = 4$.

Example C

A rocket is taking off vertically at a distance of 600 ft from an observer. If, when the angle of elevation is $\pi/4$ it is changing at the rate of 0.5 rad/s, how fast is the rocket ascending?

From Fig. 6–25, we see that

$$\theta = \operatorname{Arctan}(x/600)$$

Taking derivatives with respect to time, we have

$$\frac{d\theta}{dt} = \frac{1}{1 + x^2/(3.6 \times 10^5)} \frac{dx/dt}{600} = \frac{600 \, dx/dt}{3.6 \times 10^5 + x^2}$$

From the given information, we find that $d\theta/dt = 0.5$ rad/s, $x = 600$ ft when $\theta = \pi/4$. Thus,

$$0.5 = \frac{600 \, dx/dt}{7.2 \times 10^5} \qquad \text{or} \qquad \frac{dx}{dt} = 600 \text{ ft/s}$$

Example D

A particle moves so that its x- and y-coordinates are given by $x = \cos 2t$ and $y = \sin 2t$. Find the magnitude and direction of its velocity when $t = \pi/8$.

We write

$$v_x = \frac{dx}{dt} = -2 \sin 2t \qquad v_y = \frac{dy}{dt} = 2 \cos 2t$$

$$v_x|_{t=\pi/8} = -2 \sin 2\left(\frac{\pi}{8}\right) = -2\left(\frac{\sqrt{2}}{2}\right) = -\sqrt{2}$$

$$v_y|_{t=\pi/8} = 2 \cos 2\left(\frac{\pi}{8}\right) = 2\left(\frac{\sqrt{2}}{2}\right) = \sqrt{2}$$

$$v = \sqrt{v_x^2 + v_y^2} = \sqrt{2 + 2} = 2 \qquad \tan \theta = \frac{v_y}{v_x} = \frac{\sqrt{2}}{-\sqrt{2}} = -1$$

Since v_x is negative and v_y is positive, $\theta = 135°$. Note that in this example θ is the angle, in standard position, between the horizontal and the resultant velocity.

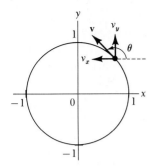

Figure 6–26

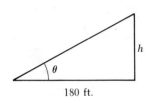

180 ft.

Figure 6–27

By plotting the curve we note that it is a circle (see Fig. 6–26). Thus, the object is moving about a circle in a counterclockwise direction. It can be determined in another way that the curve is a circle. If we square each of the expressions defining x and y, and then add these, we have the equation $x^2 + y^2 = \cos^2 2t + \sin^2 2t = 1$. This is a circle of radius 1.

Example E

At a point 180 ft from the base of a building on level ground, the angle of elevation of the top of the building is 30°. What would be the error in the height of the building due to an error of 15′ in this angle?

From Fig. 6–27 we see that $h = 180 \tan \theta$. Finding the differential of h, we have

$$dh = 180 \sec^2\theta \; d\theta$$

Since $\theta = 30°$, $\sec \theta = 2/\sqrt{3}$, or $\sec^2\theta = 4/3$. The possible error in θ is 15′ which is 0.25°, which in turn is 0.00436 rad, which is the value we must use in the calculation. Thus,

$$dh = (180)\left(\frac{4}{3}\right)(0.00436) = 1.05 \text{ ft}$$

We see that an error of 15′ in the angle can result in an error of over 1 ft in the calculation of the height.

Exercises 6–7

1. Show that the slopes of the sine and cosine curves are negatives of each other at the points of intersection.

2. Show that the tangent curve, when defined, is always increasing.

3. Show that $y = \text{Arctan } x$ is always increasing.

4. Sketch the graph of $y = \sin x + \cos x \; (0 \le x \le 2\pi)$.

5. Sketch the graph of $y = x - \tan x \; (-\pi/2 < x < \pi/2)$.

6. Sketch the graph of $y = \sin x + \sin 2x \; (0 < x < 2\pi)$.

7. Find the equation of the line tangent to the curve of $y = \sin 2x$ at $x = 5\pi/8$.

8. Find the equation of a line normal to the curve of $y = \text{Arctan}(x/2)$ at $x = 3$.

9. The displacement (in feet) a certain object travels is given as a function of time (in seconds) by $s = \sin t + \cos 2t$. Find its velocity and acceleration when $t = \pi/6$ s.

10. The displacement s (in centimeters) of an object as a function of time is given by $s = t \sin t$. Find the velocity when $t = 0.2$ s.

11. The y-component of a vector of magnitude 20 mm is given by $A_y = 20 \sin \theta$. If $d\theta/dt = 0.05$ rad/s, determine the time rate of change of A_y in millimeters per second when $\theta = 0.3$.

12. The *apparent power* P_a of an electric circuit whose power is P and impedance phase angle is θ is given by $P_a = P \sec \theta$. Given P is constant at 12 W, find the time-rate of change of P_a if θ is changing at the rate of 0.05 rad/min, when $\theta = 2\pi/9$.

13. Find the magnitude and direction of the velocity of an object which moves so that its x- and y-coordinates are given by $x = \sin 2\pi t$ and $y = \cos 2\pi t$ when $t = \frac{1}{12}$.

14. Find the magnitude and direction of the velocity of an object which moves so that the x- and y-coordinates are given by $x = \cos 2t$ and $y = 2 \sin t$ when $t = \pi/6$.

15. Find the magnitude and direction of the acceleration of the particle in Example D when $t = \pi/8$.

16. Find the magnitude and direction of the acceleration of the object in Exercise 14.

17. A person observes an object dropped from the top of a building 100 ft away. If the building is 200 ft high, how fast is the angle of elevation of the object changing after 1 s? (The distance the object drops is given by $s = 16t^2$.)

18. An airplane flying horizontally at 500 ft/s passes directly over an observer 4000 ft below. How fast is the angle of elevation changing at this instant?

19. One leg of a right triangle is 6 units long. If the acute angle of which this side is the adjacent side is increasing at the rate of 0.1 rad/s, how fast is the hypotenuse changing when it is 10 units in length?

20. In a modern hotel, where the elevators are directly observable from the lobby area (and a person can see from the elevators), a person in the lobby observes one of the elevators rising at the rate of 12 ft/s. If the person was 50 ft from the elevator when it left the lobby, how fast is the angle of elevation of the line of sight to the elevator increasing 10 s later?

21. If a block is placed on a plane inclined with the horizontal at an angle θ such that the block just moves down the plane, the coefficient of friction μ is given by $\mu = \tan \theta$. Use differentials to find the change in μ if θ changes from 20° to 21°.

22. The current (in amperes) in a certain electric circuit is given by $i = \sin^2 2t$. Find the approximate change in i between $t = 1.00$ s and $t = 1.05$ s.

23. Two sides and the included angle of a triangle are measured to be 8.00 cm, 7.50 cm, and 40.0°. What is the error caused in the calculation of the third side by an error of 0.5° in the angle?

24. A firm determined that its total weekly profit from the production of x units is $P = 10,000 \sin 0.1x$, for $0 < x < 30$ and where P is measured in dollars. What is the approximate change in profit when production changes from 20 to 22 units?

25. Use trigonometric functions to find the area of the largest rectangle with a perimeter of 12 in.

26. Find the angle between the equal sides of an isosceles triangle of greatest area for a given perimeter.

27. A wall is 6 ft high and 4 ft from a building. What is the length of the shortest pole that can touch the building and the ground beyond the wall? [*Hint:* $y = 6 \csc \theta + 4 \sec \theta$ (see Fig. 6–28).]

28. A painting 8 ft high is hung such that the lower edge is 2 ft above an observer's eye level. Assuming that the best view is obtained when the angle subtended by the painting at the eye level is a maximum, how far from the wall should the observer stand?

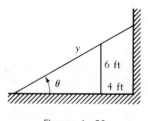

Figure 6–28

6–8 Review Exercises for Chapter 6

In Exercises 1 through 20 find the derivative of each of the given functions.

1. $y = 3 \cos(4x - 1)$

2. $y = 4 \sin(\frac{1}{2}x + 1)$

3. $y = \tan \sqrt{3 - x}$

4. $y = 5 \csc 2x^2$

5. $y = 3 \operatorname{Arctan}\left(\frac{x}{3}\right)$

6. $y = 2 \operatorname{Arcsin}(1 - 2x)$

7. $y = 2 \sin^3 \sqrt{x}$

8. $y = \cot^2 5x$

9. $y = x \sec 2x$

10. $y = (x + 1)\cos 2x$

11. $y = \dfrac{\operatorname{Arcsin} x}{x}$

12. $y = \dfrac{2 \tan^2 3x}{x}$

13. $y = \sqrt{\csc 4x + \cot 4x}$

14. $y = 3 \sin^3 2x - 2 \cos^2 3x$

15. $y = \operatorname{Arcsin}(\cos x)$

16. $y = \sin(\operatorname{Arctan} x)$

17. $y = x \operatorname{Arccos} x - \sqrt{1 - x^2}$

18. $y = x \operatorname{Arcsin}^2 x + 2\sqrt{1 - x^2}\operatorname{Arcsin} x - 2x$

19. $x \sin 2y = y \cos 2x$

20. $x + \sec^2(xy) = 1$

In Exercises 21 through 36 solve the given problems.

21. Power can be defined as the time rate of doing work. If work is being done in an electric circuit according to $W = 10 \cos 2t$, find P as a function of t.

22. A ball on the end of a string is moving in a vertical circle. The distance of its shadow from a certain point is given by

$$x = 4 \cos\left(\frac{\pi}{3}t + \frac{\pi}{6}\right)$$

Find the velocity, in ft/s, of the shadow when $t = 6$ s.

23. Find the magnitude and direction of the velocity of an object which moves so that its x- and y-coordinates are given by $x = \sin 2\pi t$ and $y = \cos 2\pi t$ when $t = \frac{1}{12}$.

24. Find the magnitude and direction of the acceleration of the object in Exercise 23.

25. Find the equation of the line tangent to the curve $y = 4 \cos^2(x^2)$ where $x = 1$.

26. Find the equation of the line normal to the curve $y = \operatorname{Arctan} x$ where $x = 1$.

27. Sketch the graph of the function $y = x - \cos x$.

28. Sketch the graph of the function $y = 4 \sin x + \cos 2x$.

29. A revolving light 300 ft from a straight wall makes 6 r/min. Find the velocity of the beam along the wall at the instant it makes an angle of 45° with the wall.

30. In optics, the index of refraction of a liquid is defined as $n = \sin i/\sin r$, where i is the angle at which the light is incident on the liquid from the air, and r is the angle of the refracted light in the liquid. If the angle of incidence is increasing at the rate of 0.05 rad/min, what is the rate of change of the angle r when $i = 10°$ for benzene, for which $n = 1.50$?

31. In the study of polarized light, under certain conditions, the intensity of transmitted light is given by $I = I_m \cos^2\theta$, where I_m is the maximum intensity and θ is the angle of transmission. If $I = 8$ units when $\theta = 60°$, find the approximate change in I if θ changes to $62°$.

32. The force which acts on a pendulum bob to restore it to its equilibrium position is given by $F = -mg \sin\theta$ where m is the mass of the bob, g is the acceleration of gravity, and θ is the angle between the string and the vertical. For a bob of 10 g, find the approximate difference in the force for angles of $3.0°$ and $3.1°$ if $g = 980$ cm/s².

33. A gutter is to be made from a sheet of metal 12 in. wide by turning up strips of width 4 in. along each side such that they make equal angles θ with the vertical. Sketch a graph of the cross-sectional area A as a function of θ (see Fig. 6–29).

34. A force P at an angle θ above the horizontal drags a 50-lb box across a level floor. The coefficient of friction between the floor and box is constant and equals 0.2. The magnitude of the force P is given by

$$P = \frac{(0.2)(50)}{(0.2)\sin\theta + \cos\theta}$$

Find θ such that P is a minimum.

35. The illuminance of light from a source varies directly as the cosine of the angle of incidence (measured from the perpendicular) and inversely as the square of the distance r from the source. How high above the center of a circle of radius 10 cm should a light be placed so that the illuminance at the circumference will be a maximum (see Fig. 6–30)?

36. A metal bracket in the shape of a Y is to be made such that its height is 10 in. and width across the top is 6 in. What shape will require the least amount of material?

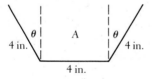

Figure 6–29

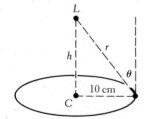

Figure 6–30

7

Derivatives of the Exponential and Logarithmic Functions

<hr>

7–1 Exponential and Logarithmic Functions

Another important function in many of the applications of technology, especially in mechanics and electricity, is the exponential function. Closely associated with the exponential function is the logarithmic function, which also has numerous applications. Therefore, in this section we shall review the properties of these functions, and in the remaining sections of this chapter we shall develop the derivatives of them.

By definition, *the exponential function is of the form*

$$y = b^x \tag{7–1}$$

where b is the **base**, *a constant, and x is the exponent, the independent variable.* Restricting values of both variables to real numbers leads us to choose the base as a positive number other than 1. We know that 1 raised to any finite power will result in an answer of 1. Negative numbers for b would result in imaginary values for y, if x were any fractional exponent with an even integer as its denominator.

Example A

$y = 2^x$ is an exponential function, where we see that x is the exponent of 2. That is, 2 raised to any given power gives us the corresponding value of y. If $x = 2$, $y = 4$. If $x = -4$, $y = \frac{1}{16}$. If $x = \frac{1}{2}$, $y = \sqrt{2} = 1.41$. If $x = \frac{3}{2}$, $y = 2\sqrt{2} = 2.83$. By the definition of the exponential function, any real value of x may be assumed.

In Eq. (7–1), since y is a function of x, it is possible to define x as a function of y. Therefore, in Eq. (7–1) we *define* x *as the* **logarithm** *of the number* y *to the base* b, which is written as

$$x = \log_b y \tag{7–2}$$

Equations (7–1) and (7–2) state precisely the same relationship, but in a different manner. Equation (7–1) is the exponential form, and Eq. (7–2) is the logarithmic form. In either case, x is a logarithm, and a logarithm is an exponent.

Example B

The equation $y = 2^x$ would be written as $x = \log_2 y$ if we put it in logarithmic form. When we choose values of y to find the corresponding values of x from this equation, we ask ourselves, "2 raised to what power gives y?" Hence if $y = 4$, we know that 2^2 is 4, and x would be 2. If $y = 8$, $2^3 = 8$, or $x = 3$.

Example C

$3^2 = 9$ in logarithmic form is $2 = \log_3 9$; $4^{-1} = \frac{1}{4}$ in logarithmic form is $-1 = \log_4\left(\frac{1}{4}\right)$. Remember, the exponent may be negative, but the base must be positive.

If $-4 = \log_b\left(\frac{1}{81}\right)$, and if we write this in exponential form, we may find b. In exponential form it becomes $b^{-4} = \frac{1}{81}$ which means that $b = 3$.

When working with functions, we normally denote the dependent variable as y and the independent variable as x. For this reason, *we define the* **logarithmic function** *to be*

$$y = \log_b x \tag{7–3}$$

Equations (7–2) and (7–3) may seem to have different meanings, due to the location of the variables. We must keep in mind that a function is defined by the operation being performed on the independent variable, and not by the letter chosen to represent it. These two equations express the same function, the logarithmic function.

We shall now show the graphs of the exponential function, Eq. (7–1), and the logarithmic function, Eq. (7–3). In this way we may demonstrate their properties. The following examples are used for this purpose.

Example D

Plot the graph of $y = 2^x$.

Assuming values for x and then finding the corresponding values for y, we obtain the values in the following table.

x	-3	-2	-1	0	1	2	3	4
y	$\frac{1}{8}$	$\frac{1}{4}$	$\frac{1}{2}$	1	2	4	8	16

From these values we plot the curve, as shown in Fig. 7–1.

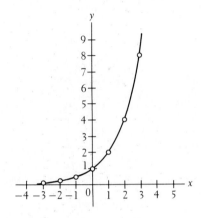

Figure 7–1

Example E

Plot the graph of $y = \log_3 x$.

We can find the points for this graph more easily if we first put the equation in exponential form: $x = 3^y$. By assuming values for y, we can find the corresponding values for x.

x	$\frac{1}{9}$	$\frac{1}{3}$	1	3	9	27
y	-2	-1	0	1	2	3

Using these values, we construct the graph seen in Fig. 7–2.

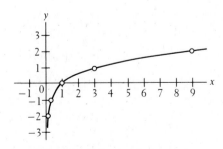

Figure 7–2

Any exponential or logarithmic curve, where $b > 1$, will be similar to those of Examples D and E. From these we can draw certain conclusions:

(1) If $0 < x < 1$, $\log_b x < 0$; $\log_b 1 = 0$; if $x > 1$, $\log_b x > 0$
(2) If $x > 1$, x increases more rapidly than $\log_b x$
(3) $b^x > 0$ for all values of x
(4) If $x > 1$, b^x increases more rapidly than x

Since a logarithm is an exponent, it must obey the laws of exponents. Those laws which are of the greatest importance at this time are listed here for reference.

$$b^u \cdot b^v = b^{u+v}, \tag{7-4}$$
$$b^u / b^v = b^{u-v}, \tag{7-5}$$
$$(b^u)^n = b^{nu} \tag{7-6}$$

We shall now show how these laws for exponents give certain useful properties to logarithms.

If we let $u = \log_b x$ and $v = \log_b y$, and write these equations in exponential form, we have $x = b^u$ and $y = b^v$. Therefore, forming the product of x and y, we obtain

$$xy = b^u b^v = b^{u+v} \qquad \text{or} \qquad xy = b^{u+v}$$

Writing this last equation in logarithmic form yields

$$u + v = \log_b xy$$

or

$$\log_b x + \log_b y = \log_b xy \tag{7-7}$$

Equation (7–7) states the property that *the logarithm of the product of two numbers is equal to the sum of the logarithms of the numbers.*

Using the same definitions of u and v to form the quotient of x and y, we have

$$\frac{x}{y} = \frac{b^u}{b^v} = b^{u-v} \qquad \text{or} \qquad \frac{x}{y} = b^{u-v}$$

Writing this last equation in logarithmic form, we have

$$u - v = \log_b\left(\frac{x}{y}\right)$$

or

$$\log_b x - \log_b y = \log_b\left(\frac{x}{y}\right) \tag{7-8}$$

Equation (7–8) states the property that *the logarithm of the quotient of two numbers is equal to the logarithm of the numerator minus the logarithm of the denominator.*

If we again let $u = \log_b x$ and write this in exponential form, we have $x = b^u$. To find the nth power of x, we write

$$x^n = (b^u)^n = b^{nu}$$

Expressing this equation in logarithmic form yields

$$nu = \log_b (x^n)$$

or

$$n \log_b x = \log_b (x^n) \tag{7–9}$$

This last equation states that *the logarithm of the nth power of a number is equal to n times the logarithm of the number.* The exponent n may be integral or fractional and, therefore, we may use Eq. (7–9) for finding powers and roots of numbers.

Example F

By Eq. (7–7) we can see that

$$\log_2 6 = \log_2 (2 \cdot 3) = \log_2 2 + \log_2 3 = 1 + \log_2 3$$

By Eq. (7–8) we have

$$\log_5 100 - \log_5 4 = \log_5 \left(\frac{100}{4}\right) = \log_5 25 = 2$$

If we have

$$\log_b y = 2 \log_b x + \log_b a$$

by Eqs. (7–9) and (7–7) we may write

$$\log_b y = \log_b x^2 + \log_b a = \log_b (ax^2)$$

Therefore,

$$y = ax^2$$

There are two bases of logarithms that are used widely. Base 10 is used primarily for calculational purposes, and base e, an irrational number equal to about 2.718, is used primarily for theoretical purposes. In the next section we shall see why e is chosen for this purpose. Tables of logarithms in both bases are found in the Appendix.

It is assumed here that the reader is familiar with logarithms to the base 10 for calculations. *The notation log x, without any base indicated, will be used for logarithms to the base 10.*

Since our primary purpose in this chapter is to develop the derivatives of the logarithmic and exponential functions, we shall be using the base e extensively. *The name* **natural logarithm** *is given to a logarithm with the base e, and the notation ln x is used for a natural logarithm.*

Since more than one base is important, it is occasionally useful to be able to change a logarithm in one base to another base. If $u = \log_b x$, $b^u = x$, and taking logarithms of each side to the base a, we have

$$u \log_a b = \log_a x, \qquad u = \frac{\log_a x}{\log_a b}$$

or

$$\log_b x = \frac{\log_a x}{\log_a b} \tag{7-10}$$

Since natural logarithms are used extensively, it is often convenient to have Eq. (7-10) written specifically for use with logarithms to the base 10 and natural logarithms. Since $\ln 10 = 2.3026$, we can write

$$\ln x = 2.3026 \log x \tag{7-11}$$

Example G

Since logarithms to the base 10 are commonly tabulated we can find the natural logarithm by use of either Eq. (7-10) or Eq. (7-11). Using Eq. (7-10) we find $\ln 20$ as follows.

$$\ln 20 = \frac{\log 20}{\log e} = \frac{\log 20}{\log 2.718} = \frac{1.3010}{0.4343} = 2.996$$

Using Eq. (7-11), we find $\ln 20$ as follows.

$$\ln 20 = 2.3026(\log 20) = 2.3026(1.3010)$$
$$= 2.996$$

Exercises 7—1

In Exercises 1 through 12 change those in exponential form to logarithmic form, and those in logarithmic form to exponential form.

1. $4^4 = 256$ 2. $3^{-2} = \frac{1}{9}$
3. $\log_8 16 = \frac{4}{3}$ 4. $\log_7 (\frac{1}{49}) = -2$
5. $8^{1/3} = 2$ 6. $81^{3/4} = 27$
7. $\log_{0.5} 16 = -4$ 8. $\log_{243} 3 = \frac{1}{5}$
9. $(12)^0 = 1$ 10. $(\frac{1}{2})^{-2} = 4$
11. $\log_8 (\frac{1}{512}) = -3$ 12. $\log_{1/32} (\frac{1}{8}) = \frac{3}{5}$

In Exercises 13 through 16 determine the value of the letter present.

13. $\log_7 y = 3$ 14. $\log_b 4 = \frac{2}{3}$ 15. $\log (10)^{0.2} = x$ 16. $\log_8 y = -\frac{2}{3}$

In Exercises 17 through 20 plot the graphs of the given functions.

17. $y = 3^x$ 18. $y = (\frac{1}{2})^x$ 19. $y = \log_2 x$ 20. $y = 2 \log_4 x$

In Exercises 21 through 28 express each as a single logarithm.

21. $\log_5 9 - \log_5 3$ 22. $\log_2 3 + \log_2 x$

23. $\log_6 x^2 + \log_6 \sqrt{x}$ **24.** $\log_7 3^4 - \log_7 9$

25. $2 \log_b 2 + \log_b y$ **26.** $3 \log_b 4 - \log_b x$

27. $2 \log_2 x + \log_2 3 - \log_2 5$ **28.** $\log_4 7 - 2 \log_4 t + \log_4 a$

In Exercises 29 through 32 evaluate the given logarithms by use of the table of logarithms to the base 10.

29. $\ln 3$ **30.** $\ln 0.9$ **31.** $\log_7 42$ **32.** $\log_{12} 122$

In Exercises 33 through 40 solve the given problems.

33. The atmospheric pressure at a given height h is given by $p = p_0 e^{-kh}$, where p_0 and k are constants. Solve for h.

34. If an amount A is deposited in a bank which pays 4% interest, compounded quarterly, the value of the investment after t years is given by $V = A(1.01)^{4t}$. By use of logarithms to the base 10, solve for t.

35. For an electric circuit containing a battery of voltage E, a resistance R, and an inductance L, the current as a function of time is given by

$$i = \frac{E}{R}(1 - e^{-Rt/L})$$

Solve for t.

36. The vapor pressure P over a liquid may be related to temperature by $\log P = (a/T) + b$, where a and b are constants. Solve for P.

37. Table 4 in the Appendix is a short table of natural logarithms. Computations can be made by means of these logarithms, but such computations require adding or subtracting multiples of 2.3026 to locate the decimal point properly. We can see this in a case such as $N = P \cdot 10^k$. Then

$$\ln N = \ln P + k \ln 10 = \ln P + k(2.3026)$$

For example,

$$\ln 2.500 = 0.9163 \quad \text{and} \quad \ln 250.0 = 0.9163 + 2(2.3026) = 5.5215$$

Find the natural logarithms of the following numbers:

    (a) 6.700   (b) 67.00   (c) 3400   (d) 0.3400

Calculate $(6.700)(3400)$ by means of these values, rounding off the answer to the nearest number from the table.

38. The **hyperbolic sine of u** is defined as

$$\sinh u = \frac{1}{2}(e^u - e^{-u})$$

and the **hyperbolic cosine of u** is defined as

$$\cosh u = \frac{1}{2}(e^u + e^{-u})$$

These functions are called *hyperbolic* functions since, if $x = \cosh u$ and $y = \sinh u$, x and y satisfy the equation of the hyperbola $x^2 - y^2 = 1$. Verify this by substituting the appropriate exponential expressions into the equation of the hyperbola and simplifying.

39. From the definitions of sinh u and cosh u (see Exercise 38), find the values of sinh 0.25 and cosh 0.25 and show that $\cosh^2 0.25 - \sinh^2 0.25 = 1$.

40. From the values of sinh 0.25 and cosh 0.25 (see Exercises 38 and 39), show that $\sinh 0.50 = 2 \sinh 0.25 \cosh 0.25$.

7—2 Derivative of the Logarithmic Function

We shall first find the derivative of the logarithmic function. We can use the derivative of the logarithmic function to find the derivative of the exponential function, which we shall do in the next section. Again we utilize the Δ-process.

If we let $y = \log_b u$, where u is a function of x, we have

$$y + \Delta y = \log_b (u + \Delta u)$$

$$\Delta y = \log_b (u + \Delta u) - \log_b u = \log_b\left(\frac{u + \Delta u}{u}\right) = \log_b\left(1 + \frac{\Delta u}{u}\right)$$

$$\frac{\Delta y}{\Delta u} = \frac{\log_b(1 + \Delta u/u)}{\Delta u} = \frac{1}{u} \cdot \frac{u}{\Delta u} \log_b\left(1 + \frac{\Delta u}{u}\right)$$

$$= \frac{1}{u} \log_b\left(1 + \frac{\Delta u}{u}\right)^{u/\Delta u}$$

(We multiply and divide by u for purposes of evaluating the limit, as we shall now show.)

Before we can evaluate the $\lim_{\Delta u \to 0} \Delta y/\Delta u$ it is necessary to determine

$$\lim_{\Delta u \to 0}\left(1 + \frac{\Delta u}{u}\right)^{u/\Delta u}$$

We can see that the exponent becomes unbounded, but the number being raised to this exponent approaches 1. Therefore, we shall investigate this limiting value.

To find an approximate value, let us graph the function $y = (1 + x)^{1/x}$ (for purposes of graphing we let $\Delta u/u = x$). We construct a table of values and then graph the function (see Fig. 7–3).

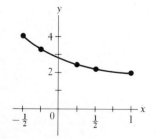

Figure 7–3

x	-0.5	-0.25	$+0.25$	$+0.50$	$+1.00$
y	4.00	3.16	2.44	2.25	2.00

Only these values are shown, since we are interested in the y-value corresponding to $x = 0$. We see from the graph that this value is approximately 2.7. Choosing very small values of x, we may obtain the following values:

x	0.1	0.01	0.001	0.0001
y	2.5937	2.7048	2.7169	2.71815

By methods developed in Chapter 12, it can be shown that this value is about 2.7182818. *The limiting value is the irrational number e*, which was the base of natural logarithms in the last section.

Returning to the derivative of the logarithmic function, we have

$$\lim_{\Delta u \to 0} \frac{\Delta y}{\Delta u} = \lim_{\Delta u \to 0} \left[\frac{1}{u} \log_b \left(1 + \frac{\Delta u}{u} \right)^{u/\Delta u} \right] = \frac{1}{u} \log_b e$$

Therefore,

$$\frac{dy}{du} = \frac{1}{u} \log_b e$$

Combining this equation with Eq. (6–28), we have

$$\frac{d(\log_b u)}{dx} = \frac{1}{u} \log_b e \frac{du}{dx} \tag{7–12}$$

At this point we see that if we choose e as the basis of a system of logarithms, the above formula becomes

$$\frac{d(\ln u)}{dx} = \frac{1}{u} \frac{du}{dx} \tag{7–13}$$

The choice of e as the base b makes $\log_e e = 1$; thus, this factor does not appear in Eq. (7–13). We can now see why the number e is chosen as the base for a system of logarithms, the natural logarithms. The notation $\ln u$ is the same as that used in Section 7–1 for natural logarithms.

Example A

Find the derivative of $y = \log 4x$.

Using Eq. (7–12), we have

$$\frac{dy}{dx} = \frac{1}{4x} \log e \,(4) \qquad (\log e = 0.4343)$$

$$= \frac{1}{x} \log e$$

Example B

Find the derivative of $y = \ln 3x^4$.

Using Eq. (7–13), we have

$$\frac{dy}{dx} = \frac{1}{3x^4} (12x^3)$$

$$= \frac{4}{x}$$

Example C

Find the derivative of $y = \ln \tan x$.

Using Eq. (7–13), along with the derivative of the tangent, we have

$$\frac{dy}{dx} = \frac{1}{\tan x} \sec^2 x = \frac{\cos x}{\sin x} \frac{1}{\cos^2 x}$$

$$= \csc x \sec x$$

Example D

Find the derivative of $y = \ln \dfrac{x - 1}{x + 1}$.

In this example it is easier to write y in the form

$$y = \ln(x - 1) - \ln(x + 1)$$

(Often we can find a derivative more simply if we use the properties of logarithms to simplify the expression.) Hence,

$$\frac{dy}{dx} = \frac{1}{x - 1} - \frac{1}{x + 1} = \frac{x + 1 - x + 1}{(x - 1)(x + 1)}$$

$$= \frac{2}{x^2 - 1}$$

Example E

Find the derivative of $y = \ln(1 - 2x)^3$.

First we may rewrite the equation as $y = 3 \ln(1 - 2x)$. Then

$$\frac{dy}{dx} = 3\left(\frac{1}{1 - 2x}\right)(-2) = \frac{-6}{1 - 2x}$$

Be careful to distinguish this function from the function $y = [\ln(1 - 2x)]^3$ [which can also be written as $y = \ln^3(1 - 2x)$]. In this latter function it is the logarithm which is being cubed, whereas in the first function the quantity $(1 - 2x)$ is being cubed. The derivative in the latter case would be

$$\frac{dy}{dx} = 3[\ln(1 - 2x)]^2\left(\frac{1}{1 - 2x}\right)(-2) = -\frac{6[\ln(1 - 2x)]^2}{1 - 2x}$$

Exercises 7–2

In Exercises 1 through 24 find the derivatives of the given functions.

1. $y = \log x^2$

2. $y = \log_2 6x$

3. $y = 2 \log_5(3x + 1)$

4. $y = 3 \log_7(x^2 + 1)$

5. $y = \ln(1 - 3x)$

6. $y = 2 \ln(3x^2 - 1)$

7. $y = 2 \ln \tan 2x$

8. $y = \ln \sin x$

9. $y = \ln \sqrt{x}$

10. $y = \ln \sqrt{4x - 3}$

11. $y = x \ln x$

12. $y = x^2 \ln 2x$

13. $y = \dfrac{3x}{\ln(2x + 1)}$

14. $y = \dfrac{\ln x}{x}$

15. $y = \ln(\ln x)$

16. $y = \ln \cos x^2$

17. $y = \ln \dfrac{x}{1 + x}$

18. $y = \ln(x\sqrt{x + 1})$

19. $y = \sin \ln x$

20. $y = (\ln x)^3$

21. $y = \ln(x \tan x)$

22. $y = \ln(x + \sqrt{x^2 - 1})$

23. $y = \ln \dfrac{x^2}{x + 2}$

24. $y = \sqrt{x^2 + 1} - \ln \dfrac{1 + \sqrt{x^2 + 1}}{x}$

In Exercises 25 through 32 solve the given problems.

25. Verify the values of $(1 + x)^{1/x}$ in the table where $(1 + x)^{1/x}$ is calculated for $x = 0.1, 0.01, 0.001$, and 0.0001.

26. Find the second derivative of the function $y = x^2 \ln x$.

27. Find the slope of a line tangent to the curve of $y = \ln \cos x$ at $x = \pi/4$.

28. Find the slope of a line tangent to the curve of $y = x \ln 2x$ at $x = 2$.

29. Find the derivative of $y = x^x$ by first taking logarithms of each side of the equation.

30. Find the derivative of $y = (\sin x)^x$ by first taking logarithms of each side of the equation.

31. If the loudness b (in decibels) of a sound of intensity I is given by $b = 10 \log(I/I_0)$ where I_0 is a constant, find the expression for db/dt in terms of dI/dt.

32. The temperature T from the inner surface of a certain steam pipe to the outer surface is given by

$$T = 100 - 40 \, \frac{\ln(r/5)}{\ln(6/5)}$$

where r is the distance from the center of the pipe. Determine the rate of change of temperature with respect to r. (This solution is valid only for $5 < r < 6$, since these are the inner and outer radii, respectively.)

7–3 Derivative of the Exponential Function

To obtain the derivative of the exponential function, we let $y = b^u$ and then write the equation in logarithmic form:

$$u = \log_b y, \qquad \frac{du}{dx} = \frac{1}{y} \log_b e \, \frac{dy}{dx}$$

$$\frac{dy}{dx} = \frac{1}{\log_b e} y \, \frac{du}{dx} = \frac{1}{\log_b e} b^u \left(\frac{du}{dx}\right)$$

Thus,

$$\frac{d(b^u)}{dx} = \frac{1}{\log_b e} b^u \left(\frac{du}{dx}\right) \qquad (7\text{–}14)$$

If we let $b = e$, Eq. (7–14) becomes

$$\frac{d(e^u)}{dx} = e^u\left(\frac{du}{dx}\right) \qquad\qquad (7\text{–}15)$$

The simplicity of Eq. (7–15) compared with Eq. (7–14) again shows the advantage of choosing e as the basis of natural logarithms. It is for this reason that e appears so often in applications of calculus.

Example A
Find the derivative of $y = e^x$.
   Using Eq. (7–15), we have

$$\frac{dy}{dx} = e^x(1) = e^x$$

We see that the derivative of the function e^x equals itself. This exponential function is widely used in applications of calculus.

Example B
Find the derivative of $y = 2^{4x}$.
   Using Eq. (7–14), we have

$$\frac{dy}{dx} = \frac{1}{\log_2 e}2^{4x}(4)$$

$$= \frac{4}{\log_2 e}2^{4x}$$

Example C
Find the derivative of $y = \ln \cos e^{2x}$.
   Using Eq. (7–15) along with the derivatives of the logarithmic and cosine functions, we have

$$\frac{dy}{dx} = \frac{1}{\cos e^{2x}}(-\sin e^{2x})(e^{2x})(2)$$

$$= -2e^{2x}\tan e^{2x}$$

Example D
Find the derivative of $y = x\,e^{\tan x}$.
   Using Eq. (7–15) along with the derivatives of a product and the tangent, we have

$$\frac{dy}{dx} = x(e^{\tan x})\sec^2 x + e^{\tan x}$$

$$= e^{\tan x}(x\,\sec^2 x + 1)$$

Example E

Find the derivative of $y = (e^{1/x})^2$.

In this example,

$$\frac{dy}{dx} = 2(e^{1/x})(e^{1/x})\left(-\frac{1}{x^2}\right) = \frac{-2(e^{1/x})^2}{x^2}$$

$$= \frac{-2e^{2/x}}{x^2}$$

This problem could have also been solved by first writing the function as $y = e^{2/x}$, which is an equivalent form determined by the laws of exponents. When we use this form, the derivative becomes

$$\frac{dy}{dx} = e^{2/x}\left(-\frac{2}{x^2}\right)$$

$$= \frac{-2e^{2/x}}{x^2}$$

We can see that this change in form of the function simplifies the steps necessary for finding the derivative.

Exercises 7–3

In Exercises 1 through 24 find the derivatives of the given functions.

1. $y = 3^{2x}$
2. $y = 3^{1-x}$
3. $y = 4^{6x}$
4. $y = 10^{x^2}$
5. $y = e^{6x}$
6. $y = 3e^{x^2}$
7. $y = e^{\sqrt{x}}$
8. $y = e^{2x^4}$
9. $y = xe^x$
10. $y = x^2e^{2x}$
11. $y = xe^{\sin x}$
12. $y = e^x\sin x$

13. $y = \dfrac{3e^{2x}}{x+1}$
14. $y = \dfrac{e^x}{x}$
15. $y = e^{-3x}\sin 4x$

16. $y = (\cos 2x)(e^{x^2-1})$
17. $y = e^{2x} - e^{-2x}$
18. $y = 4e^{-(2/x)}$
19. $y = e^{2x}\ln x$
20. $y = e^{x^2}\ln \cos x$
21. $y = \ln \sin 2e^{6x}$
22. $y = \tan e^{x+1}$
23. $y = 2 \text{ Arcsin } e^{2x}$
24. $y = \text{Arctan } e^{3x}$

In Exercises 25 through 32 solve the given problems.

25. Show that $y = xe^{-x}$ satisfies the equation $(dy/dx) + y = e^{-x}$.

26. Show that $y = e^{-x}\sin x$ satisfies the equation

$$\frac{d^2y}{dx^2} + 2\frac{dy}{dx} + 2y = 0$$

27. Show that the slope of the curve $y = e^x$ at $x = a$ equals the ordinate at $x = a$.

28. The Beer–Lambert law of light absorption may be expressed as $I = I_0e^{-\alpha x}$ where I/I_0 is that fraction of the incident light beam which is transmitted, α is a constant, and x is the distance the light travels through the medium. Find the rate of change of I with respect to x.

29. In a particular electric circuit, the voltage is given by $V = 100(1 - e^{-0.1t})$. Find the expression for dV/dt.

30. The displacement s of an object is given by $s = e^{-0.02t}\cos 2t$. Find the expression for the velocity of the object.

31. By referring to Exercise 38 of Section 7–1, show that

$$\frac{d}{dx}\sinh u = \cosh u \frac{du}{dx}$$

where u is a function of x.

32. By referring to Exercise 38 of Section 7–1, show that

$$\frac{d}{dx}\cosh u = \sinh u \frac{du}{dx}$$

where u is a function of x.

7–4 Applications

The following examples show applications of the logarithmic and exponential functions to curve-tracing and time-rate-of-change problems. Certain other applications are indicated in the exercises.

Example A

Sketch the graph of the function $y = x \ln x$.

The only intercept is $(1, 0)$ (there is no intercept for $x = 0$ since $\log_b 0$ is not defined). Thus,

$$\frac{dy}{dx} = x\left(\frac{1}{x}\right) + \ln x = 1 + \ln x, \qquad \frac{d^2y}{dx^2} = \frac{1}{x}$$

The first derivative is zero if $\ln x = -1$, or $x = e^{-1}$. The second derivative is positive for this value of x. Thus, there is a minimum point at $(1/e, -1/e)$. The only values of x which can be considered are positive values, since $\ln x$ is undefined for negative values. The second derivative indicates that the curve is always concave up. The graph is shown in Fig. 7–4.

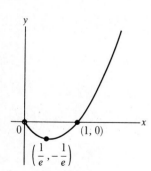

Figure 7–4

Example B

Sketch the graph of the function $y = e^{-x}\cos x$ $(0 \leq x \leq 2\pi)$.

This curve has intercepts for all values for which $\cos x$ is zero. Those values between 0 and 2π for which $\cos x = 0$ are $x = \pi/2$ and $x = 3\pi/2$. Also, $(0, 1)$ is an intercept. We next write

$$\frac{dy}{dx} = -e^{-x}\sin x - e^{-x}\cos x = -e^{-x}(\sin x + \cos x)$$

Setting the first derivative equal to zero, we obtain $\sin x + \cos x = 0$, or $\tan x = -1$. There are maximum or minimum points for $x = 3\pi/4$ and $x = 7\pi/4$ (see Fig. 7–5). We then write

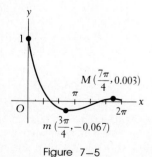

Figure 7–5

$$\frac{d^2y}{dx^2} = 2e^{-x}\sin x$$

Thus, d^2y/dx^2 is positive when $x = 3\pi/4$ and negative when $x = 7\pi/4$. Hence, $(3\pi/4, -0.067)$ is a minimum and $(7\pi/4, 0.003)$ is a maximum. From the second derivative we find that points of inflection occur for $x = 0, \pi,$ and 2π.

Example C

The population P, in terms of the population P_0 (at $t = 0$) and t, is given by $P = P_0 e^{kt}$. Show that the time-rate of change of population is proportional to the population present at time t.

By taking a derivative of P with respect to time, we have

$$\frac{dP}{dt} = kP_0 e^{kt} = kP$$

Example D

A rocket moving vertically, if the only force acting on it is due to gravity and its mass is decreasing at a constant rate r, has its velocity as a function of time given by

$$v = v_0 - gt - k \ln\left(1 - \frac{rt}{m_0}\right)$$

where v_0 is the initial velocity, m_0 is the initial mass, and k is a constant. Determine the expression for the acceleration.

Since the acceleration is the time-rate of change of the velocity, we must find dv/dt. Therefore,

$$\frac{dv}{dt} = -g - k\frac{1}{1 - \dfrac{rt}{m_0}}\left(\frac{-r}{m_0}\right) = -g + \frac{km_0}{m_0 - rt}\left(\frac{r}{m_0}\right)$$

$$= \frac{kr}{m_0 - rt} - g$$

Exercises 7-4

In Exercises 1 through 12 sketch the graphs of the given functions.

1. $y = \ln \cos x$ 2. $y = \dfrac{\ln x}{x}$ 3. $y = xe^{-x}$ 4. $y = xe^x$

5. $y = \ln \dfrac{1}{x^2 + 1}$ 6. $y = \ln \dfrac{1}{x}$ 7. $y = e^{-x^2}$ 8. $y = x - e^x$

9. $y = \ln x - x$ 10. $y = e^{-x}\sin x$

11. $y = \frac{1}{2}(e^x - e^{-x})$ (See Exercise 38 of Section 7-1.)

12. $y = \frac{1}{2}(e^x + e^{-x})$ (See Exercise 38 of Section 7-1.)

In Exercises 13 through 24 solve the given problems by finding the appropriate derivatives.

13. Find the equation of the line tangent to the curve $y = x^2 \ln x$ at the point $(1, 0)$.

14. Find the equation of the line normal to the curve of $y = \dfrac{e^{2x}}{x}$ at $x = 1$.

15. The power supply P (in watts) in a satellite is given by $P = 100e^{-0.005t}$, where t is measured in days. Find the time-rate of change of power after 100 days.

16. The number N of atoms of radium at any time t is given in terms of the number at $t = 0$, N_0, by $N = N_0 e^{-kt}$. Show that the time-rate of change of N is proportional to N.

17. The velocity of an object moving through a resisting medium is found to obey the relation $v = 40(1 - e^{-0.05t})$. Find the acceleration when $t = 10$.

18. The charge on a capacitor in a circuit containing the capacitor of capacitance C, a resistance R, and voltage source of voltage E is given by $q = CE(1 - e^{-t/RC})$. Show that this equation satisfies the equation

$$R\frac{dq}{dt} + \frac{q}{C} = E$$

19. Assuming that force is proportional to acceleration, show that a particle moving along the x-axis, so that its displacement $x = ae^{kt} + be^{-kt}$, has a force acting on it which is proportional to its displacement.

20. The radius of curvature at any point on a curve is given by

$$R = \frac{[1 + (dy/dx)^2]^{3/2}}{d^2y/dx^2}$$

Determine the radius of curvature of $y = \ln \sec x$ when $x = \pi/4$.

21. An object on the end of a spring is moving so that its displacement in centimeters from the equilibrium position is given by

$$y = e^{-0.5t}(0.4 \cos 6t - 0.2 \sin 6t)$$

Find the expression for the velocity of the object. What is the velocity when $t = \pi/12$ s? The motion described by this equation is called *damped harmonic motion*. The general shape of the curve of displacement as a function of time is that of the curve in Example B.

22. In an electronic device, the maximum current density i_m as a function of the temperature T is given by $i_m = AT^2 e^{k/T}$, where A and k are constants. If the temperature is changing with time, find the expression for the time rate of change of i_m.

23. In developing the theory dealing with the friction between a pulley wheel and the pulley belt, the ratio of the tensions in the belt on either side of the wheel is given by

$$\frac{T_2}{T_1} = e^{k/\sin(\theta/2)}$$

where k is a constant and θ is the angle of the opening of the pulley wheel. Find the expression for a small change in the ratio of tensions for a small change in the angle θ.

24. Find the area of the largest rectangle which can be inscribed in the first quadrant under the curve of $y = e^{-x}$.

7—5 Review Exercises for Chapter 7

In Exercises 1 through 20 find the derivative of each of the given functions.

1. $y = 3 \ln(x^2 + 1)$

2. $y = \ln(1 - 5x)^3$

3. $y = (e^{x-3})^2$

4. $y = e^{\sqrt{1-x}}$

5. $y = \ln(\csc x^2)$

6. $y = \ln(3 + \sin x^2)$

7. $y = [\ln(3 + \sin x)]^2$

8. $y = \ln(3 + \sin x)^2$

9. $y = e^{\sin 2x}$

10. $y = 3 e^{\sec 3x}$

11. $y = x^3 e^x$

12. $y = 2x\, e^{4x}$

13. $y = x^2 \ln x$

14. $y = (x + 1)\ln(x^3 + 1)$

15. $y = \dfrac{e^{2x}}{x^2 + 1}$

16. $y = \dfrac{\ln x}{2x + 1}$

17. $y = e^{-x}\sec x$

18. $y = e^{3x}\ln x$

19. $x^2 \ln y = y + x$

20. $y = x^2(e^{\cos^2 x})^2$

In Exercises 21 through 40 solve the given problems by finding the appropriate derivatives.

21. Find the equation of the line tangent to the curve $y = \ln \cos x$ where $x = \pi/6$.

22. Find the equation of the line normal to the curve $y = e^{x^2}$ where $x = \frac{1}{2}$.

23. Sketch the graph of $y = \ln(1 + x)$.

24. Sketch the graph of $y = x(\ln x)^2$.

25. Fans are exhausting a room such that the number of cubic feet of carbon dioxide in the room after t minutes is given by $n = 3(e^{-0.05t} + 1)$. Find the time rate of change of n with respect to t. Show that the result can be written as $0.05(3 - n)$.

26. A businessman determined that the gross income I he received by selling x items per week was $I = 100xe^{-x/10}$. How many items should he sell per week in order that I is a maximum?

27. If inflation makes the dollar worth 5% less each year, then the value of $100 in t years will be $V = 100(0.95)^t$. What is the approximate change in the value during the fourth year?

28. An analysis of samples of air for a city showed that the number of parts per million (PPM) of sulfur dioxide on a certain day was PPM $= 0.05 \ln(2 + 24t - t^2)$, where t was the hour of the day. What was the approximate change in the amount of sulfur dioxide in the air between 10 AM and noon?

29. For a circuit containing a capacitance C and a resistance R, the current as a function of time is given by $i = i_0 e^{-t/RC}$. If $i_0 = 10$ A, $R = 2\ \Omega$, and $C = 50\ \mu$F, find the approximate change in the current between 200 μs and 210 μs.

30. Under certain circumstances, 2% of a substance changes chemically each minute. If there are originally 100 g present, the amount at any time t which is present is $A = 100(0.98)^t$. What is the approximate change in the amount present during the fourth second?

31. Under certain conditions the work done by a gas in an expansion from a pressure P_1 to a pressure P_2 is given by $W = k \ln(P_1/P_2)$, where k is a constant depending on the gas. If P_2 varies with time at the rate of 2 kPa/min, find the rate (in joules per minute) at which the work changes when $P_2 = 300$ kPa, if $k = 2400$ J/mol and $P_1 = 200$ kPa.

32. The charge q on a certain capacitor as a function of time is given by

$$q = e^{-0.1t}(0.2 \sin 100t + \cos 100t).$$

The current i in the circuit is the time-rate of change of charge. Find the expression for i as a function of t.

33. An object is dropped from a weather balloon. The distance (in feet) it falls, assuming a resisting force of the air on the object, is given by

$$y = 320(t + 10e^{-0.1t} - 10).$$

Find the velocity after 10 s.

34. The velocity of an object moving through a resisting medium is found to obey the relation $v = 40(1 - e^{-0.05t})$. Find the acceleration when $t = 10$.

35. Find the area of the largest rectangle which can be inscribed under the curve of $y = e^{-x^2}$ in the first quadrant.

36. A revolving light 300 ft from a straight wall makes 6 r/min. Find the velocity of the beam along the wall at the instant it makes an angle of $45°$ with the wall.

37. Show, by graphical differentiation, that $de^x/dx = e^x$.

38. Show, by graphical differentiation, that $d \ln x/dx = 1/x$.

39. Find the acceleration of an object whose x- and y-components of displacement are given by $x = e^{-t}\sin 2t$ and $y = e^{-t}\cos 2t$ when $t = \pi/4$.

40. The curve formed by a uniform cable hanging from two points under its own weight is called the *catenary*. The equation of this curve is

$$y = \frac{H}{w} \cosh \frac{wx}{H}$$

where w and H are constants. Show that the equation of the catenary satisfies the equation

$$\frac{d^2y}{dx^2} = \frac{w}{H} \sqrt{1 + \left(\frac{dy}{dx}\right)^2}$$

(see Exercise 38 of Section 7–1, and Exercises 31 and 32 of Section 7–3).

8

Methods of Integration

8–1 The General Power Formula

Having developed the derivatives of the basic transcendental functions, we can now expand considerably the functions which we can integrate. Many algebraic functions which we were unable to integrate previously will now lend themselves to integration. The first integration formula to be discussed is the general power formula.

For reference, we repeat here the general power formula for integration, Eq. (4–5):

$$\int u^n du = \frac{u^{n+1}}{n+1} + C \qquad n \neq -1 \tag{8–1}$$

This formula will now be applied to transcendental functions as well as algebraic functions. When we are applying the general power formula, we must properly recognize the quantities u, n, and du. This requires familiarity with the differential forms presented in Chapters 2, 6, and 7. The following examples illustrate the use of the general power formula.

Example A

Integrate: $\int \sin^3 x \cos x \, dx$.

Since $d(\sin x) = \cos x \, dx$, we note that this integral fits the form of Eq. (8–1) for $u = \sin x$. Thus, with $u = \sin x$, we have $du = \cos x \, dx$, which means that this integral is of the form $\int u^3 du$. Therefore, the integration can now be completed.

$$\int \sin^3 x \cos x \, dx = \int \sin^3 x (\cos x \, dx)$$

$$= \frac{1}{4} \sin^4 x + C$$

We note here that the factor $\cos x$ is a necessary part of the du in order to have the proper form of integration, and therefore does not appear in the final result.

Example B

Integrate: $\int \sqrt{1 + \tan x} \, \sec^2 x \, dx$.

Here we note that $d(\tan x) = \sec^2 x \, dx$, which means that the integral fits the form of Eq. (8–1) with $u = 1 + \tan x$, $du = \sec^2 x \, dx$, and $n = \frac{1}{2}$. The integral is of the form $\int u^{1/2} du$. Thus,

$$\int \sqrt{1 + \tan x} \, (\sec^2 x \, dx) = \frac{2}{3} (1 + \tan x)^{3/2} + C$$

Example C

Integrate: $\int \ln x \left(\frac{dx}{x} \right)$.

By noting that $d(\ln x) = \frac{dx}{x}$, we have $u = \ln x$, $du = \frac{dx}{x}$, and $n = 1$. This means that the integral is of the form $\int u \, du$. Thus,

$$\int \ln x \left(\frac{dx}{x} \right) = \frac{1}{2} (\ln x)^2 + C$$

Example D

Integrate: $\int (2 + 3e^{2t})^3 e^{2t} dt$.

In this case $u = 2 + 3e^{2t}$, $du = 6e^{2t} dt$, and $n = 3$. This means that we must introduce a 6 to complete the differential in order that the integral fits the proper form. Also, a $\frac{1}{6}$ must then be placed before the integral. This means that the integration is as follows:

$$\int (2 + 3e^{2t})^3 e^{2t} dt = \frac{1}{6} \int (2 + 3e^{2t})^3 (6e^{2t} dt)$$

$$= \frac{1}{6} \cdot \frac{1}{4} (2 + 3e^{2t})^4 + C$$

$$= \frac{1}{24} (2 + 3e^{2t})^4 + C$$

Example E

Find the value of

$$\int_0^{0.5} \frac{\text{Arcsin } x}{\sqrt{1 - x^2}} \, dx; \quad n = 1, \quad u = \text{Arcsin } x, \quad \text{and} \quad du = \frac{dx}{\sqrt{1 - x^2}}$$

Therefore,

$$\int_{0.}^{0.5} \frac{\text{Arcsin } x}{\sqrt{1 - x^2}} \, dx = \frac{(\text{Arcsin } x)^2}{2} \Big|_0^{0.5} = \frac{(\pi/6)^2}{2} - 0 = \frac{\pi^2}{72}$$

Exercises 8–1

In Exercises 1 through 24 integrate each of the given functions.

1. $\displaystyle\int \sin^4 x \, \cos x \, dx$

2. $\displaystyle\int \cos^5 x \, (-\sin x \, dx)$

3. $\displaystyle\int \sqrt{\cos x} \, \sin x \, dx$

4. $\displaystyle\int \sin^{1/3} x \, \cos x \, dx$

5. $\displaystyle\int \tan^2 x \, \sec^2 x \, dx$

6. $\displaystyle\int \sec^3 x \, (\sec x \, \tan x) \, dx$

7. $\displaystyle\int_0^{\pi/8} \cos 2x \, \sin 2x \, dx$

8. $\displaystyle\int_{\pi/6}^{\pi/4} \sqrt{\cot x} \, \csc^2 x \, dx$

9. $\displaystyle\int (\text{Arcsin } x)^3 \left(\frac{dx}{\sqrt{1 - x^2}} \right)$

10. $\displaystyle\int \frac{(\text{Arccos } 2x)^4 \, dx}{\sqrt{1 - 4x^2}}$

11. $\displaystyle\int \frac{\text{Arctan } 5x}{1 + 25x^2} \, dx$

12. $\displaystyle\int \frac{\text{Arcsin } 4x \, dx}{\sqrt{1 - 16x^2}}$

13. $\displaystyle\int [\ln(x + 1)]^2 \, \frac{dx}{x + 1}$

14. $\displaystyle\int (3 + \ln 2x)^3 \, \frac{dx}{x}$

15. $\displaystyle\int_0^{1/2} \frac{\ln(2x + 3)}{2x + 3} \, dx$

16. $\displaystyle\int_1^e \frac{(1 - 2 \ln x) \, dx}{x}$

17. $\displaystyle\int (4 + e^x)^3 e^x \, dx$

18. $\displaystyle\int \sqrt{1 - e^{-x}} \, (e^{-x} dx)$

19. $\displaystyle\int (2e^{2x} - 1)^{1/3} e^{2x} \, dx$

20. $\displaystyle\int \frac{(1 + 3e^{-2x})^4 \, dx}{e^{2x}}$

21. $\displaystyle\int (1 + \sec^2 x)^4 \, (\sec^2 x \, \tan x \, dx)$

22. $\displaystyle\int (e^x + e^{-x})^{1/4} (e^x - e^{-x}) \, dx$

23. $\displaystyle\int_0^{\pi/6} \frac{\tan x}{\cos^2 x} \, dx$

24. $\displaystyle\int_{\pi/3}^{\pi/2} \frac{\sin \theta \, d\theta}{\sqrt{1 + \cos \theta}}$

In Exercises 25 through 28 solve the given problems by integration.

25. Find the area under the curve $y = \sin x \cos x$ from $x = 0$ to $x = \pi/2$.

26. In developing the expression for the total pressure P on a wall due to molecules with mass m and velocity v striking the wall, the following equation is found: $P = mnv^2 \int_0^{\pi/2} \sin \theta \cos^2\theta \ d\theta$. The symbol n represents the number of molecules per unit volume and θ represents the angle between a perpendicular to the wall and the direction of the molecule. Find the final expression for P.

27. The general expression for the slope of a given curve is $(\ln x)^2/x$. If the curve passes through $(1, 2)$, find its equation.

28. The voltage across a certain capacitor is given by $V = 10^6 \int e^t dt/\sqrt{3 + e^t}$. Determine the function which gives the voltage in terms of time.

8–2 The Basic Logarithmic Form

The general power formula for integration, Eq. (8–1), is valid for all values of n except $n = -1$. If n were set equal to -1, this would cause the result to be undefined. When we obtained the derivative of the logarithmic function, we found that

$$\frac{d(\ln u)}{dx} = \frac{1}{u}\frac{du}{dx}$$

which means the differential of the logarithmic form is $d(\ln u) = du/u$. Reversing the process, we then determine that $\int du/u = \ln u + C$. In other words, when the exponent of the expression being integrated is -1, it is a logarithmic form.

Logarithms are defined only for positive numbers. Thus, $\int du/u = \ln u + C$ is valid if $u > 0$. If $u < 0$, then $-u > 0$. In this case $d(-u) = -du$, or $\int (-du)/(-u) = \ln(-u) + C$. However, $\int du/u = \int (-du)/(-u)$. These results can be combined into the single form

$$\int \frac{du}{u} = \ln |u| + C \tag{8–2}$$

Example A

Integrate: $\displaystyle\int \frac{dx}{x + 1}$.

Since $d(x + 1) = dx$, this integral fits the form of Eq. (8–2) with $u = x + 1$ and $du = dx$. Therefore, we have

$$\int \frac{dx}{x + 1} = \ln |x + 1| + C$$

Example B

Integrate: $\displaystyle\int \frac{\cos x}{\sin x}\, dx.$

We note that $d(\sin x) = \cos x\, dx$. This means that this integral fits the form of Eq. (8–2) with $u = \sin x$ and $du = \cos x\, dx$. Thus,

$$\int \frac{\cos x}{\sin x}\, dx = \int \frac{\cos x\, dx}{\sin x}$$

$$= \ln|\sin x| + C$$

Example C

Integrate: $\displaystyle\int \frac{x\, dx}{4 - x^2}.$

This integral fits the form of Eq. (8–2) with $u = 4 - x^2$ and $du = -2x\, dx$. This means that we must introduce a factor of -2 into the numerator and a factor of $-\frac{1}{2}$ before the integral. Therefore,

$$\int \frac{x\, dx}{4 - x^2} = -\frac{1}{2}\int \frac{-2x\, dx}{4 - x^2}$$

$$= -\frac{1}{2}\ln|4 - x^2| + C$$

We should note that if the quantity $4 - x^2$ were raised to any power other than that in the example, we would have to employ the general power formula for integration. For example,

$$\int \frac{x\, dx}{(4 - x^2)^2} = -\frac{1}{2}\int \frac{-2x\, dx}{(4 - x^2)^2} = -\frac{1}{2}\frac{(4 - x^2)^{-1}}{-1} + C$$

$$= \frac{1}{2(4 - x^2)} + C$$

Example D

Integrate: $\displaystyle\int \frac{e^{4x}dx}{1 + 3e^{4x}};\ u = 1 + 3e^{4x},\ du = 12e^{4x}dx.$

We write

$$\int \frac{e^{4x}dx}{1 + 3e^{4x}} = \frac{1}{12}\int \frac{12e^{4x}dx}{1 + 3e^{4x}} = \frac{1}{12}\ln|1 + 3e^{4x}| + C$$

$$= \frac{1}{12}\ln(1 + 3e^{4x}) + C \qquad (\text{since } 1 + 3e^{4x} > 0)$$

Example E

Evaluate: $\displaystyle\int_0^{\pi/8} \frac{\sec^2 2x}{1 + \tan 2x}\, dx;\ u = 1 + \tan 2x,\ du = 2 \sec^2 2x\, dx.$

We write

$$\int_0^{\pi/8} \frac{\sec^2 2x}{1 + \tan 2x}\, dx = \frac{1}{2}\int_0^{\pi/8} \frac{2 \sec^2 2x\, dx}{1 + \tan 2x}$$

$$= \frac{1}{2} \ln|1 + \tan 2x|\, \Big|_0^{\pi/8} = \frac{1}{2}\ln|2| - \frac{1}{2}\ln|1|$$

$$= \frac{1}{2}\ln 2 - \frac{1}{2}\ln 1 = \frac{1}{2}\ln 2 - 0 = \frac{1}{2}\ln 2$$

Exercises 8–2

In Exercises 1 through 24 integrate each of the given functions.

1. $\displaystyle\int \frac{dx}{1 + 4x}$

2. $\displaystyle\int \frac{dx}{1 - 4x}$

3. $\displaystyle\int \frac{x\, dx}{4 - 3x^2}$

4. $\displaystyle\int \frac{x^2 dx}{1 - x^3}$

5. $\displaystyle\int \frac{\csc^2 2x}{\cot 2x}\, dx$

6. $\displaystyle\int \frac{\sin x}{\cos x}\, dx$

7. $\displaystyle\int_0^{\pi/2} \frac{\cos x\, dx}{1 + \sin x}$

8. $\displaystyle\int_0^{\pi/4} \frac{\sec^2 x\, dx}{4 + \tan x}$

9. $\displaystyle\int \frac{e^{-x}}{1 - e^{-x}}\, dx$

10. $\displaystyle\int \frac{e^{3x}}{1 - e^{3x}}\, dx$

11. $\displaystyle\int \frac{1 + e^x}{x + e^x}\, dx$

12. $\displaystyle\int \frac{e^x - e^{-x}}{e^x + e^{-x}}\, dx$

13. $\displaystyle\int \frac{\sec x \tan x\, dx}{1 + 4 \sec x}$

14. $\displaystyle\int \frac{\sin 2x}{1 - \cos^2 x}\, dx$

15. $\displaystyle\int_1^3 \frac{1 + x}{4x + 2x^2}\, dx$

16. $\displaystyle\int_1^2 \frac{4x + 6x^2}{x^2 + x^3}\, dx$

17. $\displaystyle\int \frac{dx}{x \ln x}$

18. $\displaystyle\int \frac{dx}{x(1 + 2 \ln x)}$

19. $\displaystyle\int \frac{2 + \sec^2 x}{2x + \tan x}\, dx$

20. $\displaystyle\int \frac{x + \cos 2x}{x^2 + \sin 2x}\, dx$

21. $\displaystyle\int \frac{dx}{\sqrt{1 - 2x}}$

22. $\displaystyle\int \frac{x\, dx}{(1 + x^2)^2}$

23. $\displaystyle\int_0^{\pi/12} \frac{\sec^2 3x}{4 + \tan 3x}\, dx$

24. $\displaystyle\int_1^2 \frac{x^2 + 1}{x^3 + 3x}\, dx$

In Exercises 25 through 32 solve the given problems by integration.

25. Find the area bounded by $y(x + 1) = 1$, $x = 0$, $y = 0$, and $x = 2$.

26. Find the area bounded by $xy = 1$, $x = 1$, $x = 2$, and $y = 0$.

27. Find the volume generated by rotating the area bounded by $y = 1/(x^2 + 1)$, $x = 0$, $x = 1$, $y = 0$ about the y-axis. (Use shells.)

28. Find the average value of the function $xy = 4$ from $x = 1$ to $x = 2$.

29. If the current (in amperes), as a function of time, in a certain electric circuit is given by $i = t/(1 + 2t^2)$, find the total charge to pass a given point in the circuit in the first two seconds.

30. The electric current in a certain inductor is given by

$$i = 5 \int \frac{e^t \cos e^t \, dt}{\sin e^t}$$

Find the function which expresses the current in terms of the time.

31. Under certain conditions an object is subject to a force (in pounds) given by $F = 1/(1 + 3x)$. Find the work done on the object in moving it 5 ft from the origin.

32. Conditions are often such that a force proportional to the velocity tends to retard the motion of an object. Under such conditions, the acceleration of a certain object moving down an inclined plane is given by $20 - v$. This leads to the equation $dv/(20 - v) = dt$. If the object starts from rest, find the expression for the velocity as a function of time.

8–3 The Exponential Form

In deriving the derivative for the exponential function we obtained the result $de^u/dx = e^u(du/dx)$. This means that the differential of the exponential form is $d(e^u) = e^u du$. Reversing this form to find the proper form of the integral for the exponential function, we have

$$\int e^u du = e^u + C \qquad\qquad (8\text{–}3)$$

Example A
Integrate: $\int e^{5x} \, dx$.

Since $d(5x) = 5 \, dx$, this integral can be put in the form of Eq. (8–3) with $u = 5x$ and $du = 5 \, dx$. This means that we place a 5 with the dx and a $\frac{1}{5}$ before the integral. Therefore,

$$\int e^{5x} dx = \frac{1}{5} \int e^{5x}(5 \, dx)$$

$$= \frac{1}{5} e^{5x} + C$$

Example B
Integrate: $\int x e^{x^2} \, dx$.

Since $d(x^2) = 2x \, dx$, we can write this integral in the form of Eq. (8–3) with $u = x^2$ and $du = 2x \, dx$. Thus,

$$\int x e^{x^2} dx = \frac{1}{2} \int e^{x^2}(2x \, dx)$$

$$= \frac{1}{2} e^{x^2} + C$$

Example C

Integrate: $\int \dfrac{dx}{e^{3x}}.$

This integral can be put in proper form by writing it as $\int e^{-3x} dx$. In this form $u = -3x$, $du = -3\ dx$. Thus,

$$\int \frac{dx}{e^{3x}} = \int e^{-3x} dx = -\frac{1}{3} \int e^{-3x}(-3\ dx)$$

$$= -\frac{1}{3} e^{-3x} + C$$

Example D

Evaluate: $\int_0^{\pi/2} (\sin\ 2x)\ (e^{\cos 2x})\ dx;\ u = \cos\ 2x,\ du = -2 \sin\ 2x\ dx.$

We write

$$\int_0^{\pi/2} (\sin\ 2x)\ (e^{\cos 2x})\ dx = -\frac{1}{2} \int_0^{\pi/2} (e^{\cos 2x})(-2 \sin\ 2x\ dx)$$

$$= -\frac{1}{2} e^{\cos 2x} \Big|_0^{\pi/2} = -\frac{1}{2}\Big(\frac{1}{e} - e\Big) = 1.175$$

Example E

Find the equation of the curve for which $\dfrac{dy}{dx} = \dfrac{e^{\sqrt{x+1}}}{\sqrt{x+1}}$ if the curve passes through $(0, 1)$.

The solution of this problem requires that we integrate the given function and then evaluate the constant of integration. Hence,

$$dy = \frac{e^{\sqrt{x+1}}}{\sqrt{x+1}} dx, \quad \int dy = \int \frac{e^{\sqrt{x+1}}}{\sqrt{x+1}} dx$$

For purposes of integrating the righthand side,

$$u = \sqrt{x+1} \quad \text{and} \quad du = \frac{1}{2\sqrt{x+1}} dx$$

$$y = 2 \int e^{\sqrt{x+1}} \Big(\frac{1}{2\sqrt{x+1}} dx\Big) = 2e^{\sqrt{x+1}} + C$$

Letting $x = 0$ and $y = 1$, we have $1 = 2e + C$, or $C = 1 - 2e$. This means that the equation is

$$y = 2e^{\sqrt{x+1}} + 1 - 2e$$

Exercises 8–3

In Exercises 1 through 20 integrate each of the given functions.

1. $\int e^{7x}(7\ dx)$

2. $\int e^{x^4}(4x^3\,dx)$

3. $\int e^{2x+5}\,dx$

4. $\int e^{-4x}\,dx$

5. $\int_0^2 e^{x/2}\,dx$

6. $\int_1^2 3e^{4x}\,dx$

7. $\int x^2 e^{x^3}\,dx$

8. $\int xe^{-x^2}\,dx$

9. $\int \dfrac{e^{\sqrt{x}}}{\sqrt{x}}\,dx$

10. $\int 4x^3 e^{2x^4}\,dx$

11. $\int (\sec x \tan x)\,e^{2\sec x}\,dx$

12. $\int (\sec^2 x)\,e^{\tan x}\,dx$

13. $\int \dfrac{(3 - e^x)\,dx}{e^{2x}}$

14. $\int (e^x - e^{-x})^2\,dx$

15. $\int \dfrac{e^{\text{Arctan } x}}{x^2 + 1}\,dx$

16. $\int \dfrac{e^{\text{Arcsin } 2x}\,dx}{\sqrt{1 - 4x^2}}$

17. $\int \dfrac{e^{\cos 3x}\,dx}{\csc 3x}$

18. $\int \dfrac{e^{2/x}\,dx}{x^2}$

19. $\int_0^\pi (\sin 2x)\,e^{\cos^2 x}\,dx$

20. $\int_{-1}^1 \dfrac{dx}{e^{2-3x}}$

In Exercises 21 through 28 solve the given problems by integration.

21. Find the area bounded by $y = e^x$, $x = 0$, $y = 0$, and $x = 2$.
22. Prove that the area bounded by $x = a$, $x = b$, $y = 0$, and $y = e^x$ equals the difference of the ordinates at $x = b$ and $x = a$.
23. Find the volume generated by rotating the area bounded by $y = e^{x^2}$, $x = 1$, $y = 0$, and $x = 2$ about the y-axis.
24. Find the equation of the curve for which $dy/dx = \sqrt{e^{x+3}}$ if the curve passes through $(1, 0)$.
25. Find the average value of the function $y = e^{2x}$ from $x = 0$ to $x = 4$.
26. Find the moment of inertia with respect to the y-axis of the first quadrant area bounded by $y = e^{x^3}$, $x = 1$, and the axes.
27. If the current in a certain electric circuit is given by $i = 2e^{-30t}$, find the expression for the voltage across a 250 μF capacitor as a function of time. The initial voltage across the capacitor is 60 V.
28. A particle moves such that its velocity is given by $v = e^{\sin 2x}\cos 2x\ dx$. Find the equation for the displacement as a function of the time if $s = 2$ when $t = 0$.

8–4 Basic Trigonometric Forms

In this section we shall discuss the integrals of the six trigonometric functions and the trigonometric integrals which arise directly from reversing the formulas for differentiation. Other trigonometric forms will be discussed later.

By directly reversing the differentiation formulas, the following integral formulas are obtained:

$$\int \sin u \; du = -\cos u + C \tag{8-4}$$

$$\int \cos u \; du = \sin u + C \tag{8-5}$$

$$\int \sec^2 u \; du = \tan u + C \tag{8-6}$$

$$\int \csc^2 u \; du = -\cot u + C \tag{8-7}$$

$$\int \sec u \tan u \; du = \sec u + C \tag{8-8}$$

$$\int \csc u \cot u \; du = -\csc u + C \tag{8-9}$$

Example A

Integrate: $\int \sin 3x \; dx$; $u = 3x$, $du = 3 \; dx$.
We have

$$\int \sin 3x \; dx = \frac{1}{3} \int \sin 3x (3 \; dx)$$

$$= -\frac{1}{3}\cos 3x + C$$

Example B

Integrate: $\int x \sec^2 (x^2) \, dx$; $u = x^2$, $du = 2x \; dx$.
We write

$$\int x \sec^2 (x^2) \, dx = \frac{1}{2} \int \sec^2 (x^2) \cdot (2x \; dx)$$

$$= \frac{1}{2}\tan(x^2) + C$$

Example C

Integrate: $\displaystyle \int \frac{\tan 2x}{\cos 2x} \, dx$.

By using the basic identity $\sec \theta = 1/\cos \theta$, we can transform this integral to form $\int \sec 2x \tan 2x \; dx$. In this form $u = 2x$, $du = 2 \; dx$. Therefore,

$$\int \frac{\tan 2x}{\cos 2x} \, dx = \int \sec 2x \tan 2x \; dx = \frac{1}{2} \int \sec 2x \tan 2x (2 \; dx)$$

$$= \frac{1}{2}\sec 2x + C$$

To find the integrals for the other trigonometric functions, we must change them to a form for which the integral can be determined by methods previously discussed. We can accomplish this by use of the basic trigonometric relations.

The formula for $\int \tan u \, du$ is found by expressing the integral in the form $\int [(\sin u)/(\cos u)] \, du$. We recognize this as being a logarithmic form, where the u of the logarithmic form is $\cos u$ in this integral. The differential of $\cos u$ is $-\sin u \, du$. Therefore, we have

$$\int \tan u \, du = \int \frac{\sin u}{\cos u} \, du = -\int \frac{-\sin u \, du}{\cos u} = -\ln|\cos u| + C$$

The formula for $\int \cot u \, du$ is found by writing it in the form

$$\int \frac{\cos u}{\sin u} \, du$$

In this manner we obtain the following result:

$$\int \cot u \, du = \int \frac{\cos u}{\sin u} \, du = \int \frac{\cos u \, du}{\sin u} = \ln|\sin u| + C$$

The formula for $\int \sec u \, du$ is found by writing it in the form

$$\int \frac{\sec u (\sec u + \tan u)}{\sec u + \tan u} \, du$$

In this form we recognize this as being also a logarithmic form, since

$$d(\sec u + \tan u) = (\sec u \tan u + \sec^2 u) \, du$$

The right side of this equation is the expression appearing in the numerator of the integral. Thus,

$$\int \sec u \, du = \int \frac{\sec u (\sec u + \tan u) \, du}{\sec u + \tan u} = \int \frac{\sec u \tan u + \sec^2 u}{\sec u + \tan u} \, du$$

$$= \ln|\sec u + \tan u| + C$$

To obtain the formula for $\int \csc u \, du$, we write it in the form

$$\int \frac{\csc u (\csc u - \cot u) \, du}{\csc u - \cot u}$$

Thus, we have

$$\int \csc u \, du = \int \frac{\csc u (\csc u - \cot u)}{\csc u - \cot u} \, du$$

$$= \int \frac{(-\csc u \cot u + \csc^2 u) \, du}{\csc u - \cot u}$$

$$= \ln|\csc u - \cot u| + C$$

Summarizing these results, we have the following integrals:

$$\int \tan u \, du = -\ln |\cos u| + C \qquad (8\text{--}10)$$

$$\int \cot u \, du = \ln |\sin u| + C \qquad (8\text{--}11)$$

$$\int \sec u \, du = \ln |\sec u + \tan u| + C \qquad (8\text{--}12)$$

$$\int \csc u \, du = \ln |\csc u - \cot u| + C \qquad (8\text{--}13)$$

Example D

Integrate: $\int \tan 4x \, dx$; $u = 4x$, $du = 4 \, dx$.

We write

$$\int \tan 4x \, dx = \frac{1}{4} \int \tan 4x \, (4 \, dx) = -\frac{1}{4} \ln |\cos 4x| + C$$

Example E

Integrate: $\displaystyle\int \frac{\sec e^{-x} dx}{e^x}$; $u = e^{-x}$, $du = -e^{-x} dx$.

We have

$$\int \frac{\sec e^{-x} dx}{e^x} = -\int (\sec e^{-x})(-e^{-x} dx)$$

$$= -\ln |\sec e^{-x} + \tan e^{-x}| + C$$

Example F

Evaluate: $\displaystyle\int_{\pi/6}^{\pi/4} \frac{1 + \cos x}{\sin x} \, dx$.

We write

$$\int_{\pi/6}^{\pi/4} \frac{1 + \cos x}{\sin x} \, dx = \int_{\pi/6}^{\pi/4} \csc x \, dx + \int_{\pi/6}^{\pi/4} \cot x \, dx$$

$$= \ln |\csc x - \cot x| \Big|_{\pi/6}^{\pi/4} + \ln |\sin x| \Big|_{\pi/6}^{\pi/4}$$

$$= \ln |\sqrt{2} - 1| - \ln |2 - \sqrt{3}| + \ln \left|\frac{1}{2}\sqrt{2}\right| - \ln \left|\frac{1}{2}\right|$$

$$= \ln \frac{\left(\frac{1}{2}\sqrt{2}\right)(\sqrt{2} - 1)}{\left(\frac{1}{2}\right)(2 - \sqrt{3})} = \ln \frac{2 - \sqrt{2}}{2 - \sqrt{3}} = \ln \frac{0.586}{0.268}$$

$$= \ln 2.19 = 0.784$$

Exercises 8—4

In Exercises 1 through 24 integrate each of the given functions.

1. $\int \cos 2x \, dx$

2. $\int \sin(2 - x) \, dx$

3. $\int \sec^2 3x \, dx$

4. $\int \csc 2x \cot 2x \, dx$

5. $\int \sec \frac{1}{2}x \tan \frac{1}{2}x \, dx$

6. $\int e^x \csc^2(e^x) \, dx$

7. $\int x^2 \cot x^3 \, dx$

8. $\int \tan \frac{1}{2}x \, dx$

9. $\int x \sec x^2 \, dx$

10. $\int 2 \csc 3x \, dx$

11. $\int \frac{\sin(1/x)}{x^2} \, dx$

12. $\int \frac{dx}{\sin 4x}$

13. $\int_0^{\pi/6} \frac{dx}{\cos^2 2x}$

14. $\int_0^1 e^x \cos e^x \, dx$

15. $\int \frac{\sec 5x}{\cot 5x} \, dx$

16. $\int \frac{\sin 2x}{\cos^2 x} \, dx$

17. $\int \sqrt{\tan^2 2x + 1} \, dx$

18. $\int (1 + \cot x)^2 \, dx$

19. $\int_0^{\pi/9} \sin 3x(\csc 3x + \sec 3x) \, dx$

20. $\int_{\pi/4}^{\pi/3} (1 + \sec x)^2 \, dx$

21. $\int \frac{1 - \sin x}{1 + \cos x} \, dx$

22. $\int \frac{1 + \sec^2 x}{x + \tan x} \, dx$

23. $\int \frac{1 + \sin 2x}{\tan 2x} \, dx$

24. $\int \frac{1 - \cot^2 x}{\cos^2 x} \, dx$

In Exercises 25 through 32 solve the given problems by integration.

25. Find the area bounded by $y = \tan x$, $x = \pi/4$, and $y = 0$.
26. Find the area under the curve $y = \sin x$ from $x = 0$ to $x = \pi$.
27. Find the volume generated by rotating the area bounded by $y = \sec x$, $x = 0$, $x = \pi/3$, and $y = 0$ about the x-axis.
28. Find the volume generated by rotating the area bounded by $y = \cos x^2$, $x = 0$, $y = 0$, and $x = 1$ about the y-axis.
29. Under certain conditions the expression for the velocity as a function of time for a particle is $v = 4 \cos 3t$. Determine the distance as a function of the time, if $s = 0$ when $t = 0$.
30. If the current in a certain electric circuit is $i = 110 \cos 377t$, find the expression for the voltage across a 500 μF capacitor as a function of time. The initial voltage is zero. Show that the voltage across the capacitor is 90° out of phase with the current.
31. If the area bounded by $y = \sin(x^2)$, $y = 0$, and $x = 1$ is 0.3103, find the x-coordinate of the centroid of this area.
32. A force is given as a function of the distance from the origin by

$$F = \frac{\sin x}{2 + \cos x}$$

Express the work done by this force as a function of x, if $W = 0$ when $x = 0$.

8–5 Other Trigonometric Forms

The basic trigonometric relations developed in Chapter 6 provide the means by which many other integrals involving trigonometric functions may be integrated. By use of the square relations, Eqs. (6–7), (6–8), and (6–9), and the equations for the cosine of the double angle, Eq. (6–21), it is possible to transform integrals involving powers of the trigonometric functions into integrable form. We repeat these equations here for reference:

$$\cos^2 x + \sin^2 x = 1 \tag{8–14}$$
$$1 + \tan^2 x = \sec^2 x \tag{8–15}$$
$$1 + \cot^2 x = \csc^2 x \tag{8–16}$$
$$2 \cos^2 x = 1 + \cos 2x \tag{8–17}$$
$$2 \sin^2 x = 1 - \cos 2x \tag{8–18}$$

To integrate a product of powers of the sine and cosine, we use Eq. (8–14) *if at least one of the powers is odd.* The method is based on transforming the integral so that it is made up of powers of either the sine or cosine and the first power of the other. In this way this first power becomes a factor of *du.*

Example A

Integrate: $\int \sin^3 x \cos^2 x \, dx$.

Since $\sin^3 x = \sin^2 x \sin x = (1 - \cos^2 x)\sin x$, it is possible to write this integral with powers of cos x along with sin x dx. Thus, sin x dx becomes the necessary du of the integral. Therefore,

$$\int \sin^3 x \cos^2 x \, dx = \int (1 - \cos^2 x)(\sin x)(\cos^2 x)\,dx$$

$$= \int (\cos^2 x - \cos^4 x)(\sin x \, dx)$$

$$= \int \cos^2 x (\sin x \, dx) - \int \cos^4 x (\sin x \, dx)$$

$$= -\int \cos^2 x (-\sin x \, dx) + \int \cos^4 x (-\sin x \, dx)$$

$$= -\frac{1}{3}\cos^3 x + \frac{1}{5}\cos^5 x + C$$

Example B

Integrate: $\int \cos^5 2x \, dx$.

Since $\cos^5 2x = \cos^4 2x \cos 2x = (1 - \sin^2 2x)^2 \cos 2x$, it is possible to write this integral with powers of sin $2x$ along with cos $2x$ dx. Thus, with the introduction of a factor of 2, $(\cos 2x)(2\ dx)$ is the necessary du of the integral. Thus,

$$\int \cos^5 2x\ dx = \int (1 - \sin^2 2x)^2 \cos 2x\ dx$$

$$= \int (1 - 2\sin^2 2x + \sin^4 2x)\cos 2x\ dx$$

$$= \int \cos 2x\ dx - \int 2\sin^2 2x \cos 2x\ dx + \int \sin^4 2x \cos 2x\ dx$$

$$= \frac{1}{2}\int \cos 2x(2\ dx) - \int \sin^2 2x(2\cos 2x\ dx)$$

$$+ \frac{1}{2}\int \sin^4 2x(2\cos 2x\ dx)$$

$$= \frac{1}{2}\sin 2x - \frac{1}{3}\sin^3 2x + \frac{1}{10}\sin^5 2x + C$$

In products of powers of the sine and cosine, if the powers to be integrated are even, we use Eqs. (8–17) and (8–18) to transform the integral. Those most commonly met are $\int \cos^2 u\ du$ and $\int \sin^2 u\ du$.

Example C
Integrate: $\int \sin^2 2x\ dx$.

Using Eq. (8–18) in the form $\sin^2 2x = \frac{1}{2}(1 - \cos 4x)$, this integral can be transformed into a form which can be integrated. (Here we note the x of Eq. (8–18) is treated as $2x$ for this integral.) Therefore, we write

$$\int \sin^2 2x\ dx = \int \left[\frac{1}{2}(1 - \cos 4x)\right]dx = \frac{1}{2}\int dx - \frac{1}{8}\int \cos 4x(4\ dx)$$

$$= \frac{x}{2} - \frac{1}{8}\sin 4x + C$$

To integrate even powers of the secant, powers of the tangent, or products of the secant and tangent, we use Eq. (8–15) to transform the integral. In transforming, the forms we look for are powers of the tangent with $\sec^2 x$, which becomes part of du, or powers of the secant along with $\sec x \tan x$, which becomes part of du in this case. Similar transformations are made when we integrate powers of the cotangent and cosecant, with the use of Eq. (8–16).

Example D

Integrate: $\int \tan^5 x \, dx$.

Since $\tan^5 x = \tan^3 x \tan^2 x = \tan^3 x (\sec^2 x - 1)$, we can write this integral with powers of tan x along with $\sec^2 x \, dx$. Thus, $\sec^2 x \, dx$ becomes the necessary du of the integral. It is necessary to replace $\tan^2 x$ with $\sec^2 x - 1$ twice during the integration. Therefore,

$$\int \tan^5 x \, dx = \int \tan^3 x (\sec^2 x - 1) \, dx$$

$$= \int \tan^3 x (\sec^2 x \, dx) - \int \tan^3 x \, dx$$

$$= \frac{1}{4}\tan^4 x - \int \tan x (\sec^2 x - 1) \, dx$$

$$= \frac{1}{4}\tan^4 x - \int \tan x (\sec^2 x \, dx) + \int \tan x \, dx$$

$$= \frac{1}{4}\tan^4 x - \frac{1}{2}\tan^2 x - \ln|\cos x| + C$$

Example E

Integrate: $\int \sec^3 x \tan x \, dx$.

By writing $\sec^3 x \tan x$ as $\sec^2 (\sec x \tan x)$, we can use the sec x tan x dx as the du of the integral. Thus,

$$\int \sec^3 x \tan x \, dx = \int (\sec^2 x)(\sec x \tan x \, dx)$$

$$= \frac{1}{3}\sec^3 x + C$$

Example F

Integrate: $\int \csc^4 2x \, dx$.

By writing $\csc^4 2x = \csc^2 2x \csc^2 2x = \csc^2 2x(1 + \cot^2 2x)$, we can write this integral with powers of cot $2x$ along with $\csc^2 2x \, dx$, which becomes part of the necessary du of the integral. Thus,

$$\int \csc^4 2x \, dx = \int \csc^2 2x(1 + \cot^2 2x) \, dx$$

$$= \frac{1}{2}\int \csc^2 2x(2 \, dx) - \frac{1}{2}\int \cot^2 2x(-2 \csc^2 2x \, dx)$$

$$= -\frac{1}{2}\cot 2x - \frac{1}{6}\cot^3 2x + C$$

Example G

Integrate: $\displaystyle\int_0^{\pi/4} \frac{\tan^3 x}{\sec^3 x}dx$.

This integral requires the use of several trigonometric relationships to obtain integrable forms.

$$\int_0^{\pi/4} \frac{\tan^3 x}{\sec^3 x} dx = \int_0^{\pi/4} \frac{(\sec^2 x - 1)\tan x}{\sec^3 x} dx$$

$$= \int_0^{\pi/4} \frac{\tan x}{\sec x} dx - \int_0^{\pi/4} \frac{\tan x \, dx}{\sec^3 x}$$

$$= \int_0^{\pi/4} \sin x \, dx - \int_0^{\pi/4} \frac{\sec x \tan x \, dx}{\sec^4 x}$$

$$= -\cos x + \frac{1}{3}\cos^3 x \Big|_0^{\pi/4}$$

$$= -\frac{\sqrt{2}}{2} + \frac{1}{3}\left(\frac{\sqrt{2}}{2}\right)^3 - \left(-1 + \frac{1}{3}\right) = \frac{8 - 5\sqrt{2}}{12}$$

Example H

The *root-mean-square value of a function* with respect to x is defined by

$$y_{\text{rms}} = \sqrt{\frac{1}{T}\int_0^T y^2 dx} \tag{8–19}$$

Usually the value of T which is of importance is the period of the function. Find the root-mean-square value of the current in an electric circuit for one period, if $i = 3 \cos \pi t$.

The period of the current is $2\pi/\pi = 2$ s. Therefore, we must find the square root of the integral

$$\frac{1}{2}\int_0^2 (3 \cos \pi t)^2 dt = \frac{9}{2}\int_0^2 \cos^2 \pi t \, dt$$

Hence, we have

$$\frac{9}{2}\int_0^2 \cos^2 \pi t \, dt = \frac{9}{4}\int_0^2 (1 + \cos 2\pi t) \, dt$$

after substituting $\frac{1}{2}(1 + \cos 2\pi t)$ for $\cos^2 \pi t$. Evaluating, we have

$$\frac{9}{4}\int_0^2 (1 + \cos 2\pi t) \, dt = \frac{9}{4}t \Big|_0^2 + \frac{9}{8\pi}\int_0^2 \cos 2\pi t (2\pi \, dt)$$

$$= \frac{9}{2} + \frac{9}{8\pi}\sin 2\pi t \Big|_0^2 = \frac{9}{2}$$

Thus, the root-mean-square current is

$$i_{\text{rms}} = \sqrt{\frac{9}{2}} = \frac{3\sqrt{2}}{2} = 2.12 \text{ A}$$

This value of the current, often referred to as the *effective current,* is the value of the direct current which would develop the same quantity of heat in the same time.

Exercises 8—5

In Exercises 1 through 28 integrate each of the given functions.

1. $\displaystyle\int \sin^2 x \cos x \, dx$

2. $\displaystyle\int \sin x \cos^5 x \, dx$

3. $\displaystyle\int \sin^3 2x \, dx$

4. $\displaystyle\int \cos^3 x \, dx$

5. $\displaystyle\int \sin^2 x \cos^3 x \, dx$

6. $\displaystyle\int \sin^3 x \cos^6 x \, dx$

7. $\displaystyle\int_0^{\pi/4} \sin^5 x \, dx$

8. $\displaystyle\int_{\pi/3}^{\pi/2} \sqrt{\cos x} \, \sin^3 x \, dx$

9. $\displaystyle\int \sin^2 x \, dx$

10. $\displaystyle\int \cos^2 2x \, dx$

11. $\displaystyle\int \cos^2 3x \, dx$

12. $\displaystyle\int \sin^2 4x \, dx$

13. $\displaystyle\int \tan^3 x \, dx$

14. $\displaystyle\int \cot^3 x \, dx$

15. $\displaystyle\int \tan x \sec^4 x \, dx$

16. $\displaystyle\int \cot 4x \csc^4 4x \, dx$

17. $\displaystyle\int \tan^4 2x \, dx$

18. $\displaystyle\int \cot^4 x \, dx$

19. $\displaystyle\int \tan^3 3x \sec^3 3x \, dx$

20. $\displaystyle\int \sqrt{\tan x} \, \sec^4 x \, dx$

21. $\displaystyle\int (\sin x + \cos x)^2 \, dx$

22. $\displaystyle\int (\tan 2x + \cot 2x)^2 \, dx$

23. $\displaystyle\int \frac{1 - \cot x}{\sin^4 x} \, dx$

24. $\displaystyle\int \frac{1 + \sin x}{\cos^4 x} \, dx$

25. $\displaystyle\int_{\pi/6}^{\pi/4} \cot^5 x \, dx$

26. $\displaystyle\int_{\pi/6}^{\pi/3} \frac{dx}{1 + \sin x}$

27. $\displaystyle\int \sec^6 x \, dx$

28. $\displaystyle\int \tan^7 x \, dx$

In Exercises 29 through 36 solve the given problems by integration.

29. Find the volume generated by rotating the area bounded by $y = \sin x$ and $y = 0$, from $x = 0$ to $x = \pi$, about the x-axis.

30. Find the volume generated by rotating the area bounded by $y = \tan^3(x^2)$, $y = 0$, and $x = \pi/4$ about the y-axis.

31. In determining the rate of radiation by an accelerated charge, the following integral must be evaluated: $\int_0^\pi \sin^3\theta \, d\theta$. Find the value of the integral.

32. Find the length of the curve $y = \ln \cos x$ from $x = 0$ to $x = \pi/3$.

33. Find the root-mean-square value of the current for one period if $i = 2 \sin t$.

34. Find the root-mean-square value of the voltage for one period if the voltage is given by $V = 100 \sin 120\pi t$.

35. For a current $i = i_0 \sin \omega t$, show that the root-mean-square value of the current for one period is $i_0 / \sqrt{2}$.

36. Show that $\int \sec^2 x \tan x \, dx$ can be integrated in two ways. Explain the difference in the answers.

8—6 Inverse Trigonometric Forms

Referring to Eq. (6–37), we can find the differential $\text{Arcsin}(u/a)$, where a is a constant:

$$d\left(\text{Arcsin}\,\frac{u}{a}\right) = \frac{1}{\sqrt{1 - (u/a)^2}} \frac{du}{a} = \frac{a}{\sqrt{a^2 - u^2}} \frac{du}{a} = \frac{du}{\sqrt{a^2 - u^2}}$$

Reversing this differentiation formula, we have the important integration formula:

$$\int \frac{du}{\sqrt{a^2 - u^2}} = \text{Arcsin } \frac{u}{a} + C \tag{8-20}$$

By finding the differential of $\text{Arctan}(u/a)$, and then reversing the equation, we derive another important integration formula:

$$d\left(\text{Arctan } \frac{u}{a}\right) = \frac{1}{1 + (u/a)^2} \frac{du}{a} = \frac{a^2}{a^2 + u^2} \frac{du}{a} = \frac{a \, du}{a^2 + u^2}$$

Thus,

$$\int \frac{du}{a^2 + u^2} = \frac{1}{a} \text{Arctan } \frac{u}{a} + C \tag{8-21}$$

This shows one of the principal uses of the inverse trigonometric functions: they provide a solution to the integration of important algebraic functions.

Example A

Integrate: $\displaystyle\int \frac{dx}{\sqrt{9 - x^2}}$.

This integral fits the form of Eq. (8–20) with $u = x$, $du = dx$, and $a = 3$. Thus,

$$\int \frac{dx}{\sqrt{9 - x^2}} = \text{Arcsin } \frac{x}{3} + C$$

Example B

Integrate: $\displaystyle\int \frac{dx}{4x^2 + 25}$.

This integral fits the form of Eq. (8–21) with $u = 2x$, $du = 2 \, dx$, and $a = 5$. Thus, we must insert a 2 in the numerator and a $\frac{1}{2}$ before the integral. Therefore,

$$\int \frac{dx}{4x^2 + 25} = \frac{1}{2} \int \frac{2 \, dx}{(2x)^2 + (5)^2}$$

$$= \frac{1}{2} \cdot \frac{1}{5} \text{Arctan } \frac{2x}{5} + C$$

$$= \frac{1}{10} \text{Arctan } \frac{2x}{5} + C$$

Example C

Integrate: $\displaystyle\int_{-1}^{3} \frac{dx}{x^2 + 6x + 13}$.

At first glance it does not appear that this integral fits any of the forms presented up to this point. However, by writing the denominator in the form $(x^2 + 6x + 9) + 4 = (x + 3)^2 + 2^2$, we recognize that $u = x + 3$, $du = dx$, and $a = 2$. Thus,

$$\int_{-1}^{3} \frac{dx}{x^2 + 6x + 13} = \int_{-1}^{3} \frac{dx}{(x + 3)^2 + 2^2} = \frac{1}{2} \operatorname{Arctan} \frac{x + 3}{2} \Big|_{-1}^{3}$$

$$= \frac{1}{2}(\operatorname{Arctan} 3 - \operatorname{Arctan} 1) = \frac{1}{2}(1.249 - 0.785)$$

$$= 0.232$$

Now we can see the use of completing the square when we are transforming integrals into proper form. We also see the use of inverse trigonometric functions in evaluating integrals.

Example D

Integrate: $\displaystyle\int \frac{2x + 5}{x^2 + 9} dx$.

By writing this integral as the sum of two integrals, we may integrate each of these separately:

$$\int \frac{2x + 5}{x^2 + 9} dx = \int \frac{2x\, dx}{x^2 + 9} + \int \frac{5\, dx}{x^2 + 9}$$

The first of these integrals is seen to be a logarithmic form, and the second the inverse tangent form. For the first, $u = x^2 + 9$, $du = 2x\, dx$. For the second, $u = x$, $du = dx$, $a = 3$. Thus,

$$\int \frac{2x\, dx}{x^2 + 9} + 5 \int \frac{dx}{x^2 + 9} = \ln|x^2 + 9| + \frac{5}{3} \operatorname{Arctan} \frac{x}{3} + C$$

The inverse trigonometric integral forms of this section show very well the importance of proper recognition of the form of the integral. It is important that these forms are not confused with those of the general power rule or the logarithmic form. Consider the following example.

Example E

The integral $\displaystyle\int \frac{dx}{\sqrt{1 - x^2}}$ is of the inverse sine form, with $u = x$, $du = dx$, and $a = 1$. Thus,

$$\int \frac{dx}{\sqrt{1 - x^2}} = \operatorname{Arcsin} x + C$$

The integral $\int \dfrac{x \, dx}{\sqrt{1 - x^2}}$ is not of the inverse sine form due to the factor of x in the numerator. It is integrated by use of the general power rule, with $u = 1 - x^2$, $du = -2x \, dx$, and $n = -\frac{1}{2}$. Thus,

$$\int \frac{x \, dx}{\sqrt{1 - x^2}} = -\sqrt{1 - x^2} + C$$

In the same way, the integral $\int \dfrac{dx}{1 + x^2}$ is of the inverse tangent form, whereas the integral $\int \dfrac{x \, dx}{1 + x^2}$ is of logarithmic form due to the factor of x.

Exercises 8–6

In Exercises 1 through 24 integrate each of the given functions.

1. $\displaystyle\int \frac{dx}{\sqrt{4 - x^2}}$

2. $\displaystyle\int \frac{dx}{\sqrt{49 - x^2}}$

3. $\displaystyle\int \frac{dx}{64 + x^2}$

4. $\displaystyle\int \frac{dx}{4 + x^2}$

5. $\displaystyle\int \frac{dx}{\sqrt{1 - 16x^2}}$

6. $\displaystyle\int \frac{dx}{\sqrt{9 - 4x^2}}$

7. $\displaystyle\int_0^2 \frac{dx}{1 + 9x^2}$

8. $\displaystyle\int_1^3 \frac{dx}{49 + 4x^2}$

9. $\displaystyle\int \frac{2 \, dx}{\sqrt{4 - 5x^2}}$

10. $\displaystyle\int \frac{4x \, dx}{\sqrt{3 - 2x^2}}$

11. $\displaystyle\int \frac{8x \, dx}{9x^2 + 16}$

12. $\displaystyle\int \frac{3 \, dx}{25 + 16x^2}$

13. $\displaystyle\int_1^2 \frac{dx}{5x^2 + 7}$

14. $\displaystyle\int_0^1 \frac{x \, dx}{1 + x^4}$

15. $\displaystyle\int \frac{e^x \, dx}{\sqrt{1 - e^{2x}}}$

16. $\displaystyle\int \frac{\sec^2 x \, dx}{\sqrt{1 - \tan^2 x}}$

17. $\displaystyle\int \frac{dx}{x^2 + 2x + 2}$

18. $\displaystyle\int \frac{2 \, dx}{x^2 + 8x + 17}$

19. $\displaystyle\int \frac{4 \, dx}{\sqrt{-4x - x^2}}$

20. $\displaystyle\int \frac{dx}{\sqrt{2x - x^2}}$

21. $\displaystyle\int_{\pi/6}^{\pi/2} \frac{\cos 2x}{1 + \sin^2 2x} \, dx$

22. $\displaystyle\int_{-4}^0 \frac{dx}{x^2 + 4x + 5}$

23. $\displaystyle\int \frac{2 - x}{\sqrt{4 - x^2}} \, dx$

24. $\displaystyle\int \frac{3 - 2x}{1 + 4x^2} \, dx$

In Exercises 25 through 32 solve the given problems by integration.

25. Find the area bounded by the curve $y(1 + x^2) = 1$, $x = 0$, $y = 0$, and $x = 2$.

26. Find the area bounded by $y^2(4 - x^2) = 1$, $x = 0$, $y = 0$, and $x = 1$.

27. Find the volume generated by rotating the area bounded by $y = 1/\sqrt{16 + x^2}$, $x = 0$, $y = 0$, and $x = 3$ about the x-axis.

28. The electric current in a certain inductor is given by $i = 8 \displaystyle\int \frac{dt}{100 + t^2}$. Determine the function for i in terms of t.

29. In dealing with the theory for simple harmonic motion, it is necessary to solve the following equation:

$$\frac{dx}{\sqrt{A^2 - x^2}} = \sqrt{\frac{k}{m}}\ dt \quad \text{with } k, m, \text{ and } A \text{ as constants}$$

Determine the solution to this equation if $x = x_0$ when $t = 0$.

30. Determine the function of distance in terms of time if the velocity of an object is given by $v = 1/(4 + t^2)$ and $s = 0$ when $t = 0$.

31. Find the moment of inertia with respect to the y-axis for the area bounded by $y = 1/(1 + x^6)$, the x-axis, $x = 1$, and $x = 2$.

32. Find the length of arc along the curve $y = \sqrt{1 - x^2}$ between $x = 0$ and $x = 1$.

8–7 Integration by Parts

There are many methods of transforming integrals into a form which can be integrated by one of the basic formulas. In the preceding section we saw that completing the square and trigonometric identities can be used for this purpose. In this section and the following one, we shall develop two general methods. The method of integration by parts is discussed in this section.

Since the derivative of a product of functions is found by use of the formula

$$\frac{d(uv)}{dx} = u\frac{dv}{dx} + v\frac{du}{dx}$$

the differential of a product of functions is given by $d(uv) = u\ dv + v\ du$. Integrating both sides of this equation we have $uv = \int u\ dv + \int v\ du$. Solving for $\int u\ dv$, we obtain

$$\int u\ dv = uv - \int v\ du \qquad (8\text{–}22)$$

Integration by use of Eq. (8–22) is called **integration by parts**.

Example A

Integrate: $\int x \sin x\ dx$.

This integral does not fit any of the previous forms which have been discussed, since neither x nor $\sin x$ can be made a factor of a proper du. However, by choosing $u = x$, and $dv = \sin x\ dx$, the above formula may be used. Thus,

$$u = x, \qquad dv = \sin x\ dx$$
$$du = dx, \qquad v = -\cos x$$

(The constant of integration will be included in the final result.) Hence,

$$\int (x)(\sin x\ dx) = -x \cos x - \int (-\cos x)\ dx$$
$$= -x \cos x + \sin x + C$$

Other choices of u and dv may be made in the preceding example, but they do not prove useful. If $u = \sin x$ and $dv = x\, dx$, v becomes $x^2/2$, which makes the integral $\int v\, du$ more complex than the original problem. The choice of $u = x \sin x$ makes the expression for du/dx more complex than u itself. There are no set rules which may be stated for the best choice of u and dv, but there are two guidelines which may be stated: (1) *The quantity u is normally chosen such that du/dx is of simpler form than u.* (2) *The differential dv is normally chosen such that $\int dv$ is easily obtained.* Working examples, and thereby gaining experience in methods of integrating, is the best way to determine when this method should be used and how to employ it.

Example B

Integrate: $\int_0^1 xe^{-x}dx$. Let $u = x$, $dv = e^{-x}dx$, $du = dx$, $v = -e^{-x}$.
We write

$$\int_0^1 xe^{-x}dx = -xe^{-x}\Big|_0^1 + \int_0^1 e^{-x}dx = -xe^{-x} - e^{-x}\Big|_0^1$$

$$= -e^{-1} - e^{-1} + 1$$

$$= 1 - \frac{2}{e}$$

Example C

Integrate: $\int \text{Arcsin } x\, dx$. Let $u = \text{Arcsin } x$, $dv = dx$, $du = \dfrac{dx}{\sqrt{1 - x^2}}$, $v = x$.
We obtain

$$\int \text{Arcsin } x\, dx = x \text{ Arcsin } x - \int \frac{x\, dx}{\sqrt{1 - x^2}}$$

$$= x \text{ Arcsin } x + \frac{1}{2}\int \frac{-2x\, dx}{\sqrt{1 - x^2}}$$

$$= x \text{ Arcsin } x + \sqrt{1 - x^2} + C$$

Example D

Integrate: $\int \sqrt{x} \ln x\, dx$. Let $u = \ln x$, $dv = x^{1/2}dx$, $du = \dfrac{1}{x}dx$, $v = \frac{2}{3}x^{3/2}$.
We write

$$\int \sqrt{x} \ln x\, dx = \frac{2}{3}x^{3/2}\ln x - \frac{2}{3}\int x^{1/2}dx$$

$$= \frac{2}{3}x^{3/2}\ln x - \frac{4}{9}x^{3/2} + C$$

Example E

Integrate: $\int e^x \sin x \, dx$. Let $u = \sin x$, $dv = e^x dx$, $du = \cos x \, dx$, $v = e^x$. We write

$$\int e^x \sin x \, dx = e^x \sin x - \int e^x \cos x \, dx$$

At first glance it appears that we have made no progress in applying the method of integration by parts. We note, however, that when we integrated $\int e^x \sin x \, dx$, part of the result was a term of $\int e^x \cos x \, dx$. This implies that if $\int e^x \cos x \, dx$ were integrated, a term of $\int e^x \sin x \, dx$ might result. Thus, the method of integration by parts is now applied to the integral $\int e^x \cos x \, dx$:

$$u = \cos x, \qquad dv = e^x dx, \qquad du = -\sin x \, dx, \qquad v = e^x$$

And so $\int e^x \cos x \, dx = e^x \cos x + \int e^x \sin x \, dx$. Substituting this expression into the expression for $\int e^x \sin x \, dx$, we obtain

$$\int e^x \sin x \, dx = e^x \sin x - \left(e^x \cos x + \int e^x \sin x \, dx \right)$$

$$= e^x \sin x - e^x \cos x - \int e^x \sin x \, dx$$

$$2 \int e^x \sin x \, dx = e^x (\sin x - \cos x) + 2C$$

$$\int e^x \sin x \, dx = \frac{e^x}{2} (\sin x - \cos x) + C$$

Thus, by combining integrals of like form, we obtain the desired result.

Exercises 8 – 7

In Exercises 1 through 16 integrate each of the given functions.

1. $\int x \cos x \, dx$

2. $\int x \sin 2x \, dx$

3. $\int x e^{2x} dx$

4. $\int x e^x dx$

5. $\int x \sec^2 x \, dx$

6. $\int_0^{\pi/4} x \sec x \tan x \, dx$

7. $\int \text{Arctan } x \, dx$

8. $\int \ln x \, dx$

9. $\int \frac{x \, dx}{\sqrt{1 - x}}$

10. $\int x \sqrt{x + 1} \, dx$

11. $\int x \ln x \, dx$

12. $\int x^2 \ln 4x \, dx$

13. $\int x^2 \sin 2x \, dx$

14. $\int x^2 e^{2x} dx$

15. $\int_0^{\pi/2} e^x \cos x \, dx$

16. $\int e^{-x} \sin 2x \, dx$

In Exercises 17 through 24 solve the given problems by integration.

17. Find the area bounded by $y = xe^{-x}$, $y = 0$, and $x = 2$.

18. Find the volume generated by rotating the area bounded by $y = \sin x$ and $y = 0$ (from $x = 0$ to $x = \pi$) about the y-axis.

19. Find the x-coordinate of the centroid of the area bounded by $y = \cos x$ and $y = 0$ for $0 \leq x \leq \pi/2$.

20. Find the moment of inertia with respect to its axis of the volume generated by rotating the area bounded by $y = e^x$, $x = 1$, and the coordinate axes about the y-axis.

21. Find the root-mean-square value of the function $y = \sqrt{\text{Arcsin } x}$ between $x = 0$ and $x = 1$. (See Example H of Section 8–5.)

22. The general expression for the slope of a curve is $dy/dx = x^3\sqrt{1 + x^2}$. Find the equation of the curve if it passes through the origin.

23. The current in a given circuit is given by $i = e^{-2t}\cos t$. Find an expression for the amount of charge which passes a given point in the circuit as a function of the time, if $q_0 = 0$.

24. A particle moves such that its velocity v is given by $v = t\sqrt{t + 1}$. Find the expression for the displacement s as a function of time if $s = 0$ when $t = 0$.

8–8 Integration by Trigonometric Substitution

Trigonometric relations not only provide a means of transforming trigonometric integrals, but they can also be used to transform algebraic integrals. Substitutions based on Eqs. (8–14), (8–15), and (8–16) prove to be particularly useful for integrals involving radicals. The method is illustrated in the following examples.

Example A

Integrate: $\displaystyle\int \frac{dx}{x^2\sqrt{1 - x^2}}$.

If we let $x = \sin\theta$, the radical becomes $\sqrt{1 - \sin^2\theta} = \cos\theta$. Therefore, by making this substitution, the integral can be transformed into a trigonometric integral. We must be careful to replace all factors of the integral by proper expressions in terms of θ; $x = \sin\theta$, $dx = \cos\theta\, d\theta$.

And so we obtain

$$\int \frac{dx}{x^2\sqrt{1 - x^2}} = \int \frac{\cos\theta\, d\theta}{\sin^2\theta\sqrt{1 - \sin^2\theta}} = \int \frac{\cos\theta\, d\theta}{\sin^2\theta\,\cos\theta} = \int \csc^2\theta\, d\theta$$

$$= -\cot\theta + C$$

We have now performed the integration, but the answer is expressed in terms of a variable different from the original integral. We must express the result in terms of x. Making a right triangle with an angle θ such that

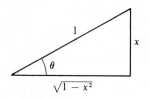

Figure 8–1

$\sin \theta = x/1$ (see Fig. 8–1), we may express any of the trigonometric functions in terms of x. (This is the method used with inverse trigonometric functions.) Thus,

$$\cot \theta = \frac{\sqrt{1 - x^2}}{x}$$

Therefore, the result of the integration becomes

$$\int \frac{dx}{x^2 \sqrt{1 - x^2}} = -\cot \theta + C = -\frac{\sqrt{1 - x^2}}{x} + C$$

Example B

Integrate: $\int \dfrac{dx}{\sqrt{x^2 + 4}}$. Let $x = 2 \tan \theta$, $dx = 2 \sec^2\theta \, d\theta$.

We write

$$\int \frac{dx}{\sqrt{x^2 + 4}} = \int \frac{2 \sec^2\theta \, d\theta}{\sqrt{4 \tan^2\theta + 4}}$$

$$= \int \frac{2 \sec^2\theta \, d\theta}{2\sqrt{\tan^2\theta + 1}} = \int \frac{2 \sec^2\theta \, d\theta}{2 \sec \theta}$$

$$= \int \sec \theta \, d\theta = \ln |\sec \theta + \tan \theta| + C$$

$$= \ln \left| \frac{\sqrt{x^2 + 4}}{2} + \frac{x}{2} \right| + C$$

$$= \ln \left| \frac{\sqrt{x^2 + 4} + x}{2} \right| + C$$

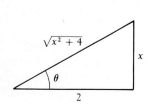

Figure 8–2

See Fig. 8–2. This answer is acceptable, but there is another form. By using the properties of logarithms, we have

$$\ln \left| \frac{\sqrt{x^2 + 4} + x}{2} \right| + C = \ln |\sqrt{x^2 + 4} + x| + (C - \ln 2)$$

$$= \ln | \sqrt{x^2 + 4} + x| + C'$$

Here C' is another arbitrary constant, which combines all constants of the previous expression. Combining constants in this manner is a common practice in integration problems.

Example C

Integrate: $\int \dfrac{dx}{x\sqrt{x^2 - 9}}$. Let $x = 3 \sec \theta$, $dx = 3 \sec \theta \tan \theta \, d\theta$.

$$\int \frac{dx}{x\sqrt{x^2 - 9}} = \int \frac{3 \sec \theta \tan \theta \, d\theta}{3 \sec \theta \sqrt{9 \sec^2\theta - 9}}$$

$$= \int \frac{\tan \theta \, d\theta}{3 \sqrt{\sec^2\theta - 1}} = \int \frac{\tan \theta \, d\theta}{3 \tan \theta}$$

$$= \frac{1}{3} \int d\theta = \frac{1}{3}\theta + C = \frac{1}{3}\text{Arcsec}\frac{x}{3} + C$$

In this solution it is not necessary to refer to a triangle to express the integral in terms of x. This solution is found by solving $x = 3 \sec \theta$ for θ, as indicated.

From the preceding examples we can see that by making the proper substitution in terms of a trigonometric function, we can solve integrals of algebraic functions by means of simpler equivalent trigonometric forms, due to the properties of the square relations of the trigonometric functions. In summary, for the indicated radical form the following trigonometric substitutions are used.

For	$\sqrt{a^2 - x^2}$	use	$x = a \sin \theta$
For	$\sqrt{a^2 + x^2}$	use	$x = a \tan \theta$
For	$\sqrt{x^2 - a^2}$	use	$x = a \sec \theta$

Example D

Evaluate: $\displaystyle\int_1^4 \frac{\sqrt{9 + x^2}}{x}dx.$

Since we have the form $\sqrt{a^2 + x^2}$, where $a = 3$, we make the substitution $x = 3 \tan \theta$. This means that $dx = 3 \sec^2\theta \, d\theta$. Thus,

$$\int \frac{\sqrt{9 + x^2}}{x} dx = \int \frac{\sqrt{9 + 9 \tan^2\theta}}{3 \tan \theta} (3 \sec^2\theta \, d\theta)$$

$$= 3 \int \frac{\sqrt{1 + \tan^2\theta}}{\tan \theta} \sec^2\theta \, d\theta = 3 \int \frac{\sec^3\theta}{\tan \theta} d\theta$$

$$= 3 \int \frac{1 + \tan^2\theta}{\tan \theta} \sec \theta \, d\theta$$

$$= 3\left(\int \frac{\sec \theta}{\tan \theta} d\theta + \int \tan \theta \sec \theta \, d\theta \right)$$

$$= 3\left(\int \csc \theta \, d\theta + \int \tan \theta \sec \theta \, d\theta \right)$$

$$= 3[\ln |\csc \theta - \cot \theta| + \sec \theta]$$

$$= 3\left[\ln \left| \frac{\sqrt{x^2 + 9}}{x} - \frac{3}{x} \right| + \frac{\sqrt{x^2 + 9}}{3}\right]$$

$$= 3\left[\ln \left| \frac{\sqrt{x^2 + 9} - 3}{x} \right| + \frac{\sqrt{x^2 + 9}}{3}\right]$$

Limits have not been included, due to the changes in variables. The actual evaluation may now be completed:

$$\int_1^4 \frac{\sqrt{9 + x^2}}{x}dx = 3\left[\ln\left|\frac{\sqrt{x^2 + 9} - 3}{x}\right| + \frac{\sqrt{x^2 + 9}}{3}\right]_1^4$$

$$= 3\left[\left(\ln\frac{1}{2} - \ln\frac{\sqrt{10} - 3}{1}\right) + \left(\frac{5}{3} - \frac{\sqrt{10}}{3}\right)\right]$$

$$= 3\left[\ln\frac{1}{2(\sqrt{10} - 3)} + \frac{5 - \sqrt{10}}{3}\right]$$

$$= 3\ln\frac{\sqrt{10} + 3}{2} + \frac{5 - \sqrt{10}}{1} = 5.21$$

Exercises 8—8

In Exercises 1 through 16 integrate each of the given functions.

1. $\displaystyle\int \frac{\sqrt{1 - x^2}}{x^2}dx$

2. $\displaystyle\int \frac{dx}{(x^2 + 9)^{3/2}}$

3. $\displaystyle\int \frac{dx}{\sqrt{x^2 - 4}}$

4. $\displaystyle\int \frac{\sqrt{x^2 - 25}}{x}dx$

5. $\displaystyle\int \frac{dx}{x^2\sqrt{x^2 + 9}}$

6. $\displaystyle\int \frac{dx}{x\sqrt{4 - x^2}}$

7. $\displaystyle\int \frac{dx}{(4 - x^2)^{3/2}}$

8. $\displaystyle\int \frac{x^3\,dx}{\sqrt{9 + x^2}}$

9. $\displaystyle\int_0^{0.5} \frac{x^3\,dx}{\sqrt{1 - x^2}}$

10. $\displaystyle\int_4^5 \frac{\sqrt{x^2 - 16}}{x^2}dx$

11. $\displaystyle\int \frac{dx}{\sqrt{x^2 + 2x + 2}}$

12. $\displaystyle\int \frac{dx}{\sqrt{x^2 + 2x}}$

13. $\displaystyle\int \frac{dx}{x\sqrt{4x^2 - 9}}$

14. $\displaystyle\int \sqrt{16 - x^2}\,dx$

15. $\displaystyle\int \frac{dx}{\sqrt{e^{2x} - 1}}$

16. $\displaystyle\int \frac{\sec^2 x\,dx}{(4 - \tan^2 x)^{3/2}}$

In Exercises 17 through 22 solve the given problems by integration.

17. Find the area of a circle of radius 1 by integration.

18. Find the area bounded by $y = \dfrac{1}{x^2\sqrt{x^2 - 1}}$, $x = \sqrt{2}$, $x = \sqrt{5}$, and $y = 0$.

19. Find the moment of inertia with respect to the y-axis of the first-quadrant area under the circle $x^2 + y^2 = a^2$ in terms of its mass.

20. Find the moment of inertia of a sphere of radius a with respect to its axis in terms of its mass.

21. Find the volume generated by rotating the area bounded by $y = \dfrac{\sqrt{x^2 - 16}}{x^2}$, $y = 0$, and $x = 5$ about the y-axis.

22. Find the length of arc along the curve of $y = \ln x$ from $x = 1$ to $x = 3$.

8—9 Integration by Use of Tables

In this chapter we have introduced certain basic integrals, and have also brought in some methods of reducing other integrals to these basic forms. Often this transformation and integration require a number of steps to be performed, and therefore integrals are tabulated for reference. The integrals found in tables have been derived by using the methods introduced thus far, as well as many other methods which can be used. Therefore, an understanding of the basic forms and some of the basic methods is very useful in finding integrals from tables. Such an understanding forms a basis for proper recognition of the forms which are used in the tables, as well as the types of results which may be expected. Therefore, *the use of the tables depends on the proper recognition of the form, and the variables and constants of the integral.* The following examples illustrate the use of the table of integrals found in Table 6 in Appendix C.

Example A

Integrate: $\displaystyle\int \frac{x\,dx}{\sqrt{2+3x}}$.

We first note that this integral fits the form of formula 6 of Table 6, with $u = x$, $a = 2$, and $b = 3$. Therefore,

$$\int \frac{x\,dx}{\sqrt{2+3x}} = -\frac{2(4-3x)\sqrt{2+3x}}{27} + C$$

Example B

Integrate: $\displaystyle\int \frac{\sqrt{4-9x^2}}{x}\,dx$.

This fits the form of formula 18, with proper identification of constants; $u = 3x$, $du = 3\,dx$, $a = 2$. Hence,

$$\int \frac{\sqrt{4-9x^2}}{x}\,dx = \int \frac{\sqrt{4-9x^2}}{3x}\,3\,dx$$

$$= \sqrt{4-9x^2} - 2\ln\!\left(\frac{2+\sqrt{4-9x^2}}{3x}\right) + C$$

Example C

Integrate: $\int \sec^3 2x\,dx$.

This fits the form of formula 37; $n = 3$, $u = 2x$, $du = 2\,dx$. And so

$$\int \sec^3 2x\,dx = \frac{1}{2}\int \sec^3 2x(2\,dx)$$

$$= \frac{1}{2}\frac{\sec 2x \tan 2x}{2} + \frac{1}{2}\cdot\frac{1}{2}\int \sec 2x(2\,dx)$$

[To complete this integral, we must use the basic form of Eq. (8–12).] Thus, we complete it by

$$\int \sec^3 2x\,dx = \frac{\sec 2x \tan 2x}{4} + \frac{1}{4}\ln|\sec 2x + \tan 2x| + C$$

Example D

Evaluate: $\int_1^e x^2 \ln 2x \, dx$.

This integral fits the form of formula 46 if $u = 2x$. Thus, we have

$$\frac{1}{8}\int_1^e (2x)^2 \ln 2x (2 \, dx) = \frac{1}{8}(2x)^3\left[\frac{\ln 2x}{3} - \frac{1}{9}\right]_1^e$$

$$= e^3\left(\frac{\ln 2e}{3} - \frac{1}{9}\right) - \left(\frac{\ln 2}{3} - \frac{1}{9}\right)$$

$$= e^3\left(\frac{\ln 2}{3} + \frac{\ln e}{3} - \frac{1}{9}\right) - \left(\frac{\ln 2}{3} - \frac{1}{9}\right)$$

$$= \frac{\ln 2}{3}(e^3 - 1) + \frac{1}{9}(2e^3 + 1) = 8.99$$

We summarize briefly here the approach to integrating a function. There are two methods of approaching a problem in order to obtain the exact result for the integral.

(1) Write the integral such that it fits an integral form. Either a basic form as developed in this chapter, or a form from a table of integrals may be used.

(2) Use a technique of transforming the integral such that an integral form may be used. Techniques include integration by parts, trigonometric substitution, and others which may be found in other sources.

Also, definie integrals may be approximated by methods such as the trapezoidal rule and Simpson's rule. There are other approximation methods, one of which is developed in Chapter 12.

Exercises 8—9

In Exercises 1 through 32 integrate each function by use of Table 6 in Appendix C.

1. $\int \dfrac{3x \, dx}{2 + 5x}$

2. $\int \dfrac{4x \, dx}{(1 + x)^2}$

3. $\int 5x\sqrt{2 + 3x} \, dx$

4. $\int \dfrac{dx}{x^2 - 4}$

5. $\int \sqrt{4 - x^2} \, dx$

6. $\int_0^{\pi/3} \sin^3 x \, dx$

7. $\int \sin 2x \sin 3x \, dx$

8. $\int \text{Arcsin } 3x \, dx$

9. $\int \dfrac{\sqrt{4x^2 - 9}}{x} dx$

10. $\int \dfrac{(9x^2 + 16)^{3/2}}{x} dx$

11. $\int \cos^5 4x \, dx$

12. $\int \tan^2 x \, dx$

13. $\int \text{Arctan } x^2 (x \, dx)$

14. $\int xe^{4x} dx$

15. $\int_1^2 (4 - x^2)^{3/2} dx$

16. $\int \dfrac{dx}{9 - 16x^2}$

17. $\int \dfrac{dx}{x\sqrt{4x^2 + 1}}$

18. $\int \dfrac{\sqrt{4 + x^2}}{x} dx$

19. $\displaystyle \int \frac{dx}{x\sqrt{1-4x^2}}$

20. $\displaystyle \int \frac{dx}{x(1+4x)^2}$

21. $\displaystyle \int \sin x \cos 5x \, dx$

22. $\displaystyle \int_0^2 x^2 e^{3x} dx$

23. $\displaystyle \int x^5 \cos x^3 \, dx$

24. $\displaystyle \int \sin^3 x \cos^2 x \, dx$

25. $\displaystyle \int \frac{x \, dx}{(1-x^4)^{3/2}}$

26. $\displaystyle \int \frac{dx}{x(1-4x)}$

27. $\displaystyle \int_1^3 \frac{\sqrt{3+5x^2} \, dx}{x}$

28. $\displaystyle \int_0^1 \frac{\sqrt{9-4x^2}}{x} dx$

29. $\displaystyle \int x^3 \ln x^2 \, dx$

30. $\displaystyle \int \frac{x \, dx}{x^2 \sqrt{x^4-9}}$

31. $\displaystyle \int \frac{x^2 \, dx}{(x^6-1)^{3/2}}$

32. $\displaystyle \int x^7 \sqrt{x^4+4} \, dx$

In Exercises 33 through 38 evaluate the integrals by use of Table 6.

33. Find the length of arc of the curve $y = x^2$ from $x = 0$ to $x = 1$.

34. Find the moment of inertia with respect to its axis of the volume generated by rotating the area bounded by $y = \ln x$, $x = e$, and the x-axis about the y-axis.

35. Find the force on the area bounded by $x = 1/\sqrt{1+y}$, $y = 0$, $y = 3$, and the y-axis, if the surface of the water is at the upper edge of the area.

36. Under certain conditions the velocity as a function of time for a particle is given by $v = t/\sqrt{1+2t}$. Determine the displacement as a function of time if $s = 6$ when $t = 0$.

37. Find the volume of the solid generated by rotating the area bounded by $y = e^x \sin x$ and the x-axis between $x = 0$ and $x = \pi$ about the x-axis.

38. The electric current in a certain circuit as a function of time is given by $i = 2/\sqrt{t^2+100}$. Find the expression for the charge which passes a given point if $q = 0$ when $t = 0$.

8–10 Review Exercises for Chapter 8

In Exercises 1 through 32 integrate the given functions without the use of Table 6.

1. $\displaystyle \int e^{-2x} dx$

2. $\displaystyle \int e^{\cos 2x} \sin x \cos x \, dx$

3. $\displaystyle \int \frac{dx}{x(\ln 2x)^2}$

4. $\displaystyle \int x^{1/3} \sqrt{x^{4/3}+1} \, dx$

5. $\displaystyle \int \frac{\cos x \, dx}{1+\sin x}$

6. $\displaystyle \int \frac{\sec^2 x \, dx}{2+\tan x}$

7. $\displaystyle \int \frac{2 \, dx}{25+49x^2}$

8. $\displaystyle \int \frac{dx}{\sqrt{1-4x^2}}$

9. $\displaystyle \int_0^{\pi/2} \cos^3 2x \, dx$

10. $\displaystyle \int_0^{\pi/8} \sec^3 2x \tan 2x \, dx$

11. $\displaystyle \int_0^2 \frac{x \, dx}{4+x^2}$

12. $\displaystyle \int_1^e \frac{\ln x^2 \, dx}{x}$

13. $\displaystyle\int \sec^4 3x \tan 3x \, dx$

14. $\displaystyle\int \frac{\sin^3 x \, dx}{\sqrt{\cos x}}$

15. $\displaystyle\int \tan 3x \, dx$

16. $\displaystyle\int \sec 4x \, dx$

17. $\displaystyle\int \sec^4 3x \, dx$

18. $\displaystyle\int \frac{\sin^2 2x \, dx}{1 + \cos 2x}$

19. $\displaystyle\int \frac{dx}{\sqrt{4x^2 - 9}}$

20. $\displaystyle\int \frac{x^2 \, dx}{\sqrt{9 - x^2}}$

21. $\displaystyle\int \frac{e^{2x} dx}{\sqrt{e^{2x} + 1}}$

22. $\displaystyle\int \frac{(4 + \ln 2x)^3 \, dx}{x}$

23. $\displaystyle\int \sin^2 3x \, dx$

24. $\displaystyle\int \sin^4 x \, dx$

25. $\displaystyle\int x \csc^2 2x \, dx$

26. $\displaystyle\int x \operatorname{Arctan} x \, dx$

27. $\displaystyle\int e^{2x} \cos e^{2x} dx$

28. $\displaystyle\int \frac{dx}{x^2 + 6x + 10}$

29. $\displaystyle\int_1^e \frac{(\ln x)^2 \, dx}{x}$

30. $\displaystyle\int_1^3 \frac{dx}{x^2 - 2x + 5}$

31. $\displaystyle\int \frac{x^2 - 1}{x + 2} dx$

32. $\displaystyle\int [\cos(\ln x)] \frac{dx}{x}$

In Exercises 33 through 44 solve the given problems by integration.

33. The change in the thermodynamic entity of entropy ΔS may be expressed as $\Delta S = \int (c_v/T) \, dT$, where c_v is the heat capacity at constant volume and T is the temperature. For increased accuracy, c_v is often given by the equation $c_v = a + bT + cT^2$, where a, b, and c are constants. Express ΔS as a function of temperature.

34. A second-order chemical reaction leads to the equation

$$dt = \frac{k_1 \, dx}{a - x} + \frac{k_2 \, dx}{b - x}$$

where k_1 and k_2 are constants, a and b are initial concentrations, t is the time, and x is the decrease in concentration. Solve for t as a function of x.

35. Assuming a resisting force, the velocity (in feet per second) of a falling object in terms of time (in seconds) is given by

$$\frac{dv}{32 - 0.1v} = dt$$

If $v = 0$ when $t = 0$, find v as a function of t.

36. Find the area bounded by $y = x/(1 + x)^2$, the x-axis, and the line $x = 4$.

37. Find the equation of the curve for which $dy/dx = \sec^4 x$, if the curve passes through the origin.

38. Find the distance from the origin after $\frac{1}{2}$ second of the particle for which the velocity as a function of time is given by $v = 4 \cos 3t$. (That is, integrate the function with limits of 0 and $\frac{1}{2}$, or substitute into the function derived in Exercise 29 of Section 8–4). Compare the results of these two methods.)

39. Find the area inside the circle $x^2 + y^2 = 25$ and to the right of the line $x = 3$.

40. Find the area bounded by $y = x\sqrt{x + 4}$, $y = 0$, and $x = 5$.

41. Find the volume generated by rotating the area bounded by $y = xe^x$, $y = 0$, and $x = 2$ about the y-axis.

42. Find the volume generated by rotating about the y-axis the area bounded by $y = x + \sqrt{x + 1}$, $x = 3$, and the axes.

43. Find the centroid of the area bounded by $y = \ln x$, $x = 2$, and the x-axis.

44. The power delivered to an electric circuit is given by $P = ei$, where e and i are, respectively, the instantaneous voltage and the instantaneous current in the circuit. The mean power, averaged over a period $2\pi/\omega$, is given by

$$P_{av} = \frac{\omega}{2\pi} \int_0^{2\pi/\omega} ei \, dt$$

If $e = 20 \cos 2t$ and $i = 3 \sin 2t$, find the average power over a period of $\pi/4$.

9

Introduction to Partial Derivatives and Double Integrals

9—1 Functions of Two Variables

To this point we have been dealing with functions which have a single independent variable. There are, however, numerous situations in science, engineering, and technology in which functions with more than one independent variable are used. Although some of these applications involve three or more independent variables, many involve only two, and we shall concern ourselves primarily with these. Therefore, in this section we shall establish the significance of a function of two variables, and in the following sections of the chapter show how the methods of the earlier chapters may be used with them.

Example A

The total surface area of a right circular cylinder is a function of the radius and the height of the cylinder. That is, the area will change if either or both of these change. The formula for the total surface area is

$$A = 2\pi r^2 + 2\pi rh$$

We say that A is a function of r and h.

We define a function of two variables as follows. *If z is uniquely determined for given values of x and y, then z is a function of x and y.* Notation similar to that which is used for one independent variable is employed in this case. This notation is $z = f(x, y)$, where both x and y are independent variables. This is the meaning of a function of two variables.

Similar to the case with one independent variable, $f(a, b)$ means "the value of the function when $x = a$ and $y = b$."

Example B

If $f(x, y) = 3x^2 + 2xy - y^3$, find $f(-1, 2)$. Substituting, we have

$$f(-1, 2) = 3(-1)^2 + 2(-1)(2) - (2)^3 = 3 - 4 - 8 = -9$$

Example C

If $f(x, y) = 2xy^2 - y$, find $f(x, 2x) - f(x, x^2)$.

We note that in each evaluation the x factor remains as x, but that we are to substitute $2x$ for y and subtract the function for which x^2 is substituted for y.

$$
\begin{aligned}
f(x, 2x) - f(x, x^2) &= [2x(2x)^2 - (2x)] - [2x(x^2)^2 - x^2] \\
&= [8x^3 - 2x] - [2x^5 - x^2] \\
&= 8x^3 - 2x - 2x^5 + x^2 \\
&= -2x^5 + 8x^3 + x^2 - 2x
\end{aligned}
$$

This type of difference of functions is important later in this chapter.

Restricting values of the function to real numbers means that certain restrictions may be placed on the independent variables. *Values of either x or y or both which lead to division by zero or to imaginary values for z are not permissible.* The following example illustrates this point.

Example D

If

$$f(x, y) = \frac{3xy}{(x - y)(x + 3)}$$

all values of x and y are permissible, except those for which $x = y$ and $x = -3$. Each of these would indicate division by zero.

If $f(x, y) = \sqrt{4 - x^2 - y^2}$, neither x nor y may be more than 2 in absolute value, and the sum of their squares may not exceed 4. Otherwise, imaginary values of the function would result.

One of the primary difficulties students have with functions is setting them up from stated conditions. Although many examples of this have been encountered in our previous work, an example of setting up a function of two variables should prove helpful. Although no general rules can be given for this procedure, a careful analysis of the statement should lead to the desired function. The following example illustrates setting up such a function.

Example E

An open rectangular metal box is to be constructed such that it will contain 2 ft³. The cost of sheet metal is c cents per square foot. Express the cost of this material as a function of the length and width of the box.

The cost in question depends on the surface area of the sides of the box. An "open" box is one which has no top. Thus, the surface area of the box is

$$S = lw + 2lh + 2hw$$

where l is the length, w the width, and h the height of the box. However, this expression contains three independent variables. Using the condition that the volume of the box is 2 ft³, we have $lwh = 2$. Since we wish to have only l and w, we solve this expression for h, and find that $h = 2/lw$. Substituting this in the expression for the surface area we have

$$S = lw + 2l\left(\frac{2}{lw}\right) + 2\left(\frac{2}{lw}\right)w$$

$$= lw + \frac{4}{w} + \frac{4}{l}$$

Since the cost C is given by $C = cS$, we have the expression for the cost as

$$C = c\left(lw + \frac{4}{w} + \frac{4}{l}\right)$$

Exercises 9—1

1. Express the volume of a right circular cylinder as a function of the radius and height.

2. Express the volume of a right circular cone as a function of the radius of the base and the height.

3. Express the area of a triangle as a function of the base and the altitude.

4. Express the length of a diagonal of a rectangle as a function of the length and width.

5. A cylindrical can is to be made to contain a volume V. Express the total surface area of the can as a function of V and the radius of the can.

6. A right circular cylinder is to be inscribed in a sphere of radius r. Express the volume of the cylinder as a function of the height h of the cylinder and r.

7. The angle between two sides a and b of a triangle is 30°. Express the area of the triangle as a function of these sides.

8. A computer leasing firm charges a monthly fee F based on the length of time a corporation has used the service plus $100 for every hour the computer is used during the month. Express the total monthly charge T as a function of F and the number of hours h the computer is used.

In Exercises 9 through 24 evaluate the given functions.

9. $f(x, y) = 2x - 6y$, find $f(0, -4)$

10. $F(x, y) = x^2 - 5y + y^2$, find $F(2, -2)$

11. $g(r, s) = r - 2rs - r^2s$, find $g(-2, 1)$

12. $\phi(x, t) = 2x^3 - x^2t - 2t^2$, find $\phi(-3, -1)$

13. $Y(y, t) = \dfrac{2 - 3y}{t - 1} + 2y^2t$, find $Y(y, 2)$

14. $f(r, t) = r^3 - 3r^2t + 3rt^2 - t^3$, find $f(3, t)$

15. $X(x, t) = -6xt + xt^2 - t^3$, find $X(x, -t)$

16. $g(y, z) = 2yz^2 - 6y^2z - y^2z^2$, find $g(y, 2y)$

17. $H(p, q) = p - \dfrac{p - 2q^2 - 5q}{p + q}$, find $H(p, q + k)$

18. $\phi(x, y) = \dfrac{3 - x}{y - x} - xy - xy^2$, find $\phi(x + h, -1)$

19. $f(x, y) = x^2 - 2xy - 4x$, find $f(x + h, y + k) - f(x, y)$

20. $g(y, z) = 4yz - z^3 + 4y$, find $g(y + 1, z + 2) - g(y, z)$

21. $f(x, y) = xy + x^2 - y^2$, find $f(x, x) - f(x, 0)$

22. $f(x, y) = 4x^2 - xy - 2y$, find $f(x, x^2) - f(x, 1)$

23. $g(y, z) = 3y^3 - y^2z + 5z^2$, find $g(3z^2, z) - g(z, z)$

24. $X(x, t) = 2x - \dfrac{t^2 - 2x^2}{x}$, find $X(2t, t) - X(2t^2, t)$

In Exercises 25 through 28 determine which values of x or y are not permissible, if any.

25. $f(x, y) = \dfrac{\sqrt{y}}{2x}$

26. $f(x, y) = \dfrac{x^2 - 4y^2}{x^2 + 9}$

27. $f(x, y) = \sqrt{x^2 + y^2 - x^2y - y^3}$

28. $f(x, y) = \dfrac{1}{xy - y}$

In Exercise 29 through 36 solve the given problems.

29. The voltage v across a resistor R in an electric circuit is given by $v = iR$, where i is the current. What is the voltage if the current is 3 A and the resistance is 6 Ω?

30. The centripetal acceleration of an object moving in a circular path is $a = v^2/R$, where v is the velocity of the object and R is the radius of the circle. What is the centripetal acceleration of an object moving at 6 ft/s in a circular path whose radius is 4 ft?

31. The pressure of a gas as a function of its volume and temperature is given by $p = nRT/V$, where n and R are constants. If $n = 3$ mol and $R = 8.31$ J/mol·K, find the pressure (in pascals) for $T = 300$ K and $V = 50$ m³.

32. The power supplied to a resistance R in an electric circuit by a current i is given by $P = i^2R$. Find the power (in watts) supplied by a current of 4 A to a resistance of 2.4 Ω.

33. The reciprocal of the image distance from a lens as a function of the object distance p and the focal length f of the lens is given by

$$\frac{1}{q} = \frac{1}{f} - \frac{1}{p}$$

Find the image distance of an object 20 cm from a lens whose focal length is 5 cm.

34. The pressure exerted by a force on an area is given by $p = F/A$. If a given force is doubled on an area which is $\frac{2}{3}$ a given area, what is the ratio of the first pressure to the second?

35. The crushing load of a pillar varies as the fourth power of its radius and inversely as the square of its length. Express L as a function of r and l for a pillar 20 ft tall and 1 ft in diameter which is crushed by a load of 20 tons.

36. The resonant frequency of an electric circuit containing an inductance L and capacitance C is inversely proportional to the square root of the product of the inductance and capacitance. If the resonant frequency of a circuit containing a 4 H capacitor and a 64 μF capacitor is 10 Hz, express f as a function of L and C.

9–2 Curves and Surfaces in Three Dimensions

We will now undertake a brief description of the graphical representation of a function of two variables. We shall show first a method of representation in the rectangular coordinate system in two dimensions. The following example illustrates the method.

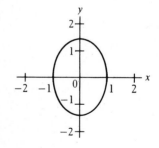

Figure 9–1

Example A

To represent $z = 2x^2 + y^2$, we will assume various values of z, and sketch the resulting equation in the xy-plane. For example, if $z = 2$ we have

$$2x^2 + y^2 = 2$$

This we recognize as an ellipse with its major axis along the y-axis and vertices at $(0, \sqrt{2})$ and $(0, -\sqrt{2})$. The ends of the minor axis are $(1, 0)$ and $(-1, 0)$ (see Fig. 9–1). However, the ellipse $2x^2 + y^2 = 2$ represents the function $z = 2x^2 + y^2$, only if $z = 2$. If $z = 4$, we have $2x^2 + y^2 = 4$, which is another ellipse. In fact, for all positive values of z, an ellipse results. Negative values for z are not possible, since neither x^2 nor y^2 may be negative. Figure 9–2 shows the ellipses obtained by using the indicated values of z.

The method of representation illustrated in Example A is useful if only a few certain values of z are to be used, or at least if the various curves do not intersect such that they cannot be distinguished. If a general representation of z as a function of x and y is desired, it is necessary to use three coordinate axes, one each for x, y, and z. The most widely applicable system of this kind is to place a third coordinate axis at right angles to each of the x- and y-axes. In this way we employ three dimensions for the representation.

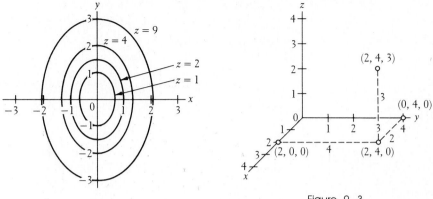

Figure 9–2

Figure 9–3

The three mutually perpendicular axes, the x-axis, the y-axis, and the z-axis, are the basis of the **rectangular coordinate system in three dimensions.** Together they form three mutually perpendicular planes in space, the xy-plane, the yz-plane, and the xz-plane. To every point in the space of the coordinate systems is associated the set of numbers (x, y, z). The point at which the axes meet is the origin. The positive directions of the axes are indicated in Fig. 9–3. *That part of space in which all the values are positive is called the first* **octant.** Numbers are not assigned to the other octants.

Example B

Represent the point (2, 4, 3) in rectangular coordinates.

It is first noted that a certain distortion is necessary to reasonably represent values of x, since the x-axis "comes out" of the plane of the page. Units $\sqrt{2}/2$ (=0.7) as long as those used on the other axes give a good representation. With this in mind, we draw a line 4 units long from the point (2, 0, 0) on the x-axis in the xy-plane. This locates the point (2, 4, 0). From this point a line 3 units long is drawn vertically upward. This locates the desired point (2, 4, 3). The point may be located by starting from (0, 4, 0), proceeding two units *parallel* to the x-axis to (2, 4, 0) and then vertically three units to (2, 4, 3). It may also be located by starting from (0, 0, 3) (see Fig. 9–3).

The remainder of this section is devoted to showing certain basic techniques by which three-dimensional figures may be drawn. We start by showing the general equation of a plane.

In Chapter 1 we showed that the graph of the equation $Ax + By + C = 0$ in two dimensions is a straight line. By the following example, we will verify that *the graph of the equation*

$$Ax + By + Cz + D = 0 \qquad (9\text{–}1)$$

is a **plane** *in three dimensions.*

Example C

Show that the graph of the equation $2x + 3y + z - 6 = 0$ in three dimensions is a plane.

If we let any of the three variables take on a specific value, we obtain a linear equation in the other two variables. For example, the point $(\frac{1}{2}, 1, 2)$ satisfies the equation, and therefore lies on the graph of the equation. For $x = \frac{1}{2}$, we have

$$3y + z - 5 = 0$$

which is the equation of a straight line. This means that all pairs of values of y and z which satisfy this equation, along with $x = \frac{1}{2}$, satisfy the given equation. Thus, for $x = \frac{1}{2}$, the straight line $3y + z - 5 = 0$ lies on the graph of the equation.

For $z = 2$, we have

$$2x + 3y - 4 = 0$$

which is also a straight line. By similar reasoning this line lies on the graph of the equation. Since two lines through a point define a plane, these lines through $(\frac{1}{2}, 1, 2)$ define a plane. This plane is the graph of the equation (see Fig. 9–4).

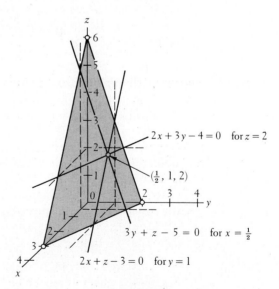

Figure 9—4

To complete the verification, it can be seen that for any point on the graph, there is a straight line parallel to one of the coordinate planes which lies on the graph. Thus, regardless of the point chosen, intersecting straight lines through the point which lies on the graph exist. Thus, the graph is a plane. Also, a similar analysis could be made for any equation of the same form.

Since we know that the graph of an equation of the form of Eq. (9–1) is a plane, its graph can be found by determining its three intercepts, and the plane can then be represented by drawing in the lines between these intercepts. If the plane passes through the origin, by letting two of the variables in turn be zero, two straight lines which define the plane are found.

Example D
Sketch the graph of $3x - y + 2z - 4 = 0$.

The intercepts of the graph of an equation are those points where it crosses the respective axes. Thus, by letting two of the variables equal zero at a time, we obtain the intercepts. For the given equation the intercepts are $(\frac{4}{3}, 0, 0)$, $(0, -4, 0)$ and $(0, 0, 2)$. These points are located (see Fig. 9–5) and lines drawn between them to represent the plane.

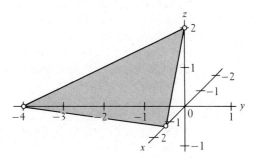

Figure 9–5

In general the graph of an equation in three variables, which is essentially equivalent to a function with two independent variables, is a surface in space. This is seen in the case of the plane, and will be verified for other cases in the examples which follow.

*The intersection of two surfaces is a **curve** in space.* This has been seen in Examples C and D, since the intersection of the given planes and the coordinate planes are lines (which in the general sense are curves). *We define the **traces** of a surface to be the curves of intersection of the surface and the coordinate planes.* The traces of a plane are those lines drawn between the intercepts to represent the plane. Many surfaces may be sketched by finding its traces and intercepts.

Example E

Find the intercepts and traces and sketch the graph of the equation $z = 4 - x^2 - y^2$.

The intercepts of the graph of the equation are $(2, 0, 0)$, $(-2, 0, 0)$, $(0, 2, 0)$, $(0, -2, 0)$ and $(0, 0, 4)$.

Since the traces of a surface lie within the coordinate planes, for each trace one of the variables is zero. Thus, by letting each variable in turn be zero, we find the trace of the surface in the plane of the other two variables. Therefore, the traces of this surface are

$$\text{in the } yz\text{-plane: } z = 4 - y^2 \text{ (a parabola)}$$
$$\text{in the } xz\text{-plane: } z = 4 - x^2 \text{ (a parabola)}$$
$$\text{in the } xy\text{-plane: } x^2 + y^2 = 4 \text{ (a circle)}$$

Using the intercepts and sketching the traces (see Fig. 9–6), we obtain the surface of the equation. This figure is called a **circular paraboloid**.

There are numerous techniques in analyzing the equation of a surface in order to obtain its graph. Another which we shall discuss here, which is closely associated to a trace, is that of a **section**. *By assuming a specific value of one of the variables, we obtain an equation in two variables, the graph of which lies in a plane parallel to the coordinate plane of the two variables.* The following example illustrates sketching a surface by use of intercepts, traces, and sections.

Example F

Sketch the graph of $4x^2 + y^2 - z^2 = 4$.

The intercepts are $(1, 0, 0)$, $(-1, 0, 0)$, $(0, 2, 0)$, and $(0, -2, 0)$. It is noted that there are no intercepts on the z-axis, for this would necessitate $z^2 = -4$.

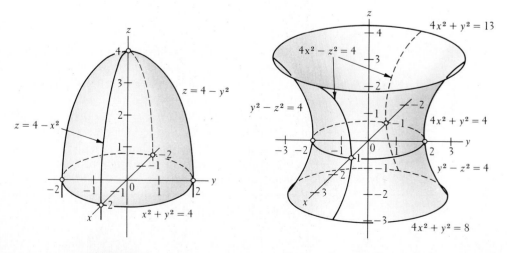

Figure 9–6   Figure 9–7

The traces are

> in the yz-plane: $y^2 - z^2 = 4$ (a hyperbola)
> in the xz-plane: $4x^2 - z^2 = 4$ (a hyperbola)
> in the xy-plane: $4x^2 + y^2 = 4$ (an ellipse)

The surface is reasonably defined by these curves, but, by assuming suitable values of z, we may indicate its shape better. For example, if $z = 3$, we have $4x^2 + y^2 = 13$, which is an ellipse. In using it we must remember that it is valid for $z = 3$ and therefore should be drawn 3 units above the xy-plane. If $z = -2$, we have $4x^2 + y^2 = 8$, which is also an ellipse. Thus, we have the sections.

> for $z = 3$: $4x^2 + y^2 = 13$ (an ellipse)
> for $z = -2$: $4x^2 + y^2 = 8$ (an ellipse)

Other sections could be found, but these are sufficient to obtain a good sketch of the graph (see Fig. 9–7). The figure is called an **elliptic hyperboloid**.

An equation with only two variables may represent a surface in space. Since only two variables are included in the equation, the surface is independent of the other variable. Another interpretation is that all sections, for all values of the variable not included, are the same. That is, *all sections parallel to the coordinate plane of the included variables are the same as the trace in that plane. The following example illustrates this type of* **cylindrical surface**.

Example G
The graph of the equation $x^2 + 4y^2 = 4$ in three dimensions is a cylindrical surface, whose sections, for all values of z, are given by the ellipse $x^2 + 4y^2 = 4$ (see Fig. 9–8).

The graph of the equation $z = 4 - x^2$ in three dimensions is a cylindrical surface, whose sections, for all values of y, are given by the parabola $z = 4 - x^2$ (see Fig. 9–9).

It must be realized that most of the figures shown extend beyond the range indicated by the traces and sections. However, these are convenient for representing and visualizing the surfaces.

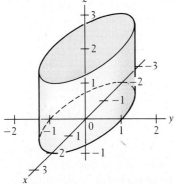

Figure 9–8   Figure 9–9

Exercises 9—2

In Exercises 1 through 4 use the method of Example A, and draw the graphs of the indicated equations for the given values of z.

1. $z = x^2 + y^2, z = 1, z = 4, z = 9$
2. $z = x - 2y, z = -2, z = 2, z = 5$
3. $z = y - x^2, z = 0, z = 2, z = 4$
4. $z = x^2 - 4y^2, z = -4, z = 1, z = 4$

Sketch the graphs of the equations given in Exercises 5 through 24 in the rectangular coordinate system in three dimensions.

5. $x + y + 2z - 4 = 0$
6. $2x - y - z + 6 = 0$
7. $4x - 2y + z - 8 = 0$
8. $3x + 3y - 2z - 6 = 0$
9. $z = y - 2x - 2$
10. $z = x - 4y$
11. $x + 2y = 4$
12. $2x - 3z = 6$
13. $x^2 + y^2 + z^2 = 4$
14. $2x^2 + 2y^2 + z^2 = 8$
15. $z = 4 - 4x^2 - y^2$
16. $z = x^2 + y^2$
17. $z = 2x^2 + y^2 + 2$
18. $x^2 + y^2 - 4z^2 = 4$
19. $x^2 - y^2 - z^2 = 9$
20. $z^2 = 9x^2 + 4y^2$
21. $x^2 + y^2 = 16$
22. $4z = x^2$
23. $y^2 + 9z^2 = 9$
24. $xy = 2$

In Exercises 25 through 30 sketch the given curves and surfaces.

25. Curves which represent a constant temperature are called *isotherms*. The temperature at a point (x, y) of a flat plate is t degrees Celsius, where $t = 4x - y^2$. In two dimensions draw the isotherms for $t = -4, 0$ and 8.

26. At a point (x, y) in the xy-plane the electric potential V is given by $V = y^2 - x^2$, where V is measured in volts. Draw the lines of equal potential for $V = -9, 0$, and 9.

27. Sketch the line in space defined by the intersection of the planes.

$$x + 2y + 3z - 6 = 0 \quad \text{and} \quad 2x + y + z - 4 = 0$$

28. Sketch the curve in space defined by the intersection of the surfaces

$$x^2 + (z - 1)^2 = 1 \quad \text{and} \quad z = 4 - x^2 - y^2$$

29. An electric charge is so distributed that the electric potential at all points on an imaginary surface is the same. Such a surface is called an *equipotential surface*. Sketch the graph of the equipotential surface whose equation is $2x^2 + 2y^2 + 3z^2 = 6$.

30. For a certain gas, the pressure, volume, and temperature are related by the equation $p = T/2V$, where p is measured in kilopascals, T in kelvins, and V in cubic meters. Sketch the p-V-T surface, by using the z-axis for p, the x-axis for V, and the y-axis for T. Use units of 100 K for T, 10 kPa for p, and 10 m³ for V. Sections must be used to sketch this surface, a *thermodynamic surface*, since none of the variables may equal zero.

9—3 Partial Derivatives

When we developed the concept of the derivative in Chapter 2, we showed that we were dealing with the rate of change of one variable with respect to another. However, only one independent variable was involved. In order to extend the derivative and its applications to functions of two (or more) variables, we find the derivative of the function with respect to one of the independent variables, while the other is held constant. If $z = f(x, y)$, this is equivalent to saying, for example, that for a specified value of y, we have the derivative of z with respect to x.

Therefore, if $z = f(x, y)$, and y is kept constant, z becomes a function of x alone. The derivative of this function with respect to x is termed the **partial derivative** of z with respect to x. Similarly, if $z = f(x, y)$ and x is kept constant, the derivative of the resulting function with respect to y is the **partial derivative** of z with respect to y.

The notation $\partial z/\partial x$ is the most common notation to denote "the partial derivative of z with respect to x," although z_x and $f_x(x, y)$ are also used. Similarly, $\partial z/\partial y$ denotes "the partial derivative of z with respect to y." Often in speaking, this is shortened to "the partial of z with respect to y."

Example A

If $z = 4x^2 + xy - y^2$, find $\partial z/\partial x$ and $\partial z/\partial y$.

In finding the partial derivative of z with respect to x, we treat y as a constant. Therefore

$$\frac{\partial z}{\partial x} = 8x + y$$

Similarly, when finding the partial derivative of z with respect to y, we treat x as a constant. Thus

$$\frac{\partial z}{\partial y} = x - 2y$$

Example B

If $y = e^x\cos t - e^{-x}\sin t$, find $\partial y/\partial x$ and $\partial y/\partial t$.

$$\frac{\partial y}{\partial x} = e^x\cos t + e^{-x}\sin t$$

$$\frac{\partial y}{\partial t} = -e^x\sin t - e^{-x}\cos t$$

Example C

If $z = \dfrac{x \ln y}{x^2 + 1}$, find $\partial z/\partial x$ and $\partial z/\partial y$.

$$\frac{\partial z}{\partial x} = \frac{(x^2 + 1)(\ln y) - (x \ln y)(2x)}{(x^2 + 1)^2} = \frac{(1 - x^2) \ln y}{(1 + x^2)^2}$$

$$\frac{\partial z}{\partial y} = \left(\frac{x}{x^2 + 1}\right)\left(\frac{1}{y}\right) = \frac{x}{y(x^2 + 1)}$$

We note that in finding $\partial z/\partial x$ it is necessary to use the quotient rule, since x appears in both numerator and denominator. However, when finding $\partial z/\partial y$ the only derivative which is necessary to find is that of $\ln y$.

To determine the geometric interpretation of a partial derivative, assume that $z = f(x, y)$ is the surface shown in Fig. 9–10. Choosing a point P on the surface, we then draw a plane through P parallel to the xz-plane. On this plane through P, the value of y is constant. The intersection of this plane and the surface is the curve as indicated. *The partial derivative of z with respect to x represents the slope of a line tangent to this curve.* When the values of the coordinates of point P are substituted into the expression which is the partial derivative, it gives the slope of the tangent line at that point. In the same way, the partial derivative of z with respect to y, evaluated at P, gives the slope of the line tangent to the curve which is found from the intersection of the surface and the plane parallel to the yz-plane through P.

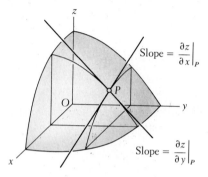

Figure 9–10

Example D
Find the slope of a line tangent to the surface $2z = x^2 + 2y^2$ and parallel to the xz-plane at the point $(2, 1, 3)$. Also find the slope of a line tangent to this surface and parallel to the yz-plane at the same point.

Finding the partial derivative of z with respect to x we have

$$2\frac{\partial z}{\partial x} = 2x \quad \text{or} \quad \frac{\partial z}{\partial x} = x$$

This derivative, evaluated at the point $(2, 1, 3)$ will give us the slope of the line tangent which is also parallel to the xz-plane. Thus, the first required slope is

$$\left.\frac{\partial z}{\partial x}\right|_{(2,1,3)} = 2$$

The partial derivative of z with respect to y, evaluated at $(2, 1, 3)$ will give us the second required slope. Thus,

$$\frac{\partial z}{\partial x} = 2y, \quad \left.\frac{\partial z}{\partial x}\right|_{(2,1,3)} = 2$$

Therefore, both slopes are 2 (see Fig. 9–11).

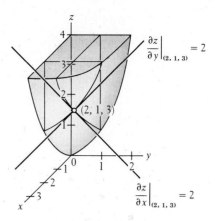

Figure 9–11

Applications of partial derivatives are found in various fields of technology. We present here an application from electricity, while others are found in the exercises.

Example E
In a parallel electric circuit with two branches having resistances r and R, the current through r can be found from

$$i = \frac{IR}{r + R}$$

where I is the total current for the two branches. Assuming I is constant at 8 A, find $\partial i/\partial r$ and evaluate it for $R = 10\ \Omega$ and $r = 0.1\ \Omega$.

Substituting for I, and finding the partial derivative, we have the following.

$$i = \frac{8R}{r + R} = 8R\,(r + R)^{-1}$$

$$\frac{\partial i}{\partial r} = (-1)\,(8R)\,(r + R)^{-2}(1)$$

$$= \frac{-8R}{(r + R)^2}$$

$$\left.\frac{\partial i}{\partial r}\right|_{\substack{r=0.1 \\ R=10}} = \frac{-8(10)}{(0.1 + 10)^2} = \frac{-80}{102} = -0.78\ \text{A}/\Omega$$

Exercises 9–3

In Exercises 1 through 16 find the partial derivative of the dependent variable or function with respect to each of the independent variables.

1. $z = 5x + 4x^2y$

2. $z = 3x^2y^3 - 3x + 4y$

3. $z = \dfrac{x^2}{y} - 2xy$

4. $f(x, y) = \dfrac{x + y}{x - y}$

5. $\phi = r\sqrt{1 + 2rs}$

6. $w = uv^2\sqrt{1 - u}$

7. $z = (x^2 + xy^3)^4$

8. $f(x, y) = (2xy - x^2)^5$

9. $z = \sin xy$

10. $y = \text{Arctan}\left(\dfrac{x}{t}\right)$

11. $y = \ln (r^2 + s)$

12. $u = e^{x+2y}$

13. $z = \sin x + \cos xy - \cos y$

14. $t = 2re^{rs^2} - \tan (2r + s)$

15. $f(x, y) = e^x\cos xy + e^{-2x}\tan y$

16. $u = \ln \dfrac{y^2}{x - y} + e^{-x} (\sin y - \cos 2y)$

In Exercises 17 through 20 evaluate the indicated partial derivatives at the given points.

17. $z = 3xy - x^2 + y$, $\dfrac{\partial z}{\partial x}\Big|_{(1, -2, -9)}$

18. $z = x\sqrt{x^2 - y^2}$, $\dfrac{\partial z}{\partial x}\Big|_{(5, 3, 20)}$

19. $z = e^y \ln xy$, $\dfrac{\partial z}{\partial y}\Big|_{(e, 1, e)}$

20. $F (r, \phi) = r^2 \sin 4\phi$, $F_\phi\left(2, \dfrac{\pi}{4}\right)$

In Exercises 21 through 30 solve the given problems.

21. Find the slope of a line, tangent to the surface $z = 9 - x^2 - y^2$, parallel to the yz-plane, which passes through $(1, 2, 4)$. Repeat the instructions for the line through $(2, 2, 1)$. Draw an appropriate figure.

22. Find $\partial z/\partial y$ for the function $z = 4x^2 - 8$. Explain your result. Draw an appropriate figure.

23. Two resistors R_1 and R_2, placed in parallel, have a combined resistance R_T given by

$$\frac{1}{R_T} = \frac{1}{R_1} + \frac{1}{R_2}$$

Find $\partial R_T/\partial R_1$.

24. For a gas, the isothermal bulk modulus B, a measure of compressibility, is given by

$$B = -V\left(\frac{\partial P}{\partial V}\right)$$

If the pressure, volume, and temperature are related by van der Waals equation

$$\left(P + \frac{a}{V^2}\right)(V - b) = RT$$

where a, b, and R are constants, find the expression for B.

25. The *mutual conductance*, measured in reciprocal ohms, of a certain electronic device is defined as $g_m = \partial i_b/\partial e_c$. Under certain circumstances, the current i_b, measured in microamperes, is given by $i_b = 50(e_b + 5e_c)^{1.5}$. Find g_m when $e_b = 200$ V and $e_c = -20$ V.

26. The *amplification factor* of the electronic device of Exercise 25 is defined as $\mu = -\partial e_b/\partial e_c$. For the device of Exercise 25, under the given conditions, find μ.

27. The *coefficient of linear expansion* α of a wire whose length is a function of the tension and temperature is given by

$$\alpha = \frac{1}{L}\left(\frac{\partial L}{\partial T}\right)$$

where L is the length and T is the temperature. If L is a function of T and the tension F given by the relation

$$L = L_0 + k_1F + k_2T + k_3FT^2$$

where k_1, k_2, k_3, and L_0 are constants, find the expression for α.

28. The fundamental frequency of vibration of a string varies directly as the square root of the tension T and inversely as the length L. If a string 60 cm long is under a tension of 65 N and has a fundamental frequency of 30 Hz, find the partial derivative of f with respect to T, and evaluate it for the given values.

29. An *isothermal process* is one during which the temperature does not change. If the pressure, volume, and temperature of an ideal gas are related by the equation

$$pV = nRT$$

where n and R are constants, find the expression for $\partial p/\partial V$, which would be the rate of change of pressure with respect to volume for an isothermal process.

30. For the ideal gas equation of Exercise 29, show that

$$\left(\frac{\partial V}{\partial T}\right)\left(\frac{\partial T}{\partial p}\right)\left(\frac{\partial p}{\partial V}\right) = -1$$

9–4 Certain Applications of Partial Derivatives

We noted some important areas of application of partial derivatives in the last section. However, as we found that we can apply the derivative to problems such as related rates, applied maximum and minimum problems, and error calculations by the use of differentials, we can also solve similar problems by the use of partial derivatives. In this section we shall show how to calculate the differential of a function of two variables and apply it to certain applied situations. Then we will discuss applied maximum and minimum problems for functions of two variables.

In Section 4–1 we defined the differential of a function of one variable as

$$dy = f'(x)\,dx$$

In a similar manner, *we define the* **total differential** *of a function* $z = f(x, y)$ *to be*

$$dz = \frac{\partial z}{\partial x}dx + \frac{\partial z}{\partial y}dy \tag{9–2}$$

It will be noted that this definition is consistent with our original definition of the differential, in that if y is constant, $dz = (\partial z/\partial x)\, dx$. Also, if x is constant, $dz = (\partial z/\partial y)\, dy$. Therefore, each term measures the differential of z, with respect to one of the independent variables, assuming the other is constant.

Example A

Find the expression for the total differential of the function $z = x^3y + x^2y^2 - 3xy^3$:

$$\frac{\partial z}{\partial x} = 3x^2y + 2xy^2 - 3y^3$$

$$\frac{\partial z}{\partial y} = x^3 + 2x^2y - 9xy^2$$

Therefore,

$$dz = (3x^2y + 2xy^2 - 3y^3)\, dx + (x^3 + 2x^2y - 9xy^2)\, dy$$

We found that we could apply the differential of a function of one variable to finding changes in the value of the function for small changes in the independent variable. This led to applied problems in error calculations. We may also apply the total differential of a function of two variables in the same manner. The following examples illustrate the method.

Example B

The power delivered to a resistor in an electric circuit is given by $P = i^2R$. Find the approximate change in power if the current changes from 4.0 to 4.1 A, and the resistance from 22.0 to 22.4 Ω.

By finding dP and evaluating the expression for $i = 4.0$, $di = 0.1$, $R = 22.0$, and $dR = 0.4$, we will have the desired result:

$$dP = \frac{\partial P}{\partial i}di + \frac{\partial P}{\partial R}dR$$

$$= (2iR)\, di + (i^2)dR$$

Evaluating we have

$$dP = [2(4.0)(22.0)](0.1) + (4.0)^2(0.4)$$
$$= 17.6 + 6.4 = 24\ \text{watts}$$

Therefore, the approximate change in the power is 24 watts.

Example C

A conical funnel, with a very small opening, was to be constructed such that the diameter across the top was 10 cm and the depth 6 cm. However, in the construction, the diameter was made only 9.8 cm and the depth 5.8 cm. What percentage error in the volume of liquid which the funnel will hold resulted?

The volume of a cone as a function of its radius and height is

$$V = \frac{1}{3}\pi r^2 h$$

Finding dV, we have

$$dV = \left(\frac{2}{3}\pi r h\right) dr + \left(\frac{1}{3}\pi r^2\right) dh$$

Using the values $r = 5.0$ cm, $dr = -0.1$ cm, $h = 6.0$ cm, and $dh = -0.2$ cm, we have

$$dV = \frac{2}{3}\pi(5.0)(6.0)(-0.1) + \frac{1}{3}\pi(25)(-0.2)$$

$$= -2\pi - \frac{5\pi}{3} = -11.5 \text{ cm}^3$$

Calculating the intended volume V, we have

$$V = \frac{1}{3}\pi(25)(6) = 50\pi = 157 \text{ cm}^3$$

The percentage error in the volume is

$$100\left(\frac{dV}{V}\right) = 100\left(\frac{11.5}{157}\right) = 7.3\%$$

We will now consider the problem of finding the maximum value or the minimum value of a function of two variables. Considering the geometric interpretation of the maximum value of a function of two variables, we note the surface indicated in Fig. 9–12.

At the maximum point M, a line tangent to the surface, *in any direction*, must be horizontal. This means that

$$\frac{\partial z}{\partial x} = 0 \qquad \text{and} \qquad \frac{\partial z}{\partial y} = 0 \qquad\qquad (9\text{--}3)$$

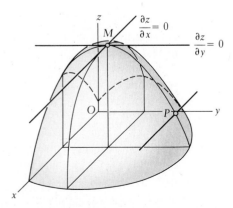

Figure 9–12

That just either of these partial derivatives be zero is not sufficient. Observe point P where $\partial z/\partial x = 0$ also. Actually, a complete analysis requires more than that shown in Eqs. (9–3), but this is beyond that which we can develop here. We shall consider only functions for which Eqs. (9–3) are sufficient to find the maximum and minimum points.

Of course, Eqs. (9–3) will hold for a minimum point as well as a maximum point. By using the first derivative test in one of the sections of the surface containing the point, the determination of whether the point is a maximum or a minimum can be made. The following example illustrates the use of Eqs. (9–3).

Example D

Examine the function $z = x^2 + 2xy + 2y^2 - 4x$ for maximum and minimum points.

First, finding the partial derivatives, we have

$$\frac{\partial z}{\partial x} = 2x + 2y - 4 \quad \text{and} \quad \frac{\partial z}{\partial y} = 2x + 4y$$

Setting these expressions equal to zero, we obtain the equations

$$2x + 2y - 4 = 0$$
$$2x + 4y = 0$$

Since these equations must hold at the same point, we solve them simultaneously. This gives us $x = 4$ and $y = -2$. Testing the section where $y = -2$, we find

$$\frac{\partial z}{\partial x} = 2x - 8$$

Thus, when $x < 4$, $\partial z/\partial x < 0$, and when $x > 4$, $\partial z/\partial x > 0$. This means that the point $(4, -2, -8)$ is a minimum point, since, as x increases, $\partial z/\partial x$ changes from negative to positive.

We shall now show an example of an applied maximum and minimum problem involving a function of two variables.

Example E

A rectangular box is to be made of material such that the base costs twice as much per square inch as the sides and top. Find the dimensions of the box of minimum cost if it is to hold 324 in.^{3}.

Assuming the length, width, and height of the box to be l, w, and h, respectively, we may write the equations for the volume and cost in terms of these quantities. The equation for the volume is

$$324 = lwh \tag{1}$$

Since the cost per square inch was not actually given, let us assume the cost for the sides and top is a cents/in.^{2}, which means that of the base is $2a$ cents/in.^{2}. Thus, the expression for the cost is

$$C = 2a(lw) + a(lw + 2wh + 2lh)$$
$$= a(3lw + 2wh + 2lh) \tag{2}$$

Solving Eq. (1) for h, and substituting this into Eq. (2), we may express C as a function of l and w:

$$C = a\left(3lw + \frac{648}{l} + \frac{648}{w}\right) \tag{3}$$

Taking the partial derivatives of C with respect to l and w, we may find the values of l and w which give the minimum cost. From these values we then can find h:

$$\frac{\partial C}{\partial w} = a\left(3l - \frac{648}{w^2}\right)$$

$$\frac{\partial C}{\partial l} = a\left(3w - \frac{648}{l^2}\right) \tag{4}$$

Setting each of these expressions equal to zero, we obtain the following equations:

$$l = \frac{216}{w^2} \qquad w = \frac{216}{l^2}$$

Substituting the expression for l into the solution for w we have

$$w = \frac{216}{\left(\frac{216}{w^2}\right)^2} = \frac{w^4}{216}$$

$$w^3 = 216 \qquad w = 6 \text{ in.}$$

Substituting, we find $l = 6$ in. and $h = 9$ in. Equation (3) shows that the maximum cost is unbounded, because of the l and w in the denominators, which means the solution we obtained must represent a minimum. Also, the first derivative test used on either of Eqs. (4) will demonstrate that the solution is a minimum.

Exercises 9–4

In Exercises 1 through 8 find the total differential of the given functions.

1. $z = 2x^2 - y^2 + 3x$ 2. $z = xy - 3yx^2 + y^3$ 3. $z = xe^y - y^2$

4. $z = x^3 - y \ln x$ 5. $z = \sin xy - y \cos x$

6. $z = \dfrac{\sec x^2}{x - y}$ 7. $z = y \operatorname{Arctan}\dfrac{y}{x^2}$ 8. $z = e^{2xy}\ln(x^2 + y^2)$

In Exercises 9 through 12 examine the given functions for maximum and minimum points.

9. $z = x^2 + y^2 - 2x - 6y + 10$ 10. $z = x^2 + xy + y^2 + 3x - 3y + 1$

11. $z = x^2 + xy + y^2 - 3y + 2$ 12. $z = x^4 - 4x + y^2$.

In Exercises 13 through 24 solve the given problems.

13. The centripetal acceleration of an object moving in a circular path is given by $a = r\omega^2$, where r is the radius of the circle and ω is the angular velocity. Find the approximate change in the centripetal acceleration if the radius

changes from 10.0 cm to 10.3 cm, and the angular velocity changes from 400 to 406 rad/min.

14. The current i through a resistor R which has a voltage V across it is $i = V/R$. Find the approximate change in current, in amperes, if the voltage changes from 220 V to 225 V and the resistance changes from 20 Ω to 21 Ω.

15. The index of refraction of a medium is defined as $n = \sin i/\sin r$, where i is the angle of incidence of light on the surface of the medium and r is the angle of refraction. In a particular experiment, n was calculated from values of $i = 45°$ and $r = 30°$. Then, an error was discovered and each angle was 1° too high. What was the approximate percentage error in the calculated value of n?

16. The electric current in a circuit containing a variable resistor R and an inductance of 10 H is a function of R and the time t given by $i = 10(1 - e^{-(Rt/10)})$. Find the approximate percentage change in current as R changes from 2.00 Ω to 2.04 Ω between $t = 10$ s and $t = 11$ s.

17. Two sides and the included angle of a triangular machine part were measured to be 2.30 cm, 2.10 cm, and 90°, respectively. The length of the third side was calculated from these values. It was then discovered that an error of 0.10 cm was made in each of the sides. What was the percentage error in the calculation of the third side?

18. If the radius and height of a right circular cylindrical container were measured to be 3.00 ft and 5.00 ft, and then the radius was found to be 3.10 ft and the height 5.20 ft, what was the error in the original value of the volume?

19. Find three positive numbers, whose sum is 120, such that their product is a maximum.

20. A flat plate is heated such that the temperature T at any point (x, y) is

$$T = x^2 + 2y^2 - x$$

Find the temperature at the coldest point.

21. An open rectangular box is to be made to contain 500 in.³. What dimensions will require the least material in building it?

22. A long metal strip 9 in. wide is to be made into a trough by bending up equal edges, making equal angles with the bottom. Find the width of the bottom and the angle between the bottom and the edges if the volume of the trough is a maximum.

23. If x, y, and z are functions of the time t, and we divide each term of Eq. (9–2) by the differential of t, dt, we obtain an equation for the **total derivative** of z with respect to t. Thus, the total derivative is

$$\frac{dz}{dt} = \frac{\partial z}{\partial x}\frac{dx}{dt} + \frac{\partial z}{\partial y}\frac{dy}{dt} \qquad (9\text{--}4)$$

Using Eq. (9–4), find dz/dt if

$$z = x^2 - xy + y^2, x = 1 + t^2, y = 1 - t^2$$

24. If the radius and height of a right circular cylinder are increasing at the rate of 0.05 ft/s and 0.20 ft/s, respectively, find the rate at which the volume is increasing when $r = 6.00$ ft and $h = 3.00$ ft, by use of Eq. (9–4).

9–5 Double Integrals

We now turn our attention to integration in the case of a function of two variables. The analysis has similarities to that of partial differentiation, in that an operation is performed while holding one of the independent variables constant.

If $z = f(x, y)$ and we wish to integrate with respect to x and y, we first consider either x or y constant, and integrate with respect to the other. After this integral is evaluated, we then integrate with respect to the variable first held constant. We shall now define this type of integral, and then give an appropriate geometric interpretation.

If $z = f(x, y)$ the **double integral** of the function over x and y is defined as

$$\int_a^b \left[\int_{g(x)}^{G(x)} f(x, y) \, dy \right] dx$$

It will be noted that the limits on the inner integral are functions of x and those on the outer integral are explicit values of x. In performing the integration, x is held constant while the inner integral is found and evaluated. This results in a function of x only. This function is then integrated and evaluated.

It is customary not to include the brackets in stating a double integral. Therefore we write

$$\int_a^b \left[\int_{g(x)}^{G(x)} f(x, y) \, dy \right] dx = \int_a^b \int_{g(x)}^{G(x)} f(x, y) \, dy \, dx \qquad (9\text{–}5)$$

Example A

Evaluate $\int_0^1 \int_{x^2}^x xy \, dy \, dx$.

Integrating and evaluating the inner integral, we have

$$\int_{x^2}^x xy \, dy = \left(x \frac{y^2}{2} \right) \Big|_{x^2}^x = x \left(\frac{x^2}{2} - \frac{x^4}{2} \right)$$

$$= \frac{1}{2}(x^3 - x^5)$$

This means

$$\int_0^1 \int_{x^2}^x xy \, dy \, dx = \int_0^1 \frac{1}{2}(x^3 - x^5) \, dx = \frac{1}{2}\left(\frac{x^4}{4} - \frac{x^6}{6} \right) \Big|_0^1$$

$$= \frac{1}{2}\left(\frac{1}{4} - \frac{1}{6} \right) - \frac{1}{2}(0) = \frac{1}{24}$$

It is also common to have an integral which is integrated over x first and then over y. The following example illustrates such a case.

Example B

Evaluate $\displaystyle\int_1^4 \int_0^{4y} \sqrt{2x + y}\, dx\, dy$.

Integrating and evaluating, we have

$$\int_1^4 \int_0^{4y} \sqrt{2x + y}\, dx\, dy = \int_1^4 \left[\frac{1}{2}\int_0^{4y}(2x + y)^{1/2}2\, dx\right] dy$$

$$= \int_1^4 \left[\frac{1}{3}(2x + y)^{3/2}\right]_0^{4y} dy$$

$$= \frac{1}{3}\int_1^4 \left[(9y)^{3/2} - (y)^{3/2}\right] dy$$

$$= \frac{1}{3}\int_1^4 (27y^{3/2} - y^{3/2})\, dy = \frac{26}{3}\int_1^4 y^{3/2}\, dy$$

$$= \frac{26}{3}\cdot\frac{2}{5}y^{5/2}\Big|_1^4 = \frac{52}{15}(32 - 1) = \frac{1612}{15}$$

To understand the geometric interpretation of a double integral, consider the surface shown in Fig. 9–13(a). Now consider an **element of volume** (as shown), of dimensions dx, dy, and z, which extends from the xy plane to the surface. We now make x a constant and sum (integrate) these elements of volume from the left boundary, $y = g(x)$, to the right boundary, $y = G(x)$. In this way we have the volume of the vertical slice as shown in Fig. 9–13(b). Now, by summing (integrating) the volumes of these slices from $x = a$ ($x = 0$ in the figure) to $x = b$, we have the complete volume as shown in Fig. 9–13(c).

Therefore, *we may interpret a double integral as the* **volume under a surface,** in the same way as the integral was interpreted as the area of a plane figure. It will be noted that this is not necessarily a volume of revolution, as discussed in Section 5–3. This means we may find the volume of a more general figure than previously possible. The following examples illustrate finding volumes by use of double integrals.

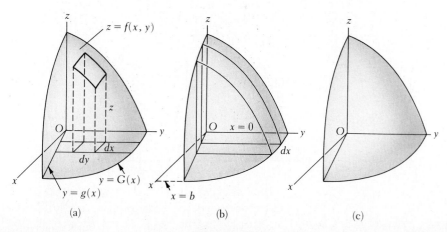

(a)   (b)   (c)

Figure 9–13

Example C

Find the volume in the first octant under the plane.

$$x + 2y + 4z - 8 = 0$$

See Fig. 9–14.

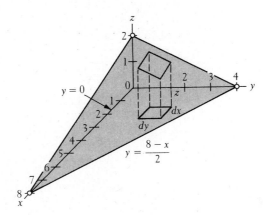

Figure 9–14

This figure is a tetrahedron, for which $V = \frac{1}{3}Bh$. Assuming the base is in the xy-plane, $B = \frac{1}{2}(4)(8) = 16$, and $h = 2$. Therefore, $V = \frac{1}{3}(16)(2) = \frac{32}{3}$ cubic units. We shall use this value to check that which we find by double integration.

To find $z = f(x, y)$, we solve the given equation for z. Thus

$$z = \frac{8 - x - 2y}{4}$$

Next, we must find the limits on y and x. Choosing to integrate over y first, we see that y goes from $y = 0$ to $y = (8 - x)/2$. This last limit is the trace of the surface in the xy-plane. Next we note that x goes from $x = 0$ to $x = 8$. Therefore, we set up and evaluate the integral:

$$V = \int_0^8 \int_0^{(8-x)/2} \left(\frac{8 - x - 2y}{4} \right) dy \, dx$$

$$= \int_0^8 \left[\frac{1}{4} \left(8y - xy - y^2 \right) \Big|_0^{(8-x)/2} \right] dx$$

$$= \frac{1}{4} \int_0^8 \left[8 \left(\frac{8 - x}{2} \right) - x \left(\frac{8 - x}{2} \right) - \left(\frac{8 - x}{2} \right)^2 \right] dx$$

$$= \frac{1}{4} \int_0^8 \left(32 - 4x - 4x + \frac{x^2}{2} - 16 + 4x - \frac{x^2}{4} \right) dx$$

$$= \frac{1}{4} \int_0^8 \left(16 - 4x + \frac{x^2}{4} \right) dx = \frac{1}{4} \left(16x - 2x^2 + \frac{x^3}{12} \right) \Big|_0^8$$

$$= \frac{1}{4} \left(128 - 128 + \frac{512}{12} \right) = \frac{1}{4} \left(\frac{128}{3} \right) = \frac{32}{3} \text{ cubic units}$$

We see that the values by the two different methods agree.

Example D

Find the volume above the xy-plane, below the surface $z = xy$, and enclosed by the cylinder $y = x^2$ and the plane $y = x$.

Constructing the figure, shown in Fig. 9–15, we now note that $z = xy$ is the desired function of x and y. Integrating over y first, the limits on y are $y = x^2$ to $y = x$. The corresponding limits on x are $x = 0$ to $x = 1$. Thus, the integral is

$$V = \int_0^1 \int_{x^2}^x xy \, dy \, dx$$

The evaluation of this integral has already been performed in Example A of this section. Therefore, we see the geometric interpretation of that integral. This means that the required volume is $\frac{1}{24}$ cubic unit.

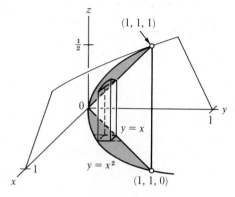

Figure 9–15

Example E

Find the volume in the first octant under the surface.

$$z = 4 - x^2 - y^2$$

and between the cylinder $x^2 = 3y$ and the plane $y = 1$ (see Fig. 9–16).

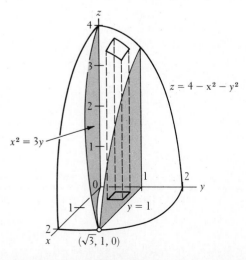

Figure 9–16

Setting up the integration such that we integrate over x first, we have

$$V = \int_0^1 \int_0^{\sqrt{3y}} (4 - x^2 - y^2)\, dx\, dy$$

$$= \int_0^1 \left[4x - \frac{x^3}{3} - y^2 x \right]_0^{\sqrt{3y}} dy$$

$$= \int_0^1 (4\sqrt{3y} - \sqrt{3}y^{3/2} - \sqrt{3}y^{5/2})\, dy$$

$$= \sqrt{3} \left[4\left(\frac{2}{3}\right) y^{3/2} - \frac{2}{5} y^{5/2} - \frac{2}{7} y^{7/2} \right]\Big|_0^1$$

$$= \sqrt{3}\left(\frac{8}{3} - \frac{2}{5} - \frac{2}{7}\right) = \frac{208\sqrt{3}}{105} \text{ cubic units}$$

If we had integrated over y first, the integral would be

$$V = \int_0^{\sqrt{3}} \int_{x^2/3}^1 (4 - x^2 - y^2)\, dy\, dx$$

Integrating and evaluating this integral, we arrive at the same result,

$$V = \frac{208\sqrt{3}}{105} \text{ cubic units}$$

Exercises 9–5

In Exercises 1 through 12 evaluate the given double integrals.

1. $\displaystyle\int_0^1 \int_0^x 2y\, dy\, dx$

2. $\displaystyle\int_0^2 \int_1^x 2x\, dy\, dx$

3. $\displaystyle\int_1^2 \int_0^{y^2} xy^2\, dx\, dy$

4. $\displaystyle\int_0^4 \int_1^{\sqrt{y}} (x - y)\, dx\, dy$

5. $\displaystyle\int_0^1 \int_0^{\sqrt{1-x^2}} y\, dy\, dx$

6. $\displaystyle\int_4^9 \int_0^x \sqrt{x - y}\, dy\, dx$

7. $\displaystyle\int_0^{\pi/6} \int_{\pi/3}^y \sin x\, dx\, dy$

8. $\displaystyle\int_0^{\sqrt{3}} \int_{x^2/3}^1 (4 - x^2 - y^2)\, dy\, dx$

9. $\displaystyle\int_1^e \int_1^y \frac{1}{x}\, dx\, dy$

10. $\displaystyle\int_{-1}^1 \int_1^{e^x} \frac{1}{xy}\, dy\, dx$

11. $\displaystyle\int_0^{\ln 3} \int_0^x e^{2x+3y}\, dy\, dx$

12. $\displaystyle\int_0^{1/2} \int_y^{y^2} \frac{dx\, dy}{\sqrt{y^2 - x^2}}$

In Exercises 13 through 22 find the indicated volumes by double integration.

13. The first octant volume under the plane $x + y + z - 4 = 0$.

14. The first octant volume under the surface $z = y^2$ and bounded by the planes $x = 2$ and $y = 3$.

15. The volume above the xy-plane and under the surface $z = 4 - x^2 - y^2$.

16. The volume above the xy-plane, below the surface $z = x^2 + y^2$, and inside the cylinder $x^2 + y^2 = 4$.

17. The first octant volume bounded by the xy-plane, the planes $x = y$, $y = 2$, and $z = 2 + x^2 + y^2$.

18. The volume bounded by the planes $x = 0$, $z = 0$, $x + 3y + 2z - 6 = 0$, and $2x = y$.

19. The first octant volume under the plane $z = x + y$ and inside the cylinder $x^2 + y^2 = 9$.

20. The volume above the xy-plane and bounded by the cylinders $x = y^2$, $y = 8x^2$, and $z = x^2 + 1$. Integrate over y first and then check by integrating over x first.

21. Draw an appropriate figure indicating a volume which is found from the integral

$$\int_0^{1/2} \int_{x^2}^1 (4 - x - 2y) \, dy \, dx$$

22. Repeat Exercise 21 for the integral

$$\int_1^2 \int_0^{2-y} \sqrt{1 + x^2 + y^2} \, dx \, dy$$

9—6 Centroids and Moments of Inertia by Double Integration

Among the important applications of double integration are those of finding centroids and moments of inertia of plane areas. In this section we shall show how double integration is applied in these cases.

Considering the area shown in Fig. 9–17, we see that *we may determine the value of this area from the double integral*

$$A = \int_a^b \int_{g(x)}^{G(x)} dy \, dx \tag{9-6}$$

This is the case, since, by summing (integrating) the elements of area $dy \, dx$ over y while holding x constant, we find the area of the vertical strip. Then by summing (integrating) from $x = a$ to $x = b$ we find the total area.

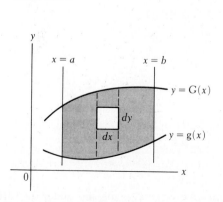

Figure 9—17

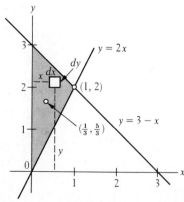

Figure 9—18

In Section 5–4, we showed that the moment of an area with respect to the y-axis is $kx\,dA$, where k is the mass per unit area, x is the moment arm, and dA is the element of area. Recalling that the x-coordinate of the centroid is the moment with respect to the y-axis, divided by the area, *we have*

$$\bar{x} = \frac{\int_a^b \int_{g(x)}^{G(x)} x\,dy\,dx}{A} \tag{9–7}$$

as *the equation by which we may find the x-coordinate of the centroid by double integration.*

Also, since the moment of the area with respect to the x-axis is $ky\,dA$, *the y coordinate of the centroid is found from*

$$\bar{y} = \frac{\int_a^b \int_{g(x)}^{G(x)} y\,dy\,dx}{A} \tag{9–8}$$

For many problems, use of Eqs. (9–7) and (9–8) is more convenient than the equivalent method in Section 5–4. One advantage is that the integration is often simplified. The following example illustrates the use of Eqs. (9–7) and (9–8).

Example A

Find the centroid of the area bounded by the lines $y = 2x$, $y = 3 - x$, and $x = 0$ (see Fig. 9–18).

$$A = \int_0^1 \int_{2x}^{3-x} dy\,dx = \int_0^1 y\Big|_{2x}^{3-x} dx$$

$$= \int_0^1 (3 - 3x)\,dx = 3x - \frac{3x^2}{2}\Big|_0^1 = \frac{3}{2}$$

$$A\bar{x} = \int_0^1 \int_{2x}^{3-x} x\,dy\,dx = \int_0^1 y\Big|_{2x}^{3-x} x\,dx$$

$$= \int_0^1 (3x - 3x^2)\,dx = \frac{3x^2}{2} - x^3\Big|_0^1 = \frac{1}{2}$$

$$A\bar{y} = \int_0^1 \int_{2x}^{3-x} y\,dy\,dx = \int_0^1 \frac{y^2}{2}\Big|_{2x}^{3-x} dx$$

$$= \frac{1}{2}\int_0^1 (9 - 6x - 3x^2)\,dx = \frac{1}{2}(9x - 3x^2 - x^3)\Big|_0^1 = \frac{5}{2}$$

Therefore, $\bar{x} = \frac{1}{2}\cdot\frac{2}{3} = \frac{1}{3}$ and $\bar{y} = \frac{5}{2}\cdot\frac{2}{3} = \frac{5}{3}$.

In many problems, integrating over x first is definitely preferable. The following example illustrates this type of problem.

Example B

Find the centroid of the area bounded by $x = 2y - y^2$ and the y-axis (see Fig. 9–19).

$$A = \int_0^2 \int_0^{2y-y^2} dx\, dy = \int_0^2 (2y - y^2)\, dy = \left[y^2 - \frac{y^3}{3} \right]\Big|_0^2 = \frac{4}{3}$$

$$A\bar{x} = \int_0^2 \int_0^{2y-y^2} x\, dx\, dy = \int_0^2 \frac{x^2}{2}\Big|_0^{2y-y^2} dy$$

$$= \frac{1}{2}\int_0^2 (4y^2 - 4y^3 + y^4)\, dy$$

$$= \frac{1}{2}\left(\frac{4}{3}y^3 - y^4 + \frac{y^5}{5} \right)\Big|_0^2 = \frac{8}{15}$$

$$A\bar{y} = \int_0^2 \int_0^{2y-y^2} y\, dx\, dy = \int_0^2 (2y^2 - y^3)\, dy = \left(\frac{2}{3}y^3 - \frac{1}{4}y^4 \right)\Big|_0^2 = \frac{4}{3}$$

Therefore, $\bar{x} = \frac{8}{15} \cdot \frac{3}{4} = \frac{2}{5}$ and $\bar{y} = \frac{4}{3} \cdot \frac{3}{4} = 1$.

From Section 5-5, we recall that the **moment of inertia** of an element of area with respect to the y-axis is $I_y = kx^2\, dA$. Therefore, using double integrals to find the moments of inertia of the area indicated in Fig. 9–17, we have (assuming $k = 1$)

$$I_y = \int_a^b \int_{g(x)}^{G(x)} x^2\, dy\, dx \tag{9-9}$$

and

$$I_x = \int_a^b \int_{g(x)}^{G(x)} y^2\, dy\, dx \tag{9-10}$$

Also, from Section 5–5, the **radius of gyration** of the area with respect to an axis was defined by $R^2 = I/A$. The following examples illustrate the

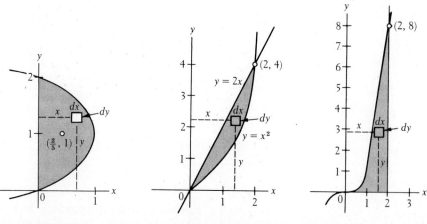

Figure 9–19   Figure 9–20   Figure 9–21

use of double integrals to find the moment of inertia and radius of gyration of an area.

Example C
Find I_y and I_x for the area bounded by $y = 2x$ and $y = x^2$ (see Fig. 9–20).

Setting up the integrals for integration over y first, we solve for I_y and I_x:

$$I_y = \int_0^2 \int_{x^2}^{2x} x^2 \, dy \, dx = \int_0^2 y \Big|_{x^2}^{2x} x^2 \, dx$$

$$= \int_0^2 (2x^3 - x^4) \, dx = \frac{x^4}{2} - \frac{x^5}{5} \Big|_0^2 = \frac{8}{5}$$

$$I_x = \int_0^2 \int_{x^2}^{2x} y^2 \, dy \, dx = \int_0^2 \frac{y^3}{3} \Big|_{x^2}^{2x} dx$$

$$= \frac{1}{3} \int_0^2 (8x^3 - x^6) \, dx = \frac{1}{3} \left(2x^4 - \frac{x^7}{7} \right) \Big|_0^2 = \frac{32}{7}$$

Example D
Find the radius of gyration with respect to the y-axis for the area bounded by $y = x^3$, $x = 2$, and $y = 0$ (see Fig. 9–21).

Since integration is more convenient over y first than over x first, we set up our integrals in this way:

$$A = \int_0^2 \int_0^{x^3} dy \, dx = \int_0^2 y \Big|_0^{x^3} dx = \int_0^2 x^3 \, dx = \frac{x^4}{4} \Big|_0^2 = 4$$

$$I_y = \int_0^2 \int_0^{x^3} x^2 \, dy \, dx = \int_0^2 y \Big|_0^{x^3} x^2 \, dx = \int_0^2 x^5 \, dx = \frac{x^6}{6} \Big|_0^2 = \frac{32}{3}$$

$$R_y^2 = \frac{32}{3} \cdot \frac{1}{4} = \frac{8}{3}, \quad R_y = \frac{2}{3} \sqrt{6} = 1.63$$

Exercises 9–6

All exercises are to be solved by the use of double integrals. In Exercises 1 through 4 find the coordinates of the centroid of the area bounded by the given curves.

1. $y = 2 - x, y = 0, x = 0$ 2. $y = x^2, y = 4, x = 0$
3. $y = 9 - x^2, y = 0, x = 0$ 4. $y = 2x, x = 1, x = 2$
5. Find I_y and I_x for the area of Exercise 1.
6. Find I_y and I_x for the area of Exercise 2.
7. Fing R_y for the area of Exercise 3.
8. Find R_x for the area of Exercise 4.
9. Find the coordinates of the centroid and R_y for the area bounded by $y = 1 - x^2$, $x = 1$, and $y = 1$.
10. Find R_y for the area bounded by $y = 3 - x$ and $y = 3 + 2x - x^2$.
11. Find the coordinates of the centroid of a right triangle with legs a and b.
12. Find the coordinates of the centroid of a quarter circle of radius a.

13. Find the moment of inertia of a rectangular area with sides a and b with respect to side a. Express the result in terms of the mass of the area.

14. Find the moment of inertia of an isosceles right triangle with respect to one of the legs a. Express the result in terms of the mass of the area.

15. Find the radius of gyration of a square of side 4 with respect to one of the sides.

16. Find the radius of gyration of a rhombus of side 2, if the angle between two adjacent sides is 45°, with respect to one of the sides.

9–7 Review Exercises for Chapter 9

In Exercises 1 through 8 find the partial derivatives of the given functions with respect to each of the independent variables.

1. $z = 5x^3y^2 - 2xy^4$

2. $z = 2x\sqrt{y} - x^2y$

3. $z = \sqrt{x^2 - 3y^2}$

4. $u = \dfrac{r}{(r - 3s)^2}$

5. $u = y \ln \sin(x^2 + 2y)$

6. $q = p \ln(r + 1) - \dfrac{rp}{r + 1}$

7. $z = \text{Arcsin}\sqrt{x + y}$

8. $z = ye^{xy}\sin(2x - y)$

In Exercises 9 through 12 evaluate each of the given double integrals.

9. $\displaystyle\int_0^3 \int_1^x (x + 2y)\, dy\, dx$

10. $\displaystyle\int_1^2 \int_0^{\pi/4} r \sec^2\theta\, d\theta\, dr$

11. $\displaystyle\int_0^1 \int_0^{2x} x^2 e^{xy}\, dy\, dx$

12. $\displaystyle\int_{\pi/4}^{\pi/2} \int_1^{\sqrt{\cos\theta}} r \sin\theta\, dr\, d\theta$

In Exercises 13 through 16 use the function $z = \sqrt{x^2 + 4y^2}$.

13. Sketch the surface representing the function.

14. Find the equation of a line tangent to the surface of the function at $(2, 1, 2\sqrt{2})$ which is parallel to the yz-plane.

15. Find the approximate change in the function as x changes from 2.00 to 2.04 and y changes from 1.00 to 1.06.

16. Find the expression for the total differential of the function.

In Exercises 17 through 20 use the function $z = e^{x+y}$.

17. Find the equation of the line tangent to the surface of the function at $(1, 1, e^2)$ which is parallel to the xz-plane.

18. Sketch the surface representing the function.

19. Find the volume in the first octant under the surface of the function, and inside the planes $x = 1$ and $y = x$.

20. Find the approximate change in the function as x changes from 1.00 to 1.01 and y changes from 1.00 to 1.02.

In Exercises 21 through 36 solve the given problems.

21. In a simple series electric circuit, with two resistors r and R connected across a voltage source E, the voltage v across r is $v = rE/(r + R)$. Assuming E to be constant, find $\partial v/\partial r$ and $\partial v/\partial R$.

22. For a gas, the volume expansivity is defined as

$$\beta = \frac{1}{V}\frac{\partial V}{\partial T}$$

where V is the volume and T the temperature of the gas. If V is a function of T and the pressure p given by

$$V = a + \frac{bT}{p} - \frac{c}{T^2}$$

where a, b and c are constants, find β.

23. In the theory dealing with transistors, the current gain α of a transistor is defined as

$$\alpha = \frac{\partial i_c}{\partial i_e}$$

where i_c is the collector current and i_e is the emitter current. If i_c is a function of i_e and the collector voltage v_c given by

$$i_c = i_e(1 - e^{-2v_c})$$

find α if v_c is 2 V.

24. Young's modulus, which measures the ratio of the stress to strain in a stretched wire, is defined as

$$Y = \frac{L}{A}\frac{\partial F}{\partial L}$$

where L is the length of the wire, A is its cross-sectional area and F is the tension on the wire. Find Y, in pascals, if

$$F = -\frac{0.01\,T}{L^2}$$

for $L = 1.10$ m, $T = 300$ K and $A = 10^{-6}$ m^2.

25. The image distance q from a lens as a function of the object distance p and the focal length f of the lens is given by

$$q = \frac{pf}{p - f}$$

Find the approximate change in q, if f changes from 20 cm to 21 cm, and if p changes from 100 cm to 105 cm.

26. The impedance Z for an alternating current circuit is given by

$$Z = \sqrt{R^2 + X^2}$$

where R is the resistance and X is the reactance. If R is measured to be 6 Ω and X to be 8 Ω, and each has a maximum possible error of 2%, what is the maximum possible error in the impedance?

27. A flat plate 5 in. square is heated such that the temperature as a function of distances x and y from intersecting edges is

$$T = 2xy - y^2 - 2x^2 + 3y + 2$$

Find the hottest point.

28. A closed rectangular box is made to contain 1000 in.^{3}. What dimensions should it have such that its surface area is a minimum?

29. Find the volume in the first octant below the plane $x + y + z - 6 = 0$ and inside the cylinder $y = 4 - x^2$.

30. Find the volume in the first octant below the surface $z = 4 - y^2$ and inside the plane $x + y = 2$.

31. Find, by double integration, the coordinates of the centroid of the area bounded by $y = 2$, $x + y = 4$, $x = 0$, and $y = 0$.

32. Find, by double integration, the coordinates of the centroid of the area bounded by $y = x$ and $y = 4x - x^2$.

33. Find, by double integration, R_y for the area of Exercise 31.

34. Find, by double integration, R_x for the area of Exercise 32.

35. The power delivered to an electric resistor is given by $P = V^2/R$, where V is the voltage across the resistor R. If V is increasing at the rate of 1 V/min and R is increasing at the rate of 2 Ω/min, find the time rate of change of power, in W/min, when $V = 60$ V and $R = 20$ Ω (see Exercise 23 of Section 9–4).

36. The kinetic energy of a moving object is given by $E = \frac{1}{2}mv^2$, where m is the mass of the object and v is its velocity. If m is decreasing at the rate of 0.01 kg/s and v is increasing at the rate of 5 m/s^2, find the time rate of change of E, in joules per second, when $m = 5$ kg and $v = 100$ m/s (see Exercise 23 of Section 9–4).

10

Polar Coordinates

10-1 Polar Coordinates

Thus far we have graphed all curves in one coordinate system. This system, the rectangular coordinate system, is probably the most useful and widely applicable system. However, for certain types of curves, other coordinate systems prove to be better adapted. These coordinate systems are widely used, especially when certain applications of higher mathematics are involved. We shall discuss one of these systems here.

Instead of designating a point by its x- and y-coordinates, we can specify its location by its radius vector and the angle which the radius vector makes with the x-axis. Thus, the r and θ that are used in the definitions of the trigonometric functions can also be used as the coordinates of points in the plane. The important aspect of choosing coordinates is that, for each set of values, there must be only one point which corresponds to this set. We can see that this condition is satisfied by the use of r and θ as coordinates. *In **polar coordinates** the origin is called the **pole**, and the positive x-axis is called the **polar axis*** (see Fig. 10–1).

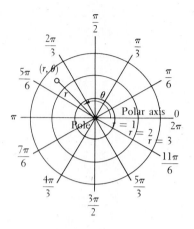

Figure 10–1

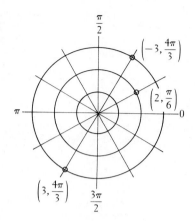

Figure 10–2

Example A

If $r = 2$ and $\theta = \pi/6$, we have the point as indicated in Fig. 10–2. The coordinates (r, θ) of this point are written as $(2, \pi/6)$ when polar coordinates are used. This point corresponds to $(\sqrt{3}, 1)$ in rectangular coordinates.

One difference between rectangular coordinates and polar coordinates is that, for each point in the plane, there are limitless possibilities for the polar coordinates of that point. For example, the point $(2, \pi/6)$ can also be represented by $(2, 13\pi/6)$ since the angles $\pi/6$ and $13\pi/6$ are coterminal. We also remove one restriction on r that we imposed in the definitions of the trigonometric functions. That is, r is allowed to take on positive and negative values. If r is considered negative, then the point is found on the opposite side of the pole from that on which it is positive.

Example B

The coordinates $(3, 4\pi/3)$ and $(3, -2\pi/3)$ represent the same point. However, the point $(-3, 4\pi/3)$ is on the opposite side of the pole, three units from the pole. Another possible set of coordinates for $(-3, 4\pi/3)$ is $(3, \pi/3)$ (see Fig. 10–2).

The relationships between the polar coordinates of a point and the rectangular coordinates of the same point come directly from the definitions of the trigonometric functions. Those most commonly used are

$$x = r \cos \theta, \qquad y = r \sin \theta \tag{10-1}$$

and

$$\tan \theta = \frac{y}{x}, \qquad r = \sqrt{x^2 + y^2} \qquad\qquad (10\text{–}2)$$

The following examples show the use of Eqs. (10–1) and (10–2) in changing coordinates in one system to coordinates in the other system. Also, they are used to transform equations from one system to the other.

Example C
Using Eqs. (10–1), we can transform the polar coordinates of $(4, \pi/4)$ into the rectangular coordinates $(2\sqrt{2}, 2\sqrt{2})$, since

$$x = 4 \cos \frac{\pi}{4} = 4 \left(\frac{\sqrt{2}}{2} \right) = 2\sqrt{2}$$

and

$$y = 4 \sin \frac{\pi}{4} = 4 \left(\frac{\sqrt{2}}{2} \right) = 2\sqrt{2}$$

Example D
Using Eqs. (10–2), we can transform the rectangular coordinates $(3, -5)$ into polar coordinates.

$$\tan \theta = -\frac{5}{3} = -1.667, \qquad \theta = 5.25 \quad (\text{or } -1.03)$$

$$r = \sqrt{3^2 + (-5)^2} = \sqrt{34} = 5.83$$

We know that θ is a fourth-quadrant angle since x is positive and y is negative. Therefore, the point $(3, -5)$ in rectangular coordinates can be expressed as the point $(5.83, 5.25)$ in polar coordinates. Other polar coordinates for the point are also possible.

Example E
Find the polar equation of the circle $x^2 + y^2 = 2x$.

Here we are given the equation of a circle in rectangular coordinates x and y and are to change it into an equation expressed in polar coordinates r and θ. Since

$$r^2 = x^2 + y^2 \qquad \text{and} \qquad x = r \cos \theta$$

we have

$$r^2 = 2r \cos \theta$$

or

$$r = 2 \cos \theta$$

as the appropriate polar equation.

Example F

Find the rectangular equation of the "rose" $r = 4 \sin 2\theta$.

Using the relation $2 \sin \theta \cos \theta = \sin 2\theta$, we have $r = 8 \sin \theta \cos \theta$. Then, using Eqs. (10–1) and (10–2), we have

$$\sqrt{x^2 + y^2} = 8\left(\frac{y}{r}\right)\left(\frac{x}{r}\right) = \frac{8xy}{r^2} = \frac{8xy}{x^2 + y^2}$$

Squaring both sides, we obtain

$$x^2 + y^2 = \frac{64x^2y^2}{(x^2 + y^2)^2} \qquad \text{or} \qquad (x^2 + y^2)^3 = 64x^2y^2$$

From this example we can see that plotting the graph from the rectangular equation would be complicated.

Exercises 10–1

In Exercises 1 through 12 plot the given points on polar coordinate paper.

1. $\left(3, \frac{\pi}{6}\right)$ 2. $(2, \pi)$ 3. $\left(\frac{5}{2}, -\frac{2\pi}{5}\right)$ 4. $\left(5, -\frac{\pi}{3}\right)$

5. $\left(-2, \frac{7\pi}{6}\right)$ 6. $\left(-5, \frac{\pi}{4}\right)$ 7. $\left(-3, -\frac{5\pi}{4}\right)$ 8. $\left(-4, -\frac{5\pi}{3}\right)$

9. $\left(0.5, -\frac{8\pi}{3}\right)$ 10. $(2.2, -6\pi)$ 11. $(2, 2)$ 12. $(-1, -1)$

In Exercises 13 through 16 find a set of polar coordinates for each of the given points expressed in rectangular coordinates.

13. $(\sqrt{3}, 1)$ 14. $(-1, -1)$ 15. $\left(-\frac{\sqrt{3}}{2}, -\frac{1}{2}\right)$ 16. $(-5, 4)$

In Exercises 17 through 20 find the rectangular coordinates corresponding to the points for which the polar coordinates are given.

17. $\left(8, \frac{4\pi}{3}\right)$ 18. $(-4, -\pi)$ 19. $\left(3, -\frac{\pi}{8}\right)$ 20. $(-1, -1)$

In Exercises 21 through 28 find the polar equation of the given rectangular equations.

21. $x = 3$ 22. $y = 2$ 23. $x^2 + y^2 = a^2$
24. $x^2 + y^2 = 4y$ 25. $y^2 = 4x$ 26. $x^2 - y^2 = a^2$
27. $x^2 + 4y^2 = 4$ 28. $y = x^2$

In Exercises 29 through 36 find the rectangular equation of each of the given polar equations.

29. $r = \sin \theta$ 30. $r = 4 \cos \theta$ 31. $r \cos \theta = 4$
32. $r \sin \theta = -2$ 33. $r = 2(1 + \cos \theta)$ 34. $r = 1 - \sin \theta$
35. $r^2 = \sin 2\theta$ 36. $r^2 = 16 \cos 2\theta$

In Exercises 37 through 40 find the required equations.

37. If we refer back to Eqs. (6–24) and (6–25), we see that the length along the arc of a circle and the area within a circular sector vary with two variables. These variables are those which we refer to as the polar coordinates. Express the arc length s and the area A in terms of rectangular coordinates. (How must we express θ?)

38. Under certain conditions, the x- and y-components of a magnetic field B are given by the equations

$$B_x = \frac{-ky}{x^2 + y^2} \quad \text{and} \quad B_y = \frac{kx}{x^2 + y^2}$$

Write these equations in terms of polar coordinates.

39. Express the equation of the cable (see Exercise 27 of Section 1–7) in polar coordinates.

40. The polar equation of the path of an artificial satellite of the earth is

$$r = \frac{4800}{1 + 0.14 \cos \theta}$$

where r is measured in miles. Find the rectangular equation of the path of this satellite. The path is an ellipse, with the earth at one of the foci.

10–2 Curves in Polar Coordinates

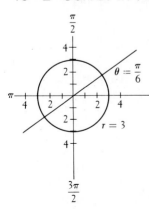

Figure 10–3

The basic method for finding a curve in polar coordinates is the same as in rectangular coordinates. We assume values of θ and then find the corresponding values of r. These points are plotted and joined, thus forming the curve which represents the function. The following examples illustrate the method.

Example A
The graph of the polar equation $r = 3$ is a circle of radius 3, with center at the pole. This is the case since $r = 3$, regardless of the value of θ. It is not really necessary to find specific points for this circle (see Fig. 10–3).

The graph of $\theta = \pi/6$ is a straight line through the pole. It represents all points for which $\theta = \pi/6$, regardless of the value of r, positive or negative (see Fig. 10–3).

Example B
Plot the graph of $r = 1 + \cos \theta$.

We find the following table of values of r corresponding to the assumed values of θ.

θ	0	$\frac{\pi}{4}$	$\frac{\pi}{2}$	$\frac{3\pi}{4}$	π	$\frac{5\pi}{4}$	$\frac{3\pi}{2}$	$\frac{7\pi}{4}$	2π
r	2	1.7	1	0.3	0	0.3	1	1.7	2

We now see that the points on the curve start repeating, and it is unnecessary to find additional points. This curve is called a **cardioid** and is shown in Fig. 10–4.

Example C
Plot the graph of $r = 1 - 2 \sin \theta$.

θ	0	$\frac{\pi}{4}$	$\frac{\pi}{2}$	$\frac{3\pi}{4}$	π	$\frac{5\pi}{4}$	$\frac{3\pi}{2}$	$\frac{7\pi}{4}$	2π
r	1	−0.4	−1	−0.4	1	2.4	3	2.4	1

Particular care should be taken in plotting the points for which r is negative. This curve is known as a **limaçon** and is shown in Fig. 10–5.

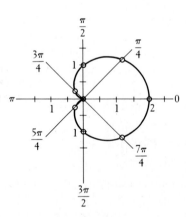

Figure 10–4

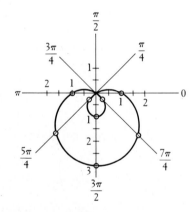

Figure 10–5

Example D
Plot the graph of $r = 2 \cos 2\theta$.

θ	0	$\frac{\pi}{12}$	$\frac{\pi}{6}$	$\frac{\pi}{4}$	$\frac{\pi}{3}$	$\frac{5\pi}{12}$	$\frac{\pi}{2}$	$\frac{7\pi}{12}$	$\frac{2\pi}{3}$	$\frac{3\pi}{4}$	$\frac{5\pi}{6}$	$\frac{11\pi}{12}$	π
r	2	1.7	1	0	−1	−1.7	−2	−1.7	−1	0	1	1.7	2

For the values of θ from π to 2π, the values of r repeat. We have a four-leaf rose (Fig. 10–6).

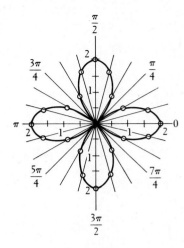

Figure 10-6

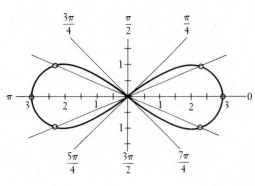

Figure 10-7

Example E

Plot the graph of $r^2 = 9 \cos 2\theta$.

θ	0	$\dfrac{\pi}{8}$	$\dfrac{\pi}{4}$	$\cdots$	$\dfrac{3\pi}{4}$	$\dfrac{7\pi}{8}$	π
r	± 3	± 2.5	0		0	± 2.5	± 3

There are no values of r corresponding to values of θ in the range $\pi/4 < \theta < 3\pi/4$, since twice these angles are in the second and third quadrants, and the cosine is negative for such angles. The value of r^2 cannot be negative. Also, the values of r repeat for $\theta > \pi$. The figure is called a **lemniscate** (Fig. 10-7).

Exercises 10-2

In Exercises 1 through 24 plot the given curves in polar coordinates.

1. $r = 4$ **2.** $r = 2$ **3.** $\theta = \dfrac{3\pi}{4}$ **4.** $\theta = \dfrac{5\pi}{3}$

5. $r = 4 \sec \theta$ **6.** $r = 4 \csc \theta$ **7.** $r = 2 \sin \theta$ **8.** $r = 3 \cos \theta$

9. $r = 1 - \cos \theta$ (cardioid)   **10.** $r = \sin \theta - 1$ (cardioid)

11. $r = 2 - \cos \theta$ (limaçon)   **12.** $r = 2 + 3 \sin \theta$ (limaçon)

13. $r = 4 \sin 2\theta$ (rose)   **14.** $r = 2 \sin 3\theta$ (rose)

15. $r^2 = 4 \sin 2\theta$ (lemniscate)   **16.** $r^2 = 2 \sin \theta$

17. $r = 2^\theta$ (spiral)   **18.** $r = 10^\theta$ (spiral)

19. $r = -4 \sin 4\theta$ (rose)   **20.** $r = -\cos 3\theta$ (rose)

21. $r = \dfrac{1}{2 - \cos \theta}$ (ellipse)   **22.** $r = \dfrac{1}{1 - \cos \theta}$ (parabola)

23. $r = 3 - \sin 3\theta$ **24.** $r = 4 \tan \theta$

In Exercises 25 through 28 sketch the indicated graphs.

25. The charged particle (see Exercise 33 of Section 1–6) is originally traveling along the line $\theta = \pi$ toward the pole. When it reaches the pole it enters the magnetic field. Its path is then described by the polar equation $r = 2a \sin \theta$, where a is the radius of the circle. Sketch the graph of this path.

26. A cam is shaped such that the equation of the edge of the upper "half" is given by $r = 2 + \cos \theta$, and the equation of the edge of the lower "half" by $r = 3/(2 - \cos \theta)$. Plot the curve which represents the shape of the cam.

27. A satellite at a height proper to make one revolution per day around the earth will have for an excellent approximation of its projection on the earth of its path the curve

$$r^2 = R^2 \cos 2\left(\theta + \frac{\pi}{2}\right)$$

where R is the radius of the earth. Sketch the path of the projection.

28. Sketch the graph of the rectangular equation

$$4(x^6 + 3x^4y^2 + 3x^2y^4 + y^6 - x^4 - 2x^2y^2 - y^4) + y^2 = 0$$

[*Hint:* The equation can be written as $4(x^2 + y^2)^3 - 4(x^2 + y^2)^2 + y^2 = 0$. Transform to polar coordinates, and then sketch the curve.]

10–3 Applications of Differentiation and Integration in Polar Coordinates

Polar coordinates are used extensively in applications and theoretical discussions in advanced mechanics, electricity, and certain other fields. However, a number of applications of differentiation and integration which we have previously discussed may be solved in polar coordinates. In this section we shall discuss an application of differentiation and one of integration. Certain other applications are in the exercises.

The application of differentiation we shall discuss is of velocity. In Section 3–2 we were able to determine the resultant velocity of an object if we knew its x- and y-components. Here we shall consider the components of velocity in the direction of increasing r and increasing θ. These components are denoted by v_r and v_θ.

The **radial component** *of velocity* v_r *and the* **transverse component** *of velocity* v_θ *are given by*

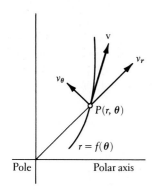

Figure 10–8

$$v_r = \frac{dr}{dt} \quad \text{and} \quad v_\theta = r\frac{d\theta}{dt} \tag{10–3}$$

See Fig. 10–8. To show that these are proper relations we recall that $x = r \cos \theta$ and $y = r \sin \theta$. Taking derivatives with respect to time, treating both r and θ as functions of time, we have

$$\frac{dx}{dt} = \cos \theta \frac{dr}{dt} - r \sin \theta \frac{d\theta}{dt}$$

$$\frac{dy}{dt} = \sin \theta \frac{dr}{dt} + r \cos \theta \frac{d\theta}{dt}$$

Since $dx/dt = v_x$ and $dy/dt = v_y$, and using Eqs. (10–3), we have

$$v_x = v_r \cos \theta - v_\theta \sin \theta$$
$$v_y = v_r \sin \theta + v_\theta \cos \theta$$

Squaring each of the sides of each of these equations and adding we have

$$v_x^2 = v_r^2 \cos^2\theta - 2v_r v_\theta \cos \theta \sin \theta + v_\theta^2 \sin^2 \theta$$
$$v_y^2 = v_r^2 \sin^2\theta + 2v_r v_\theta \cos \theta \sin \theta + v_\theta^2 \cos^2 \theta$$

$$v_x^2 + v_y^2 = v_r^2 + v_\theta^2 = v^2 \tag{10–4}$$

This analysis, along with the fact that dr/dt is the velocity in the r-direction, for constant θ, and $r(d\theta/dt)$ is the velocity in the θ-direction, for constant r, verifies that these are the correct relations. The following examples illustrate the use of Eqs. (10–3) and (10–4).

Example A
Find the radial and transverse components of velocity of an object moving such that $r = t^2 - 2$ and $\theta = \frac{1}{4}t^2$, when $t = 2$. From these values calculate the magnitude of the velocity of v when $t = 2$.

$$v_r = \frac{dr}{dt} = 2t \qquad v_\theta = r\frac{d\theta}{dt} = (t^2 - 2)\left(\frac{t}{2}\right) = \frac{1}{2}(t^3 - 2t)$$

Evaluating for $t = 2$, we have

$$v_r|_{t=2} = 4$$

$$v_\theta|_{t=2} = \frac{1}{2}(8 - 4) = 2$$

Thus,

$$v = \sqrt{16 + 4} = \sqrt{20} = 4.47$$

Example B

A particle is moving along the curve $r = 3 + 2 \sin \theta$ with a constant angular velocity, $\omega = d\theta/dt = 2$ rad/s (see Fig. 10-9). Find the magnitude of the speed in centimeters per second when $\theta = \pi/3$.

Taking derivatives of $r = 3 + 2 \sin \theta$ with respect to t, we have

$$\frac{dr}{dt} = 2 \cos \theta \frac{d\theta}{dt}$$

Therefore,

$$v_r = \frac{dr}{dt} = 2 \cos \theta \frac{d\theta}{dt}$$

$$v_\theta = r\frac{d\theta}{dt} = (3 + 2 \sin \theta)\frac{d\theta}{dt}$$

Evaluating, we have

$$v_r|_{\theta=\pi/3} = 2\left(\cos \frac{\pi}{3}\right)(2) = 4\left(\frac{1}{2}\right) = 2.00$$

$$v_\theta|_{\theta=\pi/3} = \left[3 + 2\left(\frac{\sqrt{3}}{2}\right)\right](2) = 2(3 + \sqrt{3}) = 9.46$$

$$v = \sqrt{4.00 + 89.5} = \sqrt{93.5} = 9.67 \text{ cm/s}$$

The application of integration in polar coordinates which we shall consider is the area bounded by polar curves. We recall Eq. (6-25) which states that the area of a circular sector is $A = \frac{1}{2}r^2\theta$. In Fig. 10-10, consider the element of area, with radius r and central angle $d\theta$, as shown. If these elements are summed (integrated) from $\theta = \alpha$ to $\theta = \beta$, the sum will represent the area shown. Thus, *the area is found by*

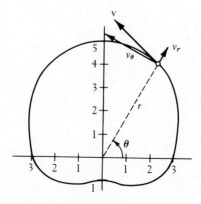

Figure 10-9

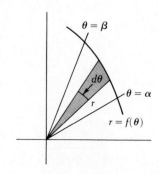

Figure 10-10

$$A = \frac{1}{2} \int_{\alpha}^{\beta} r^2 \, d\theta \tag{10–5}$$

The following examples illustrate the use of Eq. (10–5).

Example C

Find the area enclosed by one loop of the lemniscate $r^2 = \cos 2\theta$.

Employing Eq. (10–5) we substitute $r^2 = \cos 2\theta$. To determine the limits α and β, we note the symmetry of the curve (see Fig. 10–11). By choosing $\alpha = 0$ and β to be the value where r is first zero, we will find the area of the upper half of the loop. Doubling this will give the desired area. Thus, by setting $r^2 = \cos 2\theta = 0$, we find $2\theta = \pi/2$ or $\theta = \pi/4$. This means $\beta = \pi/4$. Therefore, the integral for the area is

$$A = 2\left[\frac{1}{2} \int_{0}^{\pi/4} \cos 2\theta \, d\theta\right]$$

$$= \frac{1}{2} \int_{0}^{\pi/4} \cos 2\theta (2 \, d\theta) = \frac{1}{2} \sin 2\theta \Big|_{0}^{\pi/4} = \frac{1}{2}(1 - 0) = \frac{1}{2}$$

Example D

Find the area enclosed by the curve $r = 2 - \cos \theta$.

Using Eq. (10–5) we substitute $r^2 = (2 - \cos \theta)^2 = 4 - 4 \cos \theta + \cos^2 \theta$. To determine the limits α and β, we note the sketch of the curve in Fig. 10–12.

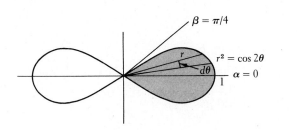

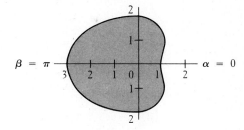

Figure 10–11   Figure 10–12

Since it is symmetrical to the polar axis, we may let $\alpha = 0$ and $\beta = \pi$ if we also multiply by 2. (Limits of $\alpha = 0$ and $\beta = 2\pi$ would also be proper.) Thus,

$$A = 2\left[\frac{1}{2} \int_{0}^{\pi} (4 - 4 \cos \theta + \cos^2 \theta) \, d\theta\right]$$

By using formula 32 of Table 6 and the basic identity $\sin 2\theta = 2 \sin \theta \cos \theta$, we have

$$A = \left[4\theta - 4 \sin \theta + \frac{\theta}{2} + \frac{1}{4} \sin 2\theta\right]_{0}^{\pi} = \left(4\pi + \frac{\pi}{2}\right) = \frac{9\pi}{2}$$

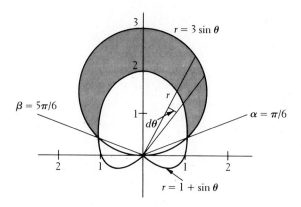

Figure 10–13

Example E

Find the area outside the cardioid $r = 1 + \sin \theta$ and inside the circle $r = 3 \sin \theta$.

Drawing the figure, we see that the desired area is the shaded portion of Fig. 10–13. By substituting $r = 3 \sin \theta$ into the equation of the cardioid, we have

$$3 \sin \theta = 1 + \sin \theta$$

$$\sin \theta = \frac{1}{2}$$

$$\theta = \frac{\pi}{6}, \frac{5\pi}{6}$$

This means the points $(3/2, \pi/6)$ and $(3/2, 5\pi/6)$ are common to each curve.

Therefore, the area inside the circle between $\theta = \pi/6$ and $\theta = 5\pi/6$ is

$$A_1 = \frac{1}{2} \int_{\pi/6}^{5\pi/6} (3 \sin \theta)^2 \, d\theta$$

The area inside the cardioid between $\theta = \pi/6$ and $\theta = 5\pi/6$ is

$$A_2 = \frac{1}{2} \int_{\pi/6}^{5\pi/6} (1 + \sin \theta)^2 \, d\theta$$

The desired area is the difference of A_1 and A_2. Thus,

$$A = \frac{1}{2} \int_{\pi/6}^{5\pi/6} [(3 \sin \theta)^2 - (1 + \sin \theta)^2] \, d\theta$$

$$= \frac{1}{2} \int_{\pi/6}^{5\pi/6} (8 \sin^2 \theta - 2 \sin \theta - 1) \, d\theta$$

$$= \frac{1}{2} \int_{\pi/6}^{5\pi/6} (4 - 4 \cos 2\theta - 2 \sin \theta - 1) \, d\theta$$

$$= \frac{1}{2} (3\theta - 2 \sin 2\theta + 2 \cos \theta) \Big|_{\pi/6}^{5\pi/6}$$

$$= \frac{1}{2}\left[3\left(\frac{5\pi}{6} - \frac{\pi}{6}\right) - 2\left(\sin\frac{5\pi}{3} - \sin\frac{\pi}{3}\right) + 2\left(\cos\frac{5\pi}{6} - \cos\frac{\pi}{6}\right)\right]$$

$$= \frac{1}{2}\left[2\pi - 2\left(-\frac{\sqrt{3}}{2} - \frac{\sqrt{3}}{2}\right) + 2\left(-\frac{\sqrt{3}}{2} - \frac{\sqrt{3}}{2}\right)\right] = \pi$$

Advantage could have been taken of the symmetry by integrating from $\pi/6$ to $\pi/2$ and doubling the result.

Exercises 10–3

In Exercises 1 through 4 r and θ are given as functions of t. Find the magnitudes of the velocities by finding v_r and v_θ for the given values of t.

1. $r = 3t, \theta = t^3, t = 1$
2. $r = 4t, \theta = \sqrt{t + 1}, t = 3$

3. $r = e^{0.1t}, \theta = \cos 2t, t = 2$
4. $r = \sin\frac{\pi t}{6}, \theta = \frac{1}{2t + 1}, t = 2$

In Exercises 5 through 12 find the magnitude of the velocity of an object moving along the given curves with a constant angular velocity $\omega = 2$ rad/s for the indicated values of θ with r measured in feet.

5. $r = 2, \theta = \pi/3$
6. $r = 4\theta, \theta = 2$

7. $r = \sin\theta, \theta = \pi/6$
8. $r = \cos 2\theta, \theta = \pi/4$

9. $r = 4 - \cos\theta, \theta = \pi/2$
10. $r = 5 + 2\sin 2\theta, \theta = \pi/6$

11. $r = \dfrac{1}{2 - \sin\theta}, \theta = \pi/6$
12. $r = 1 + \cos^2\theta, \theta = \pi/4$

In Exercises 13 through 24 find the area bounded by the given curves.

13. $\theta = \frac{1}{6}\pi, \theta = \frac{2}{3}\pi, r = 2$
14. $\theta = 0, \theta = \frac{4}{3}\pi, r = 3$

15. $\theta = 0, \theta = \frac{1}{4}\pi, r = \sec\theta$
16. $\theta = \frac{1}{6}\pi, \theta = \frac{1}{2}\pi, r\sin\theta = 2$

17. $r = 2\cos\theta$
18. $\theta = 0, \theta = \frac{1}{3}\pi, r = 4\sin\theta$

19. $\theta = 0, \theta = \frac{1}{2}\pi, r = e^{2\theta}$
20. $r = 1 - \cos\theta$

21. $r = 2\sin 3\theta$
22. $r = 3\cos 2\theta$

23. $r = 2\sin\theta, r = 2\cos\theta$ (common area)

24. $r = 2 + \cos\theta, r = 5\cos\theta$ (outside of $2 + \cos\theta$)

In Exercises 25 through 32 solve the given problems.

25. A particle moves in a counterclockwise direction on the circle $r = 2\sin\theta$. When $\theta = \pi/3, v = 10$ ft/s. Find v_r and v_θ.

26. A particle moves to the left on the parabola $r = 4/(1 + \cos\theta)$ with a constant speed of 4 m/s. Find v_r and v_θ when $\theta = \pi/2$.

27. A cam is in the shape of the curve $r = 6 - 2\cos\theta$. Find the area of one face of the cam in square centimeters.

28. Change $(x^2 + y^2)^2 - (x^2 + y^2) = 2xy$ into polar coordinates and find the area bounded by $\theta = 0, \theta = \pi/4$, and the curve.

29. Using $x = r\cos\theta$ and $y = r\sin\theta$, take derivatives of each with respect to θ and treat r as a function of θ. From these results show that

$$\frac{dy}{dx} = \frac{r'\tan\theta + r}{r' - r\tan\theta} \tag{10-6}$$

represents the slope of a line tangent to a polar curve. Here, $r' = dr/d\theta$.

30. Use Eq. (10–6) to find the slope of a line tangent to $r = \cos \theta$ at $\theta = \pi/6$.

31. The length of arc along a polar curve is given by

$$s = \int_{\alpha}^{\beta} \sqrt{r^2 + (r')^2}\, d\theta \qquad\qquad (10\text{–}7)$$

where $r' = dr/d\theta$. Using this equation find the total perimeter of $r = \cos^2(\theta/2)$.

32. Using Eq. (10–7) find the circumference of the circle $r = \sin \theta + \cos \theta$.

10–4 Review Exercises for Chapter 10

In Exercises 1 through 4 find the polar equation of each of the given rectangular equations.

1. $y = 2x$ **2.** $2xy = 1$ **3.** $x^2 - y^2 = 16$ **4.** $x^2 + y^2 = 7 - 6y$

In Exercises 5 through 8 find the rectangular equation of each of the given polar equations.

5. $r = 2 \sin 2\theta$ **6.** $r^2 = \sin \theta$

7. $r = \dfrac{4}{2 - \cos \theta}$ **8.** $r = \dfrac{2}{1 - \sin \theta}$

In Exercises 9 through 20 plot the given curves in polar coordinates.

9. $r = 4(1 + \sin \theta)$ **10.** $r = 1 + 2 \cos \theta$

11. $r = 4 \cos 3\theta$ **12.** $r = -3 \sin \theta$

13. $r = \cot \theta$ **14.** $r = \dfrac{1}{2(\sin \theta - 1)}$

15. $r = 2 \sin\left(\dfrac{\theta}{2}\right)$ **16.** $r = \theta$

17. $r = 2 + \cos 3\theta$ **18.** $r = 3 - \cos 2\theta$

19. $r^2 = \theta$ **20.** $r^2 = 4 \cos \theta$

In Exercises 21 through 24 find the area bounded by the given curves.

21. $r = 2 - \cos \theta$ **22.** $r = 1 - 2 \cos \theta$ (smaller loop)

23. Between $r = \theta$ and $\theta = \pi/2$ $(\theta \geq 0)$

24. Between $r = e^\theta$, $\theta = 0$ and $\theta = 2\pi$

In Exercises 25 through 34 solve the given problems.

25. A particle moves such that $r = t^3 - 6t$ and $\theta = t/(t + 1)$, where t is the time. Find the magnitude of the velocity of the particle when $t = 1$.

26. Find the magnitude of the velocity (in meters per second) of an object moving on the curve $r = 2 + 3 \cos 2\theta$ with a constant angular velocity $\omega = 3$ rad/s for $\theta = \pi/12$.

27. Express the equation of the artificial satellite (see Exercise 29 of Section 1–8) in polar coordinates.

28. The x- and y-coordinates of the position of a certain moving object as functions of time are given by $x = 2t$ and $y = \sqrt{8t^2 + 1}$. Find the polar equation of the path of the object.

29. In the theory dealing with an object moving through a liquid, a quantity called the velocity potential is introduced. Find v_r and v_θ if the velocity potential is given by

$$P = \frac{k \cos \theta}{r^2}$$

where k is a constant, and

$$v_r = -\frac{\partial P}{\partial r} \quad \text{and} \quad v_\theta = -\frac{1}{r}\frac{\partial P}{\partial \theta}$$

Express the result in terms of rectangular coordinates.

30. Under certain conditions the electric potential is given by

$$V = \left(\frac{a}{r^2} - br\right)\cos \theta$$

where a and b are constants. The polar components of the electric field are given by

$$E_r = -\frac{\partial V}{\partial r} \quad \text{and} \quad E_\theta = -\frac{1}{r}\frac{\partial V}{\partial \theta}$$

Find the expressions for E_r and E_θ in terms of r and θ.

31. Under a force which varies inversely as the square of the distance from an attracting object (such as the force the sun exerts on the earth), it can be shown that the equation that an object follows is given in general by

$$\frac{1}{r} = a + b \cos \theta$$

where a and b are constant for a particular path. By transforming this equation to rectangular coordinates, show that this equation represents one of the conic sections, the particular section depending on the values of a and b. It is through this kind of analysis that we know the paths of the planets and comets are conic sections.

32. The sound produced by a jet engine was measured at a distance of 100 m in all directions. The loudness of the sound (in decibels) was found to be $d = 115 + 10 \cos \theta$, where the $0°$ line for the angle θ is directed in front of the engine. Sketch the graph of d vs. θ in polar coordinates (use d as r).

33. The approximate maximum velocity attained by the artificial satellite of Exercise 40 of Section 10–1 is 5 mi/s. This velocity occurs when the satellite is closest to the earth. Calculate v_r and v_θ for this satellite at this point.

34. The angle ϕ between r and a tangent line to a curve is given by $\tan \phi = r/(dr/d\theta)$. Find the angle between r and a tangent line to the curve $r = \sin (\theta/2)$ for $\theta = \pi/2$.

11

Empirical Curve Fitting

11—1 Frequency Distributions and Measures of Central Tendency

This chapter will deal briefly with methods of analyzing empirical data. A few basic concepts from elementary statistics will be introduced, with the primary aim of being able to derive an equation of a curve to "fit" given empirical data. Such an equation determines a basic relationship between the variables of the experiment, and from which it is possible to develop and better prepare later experimentation. We shall start with basic methods of tabulating and graphing data.

When a mathematician wishes to state the relation between the area of a circle and its radius, he or she writes $A = \pi r^2$, and knows that this relation is true for all circles. When an engineer wishes to find the safe load which a particular cable is able to support, he or she cannot so simply write an equation for the relation, because the load which a cable may support depends on the diameter of the cable, the material of which it is made, the quality of material in the particular cable, any possible defects in its manufacture, and anything else which could make this cable different from all other cables. Thus, to determine the load which a cable can support, an engineer may test many cables of particular specifications. He or she will find that most of the cables can support approximately the same load, but that there is some variation, and occasionally perhaps a great variation for some reason or other. Any conclusions the person may draw will, by the nature of their source, yield probable knowledge of the cable.

The engineer, in testing a number of cables and thereby selecting a certain type of cable, is making use of the basic methods of **statistics.** That is, the engineer (1) collects data, (2) analyzes this data, and (3) interprets this data. In this section and the following section, we will discuss some of the methods of tabulating and analyzing this type of statistical information.

In statistics we are dealing with various sets of numbers. These could be measurements of length, test scores, weights of objects, or numerous other possibilities. We could deal with the entire set of numbers, but this is often too large to be practical. In such a case we deal with a selected (assumed to be representative) sample.

Example A

A company produces a machine part which is designed to be 3.80 cm long. If 10,000 are produced daily, it could be impractical to test each one to determine if it meets specifications. Therefore, it might be decided to test every tenth, or every hundredth, part.

If only ten such parts were produced daily, it might be possible to test each one for specifications.

One way of organizing data in order to develop some understanding of it is to tabulate the number of occurrences for each particular value within the set. This is called a **frequency distribution.**

Example B

A test station measured the loudness of the sound of jet aircraft taking off from a certain airport. The decibel readings of the first twenty jets were as follows: 110, 95, 100, 115, 105, 110, 120, 110, 115, 105, 90, 95, 105, 110, 100, 115, 105, 120, 95, 110.

We can see that it is difficult to determine any pattern to the readings. Therefore, we set up the following table to show the frequency distribution.

Decible reading	90	95	100	105	110	115	120
Frequency	1	3	2	4	5	3	2

We note that the distribution and pattern of readings is clearer in this form.

Just as graphs are useful in representing algebraic functions, so also are they a very convenient method of representing frequency distributions. There are several useful methods of graphing such distributions, among which the most important are the **histogram** and the **frequency polygon.** The following examples illustrate these graphical representations.

Example C
A histogram represents a particular set of data by use of rectangles. The width of the base of each rectangle is the same, and is labeled for one of the readings. The height of the rectangle represents the number of values of a given reading. In Fig. 11–1 a histogram representing the data of the frequency distribution of Example B is shown.

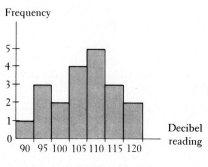

Figure 11–1

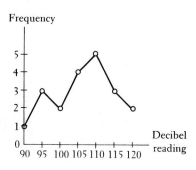

Figure 11–2

Example D
The frequency polygon is used to represent a set of data by plotting as abscissas (*x*-values) the values of the readings, and as ordinates (*y*-values) the number of occurrences of each reading (the frequency). The resulting points are joined by straight-line segments. Figure 11–2 shows a frequency polygon of the data of Example B.

Often the number of measurements is sufficiently large that it is not feasible to tabulate the number of occurrences for a particular value. In other cases such a table does not give a clear idea of the distribution. In these cases we may designate certain intervals, and tabulate the number of values within each interval. Frequency distributions, histograms, and frequency polygons can be used to represent data tabulated in this way.

Example E
A certain mathematics course had 80 students enrolled in it. After all the tests and exams were recorded, the instructors determined a numerical average (based on 100) for each student. The following is a list of the numerical grades, with the number of students receiving each.

22–1, 37–1, 40–2, 44–1, 47–1, 53–1, 55–3, 56–1, 60–3, 61–1, 63–4, 65–2, 66–1, 67–5, 68–2, 70–2, 71–4, 72–3, 74–5, 75–4, 77–7, 78–4, 79–1, 81–2, 82–4, 84–1, 85–1, 86–3, 87–2, 88–1, 90–2, 92–1, 93–2, 95–1, 97–1.

We can observe from this listing that it is difficult to see just how the grades were distributed. Therefore the instructors grouped the grades into intervals, which included five possible grades in each interval. This led to the following table.

Interval	20–24	25–29	30–34	35–39	40–44	45–49	50–54	55–59
Number in interval	1	0	0	1	3	1	1	4
Interval	60–64	65–69	70–74	75–79	80–84	85–89	90–94	95–99
Number in interval	8	10	14	16	7	7	5	2

We can see that the distribution of grades becomes much clearer in the above table. Finally, since the school graded students only with the letters A, B, C, D, and F, the instructors then grouped the grades in intervals below 60, from 60–69, from 70–79, from 80–89, and from 90–100, and assigned the appropriate letter grade to each numerical grade within each interval. This led to the following table.

Grade	A	B	C	D	F
Number receiving this grade	7	14	30	18	11

Thus, we can see the distribution according to letter grades. We can also note that, if the number of intervals were reduced much further, it would be difficult to draw any reasonable conclusions regarding the distribution of grades.

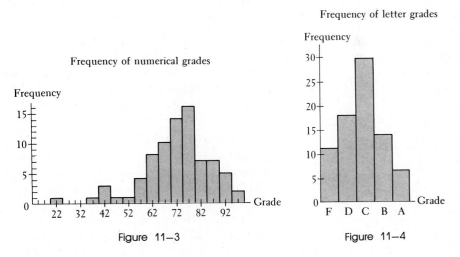

Frequency of numerical grades

Frequency of letter grades

Figure 11–3

Figure 11–4

Histograms showing the frequency of numerical grades and letter grades are shown in Figs. 11–3 and 11–4. In Fig. 11–3, each interval is represented by one number, the middle number. In Fig. 11–5 a frequency polygon of numerical grades is shown, also with each interval represented by the middle number.

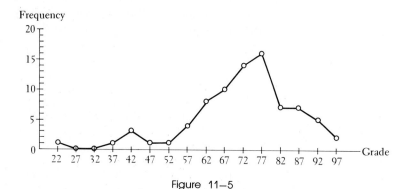

Figure 11–5

We can see from Example E that care must be taken in grouping data. If the number of intervals is too small, important characteristics of the data may not be apparent. If the number of intervals is too large, it may be difficult to analyze the data to determine the distribution and patterns.

Tables and graphical representations give a general description of data. However, it is often profitable and convenient to find representative values for the location of the center of distribution, and other numbers to give a measure of the deviation from this central value. In this way we can obtain an arithmetical description of the data. We shall now discuss the values commonly used to measure the location of the center of the distribution. These are referred to as "measures of central tendency."

The first of these measures of central tendency is the **median.** *The median is the middle number, that number for which there are as many above it as below it in the distribution.* If there is no middle number, the median is that number halfway between the two numbers nearest the middle of the distribution.

Example F

Given the numbers 5, 2, 6, 4, 7, 4, 7, 2, 8, 9, 4, 11, 9, 1, 3, we first arrange them in numerical order. This arrangement is

> 1, 2, 2, 3, 4, 4, 4, 5, 6, 7, 7, 8, 9, 9, 11.

Since there are 15 numbers, the middle number is the eighth. Since the eighth number is 5, the median is 5.

If the number 11 is not included in this set of numbers, and there are only 14 numbers in all, the median is that number halfway between the seventh and eighth numbers. Since the seventh is 4 and the eighth is 5, the median is 4.5.

Example G

In the distribution of grades given in Example E of this section, the median is 74. There are 80 grades in all, and if they are listed in numerical order we find that the 39th through the 43rd grade is 74. This means that the 40th and 41st grades are both 74. The number halfway between the 40th and 41st grades would be the median. The fact that both are 74 means that the median is also 74.

Another, and very widely applied, measure of central tendency is the **arithmetic mean.** *The mean is calculated by finding the sum of all the values and then dividing by the number of values.*

Example H

The arithmetic mean of the numbers given in Example F is found by finding the sum of all the numbers and dividing by 15. Thus, letting $\bar{x}$ (read as "x bar") represent the mean, we have

$$\bar{x} = \frac{5 + 2 + 6 + 4 + 7 + 4 + 7 + 2 + 8 + 9 + 4 + 11 + 9 + 1 + 3}{15}$$

$$= \frac{82}{15} = 5.5$$

Thus, the mean is 5.5.

If we wish to find the arithmetic mean of a large number of values, and if some of them appear more than once, the calculation can be somewhat simplified. By multiplying each number by its frequency, adding these results, and then dividing by the total number of values considered, the mean can be calculated. Letting $\bar{x}$ represent the mean of the values $x_1, x_2, \ldots, x_n$, which occur $f_1, f_2, \ldots, f_n$ times, respectively, we have

$$\bar{x} = \frac{x_1 f_1 + x_2 f_2 + \cdots + x_n f_n}{f_1 + f_2 + \cdots + f_n} \tag{11-1}$$

(This notation can be simplified by letting

$$x_1 f_1 + x_2 f_2 + \cdots + x_n f_n = \sum_{i=1}^{n} x_i f_i$$

and

$$f_1 + f_2 + \cdots + f_n = \sum_{i=1}^{n} f_i$$

This notation is often used to indicate a sum, where Σ is known as the summation sign.)

Example I

Using Eq. (11–1) to find the arithmetic mean of the numbers of Example F, we have

$$\bar{x} = \frac{1(1) + 2(2) + 3(1) + 4(3) + 5(1) + 6(1) + 7(2) + 8(1) + 9(2) + 11(1)}{15}$$

$$= \frac{82}{15} = 5.5$$

Example J
We find the arithmetic mean of the grades in Example E by

$$\bar{x} = \frac{22(1) + (37)(1) + (40)(2) + \cdots + (67)(5) + \cdots + (97)(1)}{80}$$

$$= \frac{5733}{80} = 71.7$$

Exercises 11–1

In Exercises 1 through 24 use the following sets of numbers.

A: 3, 6, 4, 2, 5, 4, 7, 6, 3, 4, 6, 4, 5, 7, 3
B: 25, 26, 23, 24, 25, 28, 26, 27, 23, 28, 25
C: 48, 53, 49, 45, 55, 49, 47, 55, 48, 57, 51, 46
D: 105, 108, 103, 108, 106, 104, 109, 104, 110, 108, 108, 104, 113, 106, 107, 106, 107, 109, 105, 111, 109, 108

In Exercises 1 through 4 set up a frequency distribution table, indicating the frequency of each number given in the indicated set.

1. Set A 2. Set B 3. Set C 4. Set D

In Exercises 5 through 8 set up a frequency distribution table, indicating the frequency of numbers for the given intervals of the given sets.

5. Intervals 2–3, 4–5, and 6–7 for set A.
6. Intervals 22–24, 25–27, and 28–30 for set B.
7. Intervals 43–45, 46–48, 49–51, 52–54, and 55–57 for set C.
8. Intervals 101–105, 106–110, and 111–115 for set D.

In Exercises 9 through 12 draw histograms for the data in the given exercise.

9. Exercise 1   10. Exercise 4   11. Exercise 7   12. Exercise 8

In Exercises 13 through 16 draw frequency polygons for the data in the given exercise.

13. Exercise 1   14. Exercise 4   15. Exercise 7   16. Exercise 8

In Exercises 17 through 20 determine the median of the numbers of the given set.

17. Set A 18. Set B 19. Set C 20. Set D

In Exercises 21 through 24 determine the arithmetic mean of the numbers of the given set.

21. Set A 22. Set B 23. Set C 24. Set D

In Exercises 25 through 36 find the indicated quantities.

25. Form a histogram for the data given in the following table, which was compiled when a particular type of cable was being tested for its breaking load.

Load interval, pounds	830–839	840–849	850–859	860–869	870–879	880–889	890–899
Number of cables breaking	4	14	48	531	85	22	2

26. Form a frequency polygon for the data given in Exercise 25.

27. A researcher, testing an electric circuit, found the following values for the current (in milliamperes) in the circuit on successive trials: 3.44, 3.46, 3.39, 3.44, 3.48, 3.40, 3.29, 3.46, 3.41, 3.37, 3.45, 3.47, 3.43, 3.38, 3.50, 3.41, 3.42, 3.47. Form a histogram for the intervals 3.25–3.29, 3.30–3.34, etc.

28. Form a histogram for the data given in Exercise 27 for the intervals 3.21–3.30, 3.31–3.40, etc.

29. Find the median of the data given in Exercise 27.

30. Find the mean of the data given in Exercise 27.

31. One hundred motorists were asked to maintain a miles per gallon record for their cars, all the same model. The records which were reported are shown in the following table.

Miles per gallon	19.0–19.4	19.5–19.9	20.0–20.4	20.5–20.9
Number reporting	4	9	14	24
Miles per gallon	21.0–21.4	21.5–21.9	22.0–22.4	22.5–22.9
Number reporting	28	16	4	1

Draw a histogram for this data.

32. Draw a frequency polygon for the data of Exercise 31.

33. Determine the arithmetic mean of the data of Exercise 31.

34. Determine the median of the data of Exercise 31.

35. Take two dice and toss them 100 times, recording at each toss the sum which appears. Draw a frequency polygon of the sum and the frequency with which it occurred. Compare this with the expected frequency as based on probability.

36. From the financial section of a newspaper, record (to the nearest dollar) the closing price of the first 50 stocks listed. Form a frequency table with intervals 0–9, 10–19, and so forth. Form a histogram of these data. Finally, find the median price.

In Exercises 37 through 40 use the following information. The **mode** of a frequency distribution, another measure of central tendency, is defined as the measure with the maximum frequency, if there is one. A set of numbers may have more than one mode. Using this definition, find the mode of the numbers in the indicated sets of numbers at the beginning of this exercise set.

37. Set A   38. Set B   39. Set C   40. Set D

11—2 Standard Deviation

In the preceding section we discussed measures of central tendency of sets of data. However, regardless of the measure which may be used, it does not tell us whether the data are grouped closely together or spread over a large range of values. Therefore, we also need some measure of the deviation, or dispersion, of the values from the median or mean. If the dispersion is small and the numbers are grouped closely together, the measure of central tendency is more reliable and descriptive of the data than the case in which the spread is greater.

There are several measures of dispersion, and we discuss in this section one which is very widely used. It is called the **standard deviation.**

The standard deviation of a set of numbers is given by the equation

$$s = \sqrt{\overline{(x - \bar{x})^2}} \tag{11-2}$$

The definition of s indicates that the following steps are to be taken in computing its value.

(1) Find the arithmetic mean $\bar{x}$ of the set of numbers.
(2) Subtract the mean from each of the numbers of the set.
(3) Square these differences.
(4) Find the arithmetic mean of these squares. (Note that a bar over any quantity signifies the mean of the set of values of that quantity.)
(5) Find the square root of this last arithmetic mean.

Defined in this way, s must be a positive number, and thus indicates a deviation from the mean, regardless of whether or not individual numbers are greater or less than the mean.

Example A

Find the standard deviation of the numbers 1, 5, 4, 2, 6, 2, 1, 1, 5, 3. The most efficient method of finding s is to make a table in which Steps (1) to (4) are indicated.

x	$x - \bar{x}$	$(x - \bar{x})^2$
1	−2	4
5	2	4
4	1	1
2	−1	1
6	3	9
2	−1	1
1	−2	4
1	−2	4
5	2	4
3	0	0
30		32

$$\bar{x} = \frac{30}{10} = 3$$

$$\overline{(x - \bar{x})^2} = \frac{32}{10} = 3.2$$

$$s = \sqrt{3.2} = 1.8$$

Example B

Find the standard deviation of the numbers given in Example F of Section 11–1.

Since several of the numbers appear more than once, it is helpful to use the frequency of each number in the table. Therefore, we have the following table and calculations.

x	f	fx	$x - \bar{x}$	$(x - \bar{x})^2$	$f(x - \bar{x})^2$
1	1	1	-4.5	20.25	20.25
2	2	4	-3.5	12.25	24.50
3	1	3	-2.5	6.25	6.25
4	3	12	-1.5	2.25	6.75
5	1	5	-0.5	0.25	0.25
6	1	6	0.5	0.25	0.25
7	2	14	1.5	2.25	4.50
8	1	8	2.5	6.25	6.25
9	2	18	3.5	12.25	24.50
11	1	11	5.5	30.25	30.25
	15	82			123.75

$$\bar{x} = \frac{82}{15} = 5.5$$

$$\overline{f(x - \bar{x})^2} = \frac{123.75}{15} = 8.25$$

$$s = \sqrt{8.25} = 2.9$$

We can see from Examples A and B that there is a great deal of calculational work required to find the standard deviation of a set of numbers. It is possible to reduce this work somewhat, and we now show how this may be done.

To find the mean of the sum of two sets of data, x_i and y_i, where the number of values in each set is the same, we can determine the mean of the x's and the mean of the y's and add the results. That is

$$\overline{x + y} = \bar{x} + \bar{y} \tag{11–3}$$

This is true by the meaning of the definition of the arithmetic mean.

$$\overline{x + y} = \frac{x_1 + x_2 + \cdots + x_n + y_1 + y_2 + \cdots + y_n}{n}$$

$$= \frac{x_1 + x_2 + \cdots + x_n}{n} + \frac{y_1 + y_2 + \cdots y_n}{n} = \bar{x} + \bar{y}$$

Also, when we find the arithmetic mean, if all the numbers contain a constant factor, this number can be factored before the mean is found. That is,

$$\overline{kx} = k\bar{x} \tag{11–4}$$

If we multiply out the quantity under the radical in Eq. (11–2), and then apply Eq. (11–3), we obtain

$$\overline{x^2 - 2x\bar{x} + \bar{x}^2} = \overline{x^2} - \overline{2x\bar{x}} + \overline{\bar{x}^2}$$

In this equation the number 2 and $\bar{x}$ are constants for any given problem. Thus, by using Eq. (11–4), we have

$$\overline{x^2} - \overline{2x\bar{x}} + \overline{\bar{x}^2} = \overline{x^2} - 2\bar{x}(\bar{x}) + \bar{x}^2(1) = \overline{x^2} - \bar{x}^2$$

Substituting this last result into Eq. (11–2), we have

$$s = \sqrt{\overline{x^2} - \bar{x}^2} \tag{11–5}$$

Equation (11–5) shows us that the standard deviation s may be found by finding the mean of the squares of the x's, subtracting the square of $\bar{x}$, and then finding the square root. This eliminates the step of subtracting $\bar{x}$ from each x, a step required by the original definition.

Example C
By using Eq. (11–5), find s for the numbers in Example A. Again, a table is the most convenient form.

x	x^2
1	1
5	25
4	16
2	4
6	36
2	4
1	1
1	1
5	25
3	9
30	122

$$\bar{x} = \frac{30}{10} = 3, \qquad \bar{x}^2 = 9$$

$$\overline{x^2} = \frac{122}{10} = 12.2$$

$$\overline{x^2} - \bar{x}^2 = 12.2 - 9 = 3.2$$

$$s = \sqrt{3.2} = 1.8$$

Example D
In an ammeter, two resistances are connected in parallel. Most of the current passing through the meter goes through the one called the shunt. In order to determine the accuracy of the resistance of shunts being made for ammeters, a manufacturer tested a sample of 100 shunts. The resistance of each, to the nearest hundredth of an ohm, is indicated in the following table. Calculate the standard deviation of the resistances of the shunts.

R (ohms)	f	fR	fR²
0.200	1	0.200	0.0400
0.210	3	0.630	0.1323
0.220	5	1.100	0.2420
0.230	10	2.300	0.5290
0.240	17	4.080	0.9792
0.250	40	10.000	2.5000
0.260	13	3.380	0.8788
0.270	6	1.620	0.4374
0.280	3	0.840	0.2352
0.290	2	0.580	0.1682
	100	24.730	6.1421

$$\overline{R} = \frac{\Sigma fR}{\Sigma f} = \frac{24.73}{100} = 0.2473$$

$$\overline{R}^2 = 0.06116$$

$$\overline{R^2} = \frac{\Sigma fR^2}{\Sigma f} = \frac{6.142}{100} = 0.06142$$

$$\overline{R^2} - \overline{R}^2 = 0.00026$$

$$s = \sqrt{0.00026} = 0.016$$

The arithmetic mean of the resistances is 0.247 Ω, with a mean deviation of 0.016 Ω.

Example E

Find the standard deviation of the grades in Example E of Section 11–1. Use the frequency distribution as grouped in intervals 20–24, 25–29, and so forth, and then assume that each value in the interval is equal to the representative value (middle value) of the interval. (This method is not exact, but when a problem involves a large number of values, the method provides a very good approximation and eliminates a great deal of arithmetic work.)

x	f	fx	fx²
22	1	22	484
27	0	0	0
32	0	0	0
37	1	37	1,369
42	3	126	5,292
47	1	47	2,209
52	1	52	2,704
57	4	228	12,996
62	8	496	30,752
67	10	670	44,890
72	14	1,008	72,576
77	16	1,232	94,864
82	7	574	47,068
87	7	609	52,983
92	5	460	42,320
97	2	194	18,818
	80	5,755	429,325

$$\overline{x} = \frac{\Sigma fx}{\Sigma f} = \frac{5,755}{80} = 71.9$$

$$\overline{x}^2 = 5,170$$

$$\overline{x^2} = \frac{\Sigma fx^2}{80} = \frac{429,325}{80} = 5,367$$

$$\overline{x^2} - \overline{x}^2 = 197$$

$$s = \sqrt{197} = 14.0$$

We have seen that the standard deviation is a measure of the dispersion of a set of data. Generally, if we make a great many measurements of a given type, we would expect to find a majority of them near the arithmetic mean. If this is the case, the distribution probably follows, at least to a reasonable extent, the **normal distribution curve** (see Fig. 11–6). This curve gives a theoretical distribution about the arithmetic mean, assuming that the number of values in the distribution becomes infinite, and its equation is

$$y = \frac{1}{\sqrt{2\pi}} e^{-x^2/2} \tag{11–6}$$

An important feature of the standard deviation is that in a normal distribution, about 68% of the values are found within the interval $\bar{x} - s$ to $\bar{x} + s$.

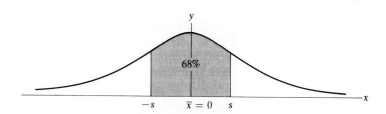

The normal distribution curve

Figure 11–6

Example F

In Example B, 9 of the 15 values, or 60%, are between 2.6 and 8.4. Here $\bar{x} = 5.5$ and $s = 2.9$. Thus, $\bar{x} - s = 2.6$ and $\bar{x} + s = 8.4$. Therefore, we see that the percentage in this interval is slightly less than in a normal distribution.

In Example E, using $\bar{x} = 72$ and $s = 14$, we have $\bar{x} - s = 58$ and $\bar{x} + s = 86$. We also note that 59 of the 80 values are in the interval from 58 to 86, which means that 74% of the values are in this interval.

Exercises 11–2

In Exercises 1 through 20 use the following sets of numbers. They are the same as those used in Exercises 11–1.

A: 3, 6, 4, 2, 5, 4, 7, 6, 3, 4, 6, 4, 5, 7, 3

B: 25, 26, 23, 24, 25, 28, 26, 27, 23, 28, 25

C: 48, 53, 49, 45, 55, 49, 47, 55, 48, 57, 51, 46

D: 105, 108, 103, 108, 106, 104, 109, 104, 110, 108, 108, 104, 113, 106, 107, 106, 107, 109, 105, 111, 109, 108

In Exercises 1 through 4 use Eq. (11–2) to find the standard deviation s for the indicated sets of numbers.

1. Set A 2. Set B 3. Set C 4. Set D

In Exercises 5 through 8 use Eq. (11–5) to find the standard deviation s for the indicated sets of numbers.

5. Set A 6. Set B 7. Set C 8. Set D

In Exercises 9 through 12 find the standard deviation s for the indicated sets of numbers.

9. The weekly salaries (in dollars) for the workers in a small factory are as follows: 250, 350, 275, 225, 175, 300, 200, 350, 275, 400, 300, 225, 250, 300.
10. The measurements of electric current in Exercise 27 of Exercises 11–1
11. The measurements of cable breaking loads in Exercise 25 of Exercises 11–1
12. The measurements of miles per gallon in Exercise 31 of Exercises 11–1

In Exercises 13 through 20 find the percentage of values in the interval $\bar{x} - s$ to $\bar{x} + s$ for the indicated sets of numbers.

13. Set A 14. Set B 15. Set C 16. Set D
17. The salaries of Exercise 9
18. The measurements of electric current in Exercise 27 of Exercises 11–1
19. The measurements of cable breaking loads in Exercise 25 of Exercises 11–1
20. The measurements of miles per gallon in Exercise 31 of Exercises 11–1

11–3 Fitting a Straight Line to a Set of Points

We have considered statistical methods for dealing with one variable. We have discussed methods of tabulating, graphing, measuring the central tendency and the deviations from this value for one variable. We now shall discuss how to obtain a relationship between two variables for which a set of points is known.

In this section we shall show a method of "fitting" a straight line to a given set of points. In the following section fitting suitable nonlinear curves to given sets of points will be discussed. Some of the reasons for doing this are (1) to express a concise relationship between the variables, (2) to use the equation for the purpose of predicting certain fundamental results, (3) to determine the reliability of certain sets of data, and (4) to use the data for testing theoretical concepts.

We shall assume for the examples and exercises of the remainder of this chapter that there is some relationship between the variables. Often when we are analyzing statistics for variables between which we think a relationship might exist, the points are so scattered as to give no reasonable idea regarding the possible functional relationship. We shall assume here that such combinations of variables have been discarded in the analysis.

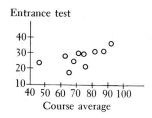

Entrance test

Figure 11–7

Example A

All the students enrolled in the mathematics course referred to in Example E of Section 11–1 took an entrance test in mathematics. To study the reliability of this test, an instructor tabulated the test scores of 10 students (selected at random), along with their course average, and made a graph of these figures (see table below, and Fig. 11–7).

Student	Entrance test score based on 40	Course average based on 100
A	29	63
B	33	88
C	22	77
D	17	67
E	26	70
F	37	93
G	30	72
H	32	81
I	23	47
J	30	74

We now ask whether or not there is a functional relationship between the test scores and the course grades. Certainly no clear-cut relationship exists, but in general we see that the higher the test score, the higher the course grade. This leads to the possibility that there might be some straight line, from which none of the points would vary too significantly. If such a line could be found, then it could be the basis of predictions as to the possible success a student might have in the course, on the basis of his or her grade on the entrance test. Assuming that such a straight line exists, the problem is to find the equation of this line. Figure 11–8 shows two such possible lines.

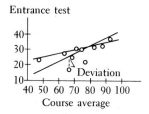

Entrance test

Figure 11–8

There are various methods of determining the line which best fits the given points. We shall employ the one which is most widely used: the **method of least squares.** *The basic principle of this method is that the sum of the squares of the deviation of all points from the best line (in accordance with this method) is the least it can be. By* **deviation** *we mean the difference between the y-value of the line and the y-value for the point (of original data) for a particular value of x.*

Example B

In Fig. 11–8 the deviation of one point is indicated. The point (67, 17) (student D of Example A) has a deviation of 8 from the indicated line of Fig. 11–8. Thus we square the value of this deviation to obtain 64. The method of least squares requires that the sum of all such squares be a minimum in order to determine the line which fits best.

Therefore, in applying this method, it is necessary to use the equation of a straight line and the coordinates of the points of the data. The deviation of these is indicated and squared. The problem then arises as to the method of determining the constants m and b in the equation of the line $y = mx + b$ for which these squares are a minimum. The following example indicates the method employed and forms a basis for determining the equations for finding m and b.

Example C
Given the points (x_1, y_1), (x_2, y_2), and (x_3, y_3), find m and b in the equation

$$y = mx + b$$

for which the sum of the squares of the deviations is a minimum.

The deviations are

$$d_1 = y_1 - (mx_1 + b)$$
$$d_2 = y_2 - (mx_2 + b)$$

and

$$d_3 = y_3 - (mx_3 + b)$$

Squaring each of these expressions and finding the sum of the squares, we have

$$\begin{aligned} S = {} & y_1^2 + y_2^2 + y_3^2 + m^2(x_1^2 + x_2^2 + x_3^2) \\ & + 3b^2 - 2m(x_1y_1 + x_2y_2 + x_3y_3) \\ & - 2b(y_1 + y_2 + y_3) + 2mb(x_1 + x_2 + x_3) \end{aligned}$$

We want S to be a minimum with respect to each of m and b. Therefore this is a maximum-minimum type of problem of Section 9–4. We shall find the partial derivatives of S with respect to m and with respect to b, set each of these equal to zero, and solve the resulting equations simultaneously:

$$\begin{aligned} \frac{\partial S}{\partial m} = {} & 2m(x_1^2 + x_2^2 + x_3^2) - 2(x_1y_1 + x_2y_2 + x_3y_3) \\ & + 2b(x_1 + x_2 + x_3) \end{aligned}$$

$$\frac{\partial S}{\partial b} = 6b - 2(y_1 + y_2 + y_3) + 2m(x_1 + x_2 + x_3)$$

Setting the second of these equations equal to zero, and solving for b, we have

$$b = \frac{(y_1 + y_2 + y_3) - m(x_1 + x_2 + x_3)}{3}$$

Observation of this equation shows it may be written as

$$b = \bar{y} - m\bar{x} \tag{1}$$

where $\bar{y}$ and $\bar{x}$ are the arithmetic means of the coordinates of the given points.

Setting $\partial S/\partial m = 0$, we may obtain

$$m(x_1^2 + x_2^2 + x_3^2) = (x_1 y_1 + x_2 y_2 + x_3 y_3) - b(x_1 + x_2 + x_3)$$

Dividing this equation through by 3, we note it may be written as

$$m\overline{x^2} = \overline{xy} - b\bar{x} \tag{2}$$

where $\overline{x^2}$ and $\overline{xy}$ are the respective arithmetic means of the coordinates of the given points.

Substituting the expression for b from Eq. (1) in Eq. (2) we have

$$m\overline{x^2} = \overline{xy} - (\bar{y} - m\bar{x})\bar{x}$$

which may be written as

$$m = \frac{\overline{xy} - \bar{x}\,\bar{y}}{\overline{x^2} - \bar{x}^2} \tag{3}$$

Evaluating the appropriate arithmetic means, m and b may be determined for the least-squares line when actual numerical values are given.

Generalizing on Example C, and using Eq. (11–5), we have the equation of the **least-squares** line as

$$y = mx + b \tag{11–7}$$

where

$$m = \frac{\overline{xy} - \bar{x}\,\bar{y}}{s_x^2} \tag{11–8}$$

and

$$b = \bar{y} - m\bar{x} \tag{11–9}$$

In Eqs. (11–8) and (11–9), the x's and y's are those of the points in the data and s_x is the standard deviation of the x-values.

Example D

Find the least-squares line of the data of Example A.

Here the y-values will be the entrance-test scores, and the x-values the course averages. The results are best obtained by tabulating the necessary quantities, as we do in the following table.

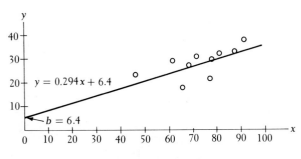

Figure 11–9

x	y	xy	x^2
63	29	1,830	3,970
88	33	2,900	7,740
77	22	1,690	5,930
67	17	1,140	4,490
70	26	1,820	4,900
93	37	3,440	8,650
72	30	2,160	5,180
81	32	2,590	6,560
47	23	1,080	2,210
74	30	2,220	5,480
732	279	20,870	55,110

$$\bar{x} = \frac{732}{10} = 73.2, \quad \overline{x^2} = 5358$$

$$\bar{y} = \frac{279}{10} = 27.9, \quad \bar{x}\,\bar{y} = 2042$$

$$\overline{xy} = \frac{20,870}{10} = 2087, \quad \overline{x^2} = \frac{55,110}{10} = 5511$$

$$s_x^2 = \overline{x^2} - \bar{x}^2 = 153$$

$$m = \frac{\overline{xy} - \bar{x}\,\bar{y}}{s_x^2} = \frac{2087 - 2042}{153}$$

$$= \frac{45}{153} = 0.294$$

$$b = \bar{y} - m\bar{x} = 27.9 - (0.294)(73.2) = 27.9 - 21.5 = 6.4$$

$$y = 0.294x + 6.4$$

Thus, the equation of the line is $y = 0.294x + 6.4$. This is the line indicated in Fig. 11–9.

We note that when plotting the least-squares line, we know two points on it. These are $(0, b)$ and $(\bar{x}, \bar{y})$. We can see that $(\bar{x}, \bar{y})$ satisfies the equation from the solution for b [Eq. (11–9)].

Example E

In an experiment to determine the relation between the load on a spring and the length of the spring, the following data were found.

Load (pounds)	0.0	1.0	2.0	3.0	4.0	5.0
Length (inches)	10.0	11.2	12.3	13.4	14.6	15.9

Find the least-squares line for this data which expresses the length as a function of the load.

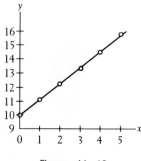

Figure 11–10

x	y	xy	x^2
0.0	10.0	0.0	0.0
1.0	11.2	11.2	1.0
2.0	12.3	24.6	4.0
3.0	13.4	40.2	9.0
4.0	14.6	58.4	16.0
5.0	15.9	79.5	25.0
15.0	77.4	213.9	55.0

$$\bar{x} = \frac{15.0}{6} = 2.50, \quad \bar{x}^2 = 6.25$$

$$\bar{y} = \frac{77.4}{6} = 12.9, \quad \bar{x}\bar{y} = 32.25$$

$$\overline{xy} = \frac{213.9}{6} = 35.65, \quad \overline{x^2} = \frac{55.0}{6} = 9.17$$

$$s_x^2 = \overline{x^2} - \bar{x}^2 = 9.17 - 6.25 = 2.92$$

$$m = \frac{35.65 - 32.25}{2.92} = 1.16, \quad b = 12.9 - (1.16)(2.50) = 10.0$$

Therefore, the least-squares line in this case is

$$y = 1.16x + 10.0$$

where y is the length of the spring and x is the load. In Fig. 11–10, the line shown is the least-squares line, and the points are the experimental data points.

Exercises 11–3

In Exercises 1 through 8 find the equation of the least-squares line for the given data. In each case graph the line and the data points on the same graph.

1. The points in the following table:

x	4	6	8	10	12
y	1	4	5	8	9

2. The points in the following table:

x	1	2	3	4	5	6	7
y	10	17	28	37	49	56	72

3. The points in the following table:

x	20	26	30	38	48	60
y	160	145	135	120	100	90

4. The points in the following table:

x	1	3	6	5	8	10	4	7	3	8
y	15	12	10	8	9	2	11	9	11	7

5. In an electrical experiment, the following data were found for the values of current and voltage for a particular element of the circuit. Find the voltage as a function of the current.

Voltage (volts)	3.00	4.10	5.60	8.00	10.50
Current (milliamperes)	15.0	10.8	9.30	3.55	4.60

6. The velocity of a falling object was found each second by use of an electric device as shown. Find the velocity as a function of time.

Velocity (meters per second)	9.70	19.5	29.5	39.4	49.2	58.9	68.6
Time (seconds)	1.00	2.00	3.00	4.00	5.00	6.00	7.00

7. In an experiment on the photoelectric effect, the frequency of the light being used was measured as a function of the stopping potential (the voltage just sufficient to stop the photoelectric current) with the results given below. Find the least-squares line for V as a function of f. The frequency for V = 0 is known as the *threshold frequency*. From the graph, determine the threshold frequency.

V (volts)	0.350	0.600	0.850	1.10	1.45	1.80
f (petahertz)	0.550	0.605	0.660	0.735	0.805	0.880

8. If a gas is cooled under conditions of constant volume, it is noted that the pressure falls nearly proportionally as the temperature. If this were to happen until there was no pressure, the theoretical temperature for this case is referred to as *absolute zero*. In an elementary experiment, the following data were found for pressure and temperature for a gas under constant volume.

P (kilopascals)	133	143	153	162	172	183
T (degrees Celsius)	0.0	20	40	60	80	100

Find the least-squares line for P as a function of T, and, from the graph, determine the value of absolute zero found in this experiment.

The linear coefficient of correlation, a measure of the relatedness of two variables, is defined by $r = m(s_x/s_y)$. Due to its definition, the values of r lie in the range $-1 \leq r \leq 1$. If r is near 1, the correlation is considered good. For values of r between $-\frac{1}{2}$ and $+\frac{1}{2}$, the correlation is poor. If r is near -1, the variables are said to be negatively correlated; that is, one increases while the other decreases.

In Exercises 9 through 12 compute r for the given sets of data.

9. Exercise 1   10. Exercise 2   11. Exercise 4   12. Example A

11—4 Fitting Nonlinear Curves to Data

If the experimental points do not appear to be on a straight line, but we recognize them as being approximately on some other type of curve, the method of least squares can be extended to use on these other curves. For example, if the points are apparently on a parabola, we could use the function $y = a + bx^2$. To use the above method, we shall extend the least-squares line to

$$y = m[f(x)] + b \tag{11–10}$$

Here, $f(x)$ must be calculated first, and then the problem can be treated as a least-squares line to find the values of m and b. Some of the functions $f(x)$ which may be considered for use are x^2, $1/x$, and 10^x.

Example A

Find the least-squares curve $y = mx^2 + b$ for the points in the following table.

y	1	5	12	24	53	76
x	0	1	2	3	4	5

The necessary quantities are tabulated for purposes of calculation.

x	y	$f(x) = x^2$	yx^2	$(x^2)^2$
0	1	0	0	0
1	5	1	5	1
2	12	4	48	16
3	24	9	216	81
4	53	16	848	256
5	76	25	1900	625
	171	55	3017	979

$$\overline{x^2} = \frac{55}{6} = 9.17, \quad \overline{x^2}^2 = 84.1$$

$$\bar{y} = \frac{171}{6} = 28.5, \quad \overline{x^2}\,\bar{y} = 261$$

$$\overline{yx^2} = \frac{3017}{6} = 503, \quad \overline{(x^2)^2} = \frac{979}{6} = 163.2$$

$$s^2_{(x^2)} = 163.2 - 84.1 = 79.1$$

$$m = \frac{503 - 261}{79.1} = \frac{242}{79.1} = 3.06 \qquad b = 28.5 - (3.06)(9.17) = 0.40$$

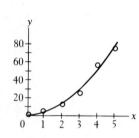

Figure 11–11

Therefore, the desired equation is $y = 3.06x^2 + 0.40$. The graph of this equation and the data points are shown in Fig. 11–11.

Example B

In a physics experiment designed to measure the pressure and volume of a gas at constant temperature, the following data were found. When the points were plotted, they were seen to approximate the hyperbola $y = c/x$. Find the least-squares approximation to this hyperbola $[y = m(1/x) + b]$ (see Fig. 11–12).

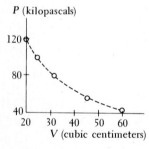

P (kilopascals)

V (cubic centimeters)

Figure 11–12

P (kPa)	V (cm³)	$x (= V)$	$y (= P)$	$f(x) = \dfrac{1}{x}$	$y\left(\dfrac{1}{x}\right)$	$\left(\dfrac{1}{x}\right)^2$
120.0	21.0	21.0	120.0	0.04762	5.714	0.002268
99.2	25.0	25.0	99.2	0.04000	3.968	0.001600
81.3	31.8	31.8	81.3	0.03145	2.557	0.000989
60.6	41.1	41.1	60.6	0.02433	1.474	0.000592
42.7	60.1	60.1	42.7	0.01664	0.711	0.000277
			403.8	0.16004	14.424	0.005726

$$\overline{\frac{1}{x}} = 0.03201 \qquad \overline{\left(\frac{1}{x}\right)^2} = 0.001025 \qquad \bar{y} = 80.76 \qquad \frac{1}{x}\bar{y} = 2.585$$

$$\overline{y\left(\frac{1}{x}\right)} = 2.885 \qquad \overline{\left(\frac{1}{x}\right)}^2 = 0.001145 \qquad s^2_{(1/x)} = 0.001145 - 0.001025$$

$$= 0.000120$$

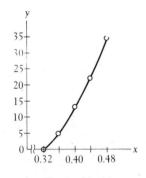

Figure 11–13

$$m = \frac{2.885 - 2.585}{0.000120} = 2500 \qquad b = 80.76 - 2500(0.03201)$$
$$= 0.74$$

Thus the equation of the hyperbola $y = m(1/x) + b$ is

$$y = \frac{2500}{x} + 0.74$$

The graph of this hyperbola and the points representing the data are shown in Fig. 11–13.

Example C

It has been found experimentally that the tensile strength of brass (a copper-zinc alloy) increases (within certain limits) with the percentage of zinc. The following table indicates the values which have been found (also see Fig. 11–14).

Tensile strength (10^5 lb/in.^{2})	0.32	0.36	0.40	0.44	0.48
Percentage of zinc	0	5	13	22	34

Fit a curve of the form $y = m(10^x) + b$ to the data. Let x = tensile strength ($\times 10^5$) and y = percentage of zinc.

x	y	$f(x) = 10^x$	$y(10^x)$	$(10^x)^2$
0.32	0	2.09	0.0	4.37
0.36	5	2.29	11.4	5.25
0.40	13	2.51	32.6	6.31
0.44	22	2.75	60.5	7.59
0.48	34	3.02	102.7	9.12
74		12.66	207.2	32.64

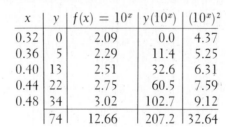

Figure 11–14

To find $f(x) = 10^x$, we may use either a calculator or logarithms. Using logarithms for $x = 0.32$, $10^{0.32}$ is found by looking up the number for which 0.32 is the logarithm. This is due to the definition of a logarithm. Values of $(10^x)^2$ are found in the same manner. When $x = 0.32$, $(10^{0.32})^2 = (10)^{0.64}$, and the antilogarithm of 0.64 is required.

$$\overline{10^x} = 2.53 \qquad\qquad \overline{10^x y} = 37.4$$
$$\overline{10^{x^2}} = 6.40 \qquad\qquad \overline{y(10^x)} = 41.4$$
$$\bar{y} = 14.8 \qquad\qquad \overline{(10^x)^2} = 6.53$$
$$s^2_{(10^x)} = 6.53 - 6.40 = 0.13$$
$$m = \frac{41.4 - 37.4}{0.13} = 31$$
$$b = 14.8 - 31(2.53) = -64$$

The equation of the curve is $y = 31(10^x) - 64$. It must be remembered that for practical purposes y must be positive. The graph of the equation is shown in Fig. 11–15, with the solid portion denoting the meaningful part of the curve. The points of the data are also shown.

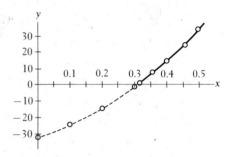

Figure 11–15

Exercises 11–4

In each of the following exercises find the indicated least-squares curve. Sketch the curve and plot the data points on the same graph.

1. For the points in the following table, find the least-squares curve $y = mx^2 + b$.

x	2	4	6	8	10
y	12	38	72	135	200

2. For the points in the following table, find the least-squares curve $y = m\sqrt{x} + b$.

x	0	4	8	12	16
y	1	9	11	14	15

3. For the points in the following table, find the least-squares curve $y = m(1/x) + b$.

x	1.10	2.45	4.04	5.86	6.90	8.54
y	9.85	4.50	2.90	1.75	1.48	1.30

4. For the points in the following table, find the least-squares curve $y = m(10^x) + b$.

x	0.00	0.200	0.500	0.950	1.325
y	6.00	6.60	8.20	14.0	26.0

5. The following data were found for the distance y that an object rolled down an inclined plane in time t. Determine the least-squares curve $y = mt^2 + b$.

y (centimeters)	6.0	23	55	98	148
t (seconds)	1.0	2.0	3.0	4.0	5.0

6. The increase in length of a certain metallic rod was measured in relation to particular increases in temperature. If y represents the increase in length for

the corresponding increase in temperature x, find the least-squares curve $y = m(x^2) + b$ for these data.

x (degrees Celsius)	50.0	100	150	200	250
y (centimeters)	1.00	4.40	9.40	16.4	24.0

7. Use the data of Exercise 6 and determine the least-squares curve $y = m(10^z) + b$ where $z = x/1000$.

8. Measurements were made of the current in an electric circuit as a function of time. The circuit contained a resistance of 5 Ω and an inductance of 10 H. The following data were found.

i (amperes)	0.00	2.52	3.45	3.80	3.92
t (seconds)	0.00	2.00	4.00	6.00	8.00

Find the least-squares curve $i = m(e^{-0.5t}) + b$ for this data. Natural logarithms may be used. From the equation, determine the value the current approaches as t approaches infinity.

9. The resonant frequency of an electric circuit containing a $4\mu F$ capacitor was measured as a function of an inductance in the circuit. The following data were found.

f (hertz)	490	360	250	200	170
L (henrys)	1.0	2.0	4.0	6.0	9.0

Find the least-squares curve $f = m(1/\sqrt{L}) + b$.

10. The displacement of a pendulum bob from its equilibrium position as a function of time gave the following results.

Displacement (centimeters)	0.00	7.80	10.0	8.10	0.00
Time (seconds)	0.00	0.90	1.60	2.20	3.15

Find the least-squares curve of the form $y = m(\sin x) + b$, expressing the displacement as y and the time as x.

11–5 Review Exercises for Chapter 11

In Exercises 1 through 4 use the following set of numbers:

2.3, 2.6, 4.2, 3.6, 3.5, 4.1, 4.8, 2.5, 3.0, 4.1, 3.8

1. Determine the median of the set of numbers.
2. Determine the arithmetic mean of the set of numbers.
3. Determine the standard deviation of the set of numbers.
4. Construct a frequency table with intervals 2.0–2.9, 3.0–3.9, and 4.0–4.9.

In Exercises 5 through 8 use the following data:

An important property of oil is its coefficient of viscosity, which gives a measure of how well it flows. In order to determine the viscosity of a certain motor oil, a refinery took samples from 12 different storage tanks and tested them at

50°C. The results (in pascal-seconds) were 0.24, 0.28, 0.29, 0.26, 0.27, 0.26, 0.25, 0.27, 0.28, 0.26, 0.26, 0.25.

5. Determine the arithmetic mean.   **6.** Determine the median.

7. Determine the standard deviation.   **8.** Make a histogram.

In Exercises 9 through 12 use the following data:

Two machine parts are considered satisfactorily assembled if their total thickness (to the nearest one-hundredth of an inch) is between or equal to 0.92 and 0.94 in. One hundred sample assemblies are tested, and the thicknesses, to the nearest one-hundredth of an inch, are given in the following table:

Total thickness	0.90	0.91	0.92	0.93	0.94	0.95	0.96
Number	3	9	31	38	12	5	2

9. Determine the arithmetic mean.   **10.** Determine the median.

11. Determine the standard deviation.   **12.** Make a frequency polygon.

In Exercises 13 through 16 use the following data:

A Geiger counter records the presence of high-energy nuclear particles. Even though no apparent radioactive source is present, a certain number of particles will be recorded. These are primarily cosmic rays, which are caused by very high-energy particles from outer space. In an experiment to measure the amount of cosmic radiation, the number of counts were recorded during 200 5-s intervals. The following table gives the number of counts, and the number of 5-s intervals having this number of counts. Draw a frequency curve for this data.

Counts	0	1	2	3	4	5	6	7	8	9	10
Intervals	3	10	25	45	29	39	26	11	7	2	3

13. Determine the median.   **14.** Determine the arithmetic mean.

15. Make a histogram.   **16.** Make a frequency polygon.

In Exercises 17 through 22 find the indicated least-squares curves.

17. In a certain experiment, the resistance of a certain resistor was measured as a function of the temperature. The data found were as follows:

R (ohms)	25.0	26.8	28.9	31.2	32.8	34.7
T (degrees Celsius)	0.0	20.0	40.0	60.0	80.0	100

Find the least-squares line for this data, expressing R as a function of T. Sketch the line and data points on the same graph.

18. The solubility of sodium chloride (table salt) (in kilograms per cubic meter of water) as a function of temperature (degrees Celsius) is measured as follows:

Solubility	357	360	366	373	384
Temperature	0.0	20.0	40.0	60.0	80.0

Find the least-squares straight line for these data.

19. The coefficient of friction of an object sliding down an inclined plane was measured as a function of the angle of incline of the plane. The following results were obtained.

Coefficient of friction	0.16	0.34	0.55	0.85	1.24	1.82	2.80
Angle (degrees)	10.0	20.0	30.0	40.0	50.0	60.0	70.0

Find the least-squares curve of the form $y = m(\tan x) + b$, expressing the coefficient of friction as y and the angle as x.

20. An electric device recorded the total distance (in centimeters) which an object fell at 0.1 s intervals. Following are the data which were found.

Distance	4.90	19.5	44.0	78.1	122.0
Time	0.100	0.200	0.300	0.400	0.500

Find the least-squares curve of the form $y = mx^2 + b$, using y to represent distance and x to represent time.

21. The period of a pendulum as a function of its length was measured, giving the following results.

Period (seconds)	1.10	1.90	2.50	2.90	3.30
Length (feet)	1.00	3.00	5.00	7.00	9.00

Find the least-squares curve of the form $y = m\sqrt{x} + b$, which expresses the period as a function of the length.

22. In an elementary experiment which measured the wavelength of sound as a function of the frequency, the following results were obtained.

Wavelength (centimeters)	140	107	81.0	70.0	60.0
Frequency (hertz)	240	320	400	480	560

Find the least-squares curve of the form $y = m(1/x) + b$ for this data, expressing wavelength as y and frequency as x.

12

Expansion of Functions in Series

___

12—1 Maclaurin Series

The values of the trigonometric functions can be determined exactly only for a few certain angles. The value of e can only be approximated in decimal form. The question arises as to how these values may be determined, particularly if a specified degree of accuracy is necessary. In this chapter we shall show how a given function may be expressed in terms of a polynomial. Once the polynomial which approximates a given function has been found, we will be able to evaluate the function to any desired accuracy. Also, a number of other uses of this polynomial form will be shown.

Example A
By using long division, we have

$$\frac{2}{2-x} = 1 + \frac{1}{2}x + \frac{1}{4}x^2 + \cdots + \left(\frac{1}{2}x\right)^{n-1} + \cdots \tag{1}$$

where n is the number of the term of the expression on the right. Since x represents a number, the right-hand side of Eq. (1) becomes a geometric progression of infinitely many terms.

From algebra we know that the sum of a geometric progression of infinitely many terms is

$$S = \frac{a}{1 - r} \tag{2}$$

where a is the first term and r is the common ratio of successive terms.
If $x = 1$, the righthand side of Eq. (1) is

$$1 + \frac{1}{2} + \frac{1}{4} + \cdots + \left(\frac{1}{2}\right)^{n-1}$$

For this progression, $r = \frac{1}{2}$ and $a = 1$, which means $S = 2$. If $x = 3$, the righthand side of Eq. (1) is

$$1 + \frac{3}{2} + \frac{9}{4} + \cdots + \left(\frac{3}{2}\right)^{n-1} + \cdots$$

and this we know does not have a defined value, since $r > 1$. Referring back to the left side of Eq. (1), we see that it also equals 2 when $x = 1$. If $x = 3$, the left side is -2. Thus, we see that the two sides agree for $x = 1$, and do not give consistent results for $x = 3$. In fact, as long as $|x| < 2$, the values will agree. From this we can conclude that the expression on the righthand side properly represents that on the left side, as long as $|x| < 2$.

From Example A, we see that an algebraic function may be properly represented by a function of the form

$$f(x) = a_0 + a_1 x + a_2 x^2 + \cdots + a_n x^n + \cdots \tag{12–1}$$

Equation (12–1) is known as a **power-series expansion** *of the function* $f(x)$. The problem now arises as to whether or not functions in general may be represented in this form. If such a representation were possible, it would provide a means of evaluating the transcendental functions for the purpose of making tables of values. Also, since a power-series expansion is in the form of a polynomial, it makes algebraic operations much simpler due to the properties of polynomials. A further study of calculus shows many other uses of power series.

At this point we shall assume that unless otherwise noted the functions with which we shall be dealing may be properly represented by a power-series expansion (it takes more advanced methods to prove that this is generally possible), for specified ranges of values of x. We shall find that the methods of calculus are very useful in developing the method of general representation. Thus, writing a general power series, along with the first few derivatives, we have

$$f(x) = a_0 + a_1 x + a_2 x^2 + a_3 x^3 + a_4 x^4 + a_5 x^5 + \cdots + a_n x^n + \cdots$$
$$f'(x) = a_1 + 2a_2 x + 3a_3 x^2 + 4a_4 x^3 + 5a_5 x^4 + \cdots + n a_n x^{n-1} + \cdots$$
$$f''(x) = 2a_2 + 2 \cdot 3a_3 x + 3 \cdot 4a_4 x^2 + 4 \cdot 5a_5 x^3 + \cdots$$
$$+ (n - 1)n a_n x^{n-2} + \cdots$$

$$f'''(x) = 2 \cdot 3a_3 + 2 \cdot 3 \cdot 4a_4 x + 3 \cdot 4 \cdot 5a_5 x^2 + \cdots$$
$$+ (n-2)(n-1)na_n x^{n-3} + \cdots$$
$$f^{iv}(x) = 2 \cdot 3 \cdot 4a_4 + 2 \cdot 3 \cdot 4 \cdot 5a_5 x + \cdots$$
$$+ (n-3)(n-2)(n-1)na_n x^{n-4} + \cdots$$

Regardless of the values of the constants a_n for any power series, if $x = 0$, the left and right sides must be equal, and all the terms on the right are zero except the first. Thus, setting $x = 0$ in each of the above equations, we have

$$f(0) = a_0 \qquad\qquad f'(0) = a_1$$
$$f''(0) = 2a_2 \qquad\qquad f'''(0) = 2 \cdot 3a_3$$
$$f^{iv}(0) = 2 \cdot 3 \cdot 4a_4$$

Solving each of these for the constants a_n, we have

$$a_0 = f(0) \qquad a_1 = f'(0) \qquad a_2 = \frac{f''(0)}{2!} \qquad a_3 = \frac{f'''(0)}{3!} \qquad a_4 = \frac{f^{iv}(0)}{4!}$$

Substituting these back into the expression for $f(x)$, we have

$$f(x) = f(0) + f'(0)x + \frac{f''(0)x^2}{2!} + \frac{f'''(0)x^3}{3!} + \cdots + \frac{f^{(n)}(0)x^n}{n!} + \cdots \qquad (12\text{--}2)$$

Equation (12–2) *is known as the* **Maclaurin series expansion** *of a function.* For a function to be represented by a Maclaurin expansion, the function and all its derivatives must exist at $x = 0$.

In developing the Maclaurin series, we introduced the **factorial notation**, $n!$. By definition,

$$n! = n(n-1)(n-2)(\cdots)(1) \qquad (12\text{--}3)$$

For example, $4! = 4 \cdot 3 \cdot 2 \cdot 1 = 24$. As we can see, this notation is convenient for use in power-series expansions. It is also used in other fields of mathematics.

Example B

Find the first four terms of the Maclaurin series expansion of $f(x) = \dfrac{2}{2-x}$.

This is written as

$$f(x) = \frac{2}{2-x} \qquad\qquad f(0) = 1$$

$$f'(x) = \frac{2}{(2-x)^2} \qquad\qquad f'(0) = \frac{1}{2}$$

$$f''(x) = \frac{4}{(2-x)^3} \qquad f''(0) = \frac{1}{2}$$

$$f'''(x) = \frac{12}{(2-x)^4} \qquad f'''(0) = \frac{3}{4}$$

$$f(x) = 1 + \frac{1}{2}x + \frac{1}{2}\left(\frac{x^2}{2!}\right) + \frac{3}{4}\left(\frac{x^3}{3!}\right) + \cdots$$

or

$$\frac{2}{2-x} = 1 + \frac{1}{2}x + \frac{1}{4}x^2 + \frac{1}{8}x^3 + \cdots$$

We see that this result agrees with that obtained by direct division.

Example C

Find the first four terms of the Maclaurin series expansion of $f(x) = e^{-x}$.

We write

$$
\begin{aligned}
f(x) &= e^{-x} & f(0) &= 1 \\
f'(x) &= -e^{-x} & f'(0) &= -1 \\
f''(x) &= e^{-x} & f''(0) &= 1 \\
f'''(x) &= -e^{-x} & f'''(0) &= -1
\end{aligned}
$$

$$f(x) = 1 + (-1)x + 1\left(\frac{x^2}{2!}\right) + (-1)\left(\frac{x^3}{3!}\right) + \cdots$$

or

$$e^{-x} = 1 - x + \frac{x^2}{2!} - \frac{x^3}{3!} + \cdots$$

Example D

Find the first three nonzero terms of the Maclaurin series expansion of $f(x) = \sin 2x$.

We have

$$
\begin{aligned}
f(x) &= \sin 2x, & f(0) &= 0, & f'(x) &= 2\cos 2x, & f'(0) &= 2 \\
f''(x) &= -4\sin 2x, & f''(0) &= 0, & f'''(x) &= -8\cos 2x, & f'''(0) &= -8 \\
f^{iv}(x) &= 16\sin 2x, & f^{iv}(0) &= 0, & f^{v}(x) &= 32\cos 2x, & f^{v}(0) &= 32
\end{aligned}
$$

$$f(x) = 0 + 2x + 0 + (-8)\frac{x^3}{3!} + 0 + 32\frac{x^5}{5!} + \cdots$$

or

$$\sin 2x = 2x - \frac{4}{3}x^3 + \frac{4}{15}x^5 - \cdots$$

*This series is called an **alternating series**, since precisely every other term is negative.*

Example E

Find the first three nonzero terms of the Maclaurin expansion of the function $f(x) = xe^{2x}$.

$$
\begin{aligned}
f(x) &= xe^{2x} & f(0) &= 0 \\
f'(x) &= e^{2x}(1 + 2x) & f'(0) &= 1 \\
f''(x) &= e^{2x}(4 + 4x) & f''(0) &= 4 \\
f'''(x) &= e^{2x}(12 + 8x) & f'''(0) &= 12
\end{aligned}
$$

$$
f(x) = 0 + x + 4\left(\frac{x^2}{2!}\right) + 12\left(\frac{x^3}{3!}\right) + \cdots
$$

or

$$
xe^{2x} = x + 2x^2 + 2x^3 + \cdots
$$

Exercises 12−1

In Exercises 1 through 12 find in each case the first three nonzero terms of the Maclaurin expansion of the given functions.

1. $f(x) = e^x$ **2.** $f(x) = \sin x$ **3.** $f(x) = \cos x$

4. $f(x) = \ln(1 + x)$ **5.** $f(x) = \sqrt{1 + x}$ **6.** $f(x) = \dfrac{1}{(1 - x)^{1/3}}$

7. $f(x) = e^{-2x}$ **8.** $f(x) = \frac{1}{2}(e^x + e^{-x})$ **9.** $f(x) = \cos 4x$

10. $f(x) = e^x \sin x$ **11.** $f(x) = \dfrac{1}{1 - x}$ **12.** $f(x) = \dfrac{1}{(1 + x)^2}$

In Exercises 13 through 20 find the first two nonzero terms of the Maclaurin expansion of the given functions.

13. $f(x) = \text{Arctan } x$ **14.** $f(x) = \cos x^2$ **15.** $f(x) = \tan x$
16. $f(x) = \sec x$ **17.** $f(x) = \ln \cos x$ **18.** $f(x) = xe^{\sin x}$
19. $f(x) = \sin^2 x$ **20.** $f(x) = e^{-x^2}$

In Exercises 21 through 24 solve the given problems.

21. Why is it not possible to find a Maclaurin expansion of the following functions?

(a) $f(x) = \csc x$,   (b) $f(x) = \sqrt{x}$,   (c) $f(x) = \ln x$.

22. By finding the Maclaurin expansion, show that

$$
(1 + x)^n = 1 + nx + \frac{n(n - 1)}{2!}x^2 + \cdots
$$

This series is the **binomial series**, which can be shown to be valid for $|x| < 1$ for all values of n.

23. If $f(x) = x^3$, show that this function is obtained when a Maclaurin expansion is found.

24. If $f(x) = x^4 + 2x^2$, show that this function is obtained when a Maclaurin expansion is found.

12–2 Certain Operations with Series

The series found in the first four exercises of Section 12–1 are of particular importance. They are used to find values of exponential functions, trigonometric functions, and logarithms. Also they may be used to develop other series. We shall give them here for reference:

$$e^x = 1 + x + \frac{x^2}{2!} + \frac{x^3}{3!} + \cdots \qquad (12\text{–}4)$$

$$\sin x = x - \frac{x^3}{3!} + \frac{x^5}{5!} - \cdots \qquad (12\text{–}5)$$

$$\cos x = 1 - \frac{x^2}{2!} + \frac{x^4}{4!} - \cdots \qquad (12\text{–}6)$$

$$\ln(1 + x) = x - \frac{x^2}{2} + \frac{x^3}{3} - \frac{x^4}{4} + \cdots \qquad (12\text{–}7)$$

In the next section we shall see how we make use of these series in the development of tables. In this section we shall not only see how new series may be developed by using the above basic series, but also we shall see certain other uses of series.

When we discussed functions in Chapter 2, we made brief mention of functions such as $f(2x)$, $f(x^2)$, and $f(-x)$. *By using this functional notation and the preceding series, we can find the series expansions of a great many other series without the use of direct expansion.* This can often save a great deal of time in finding a desired series. The examples given here illustrate the use of series for this purpose.

Example A

Find the Maclaurin expansion of e^{2x}.

From Eq. (12–4), we know the expansion of e^x. Hence,

$$f(x) = 1 + x + \frac{x^2}{2!} + \frac{x^3}{3!} + \cdots$$

Since $e^{2x} = f(2x)$, we have

$$f(2x) = 1 + (2x) + \frac{(2x)^2}{2!} + \frac{(2x)^3}{3!} + \cdots$$

or

$$e^{2x} = 1 + 2x + 2x^2 + \frac{4x^3}{3} + \cdots$$

Example B

Find the Maclaurin expansion of $\sin x^2$.

From Eq. (12–5), we know the expansion of $\sin x$. Therefore,

$$f(x) = x - \frac{x^3}{3!} + \frac{x^5}{5!} - \cdots$$

Since $\sin x^2 = f(x^2)$, we have

$$f(x^2) = (x^2) - \frac{(x^2)^3}{3!} + \frac{(x^2)^5}{5!} - \cdots$$

or

$$\sin x^2 = x^2 - \frac{x^6}{3!} + \frac{x^{10}}{5!} - \cdots$$

Direct expansion of this series is quite lengthy.

The basic algebraic operations may be applied to series in the same manner they are applied to polynomials. That is, we may add, subtract, multiply, or divide series in order to obtain other series. This is illustrated in the following example.

Example C

Multiply the series for e^x and $\cos x$ in order to obtain the series expansion for $e^x \cos x$.

Using the series expansions for e^x and $\cos x$ as shown in Eqs. (12–4) and (12–6), we have the following indicated multiplication.

$$e^x \cos x = \left(1 + x + \frac{x^2}{2!} + \frac{x^3}{3!} + \frac{x^4}{4!} + \cdots\right)\left(1 - \frac{x^2}{2!} + \frac{x^4}{4!} - \cdots\right)$$

By multiplying the series on the right, we have the following result, considering through the x^4 terms in the product.

$$e^x \cos x = 1 - \frac{x^2}{2} + \frac{x^4}{24} + x - \frac{x^3}{2} + \frac{x^2}{2} - \frac{x^4}{4} + \frac{x^3}{6} + \frac{x^4}{24} + \cdots$$

$$= 1 + x - \frac{1}{3}x^3 - \frac{1}{6}x^4 + \cdots$$

A result very important in the development of complex numbers in mathematics and in electrical theory is now demonstrated. Assuming that the Maclaurin expansions for e^x, $\sin x$, and $\cos x$ are valid for complex numbers, we have $(j = \sqrt{-1})$

$$e^{j\theta} = 1 + j\theta + \frac{(j\theta)^2}{2!} + \frac{(j\theta)^3}{3!} + \cdots = 1 + j\theta - \frac{\theta^2}{2!} - j\frac{\theta^3}{3!} + \cdots \qquad (12\text{–}8)$$

$$j \sin \theta = j\theta - j\frac{\theta^3}{3!} + \cdots \qquad (12\text{–}9)$$

$$\cos \theta = 1 - \frac{\theta^2}{2!} + \cdots \qquad (12\text{–}10)$$

When we add the terms of Eq. (12–9) to those of Eq. (12–10), the result is the series given in Eq. (12–8). Thus,

$$e^{j\theta} = \cos\theta + j\sin\theta \qquad (12\text{–}11)$$

The **rectangular form** *of a complex number is* $a + bj$. *The* **polar** *(or* **trigonometric**) **form** *is* $r(\cos\theta + j\sin\theta)$, *and the* **exponential form** *is* $re^{j\theta}$, *where* $r = \sqrt{a^2 + b^2}$. Thus, by these definitions and Eq. (12–11), we have

$$a + bj = r(\cos\theta + j\sin\theta) = re^{j\theta} \qquad (12\text{–}12)$$

where

$$r = \sqrt{a^2 + b^2} \qquad (12\text{–}13)$$

Example D

Express $2e^{4.80j}$ in rectangular form.

Using Eq. (12–12), and the fact that $4.80 = 275°$, we have

$$\begin{aligned}
2e^{4.80j} &= 2(\cos 4.80 + j\sin 4.80) \\
&= 2(\cos 275° + j\sin 275°) \\
&= 2(0.0872 - j0.9962) \\
&= 0.1744 - 1.9924j
\end{aligned}$$

An additional use of power series is now shown. Many integrals which occur in practice cannot be integrated by methods given in the preceding chapters. However, power series can be very useful in giving excellent approximations to some definite integrals.

Example E

Evaluate: $\displaystyle\int_0^1 \frac{\sin x}{x}\,dx.$

By using the series expansion of $\sin x$, and dividing each term by x, we have

$$\int_0^1 \frac{\sin x}{x}\,dx = \int_0^1 \left(1 - \frac{x^2}{3!} + \frac{x^4}{5!} - \cdots\right)dx$$

$$= \left(x - \frac{x^3}{3(3!)} + \frac{x^5}{5(5!)} - \cdots\right)\Bigg|_0^1$$

$$= 1 - \frac{1}{3(6)} + \frac{1}{5(120)} - \cdots$$

Combining the values indicated, we have $1 - 0.05556 + 0.00166 = 0.9461$. We can see that each of the terms omitted was very small. The correct answer, to 4 decimal places, is 0.9461, which is precisely the value found by using only three terms of the expansion for $\sin x$.

Example F

Evaluate: $\int_0^{0.1} e^{-x^2}\,dx$.

We write

$$e^{-x^2} = 1 + (-x^2) + \frac{(-x^2)^2}{2!} + \cdots$$

Thus,

$$\int_0^{0.1} e^{-x^2}\,dx = \int_0^{0.1}\left(1 - x^2 + \frac{x^4}{2} - \cdots\right)dx$$

$$= \left(x - \frac{x^3}{3} + \frac{x^5}{10} - \cdots\right)\Bigg|_0^{0.1}$$

$$= 0.1 - \frac{0.001}{3} + \frac{0.00001}{10} = 0.09967$$

This answer is correct to the indicated accuracy.

The question of accuracy now arises. The integrals just evaluated indicate that the more terms used, the greater the accuracy of the result. As an indication of the accuracy involved, Fig. 12–1 shows the graphs of $y = \sin x$ and the graphs of

$$y = x, \qquad y = x - \frac{x^3}{3!} \quad \text{and} \quad y = x - \frac{x^3}{3!} + \frac{x^5}{5!}$$

which are the first three approximations. We can see that each term added gives a better fit to the sin x. Also, a graphical representation of the meaning of series expansions is indicated.

We have just shown that the more terms included, the more accurate the result. For small values of x, a Maclaurin series can give good accuracy with a very few terms. When this happens, we see that the series *converges* rapidly, as we mentioned earlier. For this reason a Maclaurin series is of particular use if small values of x are being considered. If larger values of x are being considered, usually a function is expanded in a Taylor series (see Section 12–4). Of course, when we omit the later terms in any series, a certain error is inherent in the calculation.

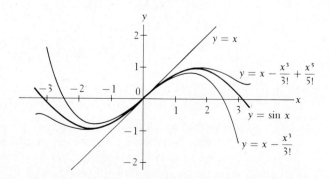

Figure 12–1

Exercises 12–2

In Exercises 1 through 8 find in each case the first four nonzero terms of the Maclaurin expansions of the given functions by using Eqs. (12–4) to (12–7).

1. $f(x) = e^{3x}$ 2. $f(x) = e^{-2x}$ 3. $f(x) = \sin \frac{1}{2}x$

4. $f(x) = \sin x^4$ 5. $f(x) = \cos 4x$ 6. $f(x) = \cos 2x^3$

7. $f(x) = \ln(1 + x^2)$ 8. $f(x) = \ln(1 - x)$

In Exercises 9 through 12 evaluate the given integrals by use of three terms of the appropriate series.

9. $\displaystyle\int_0^1 \sin x^2\, dx$ 10. $\displaystyle\int_0^{0.1} \frac{e^x - 1}{x}\, dx$

11. $\displaystyle\int_0^{0.2} \cos \sqrt{x}\, dx$ 12. $\displaystyle\int_0^1 \frac{\cos x - 1}{x}\, dx$

In Exercises 13 through 16, change the given complex numbers to the form indicated.

13. $3e^{0.5j}$ (rectangular)   14. $2(\cos 230° + j \sin 230°)$ (exponential)

15. $6 + j$ (polar)   16. $-3 + 2j$ (exponential)

In Exercises 17 through 24 determine the indicated series by the given operation.

17. Find the first four nonzero terms of the expansion of the function $f(x) = \frac{1}{2}(e^x + e^{-x})$ by adding the terms of the appropriate series. The result is the series for cosh x (see Exercise 38 of Section 7–1).

18. Find the first four nonzero terms of the expansion of the function $f(x) = \frac{1}{2}(e^x - e^{-x})$ by subtracting the terms of the appropriate series. The result is the series for sinh x (see Exercise 38 of Section 7–1).

19. Find the first three terms of the expansion for $e^x \sin x$ by multiplying the proper expansions together, term by term.

20. Find the first three nonzero terms of the expansion for $f(x) = \tan x$, by dividing the series for sin x by that for cos x.

21. Show that by differentiating term by term the expansion for sin x, the result is the expansion for cos x.

22. Show that by differentiating term by term the expansion for e^x, the result is also the expansion for e^x.

23. Show that by integrating term by term the expansion for cos x, the result is the expansion for sin x.

24. Show that by integrating term by term the expansion for $-1/(1 - x)$ (see Exercise 11 of Section 12–1), the result is the expansion for $\ln(1 - x)$.

In Exercises 25 through 30 solve the given problems.

25. Find the approximate value of the area bounded by $y = x^2 e^x$, $x = 0.2$, and the x-axis, by use of three terms of the appropriate Maclaurin series.

26. Find the approximate area bounded by $y = e^{-x^2}$, $x = 0$, $x = 1$, and $y = 0$, by use of three terms of the appropriate series.

27. Find the volume generated by rotating the area bounded by $y = \sin x$ from $x = 0$ to $x = \frac{1}{8}\pi$ and the x-axis, about the y-axis by use of two terms of the appropriate series.

28. Find the volume generated by rotating the area bounded by $y = e^{-x}$, $y = 0$, $x = 0$, and $x = 0.1$ about the y-axis, by use of three terms of the appropriate series.

29. The impedance Z of an alternating current circuit, containing a resistance R, inductance L, and capacitance C, may be defined by the equation $Z = R + j(X_L - X_C)$, where X_L and X_C are known as the inductive reactance and the capacitive reactance, respectively. Show, by use of Eq. (12–12) that the angle between Z and R is given by

$$\tan \theta = \frac{X_L - X_C}{R}$$

30. Find the magnitude of the impedance of a circuit if $R = 8\ \Omega$, $X_L = 7\ \Omega$ and $X_C = 13\ \Omega$. (Finding the magnitude of Z is equivalent to finding r for a complex number.)

12–3 Computations by Use of Series Expansions

As mentioned previously, power-series expansions can be used to compute numerical values of transcendental functions. By including a sufficient number of terms, we can calculate the values to any desired degree of accuracy. It is through such calculations that tables of values are made, and values of such numbers as e and π can be found. Also, some of the values determined by a calculator are found by use of series expansions.

Example A
Calculate the value of $e^{0.1}$.

$$e^x = 1 + x + \frac{x^2}{2!} + \cdots$$

Let $x = 0.1$. Thus,

$$e^{0.1} = 1 + 0.1 + \frac{(0.1)^2}{2} + \cdots = 1.105 \quad \text{(using 3 terms)}$$

To 5 significant digits, $e^{0.1} = 1.1052$, which indicates that our answer is valid to four significant digits.

Example B
Calculate the value of $\sin 2°$.

$$\sin x = x - \frac{x^3}{3!} + \cdots$$

Let $x = \pi/90$ (converting to radians). Hence,

$$\sin 2° = \left(\frac{\pi}{90}\right) - \frac{(\pi/90)^3}{6} + \cdots = 0.034907 - 0.000007 = 0.034900$$

This result is accurate to the number of digits shown. Here we note that the second term is much smaller than the first. In fact, a good approximation can be found by use of one term. Thus, we see the reason that $\sin \theta = \theta$ for small values of θ.

Example C

Calculate the value of $\ln(1.2)$.

$$\ln(1 + x) = x - \frac{x^2}{2} + \frac{x^3}{3} - \cdots$$

Let $x = 0.2$. Therefore,

$$\ln(1 + 0.2) = 0.2 - \frac{(0.2)^2}{2} + \frac{(0.2)^3}{3} - \cdots = 0.183$$

To 4 significant digits, $\ln(1.2) = 0.1823$. One more term is required to obtain this accuracy.

Series approximations can also be used to determine errors in calculated values due to errors in measured values. We have already discussed this topic as an application of differentials. A series solution of this kind of problem allows as close a calculation of the error as needed. With differentials, only one term can be found. Of course, for many purposes this is sufficient. Consider the following example

Example D

The velocity v attained by an object that has fallen h ft is given by

$$v = 8\sqrt{h}$$

Find the approximate error in the velocity of an object which has been found to drop 100 ft, with a possible error of 2 ft.

If we let

$$v = 8\sqrt{100 + x}$$

where x is the error in h, we may express v as a Maclaurin series in x.

$$\begin{aligned}
f(x) &= 8(100 + x)^{1/2} & f(0) &= 80 \\
f'(x) &= 4(100 + x)^{-1/2} & f'(0) &= 0.4 \\
f''(x) &= -2(100 + x)^{-3/2} & f''(0) &= -0.002
\end{aligned}$$

Therefore,

$$v = 8\sqrt{100 + x} = 80 + 0.4x - 0.001x^2 + \cdots$$

Since the actual calculated value of v, for $x = 0$, is 80, the error e in the value of v is

$$e = 0.4x - 0.001x^2 + \cdots$$

Calculating the error for $x = 2$ is

$$e = 0.4(2) - 0.001(4) = 0.800 - 0.004 = 0.796 \text{ ft/s}$$

The value 0.8 is the same as that which would be found by use of differentials, since it is a result of evaluating the first derivative term of the expansion. The additional terms are corrections to this term. The one additional term in this case indicates that the first term is a good approximation to the error. Although this problem could be done numerically, a series solution, once set up, allows us to easily calculate the error.

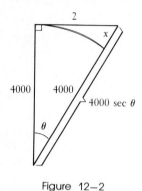

Figure 12–2

Example E

Assuming that the earth is a perfect sphere 4000 mi in radius, find how far the end of a tangent line 2 mi long is from the surface.

From Fig. 12–2 we see that

$$x = 4000 \sec \theta - 4000$$

Finding the series for $\sec \theta$, we have

$$f(\theta) = \sec \theta \qquad\qquad f(0) = 1$$
$$f'(\theta) = \sec \theta \tan \theta \qquad f'(0) = 0$$
$$f''(\theta) = \sec^3 \theta + \sec \theta \tan^2 \theta \qquad f''(0) = 1$$

Thus, the first two nonzero terms are $\sec \theta = 1 + (\theta^2/2)$. Thus,

$$x = 4000(\sec \theta - 1)$$

$$= 4000\left(1 + \frac{\theta^2}{2} - 1\right) = 2000\,\theta^2$$

The first two terms of the expansion for $\tan \theta$ are $\theta + \theta^3/3$, which means that if θ is small, which it is in this case, then $\tan \theta = \theta$. From Fig. 12–2, $\tan \theta = 2/4000$. Therefore, $\theta = 1/2000$. Substituting this value into the expression for x, we have

$$x = 2000\left(\frac{1}{2000}\right)^2 = \frac{1}{2000} = 0.0005 \text{ mi}$$

This means that the end of a two-mile line drawn tangent to the earth would only be about 2.6 ft from the surface!

Exercises 12–3

In Exercises 1 through 16 calculate the value of each of the given functions. Use the indicated number of terms of the appropriate series.

1. $e^{0.2}$ (3)   2. $e^{-0.5}$ (3)   3. $\sin(0.1)$ (2)
4. $\sin 4°$ (2)   5. e (7)   6. $\sqrt{e}$ (5)
7. $\cos 3°$ (2)   8. $\cos(0.5°)$ (2)   9. $\ln(1.4)$ (4)
10. $\ln(0.95)$ (4)   11. $\sin 1$ (4)   12. $\ln(0.5)$ (5)
13. $\sqrt{1.1}$ (3) (use the expansion for $\sqrt{1 + x}$)
14. $\sqrt{0.8}$ (3)   15. $\sqrt[3]{1.1}$ (3)   16. $\sqrt[3]{0.95}$ (3)

In Exercises 17 through 20 calculate the maximum error of the values indicated. If a series is alternating (every other term is negative) the maximum possible error in the calculated value is the value of the first term omitted.

17. The value found in Exercise 2   18. The value found in Exercise 3
19. The value found in Exercise 8   20. The value found in Exercise 9

In Exercises 21 through 26 solve the given problems by the use of series expansions.

21. The electric current in a circuit containing a resistance R, an inductance L, and a battery whose voltage is E is given by

$$i = \frac{E}{R}(1 - e^{-Rt/L})$$

where t is the time. Approximate this expression by using the first three terms of the appropriate exponential series. Under what conditions will this approximation be valid?

22. The image distance q from a certain lens as a function of the object distance p is given by $q = 20p/(p - 20)$. Find the first three nonzero terms of the expansion of the right side. From this expression calculate q for $p = 2$ cm and compare with the value found by substituting 2 in the original expression.

23. From what height can a person see a point 10 mi distant along the earth's surface?

24. The efficiency (in percent) of an internal combustion engine in terms of its compression ratio c is given approximately by $E = 100(1 - c^{-0.4})$. Determine the possible approximate error in the efficiency for a compression ratio measured to be 6.00 with a possible error of 0.50. [*Hint:* Set up a series for $(6 + x)^{-0.4}$.]

25. We can evaluate π by use of the fact that $\frac{1}{4}\pi = \text{Arctan}\frac{1}{2} + \text{Arctan}\frac{1}{3}$ (see Exercise 34 of Section 6–5), along with the use of the series expansion for Arctan x. The first three terms are Arctan $x = x - \frac{1}{3}x^3 + \frac{1}{5}x^5$. Using these terms, expand Arctan$\frac{1}{2}$ and Arctan$\frac{1}{3}$, and thereby approximate the value of π.

26. Use the fact that $\frac{1}{4}\pi = \text{Arctan}\frac{1}{7} + 2\,\text{Arctan}\frac{1}{3}$ to approximate the value of π (see Exercise 25).

12—4 Taylor's Series

If values of certain functions are determined for values of x which are not close to zero, it is usually necessary to use many terms of a Maclaurin expansion to obtain accurate values. Therefore, another type of series, called a **Taylor's series**, is used for this purpose. Also, functions for which a Maclaurin series may not be found may have a Taylor's series. Actually, a Taylor's series is a more general expansion than a Maclaurin expansion.

The basic assumption in formulating a Taylor's expansion is that a function may be expanded in a polynomial of the form

$$f(x) = c_0 + c_1(x - a) + c_2(x - a)^2 + \cdots \tag{12–14}$$

By following precisely the same line of reasoning that we did in deriving the general Maclaurin expansion, we may find the constants c_0, c_1, c_2, and so forth. That is, derivatives of Eq. (12–14) are taken, and the function and its derivatives are evaluated for $x = a$. This leads to

$$f(x) = f(a) + f'(a)(x - a) + \frac{f''(a)(x - a)^2}{2!} + \cdots \qquad (12\text{--}15)$$

This series converges rapidly for values of x which are close to a, and this is illustrated in Examples C and D. The following examples illustrate the derivation of Taylor's series.

Example A

Expand $f(x) = e^x$ in a Taylor's series with $a = 1$.

$$
\begin{aligned}
f(x) &= e^x & f(1) &= e \\
f'(x) &= e^x & f'(1) &= e \\
f''(x) &= e^x & f''(1) &= e \\
f'''(x) &= e^x & f'''(1) &= e, \text{ etc.}
\end{aligned}
$$

$$f(x) = e + e(x - 1) + e\frac{(x - 1)^2}{2!} + e\frac{(x - 1)^3}{3!} + \cdots$$

$$e^x = e\left[1 + (x - 1) + \frac{(x - 1)^2}{2} + \frac{(x - 1)^3}{6} + \cdots\right]$$

This series is of particular use in evaluating e^x for values of x near 1.

Example B

Expand $f(x) = \sqrt{x}$ in powers of $(x - 4)$.

Another way of stating this is to find the Taylor's series for $f(x) = \sqrt{x}$, with $a = 4$. Thus,

$$
\begin{aligned}
f(x) &= x^{1/2} & f(4) &= 2 \\[4pt]
f'(x) &= \frac{1}{2x^{1/2}} & f'(4) &= \frac{1}{4} \\[4pt]
f''(x) &= -\frac{1}{4x^{3/2}} & f''(4) &= -\frac{1}{32} \\[4pt]
f'''(x) &= \frac{3}{8x^{5/2}} & f'''(4) &= \frac{3}{256}, \text{ etc.}
\end{aligned}
$$

$$f(x) = 2 + \frac{1}{4}(x - 4) - \frac{1}{32}\frac{(x - 4)^2}{2!} + \frac{3}{256}\frac{(x - 4)^3}{3!} - \cdots$$

$$\sqrt{x} = 2 + \frac{(x - 4)}{4} - \frac{(x - 4)^2}{64} + \frac{(x - 4)^3}{512} - \cdots$$

This series would be used to evaluate square roots of numbers near 4.

In the last section we evaluated functions by the use of Maclaurin series. In the following examples we shall use Taylor's series to determine certain values of functions.

Example C

By use of a Taylor series, evaluate $\sqrt{4.5}$.

Using the four terms of the series found in Example B, we have

$$\sqrt{4.5} = 2 + \frac{(4.5 - 4)}{4} - \frac{(4.5 - 4)^2}{64} + \frac{(4.5 - 4)^3}{512}$$

$$= 2 + \frac{(0.5)}{4} - \frac{(0.5)^2}{64} + \frac{(0.5)^3}{512}$$

$$= 2.0000 + 0.1250 - 0.0039 + 0.0002$$

$$= 2.1213$$

The result is accurate to the number of places shown.

In Example C we see that successive terms become small rapidly. If a value of x is chosen such that $x - a$ is larger, the successive terms may not become small rapidly and many terms may be required. Therefore, we should choose a as conveniently close to the x-values which will be used. Also, we should note that a Maclaurin expansion for $\sqrt{x}$ cannot be used since the derivatives of $\sqrt{x}$ are not defined for $x = 0$.

Example D

Calculate the approximate value of $\sin 29°$ by means of the appropriate Taylor expansion.

Since the value of $\sin 30°$ is known to be $\frac{1}{2}$, if we let $a = \pi/6$ (remember, we must use values expressed in radians), when we evaluate the expansion for $x = 29°$ (when expressed in radians) the quantity $(x - a)$ is $-\pi/180$ (equivalent to $-1°$). This means that its numerical values are small and become smaller when it is raised to higher powers.

Therefore,

$$f(x) = \sin x, \qquad f\left(\frac{\pi}{6}\right) = \frac{1}{2}$$

$$f'(x) = \cos x, \qquad f'\left(\frac{\pi}{6}\right) = \frac{\sqrt{3}}{2}$$

$$f''(x) = -\sin x, \qquad f''\left(\frac{\pi}{6}\right) = -\frac{1}{2}$$

$$f(x) = \frac{1}{2} + \frac{\sqrt{3}}{2}\left(x - \frac{\pi}{6}\right) - \frac{1}{4}\left(x - \frac{\pi}{6}\right)^2 - \cdots$$

or

$$\sin x = \frac{1}{2} + \frac{\sqrt{3}}{2}\left(x - \frac{\pi}{6}\right) - \frac{1}{4}\left(x - \frac{\pi}{6}\right)^2 - \cdots$$

$$\sin 29° = \sin\left(\frac{\pi}{6} - \frac{\pi}{180}\right)$$

$$= \frac{1}{2} + \frac{\sqrt{3}}{2}\left(\frac{\pi}{6} - \frac{\pi}{180} - \frac{\pi}{6}\right) - \frac{1}{4}\left(\frac{\pi}{6} - \frac{\pi}{180} - \frac{\pi}{6}\right)^2 - \cdots$$

$$= \frac{1}{2} + \frac{\sqrt{3}}{2}\left(-\frac{\pi}{180}\right) - \frac{1}{4}\left(-\frac{\pi}{180}\right)^2 - \cdots$$

$$= 0.500000 - (0.86603)(0.017453) - (0.25000)(0.017453)^2$$
$$= 0.500000 - 0.015115 - 0.000076 = 0.48481$$

This result is accurate to the number of places shown.

Exercises 12–4

In Exercises 1 through 8 evaluate the given functions by use of the series developed in the examples of this section.

1. $e^{1.2}$ (use $e = 2.718$)   2. $e^{0.7}$ 3. $\sqrt{4.2}$ 4. $\sqrt{3.5}$
5. $\sin 31°$ 6. $\sin 28°$ 7. $\sin 29.5°$ 8. $\sqrt{3.85}$

In Exercises 9 through 16 find the first three nonzero terms of the Taylor expansion for the given function and given value of a.

9. e^{-x} $(a = 2)$ 10. $\cos x$ $(a = \pi/4)$ 11. $\sin x$ $(a = \pi/3)$

12. $\ln x$ $(a = 3)$ 13. $\sqrt[3]{x}$ $(a = 8)$ 14. $\frac{1}{x}$ $(a = 2)$

15. $\tan x$ $(a = \pi/4)$ 16. $\ln \sin x$ $(a = \pi/2)$

In Exercises 17 through 24 evaluate the given functions by use of three terms of the appropriate Taylor series.

17. $e^{-2.2}$ (use $e^{-2} = 0.1353$)   18. $\ln(3.1)$ (use $\ln 3 = 1.0986$)
19. $\sqrt{9.3}$ 20. $\sqrt{17}$ 21. $\sqrt[3]{8.3}$
22. $\tan 46°$ 23. $\sin 61°$ 24. $\cos 42°$

In Exercises 25 and 26 solve the given problems.

25. By completing the steps indicated before Eq. (12–15) in the text, complete the derivation of Eq. (12–15).
26. Calculate $e^{0.9}$ by use of four terms of the Maclaurin expansion for e^x. Also calculate $e^{0.9}$ by using the first three terms of the Taylor's expansion in Example A, using $e = 2.718$. Compare the accuracy of the values obtained with that in Table 5 in the Appendix.

12–5 Fourier Series

Many problems which are encountered in the various fields of science and technology involve functions which are periodic. A *periodic function is one for which* $F(x + P) = F(x)$, *where P is the period.* We noted that the trigonometric functions are periodic when we discussed their graphs in Chapter 6. Illustrations of applied problems which involve periodic functions are alternating-current voltages and mechanical oscillations.

Therefore, in this section we use a series made of terms of sines and cosines. This allows us to represent complicated periodic functions in terms of the simpler sines and cosines. It also provides a good approximation over a greater interval than Maclaurin and Taylor series, which give good approximations with a few terms only near a specific value. Illustrations of applications of this type of series are given in Example C and Exercise 11.

We shall assume that a function $f(x)$ may be represented by the series of sines and cosines as indicated:

$$f(x) = a_0 + a_1 \cos x + a_2 \cos 2x + \cdots + a_n \cos nx + \cdots$$
$$+ b_1 \sin x + b_2 \sin 2x + \cdots + b_n \sin nx + \cdots \quad (12\text{–}16)$$

Since all of the sines and cosines indicated in this expansion have a period of 2π (the period of any given term may be less than 2π, but all do repeat every 2π units—e.g., $\sin 2x$ has a period of π, but it also repeats every 2π), the series expansion indicated in Eq. (12–16) will also have a period of 2π. This series is called a **Fourier series.**

The principal problem to be solved is that of finding the coefficients a_n and b_n. Derivatives proved to be useful in finding the coefficients for a Maclaurin expansion. We use the properties of certain integrals to find the coefficients of a Fourier series. To utilize these properties, we multiply all terms of Eq. (12–16) by $\cos mx$, and then evaluate from $-\pi$ to π (in this way we take advantage of the period 2π). Thus, we have

$$\int_{-\pi}^{\pi} f(x) \cos mx \, dx = \int_{-\pi}^{\pi} (a_0 + a_1 \cos x + a_2 \cos 2x + \cdots)(\cos mx) \, dx$$

$$+ \int_{-\pi}^{\pi} (b_1 \sin x + b_2 \sin 2x + \cdots)(\cos mx) \, dx \quad (12\text{–}17)$$

Using the methods of integration of Chapter 8, we have the following:

$$\int_{-\pi}^{\pi} a_0 \cos mx \, dx = \frac{a_0}{m} \sin mx \Big|_{-\pi}^{\pi} = \frac{a_0}{m}(0 - 0) = 0 \quad (12\text{–}18)$$

$$\int_{-\pi}^{\pi} a_n \cos nx \cos mx \, dx =$$

$$a_n \left(\frac{\sin(n-m)x}{2(n-m)} + \frac{\sin(n+m)x}{2(n+m)} \right) \Big|_{-\pi}^{\pi} = 0 \quad n \neq m; \quad (12\text{–}19)$$

(since the sine of any multiple of π is zero),

$$\int_{-\pi}^{\pi} a_n \cos nx \cos nx \, dx = \int_{-\pi}^{\pi} a_n \cos^2 nx \, dx$$

$$= \left(\frac{a_n x}{2} + \frac{1}{2n} \sin nx \cos nx \right) \Big|_{-\pi}^{\pi} = \frac{a_n x}{2} \Big|_{-\pi}^{\pi} = \pi a_n \quad (12\text{–}20)$$

as well as

$$\int_{-\pi}^{\pi} b_n \sin nx \cos mx \, dx = b_n\left(-\frac{\cos(n-m)x}{2(n-m)} - \frac{\cos(n+m)x}{2(n+m)}\right)\Big|_{-\pi}^{\pi}$$

$$= b_n\left(-\frac{\cos(n-m)\pi}{2(n-m)} - \frac{\cos(n+m)\pi}{2(n+m)}\right.$$

$$\left. + \frac{\cos(n-m)(-\pi)}{2(n-m)} + \frac{\cos(n+m)(-\pi)}{2(n+m)}\right)$$

$$= 0 \text{ [since } \cos\theta = \cos(-\theta)\text{]}, \qquad n \neq m \qquad (12\text{--}21)$$

$$\int_{-\pi}^{\pi} b_n \sin nx \cos nx \, dx = \frac{b_n}{2n}\sin^2 nx \Big|_{-\pi}^{\pi} = 0 \qquad (12\text{--}22)$$

These integrals are seen to be zero, except for the one specific case of $\int_{-\pi}^{\pi} a_n \cos nx \cos mx \, dx$ when $n = m$, for which the result is indicated in Eq. (12–20). Using these results in Eq. (12–17), we have

$$\int_{-\pi}^{\pi} f(x)\cos nx \, dx = a_n \int_{-\pi}^{\pi} \cos^2 nx \, dx = \pi a_n$$

or

$$a_n = \frac{1}{\pi}\int_{-\pi}^{\pi} f(x)\cos nx \, dx \qquad (12\text{--}23)$$

This equation allows us to find the coefficients a_n, except a_0. We find the term a_0 by direct integration of Eq. (12–16) from $-\pi$ to π. When we perform this integration, all the sine and cosine terms integrate to zero, thereby giving the result

$$\int_{-\pi}^{\pi} f(x)\, dx = \int_{-\pi}^{\pi} a_0\, dx = a_0 x \Big|_{-\pi}^{\pi} = 2\pi a_0$$

or

$$a_0 = \frac{1}{2\pi}\int_{-\pi}^{\pi} f(x)\, dx \qquad (12\text{--}24)$$

By multiplying all terms of Eq. (12–16) by $\sin mx$ and then integrating from $-\pi$ to π, we find the coefficients b_n. We obtain the result

$$b_n = \frac{1}{\pi}\int_{-\pi}^{\pi} f(x)\sin nx \, dx \qquad (12\text{--}25)$$

We can restate our equations for the Fourier series of a function $f(x)$:

$$f(x) = a_0 + a_1 \cos x + a_2 \cos 2x + \cdots + a_n \cos nx + \cdots$$
$$+ b_1 \sin x + b_2 \sin 2x + \cdots + b_n \sin nx + \cdots \qquad (12\text{–}16)$$

where the coefficients are found by

$$a_0 = \frac{1}{2\pi} \int_{-\pi}^{\pi} f(x)\, dx \qquad (12\text{–}24)$$

$$a_n = \frac{1}{\pi} \int_{-\pi}^{\pi} f(x) \cos nx\, dx \qquad (12\text{–}23)$$

and

$$b_n = \frac{1}{\pi} \int_{-\pi}^{\pi} f(x) \sin nx\, dx \qquad (12\text{–}25)$$

Example A

Find the Fourier series for the square wave function

$$f(x) = \begin{cases} 0 & \text{for } -\pi \le x < 0 \\ 1 & \text{for } 0 \le x < \pi \end{cases}$$

(Many of the functions which we shall expand in Fourier series are discontinuous (not continuous) like this one. See Section 2–2 for a discussion of continuity.)

Since $f(x)$ is defined differently for the ranges of x indicated, it requires two integrals for each coefficient:

$$a_0 = \frac{1}{2\pi} \int_{-\pi}^{0} 0\, dx + \frac{1}{2\pi} \int_{0}^{\pi} (1)\, dx = \frac{1}{2}$$

$$a_n = \frac{1}{\pi} \int_{-\pi}^{0} 0\, dx + \frac{1}{\pi} \int_{0}^{\pi} (1) \cos nx\, dx = \frac{1}{n\pi}(\sin nx)\Big|_{0}^{\pi} = 0$$

for all values of n, since $\sin n\pi = 0$;

$$b_1 = \frac{1}{\pi} \int_{-\pi}^{0} 0\, dx + \frac{1}{\pi} \int_{0}^{\pi} \sin x\, dx = -\frac{1}{\pi}(\cos x)\Big|_{0}^{\pi}$$

$$= -\frac{1}{\pi}(-1 - 1) = \frac{2}{\pi}$$

$$b_2 = \frac{1}{\pi} \int_{-\pi}^{0} 0 \; dx + \frac{1}{\pi} \int_{0}^{\pi} \sin 2x \; dx = -\frac{1}{2\pi}(\cos 2x)\Big|_0^{\pi}$$

$$= -\frac{1}{2\pi}(1 - 1) = 0$$

$$b_3 = \frac{1}{\pi} \int_{-\pi}^{0} 0 \; dx + \frac{1}{\pi} \int_{0}^{\pi} \sin 3x \; dx = -\frac{1}{3\pi}(\cos 3x)\Big|_0^{\pi}$$

$$= -\frac{1}{3\pi}(-1 - 1) = \frac{2}{3\pi}$$

In general we can see that if n is even, $b_n = 0$, and if n is odd, then $b_n = 2/n\pi$. Thus, we have

$$f(x) = \frac{1}{2} + \frac{2}{\pi}\sin x + \frac{2}{3\pi}\sin 3x + \cdots$$

A graph of the function as defined, and the curve found by using the first three terms of the Fourier series, is shown in Fig. 12–3.

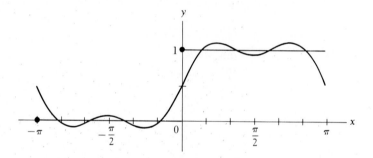

Figure 12–3

Since functions found by Fourier series have a period of 2π, they can represent functions with this period. If the function $f(x)$ were defined to be periodic with period 2π, with the same definitions as originally indicated, we would graph the function as shown in Fig. 12–4. The Fourier series representation would follow it as in Fig. 12–3. If more terms were used, the fit would be closer.

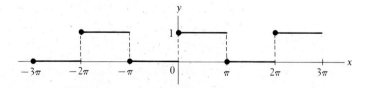

Figure 12–4

Example B

Find the Fourier series for the function

$$f(x) = \begin{cases} 1 & \text{for } -\pi \le x < 0 \\ x & \text{for } 0 \le x < \pi \end{cases}$$

For the periodic function, let $f(x + 2\pi) = f(x)$ for all x.

A graph of three periods of this function is shown in Fig. 12-5. The coefficients are

$$a_0 = \frac{1}{2\pi}\int_{-\pi}^{0} dx + \frac{1}{2\pi}\int_{0}^{\pi} x\, dx = \frac{x}{2\pi}\Big|_{-\pi}^{0} + \frac{x^2}{4\pi}\Big|_{0}^{\pi} = \frac{1}{2} + \frac{\pi}{4} = \frac{2+\pi}{4}$$

$$a_1 = \frac{1}{\pi}\int_{-\pi}^{0} \cos x\, dx + \frac{1}{\pi}\int_{0}^{\pi} x \cos x\, dx$$

$$= \frac{1}{\pi}\sin x \Big|_{-\pi}^{0} + \frac{1}{\pi}(\cos x + x \sin x)\Big|_{0}^{\pi} = -\frac{2}{\pi}$$

$$a_2 = \frac{1}{\pi}\int_{-\pi}^{0} \cos 2x\, dx + \frac{1}{\pi}\int_{0}^{\pi} x \cos 2x\, dx$$

$$= \frac{1}{2\pi}\sin 2x \Big|_{-\pi}^{0} + \frac{1}{4\pi}(\cos 2x + 2x \sin 2x)\Big|_{0}^{\pi} = 0$$

$$a_3 = \frac{1}{\pi}\int_{-\pi}^{0} \cos 3x\, dx + \frac{1}{\pi}\int_{0}^{\pi} x \cos 3x\, dx$$

$$= \frac{1}{3\pi}\sin 3x \Big|_{-\pi}^{0} + \frac{1}{9\pi}(\cos 3x + 3x \sin 3x)\Big|_{0}^{\pi} = -\frac{2}{9\pi}$$

$$b_1 = \frac{1}{\pi}\int_{-\pi}^{0} \sin x\, dx + \frac{1}{\pi}\int_{0}^{\pi} x \sin x\, dx$$

$$= -\frac{1}{\pi}\cos x \Big|_{-\pi}^{0} + \frac{1}{\pi}(\sin x - x \cos x)\Big|_{0}^{\pi} = \frac{\pi-2}{\pi}$$

$$b_2 = \frac{1}{\pi}\int_{-\pi}^{0} \sin 2x\, dx + \frac{1}{\pi}\int_{0}^{\pi} x \sin 2x\, dx$$

$$= -\frac{\cos 2x}{2\pi}\Big|_{-\pi}^{0} + \frac{\sin 2x - 2x \cos 2x}{4\pi}\Big|_{0}^{\pi} = -\frac{1}{2}$$

Thus,

$$f(x) = \frac{2+\pi}{4} - \frac{2}{\pi}\cos x - \frac{2}{9\pi}\cos 3x - \cdots$$

$$+ \left(\frac{\pi-2}{\pi}\right)\sin x - \frac{1}{2}\sin 2x + \cdots$$

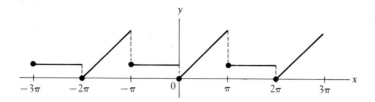

Figure 12–5

Example C

Certain electronic devices allow an electric current to pass through in only one direction. The result is that when an alternating current is applied, the current exists for only half the cycle. Figure 12–6 is a representation for such a current as a function of time. A device which causes this effect is called a half-wave rectifier. Derive the Fourier series for the half of the rectified wave for which $f(t) = \sin t$ $(0 \leq t \leq \pi)$, and $f(t) = 0$ for the other half of the cycle.

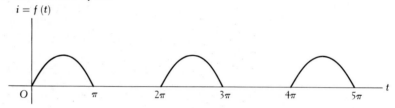

Figure 12–6

We write

$$a_0 = \frac{1}{2\pi} \int_0^\pi \sin t \, dt = \frac{1}{2\pi}(-\cos t)\Big|_0^\pi = \frac{1}{2\pi}(1+1) = \frac{1}{\pi}$$

In the previous example we evaluated each of the coefficients individually. Here we shall show how to set up a general expression for a_n and another for b_n. Once we have determined these, we can substitute values of n in the formula to obtain the individual coefficients:

$$a_n = \frac{1}{\pi} \int_0^\pi \sin t \cos nt \, dt = -\frac{1}{2\pi}\left[\frac{\cos(1-n)t}{1-n} + \frac{\cos(1+n)t}{1+n}\right]_0^\pi$$

$$= -\frac{1}{2\pi}\left[\frac{\cos(1-n)\pi}{1-n} + \frac{\cos(1+n)\pi}{1+n} - \frac{1}{1-n} - \frac{1}{1+n}\right]$$

See formula 40 of Table 6 in Appendix C. This formula is valid for all values of n except $n = 1$. Now we write

$$a_1 = \frac{1}{\pi} \int_0^\pi \sin t \cos t \, dt = \frac{1}{2\pi}\sin^2 t\Big|_0^\pi = 0$$

$$a_2 = -\frac{1}{2\pi}\left(\frac{-1}{-1} + \frac{-1}{3} - \frac{1}{-1} - \frac{1}{3}\right) = -\frac{2}{3\pi}$$

$$a_3 = -\frac{1}{2\pi}\left(\frac{1}{-2} + \frac{1}{4} - \frac{1}{-2} - \frac{1}{4}\right) = 0$$

$$a_4 = -\frac{1}{2\pi}\left(\frac{-1}{-3} + \frac{-1}{5} - \frac{1}{-3} - \frac{1}{5}\right) = -\frac{2}{15\pi}$$

$$b_n = \frac{1}{\pi}\int_0^\pi \sin t \sin nt \, dt = \frac{1}{2\pi}\left[\frac{\sin(1-n)t}{1-n} - \frac{\sin(1+n)t}{1+n}\right]_0^\pi$$

$$= \frac{1}{2\pi}\left[\frac{\sin(1-n)\pi}{1-n} - \frac{\sin(1+n)\pi}{1+n}\right]$$

See formula 39 of Table 6. This formula is valid for all values of n except $n = 1$. Thus,

$$b_1 = \frac{1}{\pi}\int_0^\pi \sin t \sin t \, dt = \frac{1}{\pi}\int_0^\pi \sin^2 t \, dt$$

$$= \frac{1}{2\pi}(t - \sin t \cos t)\Big|_0^\pi = \frac{1}{2}$$

We see that $b_n = 0$ if $n > 1$, since each is evaluated in terms of the sine of a multiple of π.

Therefore, the Fourier series for the rectified wave is

$$f(t) = \frac{1}{\pi} + \frac{1}{2}\sin t - \frac{2}{\pi}\left(\frac{1}{3}\cos 2t + \frac{1}{15}\cos 4t + \cdots\right)$$

All the types of periodic functions included in this section (as well as many others) may actually be seen on an oscilloscope when the proper signal is sent into it. In this way the oscilloscope may be used to analyze the periodic nature of such phenomena as sound waves and electric currents.

Exercises 12–5

In the following exercises find a few terms of the Fourier series for the given functions, and sketch at least three periods of the function $f(x + 2\pi) = f(x)$.

1. $f(x) = \begin{cases} 1, & -\pi \leq x < 0 \\ 0, & 0 \leq x < \pi \end{cases}$

2. $f(x) = \begin{cases} -1, & -\pi \leq x < 0 \\ 1, & 0 \leq x < \pi \end{cases}$

3. $f(x) = \begin{cases} 1, & -\pi \leq x < 0 \\ 2, & 0 \leq x < \pi \end{cases}$

4. $f(x) = \begin{cases} 0, & -\pi \leq x < 0, \, \pi/2 < x < \pi \\ 1, & 0 \leq x \leq \pi/2 \end{cases}$

5. $f(x) = \begin{cases} 0, & -\pi \leq x < 0 \\ x, & 0 \leq x < \pi \end{cases}$

6. $f(x) = x, \quad -\pi \leq x < \pi$

7. $f(x) = \begin{cases} -1, & -\pi \leq x < 0 \\ 0, & 0 \leq x < \pi/2 \\ 1, & \pi/2 \leq x < \pi \end{cases}$

8. $f(x) = x^2, \quad -\pi \leq x < \pi$

9. $f(x) = \begin{cases} -x, & -\pi \leq x < 0 \\ x, & 0 \leq x < \pi \end{cases}$

10. $f(x) = \begin{cases} 0, & -\pi \leq x < 0 \\ x^2, & 0 \leq x < \pi \end{cases}$

11. Find the Fourier expansion of the electronic device known as a *full-wave rectifier*. This is found by using as the function for the current $f(t) = -\sin t$ for $-\pi \le t \le 0$ and $f(t) = \sin t$ for $0 < t \le \pi$. See Fig. 12–7. The portion of the curve to the left of the $f(t)$-axis is dashed, since from a physical point of view we can give no significance to this part of the wave; although mathematically we can derive the proper form of the expansion by using it.

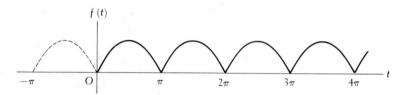

Figure 12–7

12. A function for which $f(x) = f(-x)$ is called an **even function**, and a function for which $f(x) = -f(-x)$ is called an **odd function**. From our discussion of symmetry in Section 1–3, we see that an even function is symmetrical to the *y*-axis, and an odd function is symmetrical to the origin. It can be proven that the Fourier expansion of an even function contains no sine terms, and that of an odd function contains no cosine terms. Show that the function in Exercise 9 is an even function, and the function in Exercise 6 is an odd function. Note and compare the Fourier expansions of these functions.

12–6 Review Exercises for Chapter 12

In Exercises 1 through 8 find the first three nonzero terms of the Maclaurin expansion of the given functions.

1. $f(x) = \dfrac{1}{1 + e^x}$ 2. $f(x) = \frac{1}{2}(e^x - e^{-x})$ 3. $f(x) = \sin 2x^2$ 4. $f(x) = \dfrac{1}{(1 - x)^2}$

5. $f(x) = (x + 1)^{1/3}$ 6. $f(x) = \dfrac{2}{2 - x^2}$ 7. $f(x) = \text{Arcsin } x$ 8. $f(x) = \dfrac{1}{1 - \sin x}$

In Exercises 9 through 16 calculate the value of each of the given functions. Use three terms of the appropriate series.

9. $e^{-0.2}$ 10. $\ln(1.10)$ 11. $\sqrt[3]{1.3}$ 12. $\sin 58°$

13. $\tan 43°$ 14. $\sqrt[4]{260}$ 15. $\sqrt{148}$ 16. $\cos 47°$

In Exercises 17 and 18 evaluate the given integrals by using three terms of the appropriate series.

17. $\displaystyle\int_{0.1}^{0.2} \dfrac{\cos x}{\sqrt{x}}\,dx$ 18. $\displaystyle\int_{0}^{0.1} \sqrt[3]{1 + x^2}\,dx$

In Exercises 19 and 20 find the first three nonzero terms of the Taylor expansion for the given function and the given value of a.

19. $\cos x \ (a = \pi/3)$ 20. $\ln \cos x \ (a = \pi/4)$

In Exercises 21 and 22 find a few terms of the Fourier series for the given functions. Sketch three periods of the function $f(x + 2\pi) = f(x)$.

21. $f(x) = \begin{cases} 0, & -\pi \le x < -\pi/2 \quad \text{and} \quad \pi/2 < \dot{x} < \pi \\ 1, & -\pi/2 \le x \le \pi/2 \end{cases}$

22. $f(x) = \begin{cases} -x, & -\pi \le x < 0 \\ 0, & 0 \le x < \pi \end{cases}$

In Exercises 23 and 24 determine the indicated series by the given operation.

23. If h is small, show that $\sin(x + h) - \sin(x - h) = 2h \cos x$.

24. Find the first three nonzero terms of the Maclaurin expansion of the function $\sin x + x \cos x$ by differentiating the expansion term by term for $x \sin x$.

In Exercises 25 through 30 solve the given problems.

25. On the same diagram plot the graphs of $y = e^x$, $y = 1$, $y = 1 + x$, and $y = 1 + x + \frac{1}{2}x^2$.

26. Show that the Maclaurin expansion for the polynomial $f(x) = ax^2 + bx + c$ is the polynomial itself.

27. Find the approximate area between the curve $y = \dfrac{x - \sin x}{x^2}$ and the x-axis between $x = 0$ and $x = 0.1$.

28. Find the approximate value of the moment of inertia with respect to its axis of the volume generated by rotating the area bounded by $y = \sin x$, $x = 0.3$, and the x-axis about the y-axis. Use two terms of the appropriate series.

29. A certain electric current is pulsating so that the current as a function of time is given by $f(t) = 0$ if $-\pi \le t < 0$ and $\pi/2 < t < \pi$. If $0 < t < \pi/2$, $f(t) = \sin t$. Find the Fourier expansion for this pulsating current, and plot three periods.

30. By use of series, find the greatest distance that a great circle arc 200 mi long on the earth is from its chord.

13

First-Order
Differential Equations

_____

13—1 Solutions of Differential Equations

Various relationships which occur in science and technology give rise to equations involving the rate of change of one variable with respect to another. When this type of relationship exists, we express the rate of change as a derivative. This results in an equation which contains a derivative. *Any such equation containing derivatives, or differentials, is called a* **differential equation.** It is our purpose in this section to introduce the basic meaning of the solution of a differential equation. In the sections which follow we consider certain methods of finding such solutions. We shall see that finding the solutions of differential equations is a powerful method of solving many types of applied problems.

The basic types of differential equations we shall consider are those which contain first and second derivatives. *If an equation contains only first derivatives, it is called a* **first-order** *differential equation. If the equation contains second derivatives, and possibly first derivatives, it is called a* **second-order** *differential equation. In general, the* **order** *of a differential equation is that of the highest derivative in the equation. The* **degree** *of a differential equation is the highest power of that derivative.*

Example A

The equation $\dfrac{dy}{dx} + x = y$ is a first-order differential equation. The equations

$$\frac{d^2y}{dx^2} + y = 3x^2 \quad \text{and} \quad \frac{d^2y}{dx^2} + 2\frac{dy}{dx} = x$$

are both second-order equations. The presence of dy/dx in the second equation does not affect the order.

Example B

The equation

$$\frac{d^2y}{dx^2} + \left(\frac{dy}{dx}\right)^4 - y = 6$$

is a differential equation of the second order and the first degree. That is, the highest derivative which appears is the second, and it is raised to the first power. Since the second derivative appears, the fourth power of the first derivative does not affect the degree.

In our discussion of differential equations, we shall restrict our attention to equations of the first degree.

A **solution** *of a differential equation is a relation between the variables which satisfies the differential equation. That is, when this relation is substituted into the differential equation, an algebraic identity results. A solution containing a number of independent arbitrary constants equal to the order of the differential equation is called the* **general solution** *of the equation. When specific values are given to at least one of these constants, the solution is called a* **particular solution.**

Example C

Any coefficients which are not specified numerically after like terms have been combined are independent arbitrary constants. The expression $c_1x + c_2 + c_3x$ has only two independent constants, in that the two x terms may be combined with one arbitrary constant; $c_2 + c_4x$ is an equivalent expression, with $c_4 = c_1 + c_3$.

Example D

The equation $y = c_1e^{-x} + c_2e^{2x}$ is the general solution of the differential equation

$$\frac{d^2y}{dx^2} - \frac{dy}{dx} = 2y$$

The order of this differential equation is two, and there are two independent arbitrary constants in the equation representing the solution. The equation $y = 4e^{-x}$ is a particular solution. This particular solution can be derived from the general solution by letting $c_1 = 4$ and $c_2 = 0$. Each of these solutions can be shown to satisfy the differential equation by taking two derivatives and substituting.

To solve a differential equation, we have to find some method of transforming the equation so that the terms may be integrated. This will be the topic of principal concern after this section. The purpose here is to show that a given equation is a solution of a differential equation by taking the required derivatives, and to show that an algebraic identity results after substitution.

Example E

Show that $y = c_1 \sin x + c_2 \cos x$ is the general solution of the differential equation $y'' + y = 0$.

The function and its first two derivatives are

$$y = c_1 \sin x + c_2 \cos x, \qquad y' = c_1 \cos x - c_2 \sin x$$

and

$$y'' = -c_1 \sin x - c_2 \cos x$$

Substituting these into the differential equation, we have

$$(-c_1 \sin x - c_2 \cos x) + (c_1 \sin x + c_2 \cos x) = 0 \quad \text{or} \quad 0 = 0$$

We know that this must be the general solution, since there are two independent arbitrary constants, and the order of the differential equation is two.

Example F

Show that $y = cx + x^2$ is a solution of the differential equation $xy' - y = x^2$.

Taking one derivative, $y' = c + 2x$, and substituting, we have

$$x(c + 2x) - (cx + x^2) = x^2 \quad \text{or} \quad x^2 = x^2$$

Exercises 13–1

In the following exercises show that each equation is a solution of the indicated differential equation.

1. $\dfrac{dy}{dx} = 1, \quad y = x + 3$

2. $\dfrac{dy}{dx} = 2x, \quad y = x^2 + 1$

3. $\dfrac{dy}{dx} - y = 1, \quad y = e^x - 1$

4. $\dfrac{dy}{dx} - 3 = 2x, \quad y = x^2 + 3x$

5. $xy' = 2y, \quad y = cx^2$

6. $y' = 2xy^2, \quad y = -\dfrac{1}{x^2 + c}$

7. $y' + 2y = 2x, \quad y = ce^{-2x} + x - \tfrac{1}{2}$

8. $y' - 3x^2 = 1, \quad y = x^3 + x + c$

9. $\dfrac{d^2y}{dx^2} + 4y = 0, \quad y = 3 \cos 2x$

10. $y'' + 9y = 4 \cos x, \quad 2y = \cos x$

11. $y'' - 4y' + 4y = e^{2x}, \quad y = e^{2x}\left(c_1 + c_2 x + \dfrac{x^2}{2}\right)$

12. $\dfrac{d^3y}{dx^3} = \dfrac{d^2y}{dx^2}, \quad y = c_1 + c_2 x + c_3 e^x$

13. $x^2 y' + y^2 = 0, \quad xy = cx + cy$

14. $xy' - 3y = x^2$, $y = cx^3 - x^2$ 15. $x\dfrac{d^2y}{dx^2} + \dfrac{dy}{dx} = 0$, $y = c_1 \ln x + c_2$

16. $\dfrac{d^3y}{dx^3} + 4\dfrac{d^2y}{dx^2} + 4\dfrac{dy}{dx} = 0$, $y = c_1 + c_2 e^{-2x} + xe^{-2x}$

17. $y' + y = 2 \cos x$, $y = \sin x + \cos x - e^{-x}$

18. $(x + y) - xy' = 0$, $y = x \ln x - cx$

19. $y'' + y' = 6 \sin 2x$, $y = e^{-x} - \frac{3}{5}\cos 2x - \frac{6}{5}\sin 2x$

20. $xy''' + 2y'' = 0$, $y = c_1 x + c_2 \ln x$

21. $\cos x\dfrac{dy}{dx} + \sin x = 1 - y$, $y = \dfrac{x + c}{\sec x + \tan x}$

22. $2xyy' + x^2 = y^2$, $x^2 + y^2 = cx$

23. $(y')^2 + xy' = y$, $y = cx + c^2$

24. $x^4(y')^2 - xy' = y$, $y = c^2 + \dfrac{c}{x}$

13-2 Separation of Variables

The first type of differential equation we shall solve is one of the first order and first degree. There are many methods for solving such equations, a few of which are presented in this and the following two sections. The first of these is the method of **separation of variables**.

By the definitions of order and degree, a differential equation of the first order and first degree contains the first derivative to the first power. That is, it may be written as $dy/dx = f(x, y)$. This type of equation is more commonly expressed in its differential form

$$M(x, y)\,dx + N(x, y)\,dy = 0 \qquad (13\text{--}1)$$

where $M(x, y)$ and $N(x, y)$ may represent constants, functions of either x or y, or functions of x and y.

To find the solution of an equation of the form (13–1), it is necessary to integrate. However, if $M(x, y)$ is a function of x and y, we cannot integrate this term. It is integrable only if $M(x, y)$ is a function of x only. For the same reason, the second term is integrable only if $N(x, y)$ is a function of y only. If it is possible, by some algebraic means, to rewrite Eq. (13–1) in the form

$$A(x)\,dx + B(y)\,dy = 0 \qquad (13\text{--}2)$$

where $A(x)$ is a function of x alone and $B(y)$ is a function of y alone, then we may find the solution by integrating each term and adding the constant of integration. (If division is involved, we must remember that the solution is not valid for values which make the divisor zero.) Many elementary differential equations lend themselves to this type of solution.

Example A

Solve the differential equation $dx - 4xy^3 dy = 0$; $M(x, y) = 1$ (a function of neither x nor y), $N(x, y) = -4xy^3$ (a function of both x and y).

We have to remove the x from the coefficient of dy, without introducing any factor of y in the coefficient of dx. This can be done by dividing each term of the equation by x, which results in the equation $(dx/x) - 4y^3 dy = 0$. It is now possible to integrate each term separately. Performing this integration, we have $\ln x - y^4 = c$, which is the desired solution. The c is the constant of integration, which becomes the arbitrary constant of the solution.

Example B

Solve the differential equation $xy\, dx + (x^2 + 1)\, dy = 0$.

In order to integrate each term, it is necessary to divide each term by $y(x^2 + 1)$. When this is done, we have

$$\frac{x\, dx}{x^2 + 1} + \frac{dy}{y} = 0$$

Integrating, we have $\frac{1}{2}\ln(x^2 + 1) + \ln y = c$, which is the desired solution. It is possible to make use of the properties of logarithms to make the form of this solution much neater. If we write the constant of integration as $\ln c_1$, rather than c, we have $\frac{1}{2}\ln(x^2 + 1) + \ln y = \ln c_1$. Multiplying each term by 2, and using the property of logarithms given by Eq. (7–9), we have $\ln(x^2 + 1) + \ln y^2 = \ln c_1^2$. Next, using the property of logarithms given by Eq. (7–7), we have $\ln(x^2 + 1)y^2 = \ln c_1^2$, which means $(x^2 + 1)y^2 = c_1^2$. This form of the solution is much more compact, and would be generally preferred. However, it must be pointed out that any expression representing a constant may be chosen as the constant of integration and leads to a proper result. In checking answers, we must remember that a different choice of constant will lead to a different form of the answer. Thus, two different-appearing answers may both be correct. *It often happens that there is more than one reasonable choice of a constant, and different forms of the answer may be expected.*

Example C

Solve the differential equation

$$\frac{dy}{dx} = \frac{y}{x^2 + 4}$$

The variables are separated by writing the equation in the form

$$\frac{dy}{y} = \frac{dx}{x^2 + 4}$$

Integrating, we have

$$\ln y = \frac{1}{2}\text{Arctan}\frac{x}{2} + \frac{c}{2} \quad \text{or} \quad 2\ln y = \text{Arctan}\frac{x}{2} + c$$

The choice of $\ln c$ as the constant of integration (on the left) is also reasonable. It would lead to the result $2\ln cy = \text{Arctan}(x/2)$.

Example D

Solve the differential equation $3x^2y^2\,dx + y^2\,dx + dy = 0$, subject to the condition that $x = 2$ when $y = 1$.

Dividing each term by y^2 and then combining terms, we have

$$(3x^2 + 1)\,dx + \frac{dy}{y^2} = 0$$

Integrating, the result is $x^3 + x - (1/y) = c$. Since a specified set of values is given, we can evaluate the constant of integration. Using these values, we have $8 + 2 - 1 = c$, or $c = 9$. Thus, the solution may be written as

$$x^3 + x - \frac{1}{y} = 9 \quad \text{or} \quad yx^3 + yx = 1 + 9y$$

Example E

If the function in Example B is subject to the condition that $x = 0$ when $y = e$, we have

$$\frac{1}{2}\ln(0 + 1) + \ln e = c, \quad \frac{1}{2}\ln 1 + 1 = c \quad \text{or} \quad c = 1$$

The solution is then

$$\frac{1}{2}\ln(x^2 + 1) + \ln y = 1, \quad \ln(x^2 + 1) + 2 \ln y = 2$$

$$\ln y^2(x^2 + 1) = 2 \quad \text{or} \quad y^2(x^2 + 1) = e^2$$

If we use the solution with c_1, we have $(0 + 1)e^2 = c_1^2$, or $c_1^2 = e^2$. The solution is then $y^2(x^2 + 1) = e^2$. We see, therefore, that the choice of the constant does not affect the final result, and therefore is truly arbitrary.

Exercises 13–2

In Exercises 1 through 20 solve the given differential equations.

1. $2x\,dx + dy = 0$
2. $y^2\,dy + x^3\,dx = 0$
3. $y^2\,dx + dy = 0$
4. $y\,dx + x\,dy = 0$
5. $x^2 + (x^3 + 5)y' = 0$
6. $xyy' + \sqrt{1 + y^2} = 0$
7. $e^{x^2}\,dy = x\sqrt{1 - y}\,dx$
8. $(x^3 + x^2)\,dx + (x + 1)y\,dy = 0$
9. $e^{x+y}\,dx + dy = 0$
10. $e^{2x}\,dy + e^x\,dx = 0$
11. $y \tan x\,dx + \cos^2 x\,dy = 0$
12. $\sin x \sec y\,dx = dy$
13. $yx^2\,dx = y\,dx - x^2\,dy$
14. $2y(x^3 + 1)\,dy + 3x^2(y^2 - 1)\,dx = 0$
15. $y\sqrt{1 - x^2}\,dy + 2\,dx = 0$
16. $\sqrt{1 + 4x^2}\,dy = y^3x\,dx$
17. $2 \ln x\,dx + x\,dy = 0$
18. $e^{\sin x}\,dx + \sec x\,dy = 0$

19. $y^2 e^x + (e^x + 1)\dfrac{dy}{dx} = 0$
20. $y + 1 + \sec x(\sin x + 1)\dfrac{dy}{dx} = 0$

In Exercises 21 through 24 in each case find the particular solution of the given differential equations for the indicated conditions.

21. $\dfrac{dy}{dx} + yx^2 = 0, \quad x = 0$ when $y = 1$

22. $y' = \sec y, \quad x = 0$ when $y = 0$

23. $y' = (1 - y)\cos x, \quad x = \pi/6$ when $y = 0$

24. $2y \cos y \, dy - \sin y \, dy = y \sin y \, dx, \quad x = 0$ when $y = \pi/2$

13–3 Integrable Combinations

Many differential equations cannot be solved by the method of separation of variables. Many other methods have been developed for solving such equations. One of these methods is based on the fact that certain combinations of basic differentials can be integrated together as a unit. The following differentials suggest some of these combinations which may occur:

$$d(xy) = x \, dy + y \, dx \tag{13–3}$$

$$d(x^2 + y^2) = 2(x \, dx + y \, dy) \tag{13–4}$$

$$d\left(\frac{y}{x}\right) = \frac{x \, dy - y \, dx}{x^2} \tag{13–5}$$

$$d\left(\frac{x}{y}\right) = \frac{y \, dx - x \, dy}{y^2} \tag{13–6}$$

Equation (13–3) suggests that, if the combination of $x \, dy + y \, dx$ occurs in a differential equation, we look for functions of xy. Equation (13–4) suggests that, if the combination $x \, dx + y \, dy$ occurs, we look for functions of $x^2 + y^2$. Equations (13–5) and (13–6) suggest that, if the combination $x \, dy - y \, dx$ occurs, we look for functions of y/x or x/y. The following examples illustrate the use of these combinations.

Example A

Solve the differential equation $x \, dy + y \, dx + xy \, dy = 0$.

By dividing through by xy, we have

$$\frac{x \, dy + y \, dx}{xy} + dy = 0$$

The left term is the differential of xy divided by xy. This means it integrates to $\ln xy$. Thus, we have

$$\ln xy + y = c$$

Example B

Solve the differential equation $y \, dx - x \, dy + x \, dx = 0$.

The combination of $y \, dx - x \, dy$ suggests that this equation might make use of either Eq. (13–5) or (13–6). This would require dividing each term by x^2 or y^2. If we divide by y^2, the last term cannot be integrated, but division by x^2 still allows integration of the last term. Performing this division, we obtain

$$\frac{y\,dx - x\,dy}{x^2} + \frac{dx}{x} = 0$$

This left combination is the negative of Eq. (13–5). The result then becomes

$$-\frac{y}{x} + \ln x = c$$

Example C

Solve the differential equation $(x^2 + y^2 + x)\,dx + y\,dy = 0$.

Regrouping the terms of this equation, we have

$$(x^2 + y^2)\,dx + (x\,dx + y\,dy) = 0$$

By dividing through by $x^2 + y^2$, we have

$$dx + \frac{x\,dx + ydy}{x^2 + y^2} = 0$$

The right term now can be put in the form of du/u (with $u = x^2 + y^2$) by multiplying each of the terms of the numerator by 2. The result then becomes

$$x + \frac{1}{2}\ln(x^2 + y^2) = \frac{c}{2} \quad \text{or} \quad 2x + \ln(x^2 + y^2) = c$$

Example D

Find the particular solution of the differential equation

$$(x^3 + xy^2 + 2y)\,dx + (y^3 + x^2y + 2x)\,dy = 0$$

which satisfies the condition that $x = 1$ when $y = 0$.

Regrouping the terms of the equation, we have

$$x(x^2 + y^2)\,dx + y(x^2 + y^2)\,dy + 2(y\,dx + x\,dy) = 0$$

Factoring $x^2 + y^2$ from each of the first two terms gives

$$(x^2 + y^2)(x\,dx + y\,dy) + 2(y\,dx + x\,dy) = 0$$

The first term can be integrated by multiplying the $x\,dx + y\,dy$ factor by 2. The result is

$$\frac{1}{2}\cdot\frac{1}{2}(x^2 + y^2)^2 + 2xy + \frac{c}{4} = 0 \quad \text{or} \quad (x^2 + y^2)^2 + 8xy + c = 0$$

Using the given condition gives $(1 + 0)^2 + 0 + c = 0$, or $c = -1$. The particular solution is then

$$(x^2 + y^2)^2 + 8xy = 1$$

The use of these integrable combinations depends on proper recognition of the forms. At times it may take two or three arrangements to arrive at the combination which leads to the result. Of course, many problems cannot be so arranged as to give integrable combinations in all terms.

Exercises 13−3

In Exercises 1 through 16 solve the given differential equations.

1. $x \, dy + y \, dx + x \, dx = 0$ 　　　　2. $(2y + x) \, dy + y \, dx = 0$

3. $y \, dx - x \, dy + x^3 \, dx = 2 \, dx$ 　　4. $x \, dy - y \, dx + y^2 \, dx = 0$

5. $x^3 \, dy + x^2 y \, dx + y \, dx - x \, dy = 0$ 　6. $\sec(xy) \, dx + (x \, dy + y \, dx) = 0$

7. $x^3 y^4 (x \, dy + y \, dx) = dy$ 　　　　8. $x \, dy + y \, dx + 4xy^3 \, dy = 0$

9. $\sqrt{x^2 + y^2} \, dx - 2y \, dy = 2x \, dx$ 　10. $x \, dx + (x^2 + y^2 + y) \, dy = 0$

11. $\tan(x^2 + y^2) \, dy + x \, dx + y \, dy = 0$ 　12. $(x^2 + y^3)^2 dy + 2x \, dx + 3y^2 dy = 0$

13. $y \, dy - x \, dx + (y^2 - x^2) \, dx = 0$ 　14. $e^{x+y}(dx + dy) + 4x \, dx = 0$

15. $10x \, dy + 5y \, dx + 3y \, dy = 0$ 　　16. $x^2 dy + 3xy \, dx + 2 \, dx = 0$

In Exercises 17 through 20 find the particular solutions to the given differential equations which satisfy the given conditions.

17. $2(x \, dy + y \, dx) + 3x^2 dx = 0$, 　$x = 1$ when $y = 2$

18. $x \, dx + y \, dy = 2(x^2 + y^2) \, dx$, 　$x = 1$ when $y = 0$

19. $y \, dx - x \, dy = y^3 dx + y^2 x \, dy$, 　$x = 2$ when $y = 4$

20. $e^{x/y}(x \, dy - y \, dx) = y^4 dy$, 　$x = 0$ when $y = 2$

13−4 The Linear Differential Equation of the First Order

There is one type of differential equation of the first order and first degree for which an integrable combination can always be determined. It is the linear differential equation of the first order and is of the form

$$dy + Py \, dx = Q \, dx \tag{13−7}$$

where P and Q are functions of x only. We shall find that this type of equation occurs widely in applications.

　　If each side of Eq. (13−7) is multiplied by $e^{\int P dx}$ it becomes integrable, since the left side becomes of the form du with $u = y e^{\int P dx}$ and the right side is a function of x only. This is shown by finding the differential of $y e^{\int P dx}$. Thus,

$$d(y e^{\int P dx}) = e^{\int P dx}(dy + Py \, dx)$$

(In finding the differential of $\int P \, dx$ we use the fact that, by definition, these are reverse processes. Thus, $d(\int P \, dx) = P \, dx$.) Therefore, if each side is multiplied by $e^{\int P dx}$, the left side may be immediately integrated to $y e^{\int P dx}$ and the right-side integration may be indicated. The solution becomes

$$y e^{\int P dx} = \int Q e^{\int P dx} dx + c \tag{13−8}$$

Example A

Solve the differential equation $dy + \left(\dfrac{2}{x}\right) y \, dx = 4x \, dx$. Here we note that $P = 2/x$ and $Q = 4x$.

The first expression to find is $e^{\int Pdx}$. In this case this is $e^{\int(2/x)dx} = e^{2\ln x} = e^{\ln x^2} = x^2$ (see below). This means that the left side integrates to yx^2, while the right side becomes $\int 4x(x^2)\,dx$. Thus,

$$yx^2 = \int 4x^3\,dx + c \quad \text{or} \quad yx^2 = x^4 + c$$

In finding solutions to this type of differential equation, we often have need of certain basic properties of logarithms and exponential functions. In finding the factor $e^{\int Pdx}$, we often obtain an expression of the form $e^{\ln u}$. It is easily shown that this equals u.

Let $y = e^{\ln u}$. Writing this in logarithmic form, we have $\ln y = \ln u$. Thus, $y = u$ or $u = e^{\ln u}$.

Example B
Solve the differential equation $x\,dy - 3y\,dx = x^3\,dx$.

This equation can be put in the form of Eq. (29–7) by dividing through by x. This gives $dy - (3/x)y\,dx = x^2\,dx$. Here, $P = -3/x$, $Q = x^2$, and the factor $e^{\int Pdx}$ becomes

$$e^{\int(-3/x)dx} = e^{-3\ln x} = e^{\ln x^{-3}} = x^{-3}$$

The original equation can now be written as

$$yx^{-3} = \int x^2(x^{-3})\,dx = \int x^{-1}\,dx = \ln x + c$$

A better form of the answer is $y = x^3(\ln x + c)$.

Example C
Solve the differential equation $dy + y\,dx = x\,dx$. Here, $P = 1$, $Q = x$, and

$$e^{\int Pdx} = e^{\int(1)dx} = e^x$$

Therefore,

$$ye^x = \int xe^x\,dx + c = e^x(x - 1) + c$$

or

$$y = x - 1 + ce^{-x}$$

Example D
Solve the differential equation $x^2\,dy + 2xy\,dx = \sin x\,dx$.

This equation is first written in the form of Eq. (13–7). This gives us

$$dy + \left(\frac{2}{x}\right)y\,dx = \frac{1}{x^2}\sin x\,dx \quad \text{thus} \quad e^{\int Pdx} = e^{\int(2/x)dx} = x^2$$

The solution to the equation then becomes

$$yx^2 = \int \sin x\,dx + c = -\cos x + c$$

or

$$yx^2 + \cos x = c$$

Example E

Find the particular solution of the differential equation

$$dy = (1 - 2y)x\,dx$$

such that $x = 0$ when $y = 2$.

First writing this equation in the form of Eq. (13–7), we have

$$dy + 2xy\,dx = x\,dx$$

Therefore, in this case

$$e^{\int P dx} = e^{\int 2x\,dx} = e^{x^2}$$

This leads to

$$ye^{x^2} = \int xe^{x^2}dx$$

Integrating the right side, we have

$$ye^{x^2} = \frac{1}{2}e^{x^2} + c$$

Using the condition that $x = 0$ when $y = 2$ gives

$$(2)(e^0) = \frac{1}{2}(e^0) + c$$

$$2 = \frac{1}{2} + c$$

$$c = \frac{3}{2}$$

Therefore, the required solution is

$$ye^{x^2} = \frac{1}{2}e^{x^2} + \frac{3}{2}$$

$$y = \frac{1}{2}(1 + 3e^{-x^2})$$

Exercises 13–4

In Exercises 1 through 20 solve the given differential equations.

1. $dy + y\,dx = e^{-x}dx$

2. $dy + 3y\,dx = e^{-3x}dx$

3. $dy + 2y\,dx = e^{-4x}dx$

4. $dy + y\,dx = e^{-x}\cos x\,dx$

5. $x\,dy - y\,dx = 3x\,dx$

6. $x\,dy + 3y\,dx = dx$

7. $2x\,dy + y\,dx = 8x^3\,dx$

8. $3x\,dy - y\,dx = 9x\,dx$

9. $dy + y\cot x\,dx = dx$

10. $dy + y\tan x\,dx + \sin x\,dx = 0$

11. $y' + y = 3$

12. $y' + 2y = \sin x$

13. $\dfrac{dy}{dx} = xe^{4x} + 4y$

14. $y' - 2y = 2e^{2x}$

15. $y' = x^3(1 - 4y)$ **16.** $y' = x^2y + 3x^2$

17. $x\dfrac{dy}{dx} = y + (x^2 - 1)^2$ **18.** $2x(dy - dx) + y\,dx = 0$

19. $x\,dy + (1 - 3x)y\,dx = 3x^2e^{3x}dx$ **20.** $(1 + x^2)dy + xy\,dx = x\,dx$

In Exercises 21 through 24 find the indicated particular solutions of the given differential equations.

21. $\dfrac{dy}{dx} + 2y = e^{-x}, \quad x = 0$ when $y = 1$

22. $y' - 3y = e^{4x}, \quad x = 0$ when $y = 2$

23. $y' + \dfrac{1}{2\sqrt{x}}y = e^{\sqrt{x}}, \quad x = 1$ when $y = 3$

24. $(\sin x)y' + y = \tan x, \quad x = \pi/4$ when $y = 0$

13–5 Applications

The differential equations of the first order and first degree which we have discussed thus far have a number of applications in geometry and the various fields of technology. In this section we shall illustrate some of these applications, both through the examples and the exercises.

Example A

The slope of a given curve is given by the expression $6xy$. Find the equation of the curve if it passes through the point $(2, 1)$.

Since the slope is $6xy$, we know that $dy/dx = 6xy$. Writing this differential equation as

$$dy = 6xy\,dx$$

we divide each term by y to separate the variables. Then integrating we find the general form of the solution.

$$\frac{dy}{y} = 6x\,dx$$

$$\ln y = 3x^2 + c$$

Using the fact that the curve passes through $(2, 1)$, we evaluate the constant.

$$\ln 1 = 3(2^2) + c$$
$$0 = 12 + c$$
$$c = -12$$

Thus, the required particular solution is

$$\ln y = 3x^2 - 12$$

Example B

A curve which intersects all the members of a family of curves at right angles is called an **orthogonal trajectory** *of the family.* By a family of curves is meant a specified set of curves which satisfy given conditions. Find the equations of the orthogonal trajectories of the parabolas $x^2 = cy$.

Finding the derivative of the given equation, we have $dy/dx = 2x/c$. This equation contains the constant c, which depends on the point (x, y) on the parabola. Eliminating this constant between the equations of the parabola and derivative, we have $dy/dx = 2y/x$. This expression gives us a general expression for the slope of any of the members of the family. For a curve to be perpendicular, its slope must equal the negative reciprocal of this expression, or the slope of the orthogonal trajectories must be given by

$$\left.\frac{dy}{dx}\right|_{\text{OT}} = -\frac{x}{2y}$$

Solving this differential equation gives the family of orthogonal trajectories:

$$2y\,dy = -x\,dx, \quad y^2 = -\frac{x^2}{2} + \frac{c}{2} \quad \text{or} \quad 2y^2 + x^2 = c$$

Thus, the orthogonal trajectories are a family of ellipses. Note in Fig. 13–1 that each parabola intersects each ellipse at right angles.

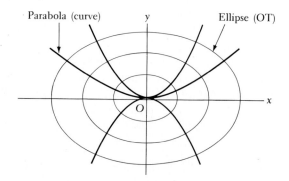

Figure 13–1

Example C

Radioactive elements decay at rates proportional to the amount present. A certain isotope of uranium decays so that half of the original amount disappears in 20 d. Determine the equation relating the amount present with the time and the fraction remaining after 50 d.

Let N_0 be the original amount and N the amount present at any time t (in days). Since we know information related to the rate of decay, we express this rate as a derivative. This means that we have the equation

$$\frac{dN}{dt} = kN$$

Separating variables, and integrating, we have

$$\frac{dN}{N} = k\ dt$$

$$\ln N = kt + \ln c$$

Evaluating the constant, we know that $N = N_0$ for $t = 0$. Thus,

$$\ln N_0 = k(0) + \ln c$$
$$c = N_0$$

Therefore, we have the solution

$$\ln N = kt + \ln N_0$$
$$\ln N - \ln N_0 = kt$$

$$\ln \frac{N}{N_0} = kt$$

$$N = N_0 e^{kt}$$

Now using the condition that half this isotope decays in 20 d, we have $N = N_0/2$ when $t = 20$ d. This gives

$$\frac{N_0}{2} = N_0 e^{20k} \quad \text{or} \quad \frac{1}{2} = (e^k)^{20} \quad \text{or} \quad e^k = \left(\frac{1}{2}\right)^{1/20}$$

Therefore, the equation relating N and t is

$$N = N_0\left(\frac{1}{2}\right)^{t/20}$$

In order to determine the fraction remaining after 50 d, we evaluate this equation for $t = 50$. This gives

$$N = N_0\left(\frac{1}{2}\right)^{5/2} = 0.177\ N_0$$

This means that 17.7% remains after 50 d.

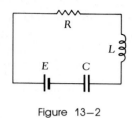

Figure 13–2

Example D
The general equation relating the current i, voltage E, inductance L, capacitance C, and resistance R of a simple electric circuit (see Fig. 13–2) is given by

$$L\frac{di}{dt} + Ri + \frac{q}{C} = E$$

where q is the charge on the capacitor. Find the general expression for the current in a circuit containing an inductance, resistance, and voltage source if $i = 0$ when $t = 0$.

The differential equation for this circuit is

$$L\frac{di}{dt} + Ri = E \tag{13–9}$$

Using the method of the linear differential equation of the first order, we have the equation

$$di + \frac{R}{L}i\ dt = \frac{E}{L}dt$$

The factor $e^{\int P\,dt}$ is $e^{\int (R/L)dt} = e^{(R/L)t}$. This gives

$$ie^{(R/L)t} = \frac{E}{L}\int e^{(R/L)t}dt = \frac{E}{R}e^{(R/L)t} + c$$

Letting the current be zero for $t = 0$, we have $c = -E/R$. The result is

$$ie^{(R/L)t} = \frac{E}{R}e^{(R/L)t} - \frac{E}{R}$$

or

$$i = \frac{E}{R}(1 - e^{-(R/L)t})$$

(We can see that $i \to E/R$ as $t \to \infty$. In practice the exponential term becomes negligible very quickly.)

Example E
Fifty gallons of brine originally containing 20 lb of salt are in a tank into which 2 gal of water run each minute with the same amount of mixture running out each minute. How much salt remains in the tank after 10 min?

Let $x =$ the number of pounds of salt in the tank at time t. Each gallon of brine contains $x/50$ lb of salt, and in time dt, $2\ dt$ gal of mixture leave the tank with $(x/50)(2\ dt)$ lb of salt. The amount of salt which is leaving may also be written as $-dx$ (the minus sign is included to show that x is decreasing). This means

$$-dx = \frac{2x\ dt}{50} \quad \text{or} \quad \frac{dx}{x} = -\frac{dt}{25}$$

This leads to $\ln x = -(t/25) + \ln c$. Using the fact that $x = 20$ when $t = 0$, we have $\ln 20 = \ln c$, or $c = 20$. Therefore,

$$x = 20e^{-t/25}$$

is the general expression for the amount of salt in the tank at time t. When $t = 10$ we have

$$x = 20e^{-10/25} = 20e^{-0.4} = 20(0.670) = 13.4$$

There are 13.4 lb of salt in the tank after 10 min.

Example F

An object moving through (or across) a resisting medium often experiences a retarding force approximately proportional to the velocity. Applying Newton's laws of motion to this leads to the equation

$$m\frac{dv}{dt} = F - kv \tag{13-10}$$

where m is the mass of the object, v is the velocity, t is the time, F is the force causing the motion, and k is a constant $(k > 0)$. The quantity kv is the retarding force.

We assume these conditions hold for a certain falling object whose mass is 1 slug (this is the unit of mass when force is expressed in pounds) and which experiences a force (its weight) of 32 lb, if the object starts from rest and the retarding force is numerically equal to 0.2 times the velocity.

Substituting in Eq. (13-10), we have

$$\frac{dv}{dt} = 32 - 0.2v$$

Separating variables, we have

$$\frac{dv}{32 - 0.2v} = dt$$

Integrating, we find

$$-5 \ln(32 - 0.2v) = t - 5 \ln c$$

Solving for v, we obtain

$$\ln(32 - 0.2v) = -\frac{t}{5} + \ln c$$

$$32 - 0.2v = ce^{-t/5}$$

$$v = 5(32 - ce^{-t/5})$$

Since the object started from rest, $v = 0$ when $t = 0$. Thus,

$$0 = 5(32 - c) \quad \text{or} \quad c = 32$$

Therefore, the expression giving the velocity as a function of time is

$$v = 160(1 - e^{-t/5})$$

Evaluating v for $t = 5$ s, we have

$$v = 160(1 - e^{-1}) = 160(1 - 0.368) = 101 \text{ ft/s}$$

Exercises 13—5

1. Find the equation of the curve for which the slope is given by $2x/y$ and which passes through $(2, 3)$.

2. Find the equation of the curve for which the slope is given by $-y/(x + y)$ and which passes through $(-1, 3)$.

3. Find the equation of the curve for which the slope is given by $y + x$ and which passes through $(0, 1)$.

4. Find the equation of the curve for which the slope is given by $-2y + e^{-x}$ and which passes through $(0, 2)$.

5. Find the equation of the orthogonal trajectories of the exponential curves $y = ce^x$.

6. Find the equation of the orthogonal trajectories of the cubic curves $y = cx^3$.

7. Find the equation of the orthogonal trajectories of the curves $y = c(\sec x + \tan x)$.

8. Find the equation of the orthogonal trajectories of the family of circles, all with centers at the origin.

9. A certain isotope decays in such a way that half the original amount disappears in 2 h. Find the proportion remaining after 3 h, by first finding the relation between the amount present and the time.

10. Radium decays in such a way that half of the original amount disappears in 1800 yr. What part of the original amount disappears in 100 yr?

11. If interest in a bank account is compounded continuously, the amount grows at a rate which is proportional to the amount which is present. A "day of deposit to day of withdrawal" account very closely approximates this situation. Determine the amount in an account after one year if $1000 is placed in an account which pays 4% interest per year, compounded continuously.

12. If interest is compounded continuously (see Exercise 11), how long will it take a bank account to double in value if the rate of interest is 5% per year?

13. If the current in an RL circuit with a voltage source E is zero when $t = 0$ (see Example D), show that $\lim_{t \to \infty} i = E/R$.

14. If a circuit contains only an inductance and a resistance, find the expression for the current in terms of the time, if $i = i_0$ when $t = 0$.

15. For the circuit in Example D, find the current after 1 ms, if $R = 10\ \Omega$, $L = 0.1$ H, and $E = 100$ V.

16. For the circuit in Exercise 14, what is the value of the current after 1 ms, if $R = 10\ \Omega$, $L = 0.1$ H, and $i_0 = 2$ A?

17. If a circuit contains only a resistance and a capacitance, find the expression relating the charge on the capacitor in terms of the time, if $i = dq/dt$ and $q = q_0$ when $t = 0$.

18. If the charge on a capacitor in an RC circuit (see Exercise 17) is 2 mC when $t = 0$, find the charge on the capacitor after 0.01 s if the resistance is 450 Ω and the capacitance is 4 μF.

19. One hundred gallons of brine originally containing 30 lb of salt are in a tank into which 5 gal of water run each minute. The same amount of mixture from the tank leaves each minute. How much salt is in the tank after 20 min?

20. Repeat Exercise 19 with the change that the water entering the tank contains 1 lb of salt per gallon.

21. An object falling under the influence of gravity has a variable acceleration given by $32 - v$, where v represents the velocity. If the object starts from rest, find an expression for the velocity in terms of the time. Also find the limiting value of the velocity (find $\lim_{t \to \infty} v$).

22. A mass of 2 slugs (weight is 64 lb) is falling under the influence of gravity. Find its velocity after 5 s if it starts from rest and experiences a retarding force numerically equal to 0.1 times the velocity.

23. A boat with a mass of 10 slugs is being towed at 8 mi/h. The tow rope is then cut and a motor which exerts a force of 20 lb on the boat is started. If the water exerts a retarding force which numerically equals twice the velocity, what is the velocity of the boat 3 min later?

24. A parachutist is falling at the rate of 200 ft/s when her parachute opens. If the air resists the fall with a force equal to $0.5v^2$, find the velocity as a function of time. The woman and her equipment have a mass of 5 slugs (weight is 160 lb).

25. A particle moves on the hyperbola $xy = 1$ such that $dx/dt = 2t$. Express x and y in terms of t if $x = 1$, $y = 1$ when $t = 0$.

26. A particle moves on the parabola $y = x^2 + x$, such that $dx/dt = 4t + 1$. Express x and y in terms of t, if both x and y are zero when $t = 0$.

27. The rate of change of air pressure with respect to height is approximately proportional to the pressure. If the pressure is 15 lb/in.^{2} when $h = 0$, and $p = 10$ lb/in.^{2} when $h = 10{,}000$ ft, find the expression relating pressure and height.

28. When a gas undergoes an adiabatic change (no gain or loss of heat), the rate of change of pressure with respect to volume is directly proportional to the pressure and inversely proportional to the volume. Express the pressure in terms of the volume.

29. Assume that the rate of depreciation of an object is proportional to its value at any time t. If a car costs $4000 new, and its value is $2000 three years later, what is its value five years after it was purchased?

30. Assume that sugar dissolves at a rate proportional to the undissolved amount. If there are initially 500 g of sugar, and 200 g remain after 4 min, how long does it take to dissolve 400 g?

31. Under certain conditions, the rate of change of temperature of an object is proportional to the difference in temperature between it and the surrounding medium. (Remember, the temperature decreases with time.) Find the temperature after an hour of an object originally 100°C, if it cools to 90°C in 5 min in air which is at 20°C.

32. The lines of equal potential in a field of force are all at right angles to the lines of force. In an electric field of force caused by charged particles, the lines of force are given by

$$x^2 + y^2 = cx$$

Find the equation of the lines of equal potential. Sketch a few members of the lines of force and those of equal potential.

13–6 Review Exercises for Chapter 13

In Exercises 1 through 16, find the general solution of each given differential equation.

1. $4xy^3 \, dx + (x^2 + 1) \, dy = 0$ 2. $\dfrac{dy}{dx} = e^{x-y}$

3. $\dfrac{dy}{dx} + 2y = e^{-2x}$ 4. $x \, dy + y \, dx = y \, dy$

5. $(x + y) \, dx + (x + y^3) \, dy = 0$ 6. $y \ln x \, dx = x \, dy$

7. $x\dfrac{dy}{dx} - 3y = x^2$ 8. $dy - 2y \, dx = (x - 2)e^x \, dx$

9. $dy = 2y \, dx + y^2 \, dx$ 10. $x^2 y \, dy = (1 + x) \csc y \, dx$

11. $\sin x \dfrac{dy}{dx} + y \cos x + x = 0$ 12. $y \, dy = (x^2 + y^2 - x) \, dx$

13. $x \, dy - y \, dx = x^3 y \, dx$ 14. $x \, dy + y \, dx = x^3 y^4 dy$

15. $y \, dy = 2(x^2 + y^2)^2 dx - x \, dx$ 16. $y' = 2 + xe^{3x}$

In Exercises 17 through 20, find the particular solution of each of the given differential equations.

17. $y' = 2y \cot x$, $x = \pi/2$ when $y = 2$
18. $x \, dy - y \, dx = y^3 \, dy$, $x = 1$ when $y = 3$
19. $y' = 4x - 2y$, $x = 0$ when $y = -2$
20. $xy^2 \, dx + e^x \, dy = 0$, $x = 0$ when $y = 2$

In Exercises 21 through 30 solve the given problems.

21. The time rate of change of volume of an evaporating substance is proportional to the surface area. Express the radius of a sphere as a function of time. Let $r = r_0$ when $t = 0$. [*Hint:* Express both V and A in terms of the radius r.]

22. An insulated tank is filled with a solution containing radioactive cobalt. Due to the radioactivity, energy is released and the temperature T (in degrees Fahrenheit) of the solution rises with the time t (in hours). The following differential equation expresses the relation between temperature and time for a specific case:

$$56,600 = 262(T - 70) + 20,200\dfrac{dT}{dt}$$

If the initial temperature is 70°F, what is the temperature 24 h later?

23. An object with a mass of 1 kg slides down a long inclined plane. The effective force of gravity is 4 N, and the motion is retarded by a force numerically equal to the velocity. If the object starts from rest, what is its velocity (in meters per second) 4 s later?

24. Under proper conditions, bacteria grow at a rate proportional to the number present. In a certain culture there were 10^5 present at a given time, and there were 3.0×10^5 present after 10 h. How many were present after 5 h?

25. Find the orthogonal trajectories of the family of curves $y = cx^5$.

26. Find the equation of the curves such that their normals at all points are in the direction with the lines connecting the points and the origin.

27. If a circuit contains a resistance R, capacitance C, and a source of voltage E, express the charge q on the capacitor as a function of time.

28. A 2-H inductor, a 40-Ω resistor, and a 20-V battery are connected in series. Find the current in the circuit as a function of time if the initial current is zero.

29. The angle ϕ between the radius vector and a curve $r = f(\theta)$ at (r, θ) is given by

$$\frac{dr}{d\theta} = \frac{r}{\tan \phi}$$

Find the polar equation of the orthogonal trajectories of the circles

$$r = a \sin \theta$$

30. A differential equation of the form

$$M(x, y)\, dx + N(x, y)\, dy = 0$$

is termed *exact* if

$$\frac{\partial M}{\partial y} = \frac{\partial N}{\partial x}$$

Determine whether or not the equation

$$3x^2 \cos^2 y\, dx - x^3 \sin 2y\, dy = 0$$

is exact.

14

Second-Order
Differential Equations

14–1 Linear Differential Equations of Higher Order

Until now we have restricted our attention to differential equations of the first order and first degree. Another important type of differential equation is the linear differential equation of the second order with constant coefficients. Before restricting our attention to this specific type, we shall briefly describe the general higher-order equation and the notation we shall use with this type of equation.

The general linear differential equation of the nth order is of the form

$$a_0 \frac{d^n y}{dx^n} + a_1 \frac{d^{n-1} y}{dx^{n-1}} + \cdots + a_{n-1} \frac{dy}{dx} + a_n y = b \qquad (14\text{–}1)$$

where the a's and b are either functions of x or are constants.

For convenience of notation, the nth derivative with respect to x will be denoted by D^n. The other derivatives are denoted in a similar manner. In this notation, the general linear differential equation becomes

$$a_0 D^n y + a_1 D^{n-1} y + \cdots + a_{n-1} Dy + a_n y = b \qquad (14\text{–}2)$$

If $b = 0$ *the general linear equation is called* **homogeneous**, *and if* $b \neq 0$ *it is called* **nonhomogeneous**. Both these types have important applications.

Although the a's may be functions of x, we shall restrict our attention to the case where they are all constants. We shall, however, consider both homogeneous and nonhomogeneous equations. Since second-order equations occur widely in applied problems, the remainder of the chapter will be devoted to this type of equation. The methods developed, however, may be applied to equations of higher order.

14–2 Second-Order Homogeneous Equations with Constant Coefficients

In accordance with the notation and terminology introduced in Section (14–1), a second-order, linear, homogeneous differential equation with constant coefficients is one of the form

$$a_0 D^2 y + a_1 Dy + a_2 y = 0 \qquad (14\text{–}3)$$

where the a's are now constants. The following example will indicate the kind of solution we should expect for this type of equation.

Example A

Solve the differential equation $D^2 y - Dy - 2y = 0$.

First we put this equation in the form $(D^2 - D - 2)y = 0$. This is another way of saying that we are to take the second derivative of y, subtract the first derivative, and finally subtract twice the function. This may now be factored as $(D - 2)(D + 1)y = 0$. (We shall not develop the algebra of the operator D. However, most such algebraic operations can be shown to be valid.) This formula tells us to find the first derivative of the function and add this to the function. Then twice this result is to be subtracted from the derivative of this result. If we let $z = (D + 1)y$, which is valid since $(D + 1)y$ is a function of x, we have $(D - 2)z = 0$. This equation is easily solved by separation of variables. Thus,

$$\frac{dz}{dx} - 2z = 0, \qquad \frac{dz}{z} - 2\,dx = 0, \qquad \ln z - 2x = \ln c_1$$

$$\ln\frac{z}{c_1} = 2x \quad \text{or} \quad z = c_1 e^{2x}$$

Replacing z by $(D + 1)y$, we have

$$(D + 1)y = c_1 e^{2x}$$

This is a linear equation of the first order. Then

$$dy + y\,dx = c_1 e^{2x} dx$$

The factor $e^{\int P\,dx}$ is $e^{\int dx} = e^x$. And so

$$ye^x = \int c_1 e^{3x} dx = \frac{c_1}{3} e^{3x} + c_2$$

or

$$y = c_1' e^{2x} + c_2 e^{-x}$$

where $c_1' = \tfrac{1}{3}c_1$. This example indicates that solutions of the form e^{mx} result for this equation.

Basing our reasoning on the result of this example, let us assume that an equation of the form (14–3) has a particular solution e^{mx}. Substituting this into Eq. (14–3) gives

$$a_0 m^2 e^{mx} + a_1 m e^{mx} + a_2 e^{mx} = 0$$

The exponential function e^{mx} is never zero, which means that the polynomial in m is zero, or

$$a_0 m^2 + a_1 m + a_2 = 0 \qquad (14\text{--}4)$$

Equation (14–4) is called the **auxiliary equation** *of Eq. (14–3). Note that it may be formed directly by inspection from Eq. (14–3).*

There are two roots of the auxiliary Eq. (14–4) and there are two arbitrary constants in the solution of Eq. (14–3). These factors lead us to the general solution of Eq. (14–3), which is

$$y = c_1 e^{m_1 x} + c_2 e^{m_2 x} \qquad (14\text{--}5)$$

where m_1 and m_2 are the solutions of Eq. (14–4). This is seen to be in agreement with the results of Example A.

Example B

Solve the differential equation $D^2 y - 5Dy + 6y = 0$. The auxiliary equation is $m^2 - 5m + 6 = 0$.

Factoring this equation, we have $(m - 3)(m - 2) = 0$, which gives the roots $m_1 = 3$ and $m_2 = 2$. Thus, the solution is

$$y = c_1 e^{3x} + c_2 e^{2x}$$

Obviously it makes no difference which constant is written with each exponential function.

Example C

Solve the differential equation $D^2 y - 6Dy = 0$. Here

$$m^2 - 6m = 0$$
$$m(m - 6) = 0$$

or

$$m_1 = 0 \quad \text{and} \quad m_2 = 6$$

The solution is $y = c_1 e^{0x} + c_2 e^{6x}$. Since $e^{0x} = 1$, we have

$$y = c_1 + c_2 e^{6x}$$

Example D

Solve the differential equation $2y'' + 7y' = 4y$.

We first rewrite this equation using the D notation for derivatives. Also, we want to write in the proper form of a homogeneous equation. This gives us

$$2D^2 y + 7Dy - 4y = 0$$

Now writing the auxiliary equation and solving it, we have

$$2m^2 + 7m - 4 = 0, \qquad (2m - 1)(m + 4) = 0$$

or

$$m_1 = \frac{1}{2} \quad \text{and} \quad m_2 = -4$$

The solution is

$$y = c_1 e^{x/2} + c_2 e^{-4x}$$

Example E

Solve the differential equation $2 \dfrac{d^2 y}{dx} + \dfrac{dy}{dx} - 7y = 0$.

First writing this equation with the D notation for derivatives, we have

$$2D^2 y + Dy - 7y = 0$$

This means that the auxiliary equation is

$$2m^2 + m - 7 = 0$$

Since this equation is not factorable, we solve it by use of the quadratic formula. This gives us

$$m = \frac{-1 \pm \sqrt{1 + 56}}{4} = \frac{-1 \pm \sqrt{57}}{4}$$

Therefore, the solution is

$$y = c_1 e^{\frac{-1 + \sqrt{57}}{4} x} + c_2 e^{\frac{-1 - \sqrt{57}}{4} x}$$

A somewhat better form can be obtained by observing that $e^{-x/4}$ is a factor of each term. Thus,

$$y = e^{-x/4} (c_1 e^{\sqrt{57}x/4} + c_2 e^{-\sqrt{57}x/4})$$

Example F

Solve the differential equation $D^2y - 2Dy - 15y = 0$, and find the particular solution which satisfies the conditions that $Dy = 2$, $y = -1$ when $x = 0$. (It is necessary to give two conditions since there are two constants to evaluate.)

We have

$$m^2 - 2m - 15 = 0, \qquad (m - 5)(m + 3) = 0$$

or

$$m_1 = 5 \quad \text{and} \quad m_2 = -3$$

$$y = c_1 e^{5x} + c_2 e^{-3x}$$

This equation is the general solution. In order to evaluate the constants c_1 and c_2, we use the given conditions to find two simultaneous equations in c_1 and c_2. These are then solved to determine the particular solution. Thus,

$$y' = 5c_1 e^{5x} - 3c_2 e^{-3x}$$

Using the given conditions in the general solution and its derivative, we have

$$c_1 + c_2 = -1, \qquad 5c_1 - 3c_2 = 2$$

The solution to this system of equations is $c_1 = -\frac{1}{8}$ and $c_2 = -\frac{7}{8}$. The particular solution becomes

$$y = -\frac{1}{8}e^{5x} - \frac{7}{8}e^{-3x}$$

or

$$8y + e^{5x} + 7e^{-3x} = 0$$

To solve the auxiliary equation for differential equations of higher order, we use methods developed in the theory of equations. With these roots, we form the solutions in the same way as we do those for second-order equations.

Exercises 14—1, 14—2

In Exercises 1 through 20 solve the given differential equations.

1. $\dfrac{d^2y}{dx^2} - \dfrac{dy}{dx} - 6y = 0$

2. $\dfrac{d^2y}{dx^2} + \dfrac{dy}{dx} = 0$

3. $\dfrac{d^2y}{dx^2} + 4\dfrac{dy}{dx} + 3y = 0$

4. $\dfrac{d^2y}{dx^2} - 2\dfrac{dy}{dx} - 8y = 0$

5. $D^2y - Dy = 0$

6. $D^2y + 7Dy + 6y = 0$

7. $3D^2y + 12y = 20Dy$

8. $4D^2y + 12Dy = 7y$

9. $3y'' + 8y' - 3y = 0$

10. $8y'' + 6y' - 9y = 0$

11. $3y'' + 2y' - y = 0$

12. $2y'' - 7y' + 6y = 0$

13. $2\dfrac{d^2y}{dx^2} - 4\dfrac{dy}{dx} + y = 0$

14. $\dfrac{d^2y}{dx^2} + \dfrac{dy}{dx} - 5y = 0$

15. $4D^2y - 3Dy - 2y = 0$

16. $2D^2y - 3Dy - y = 0$

17. $y'' = 3y' + y$

18. $5y'' - y' = 3y$

19. $y'' + y' = 8y$

20. $8y'' = y' + y$

In Exercises 21 through 24 find the particular solutions of the differential equations which satisfy the given conditions.

21. $D^2y - 4Dy - 21y = 0$, $Dy = 0$ and $y = 2$ when $x = 0$

22. $D^2y - 4Dy = 0$, $Dy = 2$ and $y = 4$ when $x = 0$

23. $D^2y - Dy - 12y = 0$, $y = 0$ when $x = 0$ and $y = 1$ when $x = 1$

24. $2D^2y + 5Dy = 0$, $y = 0$ when $x = 0$ and $y = 2$ when $x = 1$

In Exercises 25 and 26 solve the given differential equations. The auxiliary equations will be third degree, and the solutions will have three arbitrary constants.

25. $y''' - 2y'' - 3y' = 0$ **26.** $y''' - 2y'' - y' + 2y = 0$

14–3 Auxiliary Equations with Repeated or Complex Roots

The way to solve a second-order homogeneous linear differential equation was shown in the last section. However, we purposely avoided repeated and complex roots of the auxiliary equation. In this section we shall develop the solutions for such equations. The following example indicates the type of solution which results from the case of repeated roots.

Example A

Solve the differential equation $D^2y - 4Dy + 4y = 0$.

Using the method of Example A of the previous section, we have the following steps:

$$(D^2 - 4D + 4)y = 0, \qquad (D - 2)(D - 2)y = 0, \qquad (D - 2)z = 0$$

where $z = (D - 2)y$. The solution to $(D - 2)z = 0$ is found by separation of variables. And so

$$\frac{dz}{dx} - 2z = 0, \qquad \frac{dz}{z} - 2\,dx = 0$$

$$\ln z - 2x = \ln c_1 \quad \text{or} \quad z = c_1 e^{2x}$$

Substituting back, we have $(D - 2)y = c_1 e^{2x}$, which is a linear equation of the first order. Then

$$dy - 2y\,dx = c_1 e^{2x}dx, \qquad e^{\int -2\,dx} = e^{-2x}$$

This leads to

$$ye^{-2x} = c_1 \int dx = c_1 x + c_2 \quad \text{or} \quad y = c_1 xe^{2x} + c_2 e^{2x}$$

This example indicates the type of solution which results when the auxiliary equation has repeated roots. If the method of the previous section were to be used, the solution of the above example would be $y = c_1 e^{2x} + c_2 e^{2x}$. This would not be the general solution, since both terms are similar, which means that there is only one independent constant. The constants can be combined to give a solution of the form $y = ce^{2x}$, where $c = c_1 + c_2$. This solution would contain only one constant for a second-order equation.

Based on the above example, the solution to Eq. (14–3) when the auxiliary Eq. (14–4) has repeated roots is

$$y = e^{mx}(c_1 + c_2 x) \qquad\qquad (14\text{–}6)$$

where m is the double root. (In Example A, this double root is 2.)

Example B

Solve the differential equation $(D + 2)^2 y = 0$.

The auxiliary equation is $(m + 2)^2 = 0$, for which the solutions are $m = -2, -2$. Since we have repeated roots, the solution of the differential equation is

$$y = e^{-2x}(c_1 + c_2 x)$$

Example C

Solve the differential equation $\dfrac{d^2 y}{dx^2} - 10\dfrac{dy}{dx} + 25y = 0$.

Using the D notation, this equation becomes

$$D^2 y - 10Dy + 25y = 0$$

Therefore, the auxiliary equation is

$$m^2 - 10m + 25 = 0$$

Factoring, we have

$$(m - 5)^2 = 0$$

This means the roots are $m = 5, 5$. We therefore have the solution

$$y = e^{5x}(c_1 + c_2 x)$$

When the auxiliary equation has complex roots, it can be solved by the method developed in the preceding section. However, the solution can be put in a special, more useful form. Let us assume that the quadratic formula is used to find the roots of the auxiliary equation, and that these roots are complex of the form $m = \alpha \pm j\beta$. This means that the solution is of the form

$$y = c_1 e^{(\alpha + j\beta)x} + c_2 e^{(\alpha - j\beta)x} = e^{\alpha x}(c_1 e^{j\beta x} + c_2 e^{-j\beta x})$$

Using the exponential form of a complex number, Eq. (12–12), we have

$$\begin{aligned}
y &= e^{\alpha x}[c_1 \cos \beta x + jc_1 \sin \beta x + c_2 \cos(-\beta x) + jc_2 \sin(-\beta x)] \\
&= e^{\alpha x}(c_1 \cos \beta x + c_2 \cos \beta x + jc_1 \sin \beta x - jc_2 \sin \beta x) \\
&= e^{\alpha x}(c_3 \cos \beta x + c_4 \sin \beta x)
\end{aligned}$$

where $c_3 = c_1 + c_2$ and $c_4 = jc_1 - jc_2$.

Summarizing these results, we see that if the auxiliary equation has complex roots of the form $\alpha \pm j\beta$, the solution to Eq. (14–3) is

$$y = e^{\alpha x}(c_1 \sin \beta x + c_2 \cos \beta x) \tag{14–7}$$

The c_1 and c_2 here are not the same as those above. They are simply the two arbitrary constants of the solution.

Example D

Solve the differential equation $D^2y - Dy + y = 0$. Thus,

$$m^2 - m + 1 = 0 \quad \text{or} \quad m = \frac{1 \pm \sqrt{3}j}{2}$$

$$\alpha = \frac{1}{2} \quad \text{and} \quad \beta = \frac{\sqrt{3}}{2}$$

We have

$$y = e^{x/2}\left(c_1 \sin \frac{\sqrt{3}}{2}x + c_2 \cos\frac{\sqrt{3}}{2}x\right)$$

Example E

Solve the differential equation $D^2y + 4y = 0$. In this case we have $m^2 + 4 = 0$, $m_1 = 2j$, and $m_2 = -2j$; $\alpha = 0$ and $\beta = 2$.
The solution is

$$y = e^{0x}(c_1 \sin 2x + c_2 \cos 2x)$$

or

$$y = c_1 \sin 2x + c_2 \cos 2x$$

Example F

Solve the differential equation $y'' - 2y' + 12y = 0$, if $y' = 2$ and $y = 1$ when $x = 0$.
Rewriting this equation as

$$D^2y - 2Dy + 12y = 0$$

we have the auxiliary equation

$$m^2 - 2m + 12 = 0$$

Solving for m, we have

$$m = \frac{2 \pm \sqrt{4 - 48}}{2} = \frac{2 \pm 2\sqrt{11}j}{2} = 1 \pm \sqrt{11}j$$

This means that the general solution is

$$y = e^x(c_1 \cos \sqrt{11}x + c_2 \sin \sqrt{11}x)$$

Using the condition that $y = 1$ when $x = 0$, we have

$$1 = e^0(c_1 \cos 0 + c_2 \sin 0)$$

or

$$c_1 = 1$$

Since the other condition involves the value of y', we find the derivative.

$$y' = e^x(c_1 \cos \sqrt{11}x + c_2 \sin \sqrt{11}x - \sqrt{11}c_1 \sin \sqrt{11}x + \sqrt{11}c_2 \cos \sqrt{11}x)$$
$$2 = e^0(\cos 0 + c_2 \sin 0 - \sqrt{11} \sin 0 + \sqrt{11}c_2 \cos 0)$$
$$2 = 1 + \sqrt{11}c_2$$

$$c_2 = \frac{1}{11}\sqrt{11}$$

Therefore, the solution is

$$y = e^x\left(\cos \sqrt{11}x + \frac{1}{11}\sqrt{11}\sin \sqrt{11}x\right)$$

Exercises 14–3

In Exercises 1 through 24 solve the given differential equations.

1. $\dfrac{d^2y}{dx^2} - 2\dfrac{dy}{dx} + y = 0$

2. $\dfrac{d^2y}{dx^2} - 6\dfrac{dy}{dx} + 9y = 0$

3. $D^2y + 12Dy + 36y = 0$

4. $D^2y + 8Dy + 16y = 0$

5. $\dfrac{d^2y}{dx^2} + 9y = 0$

6. $\dfrac{d^2y}{dx^2} + y = 0$

7. $D^2y + Dy + 2y = 0$

8. $D^2y - 2Dy + 4y = 0$

9. $D^2y = 0$

10. $4D^2y - 12Dy + 9y = 0$

11. $4D^2y + y = 0$

12. $9D^2y + 4y = 0$

13. $16y'' - 24y' + 9y = 0$

14. $9y'' - 24y' + 16y = 0$

15. $25y'' + 2y = 0$

16. $y'' - 4y' + 5y = 0$

17. $2D^2y + 5y = 4Dy$

18. $D^2y + 4Dy + 6y = 0$

19. $25y'' + 16y = 40y'$

20. $9y'' + 0.6y' + 0.01y = 0$

21. $2D^2y - 3Dy - y = 0$

22. $D^2y - 5Dy - 4y = 0$

23. $3D^2y + 12Dy = 2y$

24. $36D^2y = 25y$

In Exercises 25 through 28 find the particular solutions of the given differential equations which satisfy the stated conditions.

25. $y'' + 2y' + 10y = 0$, $y = 0$ when $x = 0$ and $y = e^{-1}$ when $x = \pi/6$

26. $D^2y + 16y = 0$, $Dy = 0$ and $y = 2$ when $x = \pi/2$

27. $D^2y - 8Dy + 16y = 0$, $Dy = 2$ and $y = 4$ when $x = 0$

28. $4y'' + 20y' + 25y = 0$, $y = 0$ when $x = 0$ and $y = e$ when $x = -2/5$

14–4 Solutions of Nonhomogeneous Equations

To solve a nonhomogeneous linear equation of the form

$$a_0 D^2 y + a_1 Dy + a_2 y = b \tag{14–8}$$

where b is a function of x or is a constant, the solution must be such that if we substitute it into the left side, we obtain the right side. Solutions obtained from the methods of Sections 14–2 and 14–3 will give zero when substituted into the left side, but they do contain the arbitrary constants necessary in the solution. If we could find some particular solution, which when substituted into the left side produced the expression of the right, it could be added to the solution containing the arbitrary constants. Thus, the solution is of the form

$$y = y_c + y_p \tag{14–9}$$

where y_c, called the **complementary solution,** is obtained by solving the corresponding homogeneous equation and where y_p is the particular solution necessary to produce the expression b of Eq. (14–8).

Example A
The differential equation $D^2 y - Dy - 6y = e^x$ has the solution $y = c_1 e^{3x} + c_2 e^{-2x} - \frac{1}{6} e^x$, with $y_c = c_1 e^{3x} + c_2 e^{-2x}$ and $y_p = -\frac{1}{6} e^x$. The complementary solution y_c is obtained by solving the corresponding homogeneous equation $D^2 y - Dy - 6y = 0$, and we shall discuss below the method of finding y_p. We can see that y_p alone will satisfy the differential equation, but since it has no arbitrary constants, it cannot be the general solution.

By inspection of the form of the expression b on the right side of the equation, we can determine the form which the particular solution must have. Since a combination of the derivatives and the function itself must form the function b, **we assume an expression which contains all possible forms of b and its derivatives will form y_p.** The method used to find the exact form of this expression is called the **method of undetermined coefficients.**

Example B
If the function b on the right of Eq. (14–8) is $4x + e^{2x}$, we would assume the particular solution is of the form $y_p = A + Bx + Ce^{2x}$. The Bx term is included to account for the presence of the $4x$. The A term is included to account for any derivatives of the Bx term which may occur. The Ce^{2x} is included to account for the presence of the e^{2x} term. Since all the derivatives of Ce^{2x} are of this same form, it is not necessary to include any other terms for the e^{2x}.

Example C

If the expression b is of the form $x^2 + e^{-x}$, we would assume y_p to be of the form $y_p = A + Bx + Cx^2 + Ee^{-x}$.

If b is of the form $xe^x - 5$, we would assume y_p to be of the form $y_p = Ae^x + Bxe^x + C$.

If b is of the form $x \sin x$, we would assume y_p to be of the form $A \sin x + B \cos x + Cx \sin x + Ex \cos x$. All of these terms occur in the derivatives of $x \sin x$.

Once we have determined the form of the particular solution, we have to find the numerical values of the coefficients $A, B, \ldots$ *This is done by substituting the form into the differential equation, and equating coefficients of like terms.*

Example D

Solve the differential equation $D^2y - Dy - 6y = e^x$.

In this case the solution of the auxiliary equation $m^2 - m - 6 = 0$, gives us the roots $m_1 = 3$ and $m_2 = -2$. Thus,

$$y_c = c_1 e^{3x} + c_2 e^{-2x}$$

The proper form of y_p is $y_p = Ae^x$. Substituting y_p and its derivatives into the differential equation, we have

$$Ae^x - Ae^x - 6Ae^x = e^x$$

To produce equality, the coefficients of e^x must be the same on each side of the equation. Thus,

$$-6A = 1 \quad \text{or} \quad A = -\frac{1}{6}$$

This gives the complete solution $y = y_c + y_p$, where $y_p = -\frac{1}{6}e^x$. Thus,

$$y = c_1 e^{3x} + c_2 e^{-2x} - \frac{1}{6}e^x$$

See Example A.

Example E

Solve the differential equation $D^2y + 4y = x - 4e^{-x}$.

In this case we have $m^2 + 4 = 0$, which gives us $m_1 = 2j$ and $m_2 = -2j$. Therefore,

$$y_c = c_1 \sin 2x + c_2 \cos 2x$$

The proper form of the particular solution is $y_p = A + Bx + Ce^{-x}$. Finding two derivatives and then substituting gives the following result:

$$y_p' = B - Ce^{-x}, \qquad y_p'' = Ce^{-x}$$
$$(Ce^{-x}) + 4(A + Bx + Ce^{-x}) = x - 4e^{-x}$$
$$Ce^{-x} + 4A + 4Bx + 4Ce^{-x} = x - 4e^{-x}$$
$$(4A) + (4Bx) + (5Ce^{-x}) = 0 + (1)x + (-4e^{-x})$$

Equating the coefficients of the constants, x and e^{-x}, gives $4A = 0$, $4B = 1$, and $5C = -4$. Thus, $A = 0$, $B = \frac{1}{4}$, and $C = -\frac{4}{5}$. This means that the particular solution is $y_p = \frac{1}{4}x - \frac{4}{5}e^{-x}$. In turn this tells us that the complete solution is

$$y = c_1 \sin 2x + c_2 \cos 2x + \frac{1}{4}x - \frac{4}{5}e^{-x}$$

Example F

Solve the differential equation $D^2y - 3Dy + 2y = 2 \sin x$. Thus,

$$m^2 - 3m + 2 = 0, \qquad (m-1)(m-2) = 0, \qquad m_1 = 1 \quad \text{and} \quad m_2 = 2$$

We write $y_c = c_1 e^x + c_2 e^{2x}$. Then

$$y_p = A \sin x + B \cos x$$
$$Dy_p = A \cos x - B \sin x$$
$$D^2 y_p = -A \sin x - B \cos x$$
$$(-A \sin x - B \cos x) - 3(A \cos x - B \sin x) + 2(A \sin x + B \cos x)$$
$$= 2 \sin x$$

$$(A + 3B)\sin x + (B - 3A)\cos x = 2 \sin x$$
$$A + 3B = 2, \qquad -3A + B = 0$$

The solution of this system is $A = \frac{1}{5}$ and $B = \frac{3}{5}$. Thus,

$$y_p = \frac{1}{5}\sin x + \frac{3}{5}\cos x$$

Therefore,

$$y = c_1 e^x + c_2 e^{2x} + \frac{1}{5}\sin x + \frac{3}{5}\cos x$$

Exercises 14—4

In Exercises 1 through 8 solve the given differential equations. The form of y_p is indicated.

1. $D^2y - Dy - 2y = 4$ (Let $y_p = A$.)
2. $D^2y - Dy - 6y = 4x$ (Let $y_p = A + Bx$.)
3. $D^2y + y = x^2$ (Let $y_p = A + Bx + Cx^2$.)
4. $D^2y + Dy + y = e^{2x}$ (Let $y_p = Ae^{2x}$.)
5. $D^2y + 4Dy + 3y = 2 + e^x$ (Let $y_p = A + Be^x$.)
6. $D^2y - y = \sin x$ (Let $y_p = A \sin x + B \cos x$.)
7. $D^2y - Dy - 20y = 8 + e^{2x}$ (Let $y_p = A + Be^{2x}$.)
8. $D^2y + 4y = \sin x + 4$ (Let $y_p = A + B \sin x + C \cos x$.)

In Exercises 9 through 20 solve the given differential equations.

9. $\dfrac{d^2y}{dx^2} - \dfrac{dy}{dx} - 30y = 10$

10. $2\dfrac{d^2y}{dx^2} + 11\dfrac{dy}{dx} - 6y = 8x$

11. $3\dfrac{d^2y}{dx^2} + 13\dfrac{dy}{dx} - 10y = 14e^{3x}$ 12. $\dfrac{d^2y}{dx^2} + 4y = 2\sin 3x$

13. $D^2y - 4y = \sin x + 2\cos x$ 14. $2D^2y + 5Dy - 3y = e^x + 4e^{2x}$

15. $D^2y + y = 4 + \sin 2x$ 16. $D^2y - Dy + y = x + \sin x$

17. $D^2y + 5Dy + 4y = xe^x + 4$ 18. $3D^2y + Dy - 2y = 4 + 2x + e^x$

19. $y'' + 6y' + 9y = e^{2x} - e^{-2x}$ 20. $y'' + 8y' - y = x^2 + 4e^{-2x}$

In Exercises 21 through 24 find the particular solution of each differential equation for the given conditions.

21. $D^2y - Dy - 6y = 5 - e^x$, $Dy = 4$ and $y = 2$ when $x = 0$.

22. $3y'' - 10y' + 3y = xe^{-2x}$, $Dy = 0$ and $y = -1$ when $x = 0$.

23. $y'' + y = x + \sin 2x$, $Dy = 1$ and $y = 0$ when $x = \pi$.

24. $D^2y - 2Dy + y = xe^{2x} - e^{2x}$, $Dy = 4$ and $y = -2$ when $x = 0$.

14–5 Applications

Linear differential equations of the second order have many important applications. We shall restrict our attention to two of these. In this section, we shall apply second-order differential equations to solving problems in simple harmonic motion and simple electric circuits.

Example A

Simple harmonic motion may be defined as motion in a straight line for which the acceleration is proportional to the displacement and in the opposite direction. Examples of this type of motion are a weight on a spring, a simple pendulum, and an object bobbing in water. If x represents the displacement, d^2x/dt^2 is the acceleration.

Using the definition of simple harmonic motion, we have

$$\frac{d^2x}{dt^2} = -k^2x$$

(We chose k^2 for convenience of notation in the solution.) Writing this equation in the form

$$D^2x + k^2x = 0 \qquad \left(\text{here, } D = \frac{d}{dt}\right)$$

we find that its solution is $x = c_1\sin kt + c_2\cos kt$. This solution indicates an oscillating motion, which is known to be the case. If, for example, $k = 4$ and we know that $x = 2$ and $Dx = 0$ (which means the velocity is zero) for $t = 0$, we have

$$Dx = 4c_1\cos 4t - 4c_2\sin 4t$$

or (substituting into the expression for x and that for Dx)

$$2 = c_1(0) + c_2(1) \quad \text{and} \quad 0 = 4c_1(1) - 4c_2(0)$$

which gives $c_1 = 0$ and $c_2 = 2$. Therefore,

$$x = 2\cos 4t$$

is the equation relating the displacement and time; Dx is the velocity and D^2x is the acceleration.

Example B

In practice an object will in time cease to move due to unavoidable frictional forces. It is found that a "freely" oscillating object has a force retarding it which is approximately proportional to the velocity. The differential equation for this case is $D^2x = -k^2x - bDx$, which results from applying (from physics) Newton's second law of motion, which states that the net force acting on an object equals its mass times its acceleration. The term D^2x represents the acceleration, the term $-k^2x$ is a measure of the restoring force (of the spring, for example), and the term $-bDx$ is the term which represents the retarding (damping) force. This equation can be written as

$$D^2x + bDx + k^2x = 0$$

The auxiliary equation is $m^2 + bm + k^2 = 0$, for which the roots are

$$m = \frac{-b \pm \sqrt{b^2 - 4k^2}}{2}$$

If $k = 3$ and $b = 4$, $m = -2 \pm \sqrt{5}j$, which means the solution is

$$x = e^{-2t}(c_1 \sin \sqrt{5}t + c_2 \cos \sqrt{5}t) \tag{1}$$

Here, $4k^2 > b^2$, and this case is called **underdamped**.

If $k = 2$ and $b = 5$, $m = -1, -4$, which means the solution is

$$x = c_1 e^{-t} + c_2 e^{-4t} \tag{2}$$

Here $4k^2 < b^2$, and the case is called **overdamped**. It will be noted that the motion is not oscillatory, since no sine or cosine terms appear.

If $k = 2$ and $b = 4$, $m = -2, -2$, which means the solution is

$$x = e^{-2t}(c_1 + c_2 t) \tag{3}$$

Here $4k^2 = b^2$, and the case is called **critically damped**. Again the motion is not oscillatory. See Fig. 14–1 in which Eqs. (1), (2), and (3) are represented in general. Of course, the actual values depend upon c_1 and c_2, which in turn depend upon the conditions imposed on the motion.

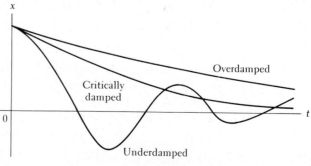

Figure 14–1

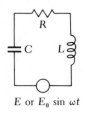

E or $E_0 \sin \omega t$

Figure 14–2

Example C

A basic relation for electric circuits is that the impressed voltage of a circuit equals the sums of the voltages across the components of the circuit. When this principle is applied to a circuit containing a resistance, inductance, capacitance, and a voltage source (see Fig. 14–2), the differential equation is

$$L\frac{d^2q}{dt^2} + R\frac{dq}{dt} + \frac{q}{C} = E \tag{14–10}$$

By definition q represents the electric charge, $dq/dt = i$ is the current and d^2q/dt^2 is the time-rate of change of current. This equation may be written as $LD^2q + RDq + q/C = E$. The auxiliary equation is $Lm^2 + Rm + 1/C = 0$. The roots are

$$m = \frac{-R \pm \sqrt{R^2 - 4L/C}}{2L} = -\frac{R}{2L} \pm \sqrt{\frac{R^2}{4L^2} - \frac{1}{LC}}$$

If we let $a = R/2L$ and $\omega = \sqrt{1/LC - R^2/4L^2}$, we have (assuming complex roots, which corresponds to realistic values of R, L, and C)

$$q_c = e^{-at}(c_1 \sin \omega t + c_2 \cos \omega t)$$

This indicates an oscillating charge, or an alternating current. However, the exponential term usually is such that the current dies out rapidly unless there is a source of voltage in the circuit.

If there is no source of voltage in the circuit of Example C, we have a homogeneous differential equation to solve. If we have a constant voltage source, the particular solution is of the form $q_p = A$. If there is an alternating voltage source, the particular solution is of the form $q_p = A \sin \omega_1 t + B \cos \omega_1 t$, where ω_1 is the angular velocity of the source. After a very short time, the exponential factor in the complementary solution makes it negligible. For this reason it is referred to as the **transient** term, and the particular solution is the **steady-state** solution. Therefore, to find the steady-state solution, we need find only the particular solution.

Example D

Find the steady-state solution for the current in a circuit containing the following elements: $C = 400 \ \mu F$, $L = 1 \ H$, $R = 10 \ \Omega$, and a voltage source of $500 \sin 100t$.

This means the differential equation to be solved is

$$\frac{d^2q}{dt^2} + 10\frac{dq}{dt} + \frac{10^4}{4}q = 500 \sin 100t$$

Since we wish to find the steady-state solution, we must find q_p from which we may find i_p by finding a derivative. Thus,

$$q_p = A \sin 100t + B \cos 100t$$

$$\frac{dq_p}{dt} = 100A \cos 100t - 100B \sin 100t$$

$$\frac{d^2q_p}{dt^2} = -10^4A \sin 100t - 10^4B \cos 100t$$

Substituting into the differential equation, we have

$$-10^4A \sin 100t - 10^4B \cos 100t + 10^3A \cos 100t - 10^3B \sin 100t$$

$$+ \frac{10^4}{4}A \sin 100t + \frac{10^4}{4}B \cos 100t = 500 \sin 100t$$

This means

$$(-0.75 \times 10^4A - 10^3B)\sin 100t + (-0.75 \times 10^4B + 10^3A)\cos 100t$$
$$= 500 \sin 100t$$

or

$$-7.5 \times 10^3A - 10^3B = 500, \qquad 10^3A - 7.5 \times 10^3B = 0$$

Solving these equations, we obtain

$$B = -8.73 \times 10^{-3} \quad \text{and} \quad A = -65.5 \times 10^{-3}$$

Therefore,

$$q_p = -65.5 \times 10^{-3}\sin 100t - 8.73 \times 10^{-3}\cos 100t$$

and

$$i_p = \frac{dq_p}{dt} = -6.55 \cos 100t + 0.87 \sin 100t$$

which is the desired solution.

It should be noted that the complimentary solutions of the mechanical and electrical cases are of the same identical form. There is also an equivalent mechanical case to that of an impressed sinusoidal voltage source in the electrical case. This arises when an external force affecting the vibrations is applied to the system. Such cases are called **forced vibrations**. Thus, we may have transient and steady-state solutions to mechanical and other nonelectrical situations.

Exercises 14–5

1. An object moves with simple harmonic motion according to the equation $D^2x + 100x = 0$; $D = d/dt$. Find the displacement x as a function of the time, subject to the conditions that $x = 4$ and $Dx = 0$ when $t = 0$. Sketch the resulting curve.

2. Solve the problem given in Exercise 1 if the motion is in accordance with the equation $D^2x + 0.2Dx + 100x = 0$. Sketch the curve.

3. What must be the value of b so that the motion given by the equation $D^2x + bDx + 100x = 0$ is critically damped?

4. What is the displacement of an object which moves according to the equation $D^2x + 4Dx + 6x = 0$ (if $x = 6$ and $Dx = 0$ when $t = 0$) after 2 s?

5. For a certain spring a 4-lb weight stretches it 1/8 ft. With this weight attached, the spring is pulled three inches longer than its equilibrium length and released. Find the equation of the resulting motion, assuming no damping.

6. Find the solution for the spring of Exercise 5 if a damping force numerically equal to the velocity is present.

7. Find the solution for the spring of Exercise 5 if no damping is present, but an external force of $4 \sin 2t$ is acting on the spring.

8. Find the solution for the spring of Exercise 5 if the damping force of Exercise 6 and the impressed force of Exercise 7 are both acting.

9. Find the equation relating the charge and time in an electric circuit with the following elements: $L = 0.2$ H, $R = 8$ Ω, $C = 1$ μF, $E = 0$. In this circuit $q = 0$ and $i = 0.5$ A when $t = 0$.

10. For a given electric circuit, $L = 0.5$ H, $R = 0$, $C = 200$ μF, and $E = 0$. Find the equation relating charge and time if $q = 100$ μC and $i = 0$ when $t = 0$.

11. For a given circuit, $L = 0.1$ H, $R = 0$, $C = 100$ μF, and $E = 100$ V. Find the equation relating the charge and time if $q = 0$ and $i = 0$ when $t = 0$.

12. Find the relation between the current and time for the circuit of Exercise 11.

13. For a given circuit, $L = 0.5$ H, $R = 10$ Ω, $C = 200$ μF, and $E = 100 \sin 200t$. Find the expression relating charge and time.

14. Find the relation between the current and time for the circuit of Exercise 13.

15. Find the steady-state current for the circuit of Exercise 13.

16. Find the steady-state current for a circuit with $L = 1$ H, $R = 5$ Ω, $C = 150$ μF, and $E = 120 \sin 100t$ V.

17. Find the steady-state solution for the charge of the circuit of Exercise 16 if the voltage source is changed to $100 \sin 400t$.

18. Find the steady-state solution for the current in an electric circuit containing the following elements: $C = 20$ μF, $L = 2$ H, $R = 20$ Ω, and $E = 200 \sin 10t$.

14–6 Review Exercises for Chapter 14

In Exercises 1 through 16 find the general solution of each of the given differential equations.

1. $2D^2y + Dy = 0$

2. $2D^2y - 5Dy + 2y = 0$

3. $y'' + 2y' + y = 0$

4. $y'' + 2y' + 2y = 0$

5. $D^2y - 2Dy + 5y = 0$

6. $4D^2y - 36Dy + 81y = 0$

7. $y'' - y' - 56y = 28$

8. $2\dfrac{d^2y}{dx^2} + \dfrac{dy}{dx} - 3y = 6$

9. $\dfrac{d^2y}{dx^2} + 6\dfrac{dy}{dx} + 9y = 3x$

10. $y'' - 3y' + 2y = 2x^2 + 1$

11. $9D^2y - 18Dy + 8y = 16 + 4x$

12. $D^2y + 9y = xe^x$

13. $y'' + y' - y = 2e^x$

14. $y'' + y = 4\cos 2x$

15. $D^2y + 9y = \sin x$

16. $y'' + y' = e^x + \cos 2x$

In Exercises 17 through 20, find the indicated particular solutions of the given differential equations.

17. $\dfrac{d^2y}{dx^2} + \dfrac{dy}{dx} + 4y = 0,\ Dy = \sqrt{15},\ y = 0$ when $x = 0$

18. $5y'' + 7y' - 6y = 0,\ y' = 10,\ y = 2$ when $x = 0$
19. $(D^2 + 4D + 4)y = 4\cos x,\ Dy = 1,\ y = 0$ when $x = 0$
20. $y'' - 2y' + y = e^{2x} + x,\ Dy = 0,\ y = 2$ when $x = 0$

In Exercises 21 through 28 solve the given problems.

21. A 64-lb weight stretches a certain spring 2 ft. With this weight suspended on it, the spring is stretched 2 ft beyond the equilibrium position and released. Find the equation of the resulting motion if the medium in which the weight is suspended retards the motion with a force equal to 16 times the velocity. Classify the motion as underdamped, critically damped, or overdamped.

22. What is the expression for the steady-state displacement x of an object which moves according to the relation $D^2x - 2Dx - 8x = 5\sin 2t$?

23. An 0.5-H inductor, a 6-Ω resistor, and a 20-mF capacitor are connected in series with a generator for which $E = 24\sin 10t$. Find the charge on the capacitor as a function of time if the initial charge and current are zero.

24. Find the general expression for the charge on the capacitor in a circuit having a resistance of 20 Ω, an inductor of 4 H, and a capacitor of 100 μF if the initial charge on the capacitor is 10 mC and the current is zero when $t = 0$.

25. Find the solution for the current for the circuit of Exercise 24 if a battery of 100 V is placed in the circuit.

26. If a circuit contains an inductor of inductance L, capacitor of capacitance C, and a sinusoidal source of voltage $E_0 \sin \omega t$, express the charge q on the capacitor as a function of the time. Assume $q = 0,\ i = 0$ when $t = 0$.

27. The approximate differential equation relating the displacement y of a beam at a horizontal distance x from one end is

$$EI\dfrac{d^2y}{dx^2} = M$$

where E is the modulus of elasticity, I is the moment of inertia of the cross section of the beam perpendicular to its axis, and M is the bending moment at the cross section. If $M = 2000x - 40x^2$ for a particular beam of length L for which $y = 0$ when $x = 0$ and when $x = L$, express y in terms of x. Consider E and I as constants.

28. When a circular disk of mass m and radius r is suspended by a wire at the center on one of its flat faces and the disk is twisted through an angle θ, torsion in the wire tends to turn the disk back in the opposite direction. The differential equation for this case is

$$\frac{1}{2}mr^2\frac{d^2\theta}{dt^2} = -k\theta$$

where k is a constant. Determine the equation of motion if $\theta = \theta_0$ and $d\theta/dt = \omega_0$ when $t = 0$.

15

Other Methods of Solving
Differential Equations

15–1 Numerical Solutions—the Increment Method

In the last two chapters we presented methods of solving certain types of differential equations. Other types of differential equations, many of which do not have exact solutions, have important applications. Numerous methods and techniques have been developed for these equations. In this chapter three of these methods are introduced. *The first of these is the* **increment method,** *a numerical method of solution.*

Numerical methods of solving differential equations often require a great deal of time and computation to arrive at the desired solution. For this reason they lend themselves very well to the use of calculators and computers, by which the calculations may be done very rapidly. Also, due to the importance of computers, numerical methods such as these are of importance in industry. The increment method, which illustrates the basic numerical technique, can be used with nearly any method of computation.

In order to find an approximate solution to a differential equation of the form $dy/dx = f(x, y)$ which passes through a known point (x_1, y_1), we approximate the increment of y, Δy, by the differential of y, dy. From Section 4–1 we recall that this approximation is good if Δx is small. The use of this approximation is illustrated in the following examples.

Example A
Find an approximate solution to the differential equation $dy/dx = x + y$ by the increment method. The curve passes through the point $(0, 1)$.

The differential equation can be written as

$$d\acute{y} = (x + y)\,\Delta x$$

(since $\Delta x = dx$). We know that $y = 1$ for $x = 0$. Let us choose $\Delta x = 0.1$. With these values we can find a good approximation for dy. Thus,

$$dy = (0 + 1)(0.1) = 0.1$$

This tells us that the curve passes (or nearly passes) through the point (0.1, 1.1). This is seen by adding 0.1 to the original value of x, 0, and then adding 0.1 to the original value of y, 1.0. Assuming this point is correct we use the values $x = 0.1$, $y = 1.1$, and $\Delta x = 0.1$ to find the next approximate point. This time

$$dy = (0.1 + 1.1)(0.1) = 0.12$$

which means that the curve passes through (0.2, 1.22). Continuing this process we find a set of points which would approximately satisfy the function which is the solution of the differential equation. Tabulating the results obtained in this manner, we have the following table.

x	y	$x + y$	dy	(correct value of y)
0.0	1.00	1.00	0.10	1.00
0.1	1.10	1.20	0.12	1.11
0.2	1.22	1.42	0.14	1.24
0.3	1.36	1.66	0.17	1.40
0.4	1.53	1.93	0.19	1.58
0.5	1.72	2.22	0.22	1.80
0.6	1.94	2.54	0.25	2.04

In this case we are able to find correct values since the differential equation can be expressed as $dy/dx - y = x$, of which the solution is $y = 2e^x - x - 1$.

In order to show that better values are obtained by using smaller values of Δx, the following table is presented in which $\Delta x = 0.05$ is also used in solving the differential equation

x	$y(\Delta x = 0.1)$	$y(\Delta x = 0.05)$	y(correct)
0.00	1.00	1.00	1.00
0.05		1.05	1.05
0.10	1.10	1.11	1.11
0.15		1.17	1.17
0.20	1.22	1.24	1.24
0.25		1.31	1.32
0.30	1.36	1.39	1.40
0.35		1.47	1.49
0.40	1.53	1.56	1.58
0.45		1.66	1.69
0.50	1.72	1.77	1.80
0.55		1.88	1.92
0.60	1.94	2.00	2.04

(see Fig. 15–1 for the graphs of the curves using these values).

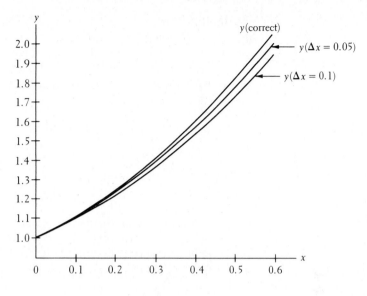

Figure 15–1

Example B

By the increment method, solve the differential equation $dy/dx = 2x^2 - y^2$ if the curve passes through the point $(1, 2)$. Use $\Delta x = 0.1$.

By following the method as outlined in Example A, the values in the following table are found

x	y	$2x^2$	y^2	$2x^2 - y^2$	dy
1.00	2.00	2.00	4.00	-2.00	-0.20
1.10	1.80	2.42	3.24	-0.82	-0.08
1.20	1.72	2.88	2.96	-0.08	-0.01
1.30	1.71	3.38	2.92	0.46	0.05
1.40	1.76	3.92	3.10	0.82	0.08
1.50	1.84	4.50	3.39	1.11	0.11
1.60	1.95	5.12	3.80	1.32	0.13

(see Fig. 15–2 for the graph of this solution).

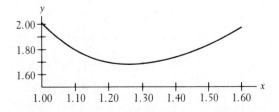

Figure 15–2

Example C

By the increment method, solve the differential equation $dy/dx = \cos x + \sin y$ if the curve passes through the point $(0, 0)$. Use $\Delta x = 0.1$ and construct the graph for $0 \le x \le 0.80$.

By following the method as outlined in Example A, the values in the following table are found

x	y	$\cos x$	$\sin y$	$\cos x + \sin y$	dy
0.00	0.00	1.00	0.00	1.00	0.10
0.10	0.10	1.00	0.10	1.10	0.11
0.20	0.21	0.98	0.21	1.19	0.12
0.30	0.33	0.96	0.32	1.28	0.13
0.40	0.46	0.92	0.44	1.36	0.14
0.50	0.60	0.88	0.56	1.44	0.14
0.60	0.74	0.83	0.67	1.50	0.15
0.70	0.89	0.76	0.78	1.54	0.15
0.80	1.04	0.70	0.86	1.56	0.16

(see Fig. 15–3 for the graph of this solution).

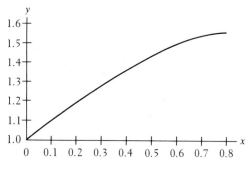

Figure 15–3

This method does have limitations. The values obtained are not particularly accurate if Δx is large, although reasonable accuracy may be obtained with a calculator and small values of Δx. Also, it is seen from Example A that the values vary more and more from the true values the farther the values of x are from that of the original given point. However, it is very useful in demonstrating the method of solving a differential equation numerically by successive approximations. Most such methods follow a similar techique.

Exercises 15–1

In Exercises 1 through 4 solve the given differential equations by the increment method for the specified values of x and Δx. Check the results against known values by solving the differential equations exactly. Plot the graphs of the solutions.

1. $\dfrac{dy}{dx} = x + 1, \quad y = 1$ when $x = 0, \quad 0 \le x \le 1, \quad \Delta x = 0.2$

2. Solve the differential equation of Exercise 1 if $\Delta x = 0.1$.

3. $\dfrac{dy}{dx} = y + e^x$, $y = 0$ when $x = 0$, $0 \le x \le 0.5$, $\Delta x = 0.1$

4. Solve the differential equation of Exercise 3 if $\Delta x = 0.05$.

In Exercises 5 through 12 solve the given differential equations by the increment method for the specified values of x and Δx. Plot the graphs of the solutions.

5. $\dfrac{dy}{dx} = xy + 1$, $y = 0$ when $x = 0$, $0 \le x \le 1$, $\Delta x = 0.1$

6. $\dfrac{dy}{dx} = x^2 + y^2$, $y = 1$ when $x = 0$, $0 \le x \le 0.8$, $\Delta x = 0.1$

7. $\dfrac{dy}{dx} = e^{xy}$, $y = 0$ when $x = 0$, $0 \le x \le 1.2$, $\Delta x = 0.2$

8. $\dfrac{dy}{dx} = \sqrt{1 + xy}$, $y = 1$ when $x = 0$, $0 \le x \le 0.5$, $\Delta x = 0.05$

9. $\dfrac{dy}{dx} = \cos(x + y)$, $y = \pi/2$ when $x = 0$, $0 \le x \le 1$, $\Delta x = 0.1$

10. $\dfrac{dy}{dx} = y + \sin x$, $y = 0$ when $x = 0.5$, $0.5 \le x \le 1.5$, $\Delta x = 0.1$

11. The equation of Example A with $\Delta x = 0.02$.

12. The equation of Example B with $\Delta x = 0.02$.

In Exercises 13 through 16 solve the given problems.

13. Evaluate the solution of the differential equation $y' = \ln(xy)$ for $x = 1.7$ if the solution passes through $(1, 1)$. Use $\Delta x = 0.1$.

14. A 3-slug object moves through a resisting medium. A constant force of 10 lb caused the motion, and a retarding force of $2v^{3/2}$ is present, where v is the velocity. The resulting differential equation is

$$3\frac{dv}{dt} = 10 - 2v^{3/2}$$

By the increment method graph the solution for v as a function of t for $0 \le t \le 2$ s, using $\Delta t = 0.2$ s, if the object starts from rest. The velocity is in feet per second.

15. An electric circuit contains a 1-H inductor, a 2-Ω resistor and a voltage source of $\sin t$. The resulting differential equation relating current and time is

$$\frac{di}{dt} + 2i = \sin t$$

Find i after 0.5 s by the increment method if the initial current is zero. Solve the equation exactly and compare the values.

16. An object is being heated such that the rate of change of temperature with respect to time is

$$\frac{dT}{dt} = \sqrt[3]{1 + t^3}$$

Find the temperature T (in degrees Celsius) for $t = 5$ min by using the increment method with $\Delta t = 1.0$ min, if the initial temperature is 0°C.

15–2 A Method of Successive Approximations

The second method of solving a differential equation to be discussed in this chapter is one based on successive approximations. In the previous section we saw how to develop a numerical solution based essentially on successive approximations. In this section we shall demonstrate a method that leads to a function which approximates the solution.

The method generally applies to a differential equation of the form $y' = f(x, y)$. If one point on the solution curve is known, the value of y is substituted into the right side, thereby giving a function of x alone. This equation can then be integrated, yielding y as a function of x, this solution being referred to as the first approximation. By substituting this expression into the right side for y, integrating again, a second approximation may be obtained. This process may be repeated as often as required, each time the solution for y being a better approximation than the last. The following example illustrates the method.

Example A

Obtain the first three approximations to the solution of the differential equation $dy = (x + y)\, dx$, if the solution curve passes through $(0, 1)$.

By substituting $y = 1$ in the right side of the differential equation, we have

$$dy = (x + 1)\, dx$$

Integrating, we obtain the first approximation

$$y_1 = \frac{x^2}{2} + x + 1$$

where the constant of integration has been evaluated as being 1 from the fact that the curve passes through $(0, 1)$. Substituting the expression for y_1 in the differential equation, we have

$$dy = \left(x + \frac{x^2}{2} + x + 1\right) dx = \left(\frac{x^2}{2} + 2x + 1\right) dx$$

$$y_2 = \frac{x^3}{6} + x^2 + x + 1$$

where the constant of integration again was evaluated to be 1, and y_2 is the second approximation. Substituting the expression for y_2 in the differential equation, we have

$$dy = \left(x + \frac{x^3}{6} + x^2 + x + 1\right) dx = \left(\frac{x^3}{6} + x^2 + 2x + 1\right) dx$$

$$y_3 = \frac{x^4}{24} + \frac{x^3}{3} + x^2 + x + 1$$

where, again, the constant of integration is 1 and y_3 is the third approximation.

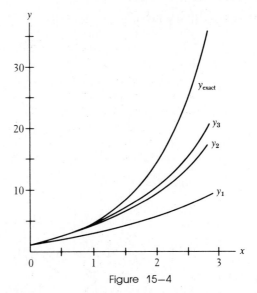

Figure 15–4

We note that this differential equation is the same as that of Example A of the previous section. Therefore, the exact solution is

$$y = 2e^x - x - 1$$

Figure 15–4 shows the graph of the exact solution and each of the three approximations. The accuracy of these approximations can be seen on the graph.

Example B

Find the first two approximations of the solution of the differential equation $y' = 2 - xy^2$ if the solution curve passes through $(0, 1)$.

Writing the equation in differential form and substituting 1 for y, we have

$$dy = (2 - x) \, dx$$

which when integrated (the constant of integration is 1) gives

$$y_1 = 1 + 2x - \frac{x^2}{2}$$

Substituting this expression for y in the differential equation, we have

$$dy = \left[2 - x\left(1 + 2x - \frac{x^2}{2} \right)^2 \right] dx$$

$$= \left(2 - x - 4x^2 - 3x^3 + 2x^4 - \frac{x^5}{4} \right) dx$$

$$y_2 = 1 + 2x - \frac{x^2}{2} - \frac{4x^3}{3} - \frac{3x^4}{4} + \frac{2x^5}{5} - \frac{x^6}{24}$$

where the constant of integration is 1 and y_2 is the desired approximation.

Example C

Find the third approximation to the solution of the differential equation $y' = y + \sin x$, if the solution curve passes through $(0, \frac{1}{2})$. Evaluate the function for $x = 1$.

Following the method as previously illustrated we have the following steps:

$$dy = (\tfrac{1}{2} + \sin x)\, dx$$

$$y_1 = \frac{x}{2} - \cos x + C_1$$

Since the curve passes through $(0, \frac{1}{2})$ we have

$$\frac{1}{2} = 0 - 1 + C_1 \quad \text{or} \quad C_1 = \frac{3}{2}$$

Thus

$$y_1 = \frac{3}{2} + \frac{x}{2} - \cos x$$

$$dy = \left(\frac{3}{2} + \frac{x}{2} - \cos x + \sin x\right) dx$$

$$y_2 = \frac{3x}{2} + \frac{x^2}{4} - \sin x - \cos x + C_2$$

Here $C_2 = \frac{3}{2}$, which means

$$y_2 = \frac{3}{2} + \frac{3x}{2} + \frac{x^2}{4} - \sin x - \cos x$$

$$dy = \left(\frac{3}{2} + \frac{3x}{2} + \frac{x^2}{4} - \cos x\right) dx$$

$$y_3 = \frac{3x}{2} + \frac{3x^2}{4} + \frac{x^3}{12} - \sin x + C_3$$

Here $C_3 = \frac{1}{2}$, which means

$$y_3 = \frac{1}{2} + \frac{3x}{2} + \frac{3x^2}{4} + \frac{x^3}{12} - \sin x$$

Evaluating y_3 for $x = 1$, we have

$$y_3 = \frac{1}{2} + \frac{3}{2} + \frac{3}{4} + \frac{1}{12} - 0.84 = 1.99$$

This function can be integrated exactly by methods developed in Chapter 13. The value found from the exact solution for $x = 1$ is 2.03. Also it should be noted that the constant of integration was not the same for each of the approximations.

The method of this section has the disadvantage that it can be applied only if the resulting differentials can be integrated. This naturally limits the equations which can be solved in this manner. However, for those to which it is applicable, it often provides a quicker and more accurate value for the solution for a specified value of x than the method of the previous section.

Exercises 15–2

In Exercises 1 through 4 find the indicated approximation of the solution of each of the given differential equations for which the solution curve passes through the given point. Solve each exactly and evaluate the solutions for the indicated value of x.

1. $y' = y^2$, y_2, $(0, 1)$, $x = 0.1$ 2. $y' = x - y$, y_2, $(0, 1)$, $x = \frac{1}{2}$

3. $y' = 2x(1 + y)$, y_3, $(0, 0)$, $x = 1$ 4. $y' = y + e^x$, y_3, $(0, 0)$, $x = 0.2$

In Exercises 5 through 8 find the indicated approximation of the solution of each of the given differential equations for which the solution curve passes through the given point.

5. $y' = x + y^2$, y_3, $(0, 0)$ 6. $y' = 1 + xy^2$, y_2, $(0, 0)$

7. $y' = y - \cos 2x$, y_2, $(0, 1)$ 8. $y' = x + y \cos x$, y_2, $(0, 1)$

In Exercises 9 through 12 solve the given problems.

9. Substitute the first few terms of the Maclaurin series for e^x in the expression $2e^x - x - 1$ and compare the results with the solution for Example A.

10. Substitute the first few terms for e^{x^2} in the expression for the exact solution to the equation of Exercise 3, and compare the results with the approximate solution.

11. A 1-slug object moves through a resisting medium such that the motion is resisted by a force numerically equal to the cube of the velocity. If a 2-lb force causes the motion, the resulting differential equation is

$$\frac{dv}{dt} = 2 - v^3$$

Find the velocity when $t = 0.3$ s if the object starts from rest. Use the second approximation of the solution.

12. Evaluate i for $t = 0.5$ s for the differential equation in Exercise 15 of Section (15–1) by finding i_2. Compare the results with those found in the indicated exercise.

15–3 Laplace Transforms

The final method of solving differential equations which we shall discuss is by means of **Laplace transforms**. As we shall see, *Laplace transforms provide an algebraic method of obtaining a **particular** solution of a differential equation from stated initial conditions.* Since this is often what is desired in practice, Laplace transforms are often preferred for the solution of differential equations in engineering and electronics. In this section we shall discuss the meaning of the transform and its operations.

These methods will be applied to solving differential equations in the following section. The treatment in this text is intended only as an introduction to the topic of Laplace transforms.

The Laplace transform of a function $f(t)$ is defined as the function $F(s)$ by the equation

$$F(s) = \int_0^\infty e^{-st} f(t)\, dt \tag{15-1}$$

By writing the transform as $F(s)$ we show that the result of integrating and evaluating is a function of s. To denote that we are dealing with "the Laplace transform of the function $f(t)$" the notation $L(f)$ is used. Thus,

$$F(s) = L(f) = \int_0^\infty e^{-st} f(t)\, dt \tag{15-2}$$

We will see that both notations are quite useful.

A note regarding the form of the integral in Eqs. (15–1) and (15–2) is in order at this point. Since the upper limit is ∞, which means that it is unbounded, this integral is one type of what is known as an **improper integral**. In evaluating this integral at the upper limit, it is necessary to find the limit of the resulting function as the upper limit approaches infinity. This may be denoted by

$$\lim_{c \to \infty} \int_0^c e^{-st} f(t)\, dt$$

where we substitute c for t in the resulting function and determine the limit as $c \to \infty$ to determine the result for the upper limit.

Example A
Find the Laplace transform of the function $f(t) = t$, $t > 0$.
By the definition of the Laplace transform

$$L(f) = L(t) = \int_0^\infty e^{-st} t\, dt$$

This may be integrated by parts or by formula (44) of Table 6 in Appendix C. Using the formula we have

$$L(t) = \int_0^\infty te^{-st}\, dt = \lim_{c \to \infty} \int_0^c te^{-st}\, dt = \lim_{c \to \infty} \left. \frac{e^{-st}(-st-1)}{s^2} \right|_0^c$$

$$= \lim_{c \to \infty} \left[\frac{e^{-sc}(-sc-1)}{s^2} \right] + \frac{1}{s^2}$$

Now, as $c \to \infty$, $e^{-sc} \to 0$ and $sc \to \infty$. However, although we cannot prove it here, $e^{-sc} \to 0$ much faster than $sc \to \infty$. We can see that this is reasonable, for $ce^{-c} = 4.5 \times 10^{-4}$ for $c = 10$ and $ce^{-c} = 3.7 \times 10^{-42}$ for $c = 100$. Thus, the value at the upper limit approaches zero, which means the limit is zero. This means that

$$L(t) = \frac{1}{s^2}$$

Example B

Find the Laplace transform of the function $f(t) = \cos at$.

By definition

$$L(f) = L(\cos at) = \int_0^\infty e^{-st}\cos at \, dt$$

Using formula (50) of Table 6 we have

$$L(\cos at) = \int_0^\infty e^{-st}\cos at \, dt = \lim_{c \to \infty} \int_0^c e^{-st}\cos at \, dt$$

$$= \lim_{c \to \infty} \frac{e^{-st}(-s\cos at + a\sin at)}{s^2 + a^2}\Big|_0^c$$

$$= \lim_{c \to \infty} \frac{e^{-sc}(-s\cos ac + a\sin ac)}{s^2 + a^2} - \left(-\frac{s}{s^2 + a^2}\right)$$

$$= 0 + \frac{s}{s^2 + a^2}$$

$$= \frac{s}{s^2 + a^2}$$

Therefore, the Laplace transform of the function $\cos at$ is

$$L(\cos at) = \frac{s}{s^2 + a^2}$$

In both examples the resulting transform was an algebraic function of s.

An important property of Laplace transforms is the **linearity property**

$$L[af(t) + bg(t)] = aL(f) + bL(g) \tag{15-3}$$

We state this property here since it determines that the transform of a sum of functions is the sum of the transforms. This is of definite importance when dealing with a sum of functions. This property is a direct result of the definition of the Laplace transform.

We now present a short table of Laplace transforms. The transforms given here are sufficient for our remaining work in this chapter. Much more complete tables are available in many standard reference sources.

Table of Laplace Transforms

	$f(t) = L^{-1}(F)$	$L(f) = F(s)$
1.	1	$\dfrac{1}{s}$
2.	$\dfrac{t^{n-1}}{(n-1)!}$	$\dfrac{1}{s^n}\,(n = 1, 2, 3, \ldots)$
3.	e^{-at}	$\dfrac{1}{s+a}$
4.	$1 - e^{-at}$	$\dfrac{a}{s(s+a)}$
5.	$\cos at$	$\dfrac{s}{s^2+a^2}$
6.	$\sin at$	$\dfrac{a}{s^2+a^2}$
7.	$1 - \cos at$	$\dfrac{a^2}{s(s^2+a^2)}$
8.	$at - \sin at$	$\dfrac{a^3}{s^2(s^2+a^2)}$
9.	$e^{-at} - e^{-bt}$	$\dfrac{b-a}{(s+a)(s+b)}$
10.	$ae^{-at} - be^{-bt}$	$\dfrac{s(a-b)}{(s+a)(s+b)}$
11.	te^{-at}	$\dfrac{1}{(s+a)^2}$
12.	$t^{n-1}e^{-at}$	$\dfrac{(n-1)!}{(s+a)^n}$
13.	$e^{-at}(1 - at)$	$\dfrac{s}{(s+a)^2}$
14.	$[(b-a)t + 1]e^{-at}$	$\dfrac{s+b}{(s+a)^2}$
15.	$\sin at - at \cos at$	$\dfrac{2a^3}{(s^2+a^2)^2}$
16.	$t \sin at$	$\dfrac{2as}{(s^2+a^2)^2}$

	$f(t) = L^{-1}(F)$	$L(f) = F(s)$
17.	$\sin at + at \cos at$	$\dfrac{2as^2}{(s^2 + a^2)^2}$
18.	$t \cos at$	$\dfrac{s^2 - a^2}{(s^2 + a^2)^2}$
19.	$e^{-at}\sin bt$	$\dfrac{b}{(s + a)^2 + b^2}$
20.	$e^{-at}\cos bt$	$\dfrac{s + a}{(s + a)^2 + b^2}$

Another Laplace transform important to the solution of a differential equation is the transform of the derivative of a function. Let us first find the Laplace transform of the first derivative of a function.

By definition

$$L(f') = \int_0^\infty e^{-st}f'(t)\,dt$$

To integrate by parts, let $u = e^{-st}$ and $dv = f'(t)\,dt$, so $du = -se^{-st}dt$ and $v = f(t)$ [the integral of the derivative of a function is the function]. Therefore,

$$L(f') = e^{-st}f(t)\Big|_0^\infty + s\int_0^\infty e^{-st}f(t)\,dt$$

$$= 0 - f(0) + sL(f)$$

It is noted that the integral in the second term on the right is the Laplace transform of $f(t)$ by definition. Therefore, the Laplace transform of the first derivative of a function is

$$L(f') = sL(f) - f(0) \tag{15–4}$$

Applying the same analysis, we may find the Laplace transform of the second derivative of a function. It is

$$L(f'') = s^2L(f) - sf(0) - f'(0) \tag{15–5}$$

Here it is necessary to integrate by parts twice to derive the result. The transforms of higher derivatives are found in a similar manner.

Equations (15–4) and (15–5) allow us to express the transform of each derivative in terms of s and the transform of the function itself.

Example C

If $f(0) = 0$ and $f'(0) = 1$, express the transform of $f''(t) - 2f'(t)$ in terms of s and the transform of $f(t)$.

By the linearity property, Eq. (15–3), we have

$$L[f''(t) - 2f'(t)] = L(f'') - 2L(f')$$
$$= [s^2 L(f) - s(0) - 1] - 2[sL(f) - 0]$$
$$= (s^2 - 2s)L(f) - 1$$

If the Laplace transform of a function is known, it is then possible to find the function by finding the **inverse transform**

$$L^{-1}(F) = f(t) \qquad\qquad (15\text{–}6)$$

where L^{-1} denotes the inverse transform.

Example D

If $F(s) = \dfrac{s}{(s^2 + a^2)}$, from transform (5) of the table we see that

$$L^{-1}(F) = L^{-1}\left(\frac{s}{s^2 + a^2}\right) = \cos at$$

or

$$f(t) = \cos at$$

Example E

If $(s^2 - 2s)L(f) - 1 = 0$, then

$$L(f) = \frac{1}{s^2 - 2s} \quad \text{or} \quad F(s) = \frac{1}{s(s - 2)}$$

Thus,

$$L^{-1}(F) = f(t) = -\frac{1}{2}(1 - e^{2t})$$

from transform (4) of the table.

Example F

If $F(s) = \dfrac{(s + 5)}{(s^2 + 6s + 10)}$, then

$$L^{-1}(F) = L^{-1}\left[\frac{s + 5}{s^2 + 6s + 10}\right]$$

It appears that this function does not fit any of the forms given. However,

$$s^2 + 6s + 10 = (s^2 + 6s + 9) + 1 = (s + 3)^2 + 1$$

By writing $F(s)$ as

$$F(s) = \frac{(s + 3) + 2}{(s + 3)^2 + 1} = \frac{s + 3}{(s + 3)^2 + 1} + \frac{2}{(s + 3)^2 + 1}$$

we can find the inverse of each term. Therefore,

$$L^{-1}(F) = e^{-3t}\cos t + 2e^{-3t}\sin t$$

or

$$f(t) = e^{-3t}(\cos t + 2 \sin t)$$

Exercises 15–3

In Exercises 1 through 4 verify the indicated transforms given in the table.

1. Transform 1   2. Transform 3   3. Transform 6   4. Transform 11

In Exercises 5 through 12 find the transforms of the given functions by use of the table.

5. $f(t) = e^{3t}$

6. $f(t) = 1 - \cos 2t$

7. $f(t) = t^3 e^{-2t}$

8. $f(t) = 2e^{-3t}\sin 4t$

9. $f(t) = \cos 2t - \sin 2t$

10. $f(t) = 2t \sin 3t + e^{-3t}\cos t$

11. $f(t) = 3 + 2t \cos 3t$

12. $f(t) = t^3 - 3te^{-t}$

In Exercises 13 through 16 express the transforms of the given expressions in terms of s and $L(f)$.

13. $y'' + y', f(0) = 0, f'(0) = 0$

14. $y'' - 3y', f(0) = 2, f'(0) = -1$

15. $2y'' - y' + y, f(0) = 1, f'(0) = 0$

16. $y'' - 3y' + 2y, f(0) = -1, f'(0) = 2$

In Exercises 17 through 24 find the inverse transform of the given functions of s.

17. $F(s) = \dfrac{2}{s^3}$

18. $F(s) = \dfrac{3}{s^2 + 4}$

19. $F(s) = \dfrac{1}{s + 5}$

20. $F(s) = \dfrac{3}{s^4 + 4s^2}$

21. $F(s) = \dfrac{1}{s^3 + 3s^2 + 3s + 1}$

22. $F(s) = \dfrac{s^2 - 1}{s^4 + 2s^2 + 1}$

23. $F(s) = \dfrac{s + 2}{(s^2 + 9)^2}$

24. $F(s) = \dfrac{s + 3}{s^2 + 4s + 13}$

15–4 Solving Differential Equations by Laplace Transforms

We will now show how certain differential equations can be solved by the use of Laplace transforms. It must be remembered that these solutions are the *particular* solutions of the equations subject to the given conditions. The necessary operations were developed in the preceding section. The following examples illustrate the method.

Example A

Solve the differential equation $2y' - y = 0$ if $y(0) = 1$. (Note that we are using y to denote the function.)

Taking transforms of each term in the equation, we have

$$L(2y') - L(y) = L(0), \qquad 2L(y') - L(y) = 0$$

$L(0) = 0$ by direct use of the definition of the transform. Now using Eq. (15–4), $L(y') = sL(y) - y(0)$, we have

$$2[sL(y) - 1] - L(y) = 0$$

Solving for $L(y)$ we obtain

$$2sL(y) - L(y) = 2, \qquad L(y) = \frac{2}{2s - 1} = \frac{1}{s - 1/2}$$

Finding the inverse transform of $F(s) = 1/(s - \frac{1}{2})$, we have

$$y = e^{t/2}$$

The reader should check this solution with that obtained by methods developed earlier. Also it should be noted that the solution was essentially an algebraic one. This points out the power and usefulness of Laplace transforms. We are able to translate a differential equation into an algebraic form, which can in turn be translated into the solution of the differential equation. Thus, we are able to solve a differential equation by use of algebra and specific algebraic forms.

Example B

Solve the differential equation $y'' + 2y' + 2y = 0$, if $y(0) = 0$ and $y'(0) = 1$.

Using the same steps as outlined in Example A, we have the following solution:

$$L(y'') + 2L(y') + 2L(y) = 0$$
$$[s^2 L(y) - sy(0) - y'(0)] + 2[sL(y) - y(0)] + 2L(y) = 0$$
$$s^2 L(y) - 1 + 2sL(y) + 2L(y) = 0$$
$$(s^2 + 2s + 2)L(y) = 1$$

$$L(y) = \frac{1}{s^2 + 2s + 2} = \frac{1}{(s + 1)^2 + 1}$$

$$y = e^{-t}\sin t$$

Example C

Solve the differential equation $y'' + y = \cos t$ if $y(0) = 1$ and $y'(0) = 2$.

Taking transforms of each term in the equation, we have

$$L(y'') + L(y) = L(\cos t)$$

The transform on the left is found by use of Eq. (15—5) and the transform on the right is found by direct use of transform (5) in the table. Therefore, we have

$$[s^2 L(y) - s(1) - 2] + L(y) = \frac{s}{s^2 + 1}$$

Solving for $L(y)$, we obtain

$$(s^2 + 1)L(y) = \frac{s}{s^2 + 1} + s + 2$$

$$L(y) = \frac{s}{(s^2 + 1)^2} + \frac{s}{s^2 + 1} + \frac{2}{s^2 + 1}$$

Finding the inverse-transform of each expression completes the solution:

$$y = \frac{t}{2}\sin t + \cos t + 2 \sin t$$

Example D

An electric circuit contains a 1 H inductor, a 10-Ω resistor, and a battery whose voltage is 6 V. Find the current as a function of the time, if the initial current is zero.

The differential equation for this circuit is

$$\frac{di}{dt} + 10i = 6$$

Following the procedures outlined in the previous examples, the solution is found:

$$L\left(\frac{di}{dt}\right) + 10L(i) = L(6), \qquad [sL(i) - 0] + 10L(i) = \frac{6}{s}$$

$$L(i) = \frac{6}{s(s + 10)}$$

$$i = 0.6(1 - e^{-10t})$$

Example E

A spring is stretched 1 ft by a weight of 16 lb (mass of $\frac{1}{2}$ slug). The medium resists the motion of the object with a force of $4v$, where v is the velocity. The differential equation describing the motion is

$$\frac{1}{2}\frac{d^2y}{dt^2} + 4\frac{dy}{dt} + 16y = 0$$

Find y, the displacement of the object, as a function of the time if $y(0) = 1$ and $dy/dt = 0$ when $t = 0$.

Clearing fractions and denoting the derivatives by y'' and y', the differential equation becomes

$$y'' + 8y' + 32y = 0$$

Following the procedure of solving this type of an equation by use of Laplace transforms, we have the following steps:

$$L(y'') + 8L(y') + 32L(y) = 0$$
$$[s^2L(y) - s(1) - 0] + 8[sL(y) - 1] + 32L(y) = 0$$
$$(s^2 + 8s + 32)L(y) = s + 8$$

$$L(y) = \frac{s + 8}{(s + 4)^2 + 4^2} = \frac{s + 4}{(s + 4)^2 + 4^2} + \frac{4}{(s + 4)^2 + 4^2}$$

$$y = e^{-4t}\cos 4t + e^{-4t}\sin 4t$$
$$= e^{-4t}(\cos 4t + \sin 4t)$$

Exercises 15−4

In the following exercises solve the given differential equations by use of Laplace transforms, where the function is subject to the given conditions.

1. $y' + y = 0$, $y(0) = 1$
2. $y' - 2y = 0$, $y(0) = 2$
3. $2y' - 3y = 0$, $y(0) = -1$
4. $y' + 2y = 1$, $y(0) = 0$
5. $y' + 3y = e^{-3t}$, $y(0) = 1$
6. $y' + 2y = te^{-2t}$, $y(0) = 0$
7. $y'' + 4y = 0$, $y(0) = 0$, $y'(0) = 1$
8. $y'' - 4y = 0$, $y(0) = 2$, $y'(0) = 0$
9. $y'' + 2y' = 0$, $y(0) = 0$, $y'(0) = 2$
10. $y'' + 2y' + y = 0$, $y(0) = 0$, $y'(0) = -2$
11. $y'' - 4y' + 5y = 0$, $y(0) = 1$, $y'(0) = 2$
12. $4y'' + 4y' + y = 0$, $y(0) = 1$, $y'(0) = 0$
13. $y'' + y = 1$, $y(0) = 1$, $y'(0) = 1$
14. $y'' + 4y = 2t$, $y(0) = 0$, $y'(0) = 0$
15. $y'' + 2y' + y = e^{-t}$, $y(0) = 1$, $y'(0) = 2$
16. $y'' + 4y = \sin 2t$, $y(0) = 0$, $y'(0) = 0$
17. A constant force of 6 lb moves a 2-slug mass through a medium which resists the motion with a force equal to v, where v is the velocity. The differential equation relating the velocity and time is

$$2\frac{dv}{dt} = 6 - v$$

Find v as a function of time if the object starts from rest.

18. A 100-Ω resistor and a 400-μF capacitor make up an electric circuit. If the initial charge on the capacitor is 10 mC, find the charge on the capacitor as a function of time after the switch is closed.

19. A 50-Ω resistor, a 4-μF capacitor, and a 40-V battery are connected in series. Find the charge as a function of the time if the initial charge on the capacitor is zero.

20. A 2-H inductor, an 80-Ω resistor, and an 8-V battery are connected in series. Find the current in the circuit as a function of time if the initial current is zero.

21. A 10-H inductor, a 40-μF capacitor, and a voltage supply whose voltage is given by 100 sin 50t are connected in series in an electric circuit. Find the current as a function of time if the initial charge on the capacitor and the initial current are zero.

22. A spring is such that it is stretched 6 in. by an 8-lb (1/4-slug) weight. The spring is stretched 6 in. below the equilibrium position with the weight on it and is given a velocity of 4 ft/s. If the spring is in a medium which resists the motion with a force equal to 4v, find the displacement y as a function of time.

15–5 Review Exercises for Chapter 15

In Exercises 1 through 8 solve the given differential equations by the increment method for the specified values of x and Δx and for which the solution curve passes through the given point.

1. $y' = x^2 - y^2$, $0 \le x \le 0.5$, $\Delta x = 0.1$, $(0, 1)$

2. $y' = \dfrac{1}{1 + x + y}$, $0 \le x \le 1$, $\Delta x = 0.2$, $(0, 0)$

3. $y' = \sqrt{x + y}$, $0 \le x \le 2$, $\Delta x = 0.2$, $(0, 1)$
4. $y' = x^2 + \sin y$, $0 \le x \le 1$, $\Delta x = 0.1$, $(0, 1)$
5. $y' = \sin x + \tan y$, $0 \le x \le 0.7$, $\Delta x = 0.1$, $(0, 0.5)$
6. $y' = e^{2x} + \sin xy$, $0 \le x \le 1$, $\Delta x = 0.1$, $(0, 0)$
7. The equation of Exercise 1 with $\Delta x = 0.02$
8. The equation of Exercise 2 with $\Delta x = 0.05$

In Exercises 9 through 12 find the indicated approximation of the solution of each of the given differential equations for which the solution curve passes through the given point.

9. $y' = y + x^2$, y_3, $(0, 1)$
10. $y' = \cos x - y$, y_3, $(0, 1)$
11. $y' = 1 + x^2 y$, y_2, $(1, 1)$
12. $y' = y^2 - \sin x$, y_2, $(0, 2)$

In Exercises 13 through 20 solve the given differential equations by use of Laplace transforms where the function is subject to the given conditions.

13. $4y' - y = 0, y(0) = 1$
14. $2y' - y = 4, y(0) = 1$
15. $y' - 3y = e^t, y(0) = 0$

16. $y' + 2y = e^{-2t}$, $y(0) = 2$

17. $y'' + y = 0$, $y(0) = 0$, $y'(0) = -4$

18. $y'' + 4y' + 5y = 0$, $y(0) = 1$, $y'(0) = 1$

19. $y'' + 9y = 3t$, $y(0) = 0$, $y'(0) = -1$

20. $y'' + 4y' + 4y = 2e^{-2t}$, $y(0) = 0$, $y'(0) = 1$

In Exercises 21 through 32 solve the given problems.

21. Find the second approximation to the solution of the differential equation $y' = 1 + ye^{-x}$ if the solution curve passes through $(0, 1)$. Evaluate y_2 for $x = 0.5$.

22. Repeat the instructions of Exercise 21 for $y' = 2x/(x^2 + e^y)$.

23. Solve $y' = 1 + ye^{-x}$ by the increment method for $0 \leq x \leq 1$ using $\Delta x = 0.1$ if the solution curve passes through $(0, 1)$. Compare the value for $x = 0.5$ with that found in Exercise 21.

24. Repeat the instructions of Exercise 23 for $y' = 2x/(x^2 + e^y)$ and compare the result with that found in Exercise 22.

25. The differential equation relating current and time for a certain electric circuit is

$$2\frac{di}{dt} + i = 12$$

Solve this equation by use of Laplace transforms given that the initial current is zero. Evaluate the current for $t = 0.3$ s.

26. A 6-H inductor and a 30-Ω resistor are connected in series. Find the current as a function of time if the initial current is 15 A. Use Laplace transforms.

27. A 0.25-H inductor, a 4-Ω resistor, and a 100-μF capacitor are connected in series. If the initial charge on the capacitor is 400 μC and the initial current is zero, find the charge on the capacitor as a function of time. Use Laplace transforms.

28. An inductor of 0.5 H, a resistor of 6 Ω, and a capacitor of 200 μF are connected in series. If the initial charge on the capacitor is 10 mC and the initial current is zero, find the charge on the capacitor as a function of time after the switch is closed. Use Laplace transforms.

29. Use the increment method to find the current for $t = 0.3$ s for the equation of Exercise 25. Use $\Delta t = 0.05$ s.

30. Find the third approximation for i as a function of t for the equation of Exercise 25. Evaluate i for $t = 0.3$ s and compare this value with those found in Exercises 25 and 29.

31. A certain spring is such that an 8-lb weight (1/4-slug mass) stretches it 6 in. An external force of $\cos 8t$ is applied to the spring. Express the displacement y of the object as a function of time if the initial displacement and velocity are zero.

32. A spring is stretched 1 m by a mass of 5 kg (force of gravity of 50 N). The medium resists the motion of the object with a force of $10v$, where v is the velocity. Find y, the displacement of the object, as a function of time if $y(0) = 1$ and $dy/dt = 0$ when $t = 0$.

Appendices

Appendix A

Formulas from Geometry

For the indicated figures, the following symbols are used: A = area, B = area of base, c = circumference, S = lateral area, V = volume.

1. **Triangle.** $A = \frac{1}{2}bh$ (Fig. 1)
2. **Pythagorean theorem.** $c^2 = a^2 + b^2$ (Fig. 2)
3. **Parallelogram.** $A = bh$ (Fig. 3)
4. **Trapezoid.** $A = \frac{1}{2}(a + b)h$ (Fig. 4)
5. **Circle.** $A = \pi r^2$, $c = 2\pi r$ (Fig. 5)
6. **Rectangular solid.** $A = 2(lw + lh + wh)$, $V = lwh$ (Fig. 6)
7. **Cube.** $A = 6e^2$, $V = e^3$ (Fig. 7)
8. **Any cylinder or prism with parallel bases.** $V = Bh$ (Fig. 8)
9. **Right circular cylinder.** $S = 2\pi rh$, $V = \pi r^2 h$ (Fig. 9)
10. **Any cone or pyramid.** $V = \frac{1}{3}Bh$ (Fig. 10)
11. **Right circular cone.** $S = \pi rs$, $V = \frac{1}{3}\pi r^2 h$ (Fig. 11)
12. **Sphere.** $A = 4\pi r^2$, $V = \frac{4}{3}\pi r^3$ (Fig. 12)

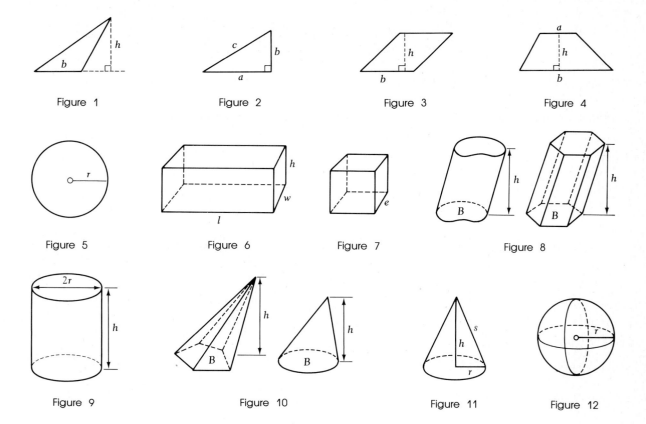

Figure 1

Figure 2

Figure 3

Figure 4

Figure 5

Figure 6

Figure 7

Figure 8

Figure 9

Figure 10

Figure 11

Figure 12

Appendix B

Units of Measurement

Table 1. Quantities and Their Associated Units

Quantity	Quantity Symbol	Unit				
		British		Metric (SI)		
		Name	Symbol	Name	Symbol	In terms of other SI units
Length	s	foot	ft	**meter**	m	
Mass	m	slug		**kilogram**	kg	
Force	F	pound	lb	newton	N	$m \cdot kg/s^2$
Time	t	second	s	**second**	s	
Area	A		ft²		m²	
Volume	V		ft³		m³	
Capacity	V	gallon	gal	liter	L	$(1\ L = 1\ dm^3)$
Velocity	v		ft/s		m/s	
Acceleration	a		ft/s²		m/s²	
Density	d, ρ		lb/ft³		kg/m³	
Pressure	p		lb/ft²	pascal	Pa	N/m²
Energy, work	E, W		ft · lb	joule	J	$N \cdot m$
Power	P	horsepower	hp	watt	W	J/s
Period	T		s		s	
Frequency	f		l/s	hertz	Hz	l/s
Angle	θ	radian	rad	radian	rad	
Electric current	I, i	ampere	A	**ampere**	A	
Electric charge	q	coulomb	C	coulomb	C	$A \cdot s$
Electric potential	V, E	volt	V	volt	V	$J/A \cdot s$
Capacitance	C	farad	F	farad	F	s/Ω
Inductance	L	henry	H	henry	H	$\Omega \cdot s$
Resistance	R	ohm	Ω	ohm	Ω	V/A
Thermodynamic Temperature	T			kelvin	K	
Temperature	T	Fahrenheit degree	°F	degree Celsius	°C	$(1°C = 1\ K)$
Quantity of heat	Q	British thermal unit	Btu	joule	J	
Amount of substance	n			**mole**	mol	
Luminous intensity	I	candlepower	cp	**candela**	cd	

Special notes:

1. The metric units used here are those of the **International System of Units** which is referred to as SI.
2. The SI base units are shown in boldface type.
3. The units symbols shown above are those which are used in the text. Many of them were adopted with the adoption of the SI system. This means, for example, that we use s rather than sec for seconds, and A rather than amp for amperes. Also, other units such as volt are not spelled out, a common practice in the past. When a given unit is used with both systems, we use the SI symbol for the unit.
4. The liter and degree Celsius are not actually SI units. However, they are recognized for use with the SI system due to their practical importance. Also, the symbol for liter has several variations. Presently L is recognized for use in the United States and Canada, l is recognized by the International Committee of Weights and Measures, and *l* is also recognized for use in several countries.
5. For *temperature intervals*, $1°C = 1$ K, whereas the temperature in kelvins equals the temperature in degrees Celsius plus 273.15.
6. Other units of time, along with their symbols, which are recognized for use with the SI system and are used in this text are: minute, min; hour, h; day, d.
7. There are many additional specialized units which are used with the SI system. However, most of those which appear in this text are shown in the table. A few of the specialized units are noted when used in the text. One which is frequently used is that for revolution, r.
8. Other common British units which are used in the text are as follows: inch, in.; yard, yd; mile, mi; ounce, oz.; ton; quart, qt; acre.
9. There are a number of units which were used with the metric system prior to the development of the SI system. However, many of these are not to be used with the SI system. Among these which were commonly used are the dyne, erg, and calorie.

Table 2. Metric Prefixes

Prefix	Factor	Symbol	Prefix	Factor	Symbol
exa	10^{18}	E	deci	10^{-1}	d
peta	10^{15}	P	centi	10^{-2}	c
tera	10^{12}	T	milli	10^{-3}	m
giga	10^{9}	G	micro	10^{-6}	μ
mega	10^{6}	M	nano	10^{-9}	n
kilo	10^{3}	k	pico	10^{-12}	p
hecto	10^{2}	h	femto	10^{-15}	f
deca	10^{1}	da	atto	10^{-18}	a

Table 3. Reduction and Conversion Factors

1 in. = 2.54 cm (exact)	1 ft^3 = 28.32 L
1 km = 0.6214 mi	1 L = 1.057 qt
1 lb = 453.6 g	1 Btu = 778.0 ft · lb
1 kg = 2.205 lb	1 hp = 550 ft · lb/s (exact)
1 lb = 4.448 N	1 hp = 746.0 W

Appendix C

Tables

Table 1   467

Table 1. Powers and Roots

No.	Sq.	Sq. Root	Cube	Cube Root	No.	Sq.	Sq. Root	Cube	Cube Root
1	1	1.000	1	1.000	51	2,601	7.141	132,651	3.708
2	4	1.414	8	1.260	52	2,704	7.211	140,608	3.733
3	9	1.732	27	1.442	53	2,809	7.280	148,877	3.756
4	16	2.000	64	1.587	54	2,916	7.348	157,464	3.780
5	25	2.236	125	1.710	55	3,025	7.416	166,375	3.803
6	36	2.449	216	1.817	56	3,136	7.483	175,616	3.826
7	49	2.646	343	1.913	57	3,249	7.550	185,193	3.849
8	64	2.828	512	2.000	58	3,364	7.616	195,112	3.871
9	81	3.000	729	2.080	59	3,481	7.681	205,379	3.893
10	100	3.162	1,000	2.154	60	3,600	7.746	216,000	3.915
11	121	3.317	1,331	2.224	61	3,721	7.810	226,981	3.936
12	144	3.464	1,728	2.289	62	3,844	7.874	238,328	3.958
13	169	3.606	2,197	2.351	63	3,969	7.937	250,047	3.979
14	196	3.742	2,744	2.410	64	4,096	8.000	262,144	4.000
15	225	3.873	3,375	2.466	65	4,225	8.062	274,625	4.021
16	256	4.000	4,096	2.520	66	4,356	8.124	287,496	4.041
17	289	4.123	4,913	2.571	67	4,489	8.185	300,763	4.062
18	324	4.243	5,832	2.621	68	4,624	8.246	314,432	4.082
19	361	4.359	6,859	2.668	69	4,761	8.307	328,509	4.102
20	400	4.472	8,000	2.714	70	4,900	8.367	343,000	4.121
21	441	4.583	9,261	2.759	71	5,041	8.426	357,911	4.141
22	484	4.690	10,648	2.802	72	5,184	8.485	373,248	4.160
23	529	4.796	12,167	2.844	73	5,329	8.544	389,017	4.179
24	576	4.899	13,824	2.884	74	5,476	8.602	405,224	4.198
25	625	5.000	15,625	2.924	75	5,625	8.660	421,875	4.217
26	676	5.099	17,576	2.962	76	5,776	8.718	438,976	4.236
27	729	5.196	19,683	3.000	77	5,929	8.775	456,533	4.254
28	784	5.292	21,952	3.037	78	6,084	8.832	474,552	4.273
29	841	5.385	24,389	3.072	79	6,241	8.888	493,039	4.291
30	900	5.477	27,000	3.107	80	6,400	8.944	512,000	4.309
31	961	5.568	29,791	3.141	81	6,561	9.000	531,441	4.327
32	1,024	5.657	32,768	3.175	82	6,724	9.055	551,368	4.344
33	1,089	5.745	35,937	3.208	83	6,889	9.110	571,787	4.362
34	1,156	5.831	39,304	3.240	84	7,056	9.165	592,704	4.380
35	1,225	5.916	42,875	3.271	85	7,225	9.220	614,125	4.397
36	1,296	6.000	46,656	3.302	86	7,396	9.274	636,056	4.414
37	1,369	6.083	50,653	3.332	87	7,569	9.327	658,503	4.431
38	1,444	6.164	54,872	3.362	88	7,744	9.381	681,472	4.448
39	1,521	6.245	59,319	3.391	89	7,921	9.434	704,969	4.465
40	1,600	6.325	64,000	3.420	90	8,100	9.487	729,000	4.481
41	1,681	6.403	68,921	3.448	91	8,281	9.539	753,571	4.498
42	1,764	6.481	74,088	3.476	92	8,464	9.592	778,688	4.514
43	1,849	6.557	79,507	3.503	93	8,649	9.644	804,357	4.531
44	1,936	6.633	85,184	3.530	94	8,836	9.695	830,584	4.547
45	2,025	6.708	91,125	3.557	95	9,025	9.747	857,375	4.563
46	2,116	6.782	97,336	3.583	96	9,216	9.798	884,736	4.579
47	2,209	6.856	103,823	3.609	97	9,409	9.849	912,673	4.595
48	2,304	6.928	110,592	3.634	98	9,604	9.899	941,192	4.610
49	2,401	7.000	117,649	3.659	99	9,801	9.950	970,299	4.626
50	2,500	7.071	125,000	3.684	100	10,000	10.000	1,000,000	4.642

Table 2. Four-Place Logarithms of Numbers

N	0	1	2	3	4	5	6	7	8	9
10	0000	0043	0086	0128	0170	0212	0253	0294	0334	0374
11	0414	0453	0492	0531	0569	0607	0645	0682	0719	0755
12	0792	0828	0864	0899	0934	0969	1004	1038	1072	1106
13	1139	1173	1206	1239	1271	1303	1335	1367	1399	1430
14	1461	1492	1523	1553	1584	1614	1644	1673	1703	1732
15	1761	1790	1818	1847	1875	1903	1931	1959	1987	2014
16	2041	2068	2095	2122	2148	2175	2201	2227	2253	2279
17	2304	2330	2355	2380	2405	2430	2455	2480	2504	2529
18	2553	2577	2601	2625	2648	2672	2695	2718	2742	2765
19	2788	2810	2833	2856	2878	2900	2923	2945	2967	2989
20	3010	3032	3054	3075	3096	3118	3139	3160	3181	3201
21	3222	3243	3263	3284	3304	3324	3345	3365	3385	3404
22	3424	3444	3464	3483	3502	3522	3541	3560	3579	3598
23	3617	3636	3655	3674	3692	3711	3729	3747	3766	3784
24	3802	3820	3838	3856	3874	3892	3909	3927	3945	3962
25	3979	3997	4014	4031	4048	4065	4082	4099	4116	4133
26	4150	4166	4183	4200	4216	4232	4249	4265	4281	4298
27	4314	4330	4346	4362	4378	4393	4409	4425	4440	4456
28	4472	4487	4502	4518	4533	4548	4564	4579	4594	4609
29	4624	4639	4654	4669	4683	4698	4713	4728	4742	4757
30	4771	4786	4800	4814	4829	4843	4857	4871	4886	4900
31	4914	4928	4942	4955	4969	4983	4997	5011	5024	5038
32	5051	5065	5079	5092	5105	5119	5132	5145	5159	5172
33	5185	5198	5211	5224	5237	5250	5263	5276	5289	5302
34	5315	5328	5340	5353	5366	5378	5391	5403	5416	5428
35	5441	5453	5465	5478	5490	5502	5514	5527	5539	5551
36	5563	5575	5587	5599	5611	5623	5635	5647	5658	5670
37	5682	5694	5705	5717	5729	5740	5752	5763	5775	5786
38	5798	5809	5821	5832	5843	5855	5866	5877	5888	5899
39	5911	5922	5933	5944	5955	5966	5977	5988	5999	6010
40	6021	6031	6042	6053	6064	6075	6085	6096	6107	6117
41	6128	6138	6149	6160	6170	6180	6191	6201	6212	6222
42	6232	6243	6253	6263	6274	6284	6294	6304	6314	6325
43	6335	6345	6355	6365	6375	6385	6395	6405	6415	6425
44	6435	6444	6454	6464	6474	6484	6493	6503	6513	6522
45	6532	6542	6551	6561	6571	6580	6590	6599	6609	6618
46	6628	6637	6646	6656	6665	6675	6684	6693	6702	6712
47	6721	6730	6739	6749	6758	6767	6776	6785	6794	6803
48	6812	6821	6830	6839	6848	6857	6866	6875	6884	6893
49	6902	6911	6920	6928	6937	6946	6955	6964	6972	6981
50	6990	6998	7007	7016	7024	7033	7042	7050	7059	7067
51	7076	7084	7093	7101	7110	7118	7126	7135	7143	7152
52	7160	7168	7177	7185	7193	7202	7210	7218	7226	7235
53	7243	7251	7259	7267	7275	7284	7292	7300	7308	7316
54	7324	7332	7340	7348	7356	7364	7372	7380	7388	7396

Table 2   469

Table 2. Continued

N	0	1	2	3	4	5	6	7	8	9
55	7404	7412	7419	7427	7435	7443	7451	7459	7466	7474
56	7482	7490	7497	7505	7513	7520	7528	7536	7543	7551
57	7559	7566	7574	7582	7589	7597	7604	7612	7619	7627
58	7634	7642	7649	7657	7664	7672	7679	7686	7694	7701
59	7709	7716	7723	7731	7738	7745	7752	7760	7767	7774
60	7782	7789	7796	7803	7810	7818	7825	7832	7839	7846
61	7853	7860	7868	7875	7882	7889	7896	7903	7910	7917
62	7924	7931	7938	7945	7952	7959	7966	7973	7980	7987
63	7993	8000	8007	8014	8021	8028	8035	8041	8048	8055
64	8062	8069	8075	8082	8089	8096	8102	8109	8116	8122
65	8129	8136	8142	8149	8156	8162	8169	8176	8182	8189
66	8195	8202	8209	8215	8222	8228	8235	8241	8248	8254
67	8261	8267	8274	8280	8287	8293	8299	8306	8312	8319
68	8325	8331	8338	8344	8351	8357	8363	8370	8376	8382
69	8388	8395	8401	8407	8414	8420	8426	8432	8439	8445
70	8451	8457	8463	8470	8476	8482	8488	8494	8500	8506
71	8513	8519	8525	8531	8537	8543	8549	8555	8561	8567
72	8573	8579	8585	8591	8597	8603	8609	8615	8621	8627
73	8633	8639	8645	8651	8657	8663	8669	8675	8681	8686
74	8692	8698	8704	8710	8716	8722	8727	8733	8739	8745
75	8751	8756	8762	8768	8774	8779	8785	8791	8797	8802
76	8808	8814	8820	8825	8831	8837	8842	8848	8854	8859
77	8865	8871	8876	8882	8887	8893	8899	8904	8910	8915
78	8921	8927	8932	8938	8943	8949	8954	8960	8965	8971
79	8976	8982	8987	8993	8998	9004	9009	9015	9020	9025
80	9031	9036	9042	9047	9053	9058	9063	9069	9074	9079
81	9085	9090	9096	9101	9106	9112	9117	9122	9128	9133
82	9138	9143	9149	9154	9159	9165	9170	9175	9180	9186
83	9191	9196	9201	9206	9212	9217	9222	9227	9232	9238
84	9243	9248	9253	9258	9263	9269	9274	9279	9284	9289
85	9294	9299	9304	9309	9315	9320	9325	9330	9335	9340
86	9345	9350	9355	9360	9365	9370	9375	9380	9385	9390
87	9395	9400	9405	9410	9415	9420	9425	9430	9435	9440
88	9445	9450	9455	9460	9465	9469	9474	9479	9484	9489
89	9494	9499	9504	9509	9513	9518	9523	9528	9533	9538
90	9542	9547	9552	9557	9562	9566	9571	9576	9581	9586
91	9590	9595	9600	9605	9609	9614	9619	9624	9628	9633
92	9638	9643	9647	9652	9657	9661	9666	9671	9675	9680
93	9685	9689	9694	9699	9703	9708	9713	9717	9722	9727
94	9731	9736	9741	9745	9750	9754	9759	9763	9768	9773
95	9777	9782	9786	9791	9795	9800	9805	9809	9814	9818
96	9823	9827	9832	9836	9841	9845	9850	9854	9859	9863
97	9868	9872	9877	9881	9886	9890	9894	9899	9903	9908
98	9912	9917	9921	9926	9930	9934	9939	9943	9948	9952
99	9956	9961	9965	9969	9974	9978	9983	9987	9991	9996

Table 3. Four-place Values of Trigonometric Functions
Angle θ in Degrees and Tenths

Degrees	Sin θ	Cos θ	Tan θ	Cot θ		Degrees	Sin θ	Cos θ	Tan θ	Cot θ	
0.0	0.0000	1.0000	0.0000	—	90.0	5.0	0.0872	0.9962	0.0875	11.43	85.0
.1	0.0017	1.0000	0.0017	573.0	.9	.1	0.0889	0.9960	0.0892	11.20	.9
.2	0.0035	1.0000	0.0035	286.4	.8	.2	0.0906	0.9959	0.0910	10.99	.8
.3	0.0052	1.0000	0.0052	191.0	.7	.3	0.0924	0.9957	0.0928	10.78	.7
.4	0.0070	1.0000	0.0070	143.2	.6	.4	0.0941	0.9956	0.0945	10.58	.6
.5	0.0087	1.0000	0.0087	114.6	.5	.5	0.0958	0.9954	0.0963	10.39	.5
.6	0.0105	0.9999	0.0105	95.49	.4	.6	0.0976	0.9952	0.0981	10.20	.4
.7	0.0122	0.9999	0.0122	81.85	.3	.7	0.0993	0.9951	0.0998	10.02	.3
.8	0.0140	0.9999	0.0140	71.62	.2	.8	0.1011	0.9949	0.1016	9.845	.2
.9	0.0157	0.9999	0.0157	63.66	.1	.9	0.1028	0.9947	0.1033	9.677	.1
1.0	0.0175	0.9998	0.0175	57.29	89.0	6.0	0.1045	0.9945	0.1051	9.514	84.0
.1	0.0192	0.9998	0.0192	52.08	.9	.1	0.1063	0.9943	0.1069	9.357	.9
.2	0.0209	0.9998	0.0209	47.74	.8	.2	0.1080	0.9942	0.1086	9.205	.8
.3	0.0227	0.9997	0.0227	44.07	.7	.3	0.1097	0.9940	0.1104	9.058	.7
.4	0.0244	0.9997	0.0244	40.92	.6	.4	0.1115	0.9938	0.1122	8.915	.6
.5	0.0262	0.9997	0.0262	38.19	.5	.5	0.1132	0.9936	0.1139	8.777	.5
.6	0.0279	0.9996	0.0279	35.80	.4	.6	0.1149	0.9934	0.1157	8.643	.4
.7	0.0297	0.9996	0.0297	33.69	.3	.7	0.1167	0.9932	0.1175	8.513	.3
.8	0.0314	0.9995	0.0314	31.82	.2	.8	0.1184	0.9930	0.1192	8.386	.2
.9	0.0332	0.9995	0.0332	30.14	.1	.9	0.1201	0.9928	0.1210	8.264	.1
2.0	0.0349	0.9994	0.0349	28.64	88.0	7.0	0.1219	0.9925	0.1228	8.144	83.0
.1	0.0366	0.9993	0.0367	27.27	.9	.1	0.1236	0.9923	0.1246	8.028	.9
.2	0.0384	0.9993	0.0384	26.03	.8	.2	0.1253	0.9921	0.1263	7.916	.8
.3	0.0401	0.9992	0.0402	24.90	.7	.3	0.1271	0.9919	0.1281	7.806	.7
.4	0.0419	0.9991	0.0419	23.86	.6	.4	0.1288	0.9917	0.1299	7.700	.6
.5	0.0436	0.9990	0.0437	22.90	.5	.5	0.1305	0.9914	0.1317	7.596	.5
.6	0.0454	0.9990	0.0454	22.02	.4	.6	0.1323	0.9912	0.1334	7.495	.4
.7	0.0471	0.9989	0.0472	21.20	.3	.7	0.1340	0.9910	0.1352	7.396	.3
.8	0.0488	0.9988	0.0489	20.45	.2	.8	0.1357	0.9907	0.1370	7.300	.2
.9	0.0506	0.9987	0.0507	19.74	.1	.9	0.1374	0.9905	0.1388	7.207	.1
3.0	0.0523	0.9986	0.0524	19.08	87.0	8.0	0.1392	0.9903	0.1405	7.115	82.0
.1	0.0541	0.9985	0.0542	18.46	.9	.1	0.1409	0.9900	0.1423	7.026	.9
.2	0.0558	0.9984	0.0559	17.89	.8	.2	0.1426	0.9898	0.1441	6.940	.8
.3	0.0576	0.9983	0.0577	17.34	.7	.3	0.1444	0.9895	0.1459	6.855	.7
.4	0.0593	0.9982	0.0594	16.83	.6	.4	0.1461	0.9893	0.1477	6.772	.6
.5	0.0610	0.9981	0.0612	16.35	.5	.5	0.1478	0.9890	0.1495	6.691	.5
.6	0.0628	0.9980	0.0629	15.89	.4	.6	0.1495	0.9888	0.1512	6.612	.4
.7	0.0645	0.9979	0.0647	15.46	.3	.7	0.1513	0.9885	0.1530	6.535	.3
.8	0.0663	0.9978	0.0664	15.06	.2	.8	0.1530	0.9882	0.1548	6.460	.2
.9	0.0680	0.9977	0.0682	14.67	.1	.9	0.1547	0.9880	0.1566	6.386	.1
4.0	0.0698	0.9976	0.0699	14.30	86.0	9.0	0.1564	0.9877	0.1584	6.314	81.0
.1	0.0715	0.9974	0.0717	13.95	.9	.1	0.1582	0.9874	0.1602	6.243	.9
.2	0.0732	0.9973	0.0734	13.62	.8	.2	0.1599	0.9871	0.1620	6.174	.8
.3	0.0750	0.9972	0.0752	13.30	.7	.3	0.1616	0.9869	0.1638	6.107	.7
.4	0.0767	0.9971	0.0769	13.00	.6	.4	0.1633	0.9866	0.1655	6.041	.6
.5	0.0785	0.9969	0.0787	12.71	.5	.5	0.1650	0.9863	0.1673	5.976	.5
.6	0.0802	0.9968	0.0805	12.43	.4	.6	0.1668	0.9860	0.1691	5.912	.4
.7	0.0819	0.9966	0.0822	12.16	.3	.7	0.1685	0.9857	0.1709	5.850	.3
.8	0.0837	0.9965	0.0840	11.91	.2	.8	0.1702	0.9854	0.1727	5.789	.2
.9	0.0854	0.9963	0.0857	11.66	.1	.9	0.1719	0.9851	0.1745	5.730	.1
5.0	0.0872	0.9962	0.0875	11.43	85.0	10.0	0.1736	0.9848	0.1763	5.671	80.0
	Cos θ	Sin θ	Cot θ	Tan θ	Degrees		Cos θ	Sin θ	Cot θ	Tan θ	Degrees

Table 3   471

Table 3. Continued

Degrees	Sin θ	Cos θ	Tan θ	Cot θ		Degrees	Sin θ	Cos θ	Tan θ	Cot θ	
10.0	0.1736	0.9848	0.1763	5.671	80.0	15.0	0.2588	0.9659	0.2679	3.732	75.0
.1	0.1754	0.9845	0.1781	5.614	.9	.1	0.2605	0.9655	0.2698	3.706	.9
.2	0.1771	0.9842	0.1799	5.558	.8	.2	0.2622	0.9650	0.2717	3.681	.8
.3	0.1788	0.9839	0.1817	5.503	.7	.3	0.2639	0.9646	0.2736	3.655	.7
.4	0.1805	0.9836	0.1835	5.449	.6	.4	0.2656	0.9641	0.2754	3.630	.6
.5	0.1822	0.9833	0.1853	5.396	.5	.5	0.2672	0.9636	0.2773	3.606	.5
.6	0.1840	0.9829	0.1871	5.343	.4	.6	0.2689	0.9632	0.2792	3.582	.4
.7	0.1857	0.9826	0.1890	5.292	.3	.7	0.2706	0.9627	0.2811	3.558	.3
.8	0.1874	0.9823	0.1908	5.242	.2	.8	0.2723	0.9622	0.2830	3.534	.2
.9	0.1891	0.9820	0.1926	5.193	.1	.9	0.2740	0.9617	0.2849	3.511	.1
11.0	0.1908	0.9816	0.1944	5.145	79.0	16.0	0.2756	0.9613	0.2867	3.487	74.0
.1	0.1925	0.9813	0.1962	5.097	.9	.1	0.2773	0.9608	0.2886	3.465	.9
.2	0.1942	0.9810	0.1980	5.050	.8	.2	0.2790	0.9603	0.2905	3.442	.8
.3	0.1959	0.9806	0.1998	5.005	.7	.3	0.2807	0.9598	0.2924	3.420	.7
.4	0.1977	0.9803	0.2016	4.959	.6	.4	0.2823	0.9593	0.2943	3.398	.6
.5	0.1994	0.9799	0.2035	4.915	.5	.5	0.2840	0.9588	0.2962	3.376	.5
.6	0.2011	0.9796	0.2053	4.872	.4	.6	0.2857	0.9583	0.2981	3.354	.4
.7	0.2028	0.9792	0.2071	4.829	.3	.7	0.2874	0.9578	0.3000	3.333	.3
.8	0.2045	0.9789	0.2089	4.787	.2	.8	0.2890	0.9573	0.3019	3.312	.2
.9	0.2062	0.9785	0.2107	4.745	.1	.9	0.2907	0.9568	0.3038	3.291	.1
12.0	0.2079	0.9781	0.2126	4.705	78.0	17.0	0.2924	0.9563	0.3057	3.271	73.0
.1	0.2096	0.9778	0.2144	4.665	.9	.1	0.2940	0.9558	0.3076	3.251	.9
.2	0.2113	0.9774	0.2162	4.625	.8	.2	0.2957	0.9553	0.3096	3.230	.8
.3	0.2130	0.9770	0.2180	4.586	.7	.3	0.2974	0.9548	0.3115	3.211	.7
.4	0.2147	0.9767	0.2199	4.548	.6	.4	0.2990	0.9542	0.3134	3.191	.6
.5	0.2164	0.9763	0.2217	4.511	.5	.5	0.3007	0.9537	0.3153	3.172	.5
.6	0.2181	0.9759	0.2235	4.474	.4	.6	0.3024	0.9532	0.3172	3.152	.4
.7	0.2198	0.9755	0.2254	4.437	.3	.7	0.3040	0.9527	0.3191	3.133	.3
.8	0.2215	0.9751	0.2272	4.402	.2	.8	0.3057	0.9521	0.3211	3.115	.2
.9	0.2233	0.9748	0.2290	4.366	.1	.9	0.3074	0.9516	0.3230	3.096	.1
13.0	0.2250	0.9744	0.2309	4.331	77.0	18.0	0.3090	0.9511	0.3249	3.078	72.0
.1	0.2267	0.9740	0.2327	4.297	.9	.1	0.3107	0.9505	0.3269	3.060	.9
.2	0.2284	0.9736	0.2345	4.264	.8	.2	0.3123	0.9500	0.3288	3.042	.8
.3	0.2300	0.9732	0.2364	4.230	.7	.3	0.3140	0.9494	0.3307	3.024	.7
.4	0.2317	0.9728	0.2382	4.198	.6	.4	0.3156	0.9489	0.3327	3.006	.6
.5	0.2334	0.9724	0.2401	4.165	.5	.5	0.3173	0.9483	0.3346	2.989	.5
.6	0.2351	0.9720	0.2419	4.134	.4	.6	0.3190	0.9478	0.3365	2.971	.4
.7	0.2368	0.9715	0.2438	4.102	.3	.7	0.3206	0.9472	0.3385	2.954	.3
.8	0.2385	0.9711	0.2456	4.071	.2	.8	0.3223	0.9466	0.3404	2.937	.2
.9	0.2402	0.9707	0.2475	4.041	.1	.9	0.3239	0.9461	0.3424	2.921	.1
14.0	0.2419	0.9703	0.2493	4.011	76.0	19.0	0.3256	0.9455	0.3443	2.904	71.0
.1	0.2436	0.9699	0.2512	3.981	.9	.1	0.3272	0.9449	0.3463	2.888	.9
.2	0.2453	0.9694	0.2530	3.952	.8	.2	0.3289	0.9444	0.3482	2.872	.8
.3	0.2470	0.9690	0.2549	3.923	.7	.3	0.3305	0.9438	0.3502	2.856	.7
.4	0.2487	0.9686	0.2568	3.895	.6	.4	0.3322	0.9432	0.3522	2.840	.6
.5	0.2504	0.9681	0.2586	3.867	.5	.5	0.3338	0.9426	0.3541	2.824	.5
.6	0.2521	0.9677	0.2605	3.839	.4	.6	0.3355	0.9421	0.3561	2.808	.4
.7	0.2538	0.9673	0.2623	3.812	.3	.7	0.3371	0.9415	0.3581	2.793	.3
.8	0.2554	0.9668	0.2642	3.785	.2	.8	0.3387	0.9409	0.3600	2.778	.2
.9	0.2571	0.9664	0.2661	3.758	.1	.9	0.3404	0.9403	0.3620	2.762	.1
15.0	0.2588	0.9659	0.2679	3.732	75.0	20.0	0.3420	0.9397	0.3640	2.747	70.0
	Cos θ	Sin θ	Cot θ	Tan θ	Degrees		Cos θ	Sin θ	Cot θ	Tan θ	Degrees

Table 3. Continued

Degrees	Sin θ	Cos θ	Tan θ	Cot θ		Degrees	Sin θ	Cos θ	Tan θ	Cot θ	
20.0	0.3420	0.9397	0.3640	2.747	70.0	25.0	0.4226	0.9063	0.4663	2.145	65.0
.1	0.3437	0.9391	0.3659	2.733	.9	.1	0.4242	0.9056	0.4684	2.135	.9
.2	0.3453	0.9385	0.3679	2.718	.8	.2	0.4258	0.9048	0.4706	2.125	.8
.3	0.3469	0.9379	0.3699	2.703	.7	.3	0.4274	0.9041	0.4727	2.116	.7
.4	0.3486	0.9373	0.3719	2.689	.6	.4	0.4289	0.9033	0.4748	2.106	.6
.5	0.3502	0.9367	0.3739	2.675	.5	.5	0.4305	0.9026	0.4770	2.097	.5
.6	0.3518	0.9361	0.3759	2.660	.4	.6	0.4321	0.9018	0.4791	2.087	.4
.7	0.3535	0.9354	0.3779	2.646	.3	.7	0.4337	0.9011	0.4813	2.078	.3
.8	0.3551	0.9348	0.3799	2.633	.2	.8	0.4352	0.9003	0.4834	2.069	.2
.9	0.3567	0.9342	0.3819	2.619	.1	.9	0.4368	0.8996	0.4856	2.059	.1
21.0	0.3584	0.9336	0.3839	2.605	69.0	26.0	0.4384	0.8988	0.4877	2.050	64.0
.1	0.3600	0.9330	0.3859	2.592	.9	.1	0.4399	0.8980	0.4899	2.041	.9
.2	0.3616	0.9323	0.3879	2.578	.8	.2	0.4415	0.8973	0.4921	2.032	.8
.3	0.3633	0.9317	0.3899	2.565	.7	.3	0.4431	0.8965	0.4942	2.023	.7
.4	0.3649	0.9311	0.3919	2.552	.6	.4	0.4446	0.8957	0.4964	2.014	.6
.5	0.3665	0.9304	0.3939	2.539	.5	.5	0.4462	0.8949	0.4986	2.006	.5
.6	0.3681	0.9298	0.3959	2.526	.4	.6	0.4478	0.8942	0.5008	1.997	.4
.7	0.3697	0.9291	0.3979	2.513	.3	.7	0.4493	0.8934	0.5029	1.988	.3
.8	0.3714	0.9285	0.4000	2.500	.2	.8	0.4509	0.8926	0.5051	1.980	.2
.9	0.3730	0.9278	0.4020	2.488	.1	.9	0.4524	0.8918	0.5073	1.971	.1
22.0	0.3746	0.9272	0.4040	2.475	68.0	27.0	0.4540	0.8910	0.5095	1.963	63.0
.1	0.3762	0.9265	0.4061	2.463	.9	.1	0.4555	0.8902	0.5117	1.954	.9
.2	0.3778	0.9259	0.4081	2.450	.8	.2	0.4571	0.8894	0.5139	1.946	.8
.3	0.3795	0.9252	0.4101	2.438	.7	.3	0.4586	0.8886	0.5161	1.937	.7
.4	0.3811	0.9245	0.4122	2.426	.6	.4	0.4602	0.8878	0.5184	1.929	.6
.5	0.3827	0.9239	0.4142	2.414	.5	.5	0.4617	0.8870	0.5206	1.921	.5
.6	0.3843	0.9232	0.4163	2.402	.4	.6	0.4633	0.8862	0.5228	1.913	.4
.7	0.3859	0.9225	0.4183	2.391	.3	.7	0.4648	0.8854	0.5250	1.905	.3
.8	0.3875	0.9219	0.4204	2.379	.2	.8	0.4664	0.8846	0.5272	1.897	.2
.9	0.3891	0.9212	0.4224	2.367	.1	.9	0.4679	0.8838	0.5295	1.889	.1
23.0	0.3907	0.9205	0.4245	2.356	67.0	28.0	0.4695	0.8829	0.5317	1.881	62.0
.1	0.3923	0.9198	0.4265	2.344	.9	.1	0.4710	0.8821	0.5340	1.873	.9
.2	0.3939	0.9191	0.4286	2.333	.8	.2	0.4726	0.8813	0.5362	1.865	.8
.3	0.3955	0.9184	0.4307	2.322	.7	.3	0.4741	0.8805	0.5384	1.857	.7
.4	0.3971	0.9178	0.4327	2.311	.6	.4	0.4756	0.8796	0.5407	1.849	.6
.5	0.3987	0.9171	0.4348	2.300	.5	.5	0.4772	0.8788	0.5430	1.842	.5
.6	0.4003	0.9164	0.4369	2.289	.4	.6	0.4787	0.8780	0.5452	1.834	.4
.7	0.4019	0.9157	0.4390	2.278	.3	.7	0.4802	0.8771	0.5475	1.827	.3
.8	0.4035	0.9150	0.4411	2.267	.2	.8	0.4818	0.8763	0.5498	1.819	.2
.9	0.4051	0.9143	0.4431	2.257	.1	.9	0.4833	0.8755	0.5520	1.811	.1
24.0	0.4067	0.9135	0.4452	2.246	66.0	29.0	0.4848	0.8746	0.5543	1.804	61.0
.1	0.4083	0.9128	0.4473	2.236	.9	.1	0.4863	0.8738	0.5566	1.797	.9
.2	0.4099	0.9121	0.4494	2.225	.8	.2	0.4879	0.8729	0.5589	1.789	.8
.3	0.4115	0.9114	0.4515	2.215	.7	.3	0.4894	0.8721	0.5612	1.782	.7
.4	0.4131	0.9107	0.4536	2.204	.6	.4	0.4909	0.8712	0.5635	1.775	.6
.5	0.4147	0.9100	0.4557	2.194	.5	.5	0.4924	0.8704	0.5658	1.767	.5
.6	0.4163	0.9092	0.4578	2.184	.4	.6	0.4939	0.8695	0.5681	1.760	.4
.7	0.4179	0.9085	0.4599	2.174	.3	.7	0.4955	0.8686	0.5704	1.753	.3
.8	0.4195	0.9078	0.4621	2.164	.2	.8	0.4970	0.8678	0.5727	1.746	.2
.9	0.4210	0.9070	0.4642	2.154	.1	.9	0.4985	0.8669	0.5750	1.739	.1
25.0	0.4226	0.9063	0.4663	2.145	65.0	30.0	0.5000	0.8660	0.5774	1.732	60.0
	Cos θ	Sin θ	Cot θ	Tan θ	Degrees		Cos θ	Sin θ	Cot θ	Tan θ	Degrees

Table 3   473

Table 3. Continued

Degrees	Sin θ	Cos θ	Tan θ	Cot θ		Degrees	Sin θ	Cos θ	Tan θ	Cot θ	
30.0	0.5000	0.8660	0.5774	1.732	60.0	35.0	0.5736	0.8192	0.7002	1.428	55.0
.1	0.5015	0.8652	0.5797	1.725	.9	.1	0.5750	0.8181	0.7028	1.423	.9
.2	0.5030	0.8643	0.5820	1.718	.8	.2	0.5764	0.8171	0.7054	1.418	.8
.3	0.5045	0.8634	0.5844	1.711	.7	.3	0.5779	0.8161	0.7080	1.412	.7
.4	0.5060	0.8625	0.5867	1.704	.6	.4	0.5793	0.8151	0.7107	1.407	.6
.5	0.5075	0.8616	0.5890	1.698	.5	.5	0.5807	0.8141	0.7133	1.402	.5
.6	0.5090	0.8607	0.5914	1.691	.4	.6	0.5821	0.8131	0.7159	1.397	.4
.7	0.5105	0.8599	0.5938	1.684	.3	.7	0.5835	0.8121	0.7186	1.392	.3
.8	0.5120	0.8590	0.5961	1.678	.2	.8	0.5850	0.8111	0.7212	1.387	.2
.9	0.5135	0.8581	0.5985	1.671	.1	.9	0.5864	0.8100	0.7239	1.381	.1
31.0	0.5150	0.8572	0.6009	1.664	59.0	36.0	0.5878	0.8090	0.7265	1.376	54.0
.1	0.5165	0.8563	0.6032	1.658	.9	.1	0.5892	0.8080	0.7292	1.371	.9
.2	0.5180	0.8554	0.6056	1.651	.8	.2	0.5906	0.8070	0.7319	1.366	.8
.3	0.5195	0.8545	0.6080	1.645	.7	.3	0.5920	0.8059	0.7346	1.361	.7
.4	0.5210	0.8536	0.6104	1.638	.6	.4	0.5934	0.8049	0.7373	1.356	.6
.5	0.5225	0.8526	0.6128	1.632	.5	.5	0.5948	0.8039	0.7400	1.351	.5
.6	0.5240	0.8517	0.6152	1.625	.4	.6	0.5962	0.8028	0.7427	1.347	.4
.7	0.5255	0.8508	0.6176	1.619	.3	.7	0.5976	0.8018	0.7454	1.342	.3
.8	0.5270	0.8499	0.6200	1.613	.2	.8	0.5990	0.8007	0.7481	1.337	.2
.9	0.5284	0.8490	0.6224	1.607	.1	.9	0.6004	0.7997	0.7508	1.332	.1
32.0	0.5299	0.8480	0.6249	1.600	58.0	37.0	0.6018	0.7986	0.7536	1.327	53.0
.1	0.5314	0.8471	0.6273	1.594	.9	.1	0.6032	0.7976	0.7563	1.322	.9
.2	0.5329	0.8462	0.6297	1.588	.8	.2	0.6046	0.7965	0.7590	1.317	.8
.3	0.5344	0.8453	0.6322	1.582	.7	.3	0.6060	0.7955	0.7618	1.313	.7
.4	0.5358	0.8443	0.6346	1.576	.6	.4	0.6074	0.7944	0.7646	1.308	.6
.5	0.5373	0.8434	0.6371	1.570	.5	.5	0.6088	0.7934	0.7673	1.303	.5
.6	0.5388	0.8425	0.6395	1.564	.4	.6	0.6101	0.7923	0.7701	1.299	.4
.7	0.5402	0.8415	0.6420	1.558	.3	.7	0.6115	0.7912	0.7729	1.294	.3
.8	0.5417	0.8406	0.6445	1.552	.2	.8	0.6129	0.7902	0.7757	1.289	.2
.9	0.5432	0.8396	0.6469	1.546	.1	.9	0.6143	0.7891	0.7785	1.285	.1
33.0	0.5446	0.8387	0.6494	1.540	57.0	38.0	0.6157	0.7880	0.7813	1.280	52.0
.1	0.5461	0.8377	0.6519	1.534	.9	.1	0.6170	0.7869	0.7841	1.275	.9
.2	0.5476	0.8368	0.6544	1.528	.8	.2	0.6184	0.7859	0.7869	1.271	.8
.3	0.5490	0.8358	0.6569	1.522	.7	.3	0.6198	0.7848	0.7898	1.266	.7
.4	0.5505	0.8348	0.6594	1.517	.6	.4	0.6211	0.7837	0.7926	1.262	.6
.5	0.5519	0.8339	0.6619	1.511	.5	.5	0.6225	0.7826	0.7954	1.257	.5
.6	0.5534	0.8329	0.6644	1.505	.4	.6	0.6239	0.7815	0.7983	1.253	.4
.7	0.5548	0.8320	0.6669	1.499	.3	.7	0.6252	0.7804	0.8012	1.248	.3
.8	0.5563	0.8310	0.6694	1.494	.2	.8	0.6266	0.7793	0.8040	1.244	.2
.9	0.5577	0.8300	0.6720	1.488	.1	.9	0.6280	0.7782	0.8069	1.239	.1
34.0	0.5592	0.8290	0.6745	1.483	56.0	39.0	0.6293	0.7771	0.8098	1.235	51.0
.1	0.5606	0.8281	0.6771	1.477	.9	.1	0.6307	0.7760	0.8127	1.230	.9
.2	0.5621	0.8271	0.6796	1.471	.8	.2	0.6320	0.7749	0.8156	1.226	.8
.3	0.5635	0.8261	0.6822	1.466	.7	.3	0.6334	0.7738	0.8185	1.222	.7
.4	0.5650	0.8251	0.6847	1.460	.6	.4	0.6347	0.7727	0.8214	1.217	.6
.5	0.5664	0.8241	0.6873	1.455	.5	.5	0.6361	0.7716	0.8243	1.213	.5
.6	0.5678	0.8231	0.6899	1.450	.4	.6	0.6374	0.7705	0.8273	1.209	.4
.7	0.5693	0.8221	0.6924	1.444	.3	.7	0.6388	0.7694	0.8302	1.205	.3
.8	0.5707	0.8211	0.6950	1.439	.2	.8	0.6401	0.7683	0.8332	1.200	.2
.9	0.5721	0.8202	0.6976	1.433	.1	.9	0.6414	0.7672	0.8361	1.196	.1
35.0	0.5736	0.8192	0.7002	1.428	55.0	40.0	0.6428	0.7660	0.8391	1.192	50.0
	Cos θ	Sin θ	Cot θ	Tan θ	Degrees		Cos θ	Sin θ	Cot θ	Tan θ	Degrees

Table 3. Continued

Degrees	Sin θ	Cos θ	Tan θ	Cot θ	
40.0	0.6428	0.7660	0.8391	1.192	50.0
.1	0.6441	0.7649	0.8421	1.188	.9
.2	0.6455	0.7638	0.8451	1.183	.8
.3	0.6468	0.7627	0.8481	1.179	.7
.4	0.6481	0.7615	0.8511	1.175	.6
.5	0.6494	0.7604	0.8541	1.171	.5
.6	0.6508	0.7593	0.8571	1.167	.4
.7	0.6521	0.7581	0.8601	1.163	.3
.8	0.6534	0.7570	0.8632	1.159	.2
.9	0.6547	0.7559	0.8662	1.154	.1
41.0	0.6561	0.7547	0.8693	1.150	49.0
.1	0.6574	0.7536	0.8724	1.146	.9
.2	0.6587	0.7524	0.8754	1.142	.8
.3	0.6600	0.7513	0.8785	1.138	.7
.4	0.6613	0.7501	0.8816	1.134	.6
.5	0.6626	0.7490	0.8847	1.130	.5
.6	0.6639	0.7478	0.8878	1.126	.4
.7	0.6652	0.7466	0.8910	1.122	.3
.8	0.6665	0.7455	0.8941	1.118	.2
.9	0.6678	0.7443	0.8972	1.115	.1
42.0	0.6691	0.7431	0.9004	1.111	48.0
.1	0.6704	0.7420	0.9036	1.107	.9
.2	0.6717	0.7408	0.9067	1.103	.8
.3	0.6730	0.7396	0.9099	1.099	.7
.4	0.6743	0.7385	0.9131	1.095	.6
.5	0.6756	0.7373	0.9163	1.091	.5
.6	0.6769	0.7361	0.9195	1.087	.4
.7	0.6782	0.7349	0.9228	1.084	.3
.8	0.6794	0.7337	0.9260	1.080	.2
.9	0.6807	0.7325	0.9293	1.076	.1
43.0	0.6820	0.7314	0.9325	1.072	47.0
.1	0.6833	0.7302	0.9358	1.069	.9
.2	0.6845	0.7290	0.9391	1.065	.8
.3	0.6858	0.7278	0.9424	1.061	.7
.4	0.6871	0.7266	0.9457	1.057	.6
.5	0.6884	0.7254	0.9490	1.054	.5
.6	0.6896	0.7242	0.9523	1.050	.4
.7	0.6909	0.7230	0.9556	1.046	.3
.8	0.6921	0.7218	0.9590	1.043	.2
.9	0.6934	0.7206	0.9623	1.039	.1
44.0	0.6947	0.7193	0.9657	1.036	46.0
.1	0.6959	0.7181	0.9691	1.032	.9
.2	0.6972	0.7169	0.9725	1.028	.8
.3	0.6984	0.7157	0.9759	1.025	.7
.4	0.6997	0.7145	0.9793	1.021	.6
.5	0.7009	0.7133	0.9827	1.018	.5
.6	0.7022	0.7120	0.9861	1.014	.4
.7	0.7034	0.7108	0.9896	1.011	.3
.8	0.7046	0.7096	0.9930	1.007	.2
.9	0.7059	0.7083	0.9965	1.003	.1
45.0	0.7071	0.7071	1.0000	1.000	45.0
	Cos θ	Sin θ	Cot θ	Tan θ	Degrees

Table 4   475

Table 4. Natural Logarithms of Numbers

n	$\ln n$	n	$\ln n$	n	$\ln n$
0.0	——— *	4.5	1.5041	9.0	2.1972
0.1	7.6974	4.6	1.5261	9.1	2.2083
0.2	8.3906	4.7	1.5476	9.2	2.2192
0.3	8.7960	4.8	1.5686	9.3	2.2300
0.4	9.0837	4.9	1.5892	9.4	2.2407
0.5	9.3069	5.0	1.6094	9.5	2.2513
0.6	9.4892	5.1	1.6292	9.6	2.2618
0.7	9.6433	5.2	1.6487	9.7	2.2721
0.8	9.7769	5.3	1.6677	9.8	2.2824
0.9	9.8946	5.4	1.6864	9.9	2.2925
1.0	0.0000	5.5	1.7047	10	2.3026
1.1	0.0953	5.6	1.7228	11	2.3979
1.2	0.1823	5.7	1.7405	12	2.4849
1.3	0.2624	5.8	1.7579	13	2.5649
1.4	0.3365	5.9	1.7750	14	2.6391
1.5	0.4055	6.0	1.7918	15	2.7081
1.6	0.4700	6.1	1.8083	16	2.7726
1.7	0.5306	6.2	1.8245	17	2.8332
1.8	0.5878	6.3	1.8405	18	2.8904
1.9	0.6419	6.4	1.8563	19	2.9444
2.0	0.6931	6.5	1.8718	20	2.9957
2.1	0.7419	6.6	1.8871	25	3.2189
2.2	0.7885	6.7	1.9021	30	3.4012
2.3	0.8329	6.8	1.9169	35	3.5553
2.4	0.8755	6.9	1.9315	40	3.6889
2.5	0.9163	7.0	1.9459	45	3.8067
2.6	0.9555	7.1	1.9601	50	3.9120
2.7	0.9933	7.2	1.9741	55	4.0073
2.8	1.0296	7.3	1.9879	60	4.0943
2.9	1.0647	7.4	2.0015	65	4.1744
3.0	1.0986	7.5	2.0149	70	4.2485
3.1	1.1314	7.6	2.0281	75	4.3175
3.2	1.1632	7.7	2.0412	80	4.3820
3.3	1.1939	7.8	2.0541	85	4.4427
3.4	1.2238	7.9	2.0669	90	4.4998
3.5	1.2528	8.0	2.0794	95	4.5539
3.6	1.2809	8.1	2.0919	100	4.6052
3.7	1.3083	8.2	2.1041		
3.8	1.3350	8.3	2.1163		
3.9	1.3610	8.4	2.1282		
4.0	1.3863	8.5	2.1401	$\ln 10$	2.3026
4.1	1.4110	8.6	2.1518	$2 \ln 10$	4.6052
4.2	1.4351	8.7	2.1633	$3 \ln 10$	6.9078
4.3	1.4586	8.8	2.1748	$4 \ln 10$	9.2103
4.4	1.4816	8.9	2.1861	$5 \ln 10$	11.5129

*Attach -10 to these logarithms.

Table 5. Exponential Functions

x	e^x	e^{-x}	x	e^x	e^{-x}
0.00	1.0000	1.0000	2.5	12.182	0.0821
0.05	1.0513	0.9512	2.6	13.464	0.0743
0.10	1.1052	0.9048	2.7	14.880	0.0672
0.15	1.1618	0.8607	2.8	16.445	0.0608
0.20	1.2214	0.8187	2.9	18.174	0.0550
0.25	1.2840	0.7788	3.0	20.086	0.0498
0.30	1.3499	0.7408	3.1	22.198	0.0450
0.35	1.4191	0.7047	3.2	24.533	0.0408
0.40	1.4918	0.6703	3.3	27.113	0.0369
0.45	1.5683	0.6376	3.4	29.964	0.0334
0.50	1.6487	0.6065	3.5	33.115	0.0302
0.55	1.7333	0.5769	3.6	36.598	0.0273
0.60	1.8221	0.5488	3.7	40.447	0.0247
0.65	1.9155	0.5220	3.8	44.701	0.0224
0.70	2.0138	0.4966	3.9	49.402	0.0202
0.75	2.1170	0.4724	4.0	54.598	0.0183
0.80	2.2255	0.4493	4.1	60.340	0.0166
0.85	2.3369	0.4274	4.2	66.686	0.0150
0.90	2.4596	0.4066	4.3	73.700	0.0136
0.95	2.5857	0.3867	4.4	81.451	0.0123
1.0	2.7183	0.3679	4.5	90.017	0.0111
1.1	3.0042	0.3329	4.6	99.484	0.0101
1.2	3.3201	0.3012	4.7	109.95	0.0091
1.3	3.6693	0.2725	4.8	121.51	0.0082
1.4	4.0552	0.2466	4.9	134.29	0.0074
1.5	4.4817	0.2231	5	148.41	0.0067
1.6	4.9530	0.2019	6	403.43	0.0025
1.7	5.4739	0.1827	7	1096.6	0.0009
1.8	6.0496	0.1653	8	2981.0	0.0003
1.9	6.6859	0.1496	9	8103.1	0.0001
2.0	7.3891	0.1353	10	22026	0.00005
2.1	8.1662	0.1225			
2.2	9.0250	0.1108			
2.3	9.9742	0.1003			
2.4	11.023	0.0907			

Table 6   477

Table 6. A Short Table of Integrals

The basic forms of Chapter 8 are not included. The constant of integration is omitted.

Forms containing $a + bu$ and $\sqrt{a + bu}$

1. $\displaystyle \int \frac{u\,du}{a + bu} = \frac{1}{b^2}[(a + bu) - a\,\ln(a + bu)]$

2. $\displaystyle \int \frac{du}{u(a + bu)} = -\frac{1}{a}\ln\frac{a + bu}{u}$

3. $\displaystyle \int \frac{u\,du}{(a + bu)^2} = \frac{1}{b^2}\left(\frac{a}{a + bu} + \ln(a + bu)\right)$

4. $\displaystyle \int \frac{du}{u(a + bu)^2} = \frac{1}{a(a + bu)} - \frac{1}{a^2}\ln\frac{a + bu}{u}$

5. $\displaystyle \int u\sqrt{a + bu}\,du = -\frac{2(2a - 3bu)(a + bu)^{3/2}}{15b^2}$

6. $\displaystyle \int \frac{u\,du}{\sqrt{a + bu}} = -\frac{2(2a - bu)\sqrt{a + bu}}{3b^2}$

7. $\displaystyle \int \frac{du}{u\sqrt{a + bu}} = \frac{1}{\sqrt{a}}\ln\left(\frac{\sqrt{a + bu} - \sqrt{a}}{\sqrt{a + bu} + \sqrt{a}}\right), \qquad a > 0$

8. $\displaystyle \int \frac{\sqrt{a + bu}}{u}\,du = 2\sqrt{a + bu} + a\int \frac{du}{u\sqrt{a + bu}}$

Forms containing $\sqrt{u^2 \pm a^2}$ and $\sqrt{a^2 - u^2}$

9. $\displaystyle \int \frac{du}{u^2 - a^2} = \frac{1}{2a}\ln\frac{u - a}{u + a}$

10. $\displaystyle \int \frac{du}{\sqrt{u^2 \pm a^2}} = \ln(u + \sqrt{u^2 \pm a^2})$

11. $\displaystyle \int \frac{du}{u\sqrt{u^2 + a^2}} = -\frac{1}{a}\ln\left(\frac{a + \sqrt{u^2 + a^2}}{u}\right)$

12. $\displaystyle \int \frac{du}{u\sqrt{u^2 - a^2}} = \frac{1}{a}\text{Arcsec}\frac{u}{a}$

13. $\displaystyle \int \frac{du}{u\sqrt{a^2 - u^2}} = -\frac{1}{a}\ln\left(\frac{a + \sqrt{a^2 - u^2}}{u}\right)$

14. $\displaystyle \int \sqrt{u^2 \pm a^2}\,du = \frac{u}{2}\sqrt{u^2 \pm a^2} \pm \frac{a^2}{2}\ln(u + \sqrt{u^2 \pm a^2})$

15. $\displaystyle \int \sqrt{a^2 - u^2}\,du = \frac{u}{2}\sqrt{a^2 - u^2} + \frac{a^2}{2}\text{Arcsin}\frac{u}{a}$

16. $\displaystyle \int \frac{\sqrt{u^2 + a^2}}{u}\,du = \sqrt{u^2 + a^2} - a\,\ln\left(\frac{a + \sqrt{u^2 + a^2}}{u}\right)$

17. $\displaystyle \int \frac{\sqrt{u^2 - a^2}}{u}\,du = \sqrt{u^2 - a^2} - a\,\text{Arcsec}\frac{u}{a}$

Table 6. Continued

18. $\int \dfrac{\sqrt{a^2 - u^2}}{u}\,du = \sqrt{a^2 - u^2} - a\,\ln\left(\dfrac{a + \sqrt{a^2 - u^2}}{u}\right)$

19. $\int (u^2 \pm a^2)^{3/2}\,du = \dfrac{u}{4}(u^2 \pm a^2)^{3/2} \pm \dfrac{3a^2 u}{8}\sqrt{u^2 \pm a^2} + \dfrac{3a^4}{8}\ln(u + \sqrt{u^2 \pm a^2})$

20. $\int (a^2 - u^2)^{3/2}\,du = \dfrac{u}{4}(a^2 - u^2)^{3/2} + \dfrac{3a^2 u}{8}\sqrt{a^2 - u^2} + \dfrac{3a^4}{8}\text{Arcsin}\dfrac{u}{a}$

21. $\int \dfrac{(u^2 + a^2)^{3/2}}{u}\,du = \dfrac{1}{3}(u^2 + a^2)^{3/2} + a^2\sqrt{u^2 + a^2} - a^3\ln\left(\dfrac{a + \sqrt{u^2 + a^2}}{u}\right)$

22. $\int \dfrac{(u^2 - a^2)^{3/2}}{u}\,du = \dfrac{1}{3}(u^2 - a^2)^{3/2} - a^2\sqrt{u^2 - a^2} + a^3\text{Arcsec}\dfrac{u}{a}$

23. $\int \dfrac{(a^2 - u^2)^{3/2}}{u}\,du = \dfrac{1}{3}(a^2 - u^2)^{3/2} - a^2\sqrt{a^2 - u^2} + a^3\ln\left(\dfrac{a + \sqrt{a^2 - u^2}}{u}\right)$

24. $\int \dfrac{du}{(u^2 \pm a^2)^{3/2}} = \pm\dfrac{u}{a^2\sqrt{u^2 \pm a^2}}$

25. $\int \dfrac{du}{(a^2 - u^2)^{3/2}} = \dfrac{u}{a^2\sqrt{a^2 - u^2}}$

26. $\int \dfrac{du}{u(u^2 + a^2)^{3/2}} = \dfrac{1}{a^2\sqrt{u^2 + a^2}} - \dfrac{1}{a^3}\ln\left(\dfrac{a + \sqrt{u^2 + a^2}}{u}\right)$

27. $\int \dfrac{du}{u(u^2 - a^2)^{3/2}} = -\dfrac{1}{a^2\sqrt{u^2 - a^2}} - \dfrac{1}{a^3}\text{Arcsec}\dfrac{u}{a}$

28. $\int \dfrac{du}{u(a^2 - u^2)^{3/2}} = \dfrac{1}{a^2\sqrt{a^2 - u^2}} - \dfrac{1}{a^3}\ln\left(\dfrac{a + \sqrt{a^2 - u^2}}{u}\right)$

Trigonometric forms

29. $\int \sin^2 u\,du = \dfrac{u}{2} - \dfrac{1}{2}\sin u \cos u$

30. $\int \sin^3 u\,du = -\cos u + \dfrac{1}{3}\cos^3 u$

31. $\int \sin^n u\,du = -\dfrac{1}{n}\sin^{n-1}u \cos u + \dfrac{n-1}{n}\int \sin^{n-2}u\,du$

32. $\int \cos^2 u\,du = \dfrac{u}{2} + \dfrac{1}{2}\sin u \cos u$

33. $\int \cos^3 u\,du = \sin u - \dfrac{1}{3}\sin^3 u$

34. $\int \cos^n u\,du = \dfrac{1}{n}\cos^{n-1}u \sin u + \dfrac{n-1}{n}\int \cos^{n-2}u\,du$

35. $\int \tan^n u\,du = \dfrac{\tan^{n-1}u}{n-1} - \int \tan^{n-2}u\,du$

Table 6   479

Table 6. Continued

36. $\displaystyle\int \cot^n u \; du = -\frac{\cot^{n-1}u}{n-1} - \int \cot^{n-2}u \; du$

37. $\displaystyle\int \sec^n u \; du = \frac{\sec^{n-2}u \tan u}{n-1} + \frac{n-2}{n-1}\int \sec^{n-2}u \; du$

38. $\displaystyle\int \csc^n u \; du = -\frac{\csc^{n-2}u \cot u}{n-1} + \frac{n-2}{n-1}\int \csc^{n-2}u \; du$

39. $\displaystyle\int \sin au \sin bu \; du = \frac{\sin(a-b)u}{2(a-b)} - \frac{\sin(a+b)u}{2(a+b)}$

40. $\displaystyle\int \sin au \cos bu \; du = -\frac{\cos(a-b)u}{2(a-b)} - \frac{\cos(a+b)u}{2(a+b)}$

41. $\displaystyle\int \cos au \cos bu \; du = \frac{\sin(a-b)u}{2(a-b)} + \frac{\sin(a+b)u}{2(a+b)}$

42. $\displaystyle\int \sin^m u \cos^n u \; du = \frac{\sin^{m+1}u \cos^{n-1}u}{m+n} + \frac{n-1}{m+n}\int \sin^m u \cos^{n-2}u \; du$

43. $\displaystyle\int \sin^m u \cos^n u \; du = -\frac{\sin^{m-1}u \cos^{n+1}u}{m+n} + \frac{m-1}{m+n}\int \sin^{m-2}u \cos^n u \; du$

Other forms

44. $\displaystyle\int u e^{au} du = \frac{e^{au}(au-1)}{a^2}$

45. $\displaystyle\int u^2 e^{au} du = \frac{e^{au}}{a^3}(a^2 u^2 - 2au + 2)$

46. $\displaystyle\int u^n \ln u \; du = u^{n+1}\left(\frac{\ln u}{n+1} - \frac{1}{(n+1)^2}\right)$

47. $\displaystyle\int u \sin u \; du = \sin u - u \cos u$

48. $\displaystyle\int u \cos u \; du = \cos u + u \sin u$

49. $\displaystyle\int e^{au}\sin bu \; du = \frac{e^{au}(a \sin bu - b \cos bu)}{a^2 + b^2}$

50. $\displaystyle\int e^{au}\cos bu \; du = \frac{e^{au}(a \cos bu + b \sin bu)}{a^2 + b^2}$

51. $\displaystyle\int \text{Arcsin } u \; du = u \text{ Arcsin } u + \sqrt{1 - u^2}$

52. $\displaystyle\int \text{Arctan } u \; du = u \text{ Arctan } u - \frac{1}{2}\ln(1 + u^2)$

Answers to Odd-numbered Exercises

Exercises 1–2, p. 4

1. (a) rational, real   (b) irrational, real   (c) imaginary   (d) imaginary   **3.** (a) 3   (b) $\dfrac{7}{2}$ (c) 4

5. (a) <   (b) >   (c) >   **7.**   **9.** The points lie on a straight line.

11. On a line parallel to the y-axis, one unit to the right.   **13.** 0   **15.** to the right of the y-axis
17. In quadrant II above a line parallel to the x-axis and one unit above.   **19.** I, III   **21.** (5, 4)
23.

Exercises 1–3, p. 9

1.   **3.**   **5.**   **7.**   **9.**

     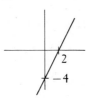   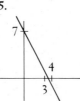

11.   **13.**   **15.**   **17.**   **19.**

21.   **23.**   **25.**   **27.**   **29.**

     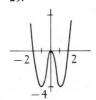

481

31.

33. (a) To y-axis　(b) Not to either axis　(c) Not to either axis

35. (a) No　(b) Yes, $x = 0$　(c) No

37.

39.

41.

43.

Exercises 1–4, p. 14

1. $2\sqrt{29}$　3. 3　5. $\sqrt{85}$　7. 7　9. $\dfrac{5}{2}$　11. undefined　13. $-\dfrac{7}{6}$　15. 0　17. $\dfrac{1}{3}\sqrt{3}$

19. $-\dfrac{1}{3}\sqrt{3}$　21. 20.0°　23. 98.5°　25. parallel　27. perpendicular　29. 8, -2　31. -3

33. two sides equal $2\sqrt{10}$　35. $m_1 = \dfrac{1}{4}$, $m_2 = \dfrac{5}{4}$　37. 10

Exercises 1–5, p. 19

1. $4x - y + 20 = 0$　3. $7x - 6y - 16 = 0$　5. $x - y + 2 = 0$　7. $y = -3$　9. $x = -3$

11. $3x - 2y - 12 = 0$　13. $x + 3y + 5 = 0$　15. $x + 7y - 18 = 0$

17.

19.

21. $y = \dfrac{3}{2}x - \dfrac{1}{2}$; $m = \dfrac{3}{2}$, $b = -\dfrac{1}{2}$　23. $y = \dfrac{5}{2}x + \dfrac{5}{2}$; $m = \dfrac{5}{2}$, $b = \dfrac{5}{2}$

25. -2　27. 1　29. $m_1 = m_2 = \dfrac{3}{2}$　31. $m_1 = 2$, $m_2 = -\dfrac{1}{2}$

33. $x + 2y - 4 = 0$　35. $3x + y - 18 = 0$　37. $s = 50t + 10$

39. $L = \dfrac{2}{3}F + 15$　41. 3.35 kJ　43. $n = \dfrac{7}{6}t + 10$; at 6:30, $n = 10$;

at 8:30, $n = 150$

45.

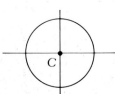

47.

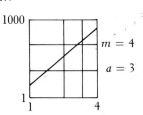

$m = 4$

$a = 3$

49.

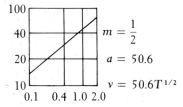

$m = \dfrac{1}{2}$

$a = 50.6$

$v = 50.6T^{1/2}$

Exercises 1–6, p. 24

1. $(2, 1)$, $r = 5$　3. $(-1, 0)$, $r = 2$　5. $x^2 + y^2 = 9$　7. $x^2 + y^2 - 4x - 4y - 8 = 0$

9. $x^2 + y^2 + 4x - 10y + 24 = 0$　11. $x^2 + y^2 - 4x - 2y - 3 = 0$　13. $x^2 + y^2 + 6x - 10y + 9 = 0$

15. $x^2 + y^2 - 4x - 10y + 4 = 0$; $x^2 + y^2 + 4x + 10y + 4 = 0$

17. $(0,0)$　19. $(1,0)$　21. $(-4,5)$　23. $(1,2)$

$r = 5$　$r = 3$　$r = 7$　$r = \dfrac{1}{2}\sqrt{22}$

25. Symmetrical to both axes and origin　27. Symmetrical to y-axis　29. $(7, 0)$, $(-1, 0)$

31. $3x^2 + 3y^2 + 4x + 8y - 20 = 0$, circle

33. $x^2 + y^2 = 2.79$ (assume center of circle at center of coordinate system)

35. $x^2 + y^2 = 576$, $x^2 + y^2 + 36y + 315 = 0$ (in inches)

Exercises 1–7, p. 29

1. $F(1,0)$, $x = -1$

3. $F(-1,0)$, $x = 1$

5. $F(0,2)$, $y = -2$

7. $F(0,-1)$, $y = 1$

9. $F\left(\frac{1}{2},0\right)$, $x = -\frac{1}{2}$

11. $F\left(0,\frac{1}{4}\right)$, $y = -\frac{1}{4}$

13. $y^2 = 12x$

15. $x^2 = 16y$

17. $x^2 = 4y$

19. $x^2 = \frac{1}{8}y$

21. $y^2 - 2y - 12x + 37 = 0$

23. $x^2 - 2x + 8y - 23 = 0$

25. $x = 2$, $y = 4$

27. $x^2 = 100y$

29.

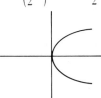

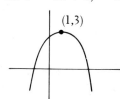

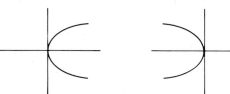

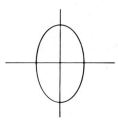

(1,3)

(3,1)

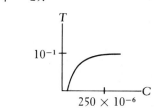

T

10^{-1}

250×10^{-6}

C

31. $y^2 = 8x$ or $x^2 = 8y$ with vertex midway between island and shore

Exercises 1–8, p. 35

1. $V(2,0)$, $V(-2,0)$
$F(\sqrt{3},0)$, $F(-\sqrt{3},0)$

3. $V(0,6)$, $V(0,-6)$
$F(0,\sqrt{11})$, $F(0,-\sqrt{11})$

5. $V(3,0)$, $V(-3,0)$
$F(\sqrt{5},0)$, $F(-\sqrt{5},0)$

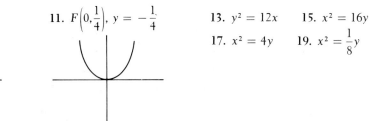

7. $V(0,7)$, $V(0,-7)$
$F(0)$, $\sqrt{45})$, $F(0, -\sqrt{45})$

9. $V(0,4)$, $V(0,-4)$
$F(0,\sqrt{14})$,
$F(0,-\sqrt{14})$

11. $V\left(\frac{5}{2},0\right)$, $V\left(-\frac{5}{2},0\right)$
$F(\sqrt{21}/2,0)$, $F(-\sqrt{21}/2,0)$

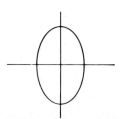

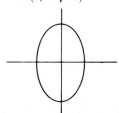

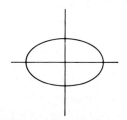

13. $144x^2 + 225y^2 = 32400$ 15. $9x^2 + 5y^2 = 45$ 17. $3x^2 + 20y^2 = 192$ 19. $4x^2 + y^2 = 20$
21. $16x^2 + 25y^2 - 32x - 50y - 359 = 0$ 23. $9x + 5y^2 - 18x - 20y - 16 = 0$
25. $2x^2 + 3y^2 - 8x - 4 = 2x^2 + 3(-y)^2 - 8x - 4$ 27. $9x^2 + 25y^2 = 22500$
29. $2.70x^2 + 2.76y^2 = 7.45 \times 10^7$ 31. major axis = 8.5 in., minor axis = 6.0 in.

Exercises 1–9, p. 41

1. $V(5,0)$, $V(-5,0)$
   $F(13,0)$, $F(-13,0)$

3. $V(0,3)$, $V(0,-3)$
   $F(0,\sqrt{10})$, $F(0,-\sqrt{10})$

5. $V(\sqrt{2},0)$, $V(-\sqrt{2},0)$
   $F(\sqrt{6},0)$, $F(-\sqrt{6},0)$

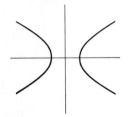
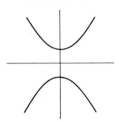
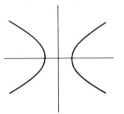

7. $V(0,\sqrt{5})$, $V(0,-\sqrt{5})$
   $F(0,\sqrt{7})$, $F(0,-\sqrt{7})$

9. $V(0,2)$, $V(0,-2)$
   $F(0,\sqrt{5})$, $F(0,-\sqrt{5})$

11. $V(2,0)$, $V(-2,0)$
    $F\left(\frac{2}{3}\sqrt{13},0\right)$, $F\left(-\frac{2}{3}\sqrt{13},0\right)$

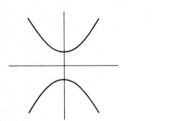
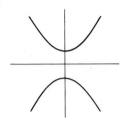
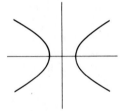

13. $16x^2 - 9y^2 = 144$ 15. $9y^2 - 25x^2 = 900$ 17. $3x^2 - y^2 = 3$ 19. $4x^2 - 5y^2 = 20$

21.
23.

25. $9x^2 - 16y^2 - 108x + 64y + 116 = .0$
27. $9x^2 - y^2 - 36x + 27 = 0$

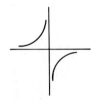

29.
R
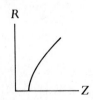

Z

31.
f

λ

33. dist(rifle to P) − dist(target to P) = constant
    (related to distance from rifle to target)

Exercises 1–10, p. 46

1. Parabola, $(-1,2)$

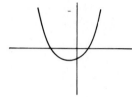

3. Hyperbola, $(1,2)$

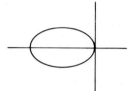

5. Ellipse, $(-1,0)$

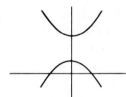

7. Parabola, $(-3,1)$

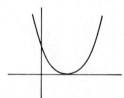

9. $y^2 - 6y - 16x - 7 = 0$ **11.** $x^2 + 6x - 4y + 17 = 0$ **13.** $16x^2 + 25y^2 + 64x - 100y - 236 = 0$
15. $16x^2 - 9y^2 + 32x + 18y - 137 = 0$
17. Parabola, $(-1,-1)$ **19.** Ellipse, $(-3,0)$ **21.** Hyperbola, $(0,4)$ **23.** Parabola, $(1,0)$

25. $x^2 - y^2 + 4x - 2y - 22 = 0$ **27.** $y^2 - 2y + 4x + 1 = 0$ **29.**   **31.** $\dfrac{(x - 0.9)^2}{3.7^2} + \dfrac{y^2}{3.6^2} = 1$

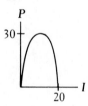

Exercises 1–11, p. 50

1. Circle   **3.** Parabola   **5.** Hyperbola   **7.** Circle   **9.** Ellipse   **11.** Hyperbola   **13.** Ellipse
15. Parabola
17. Parabola; $V(-4,0)$; $F(-4,2)$ **19.** Hyperbola; $C(+1, -2)$;   **21.** Ellipse; $C(5,0)$; $V(5, \pm 2\sqrt{2})$
    $V(+1, -2, \pm \sqrt{2})$

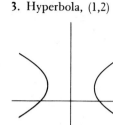
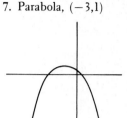

23. Parabola; $V\left(-\dfrac{1}{2}, \dfrac{5}{2}\right)$; $F\left(\dfrac{1}{2}, \dfrac{5}{2}\right)$ **25.** Hyperbola   **27.** Ellipse

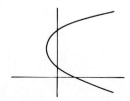

Review Exercises for Chapter 1, p. 51

1. $4x - y - 11 = 0$

3. $2x + 3y + 3 = 0$

5. $x^2 + y^2 - 2x + 4y - 5 = 0$

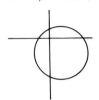

7. $y^2 = 12x$

9. $9x^2 + 25y^2 = 900$

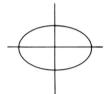

11. $144y^2 - 169x^2 = 24336$

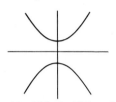

13. $(-3,0)$, $r = 4$

15. $(0,-5)$, $y = 5$

17. $V(0,4)$, $V(0,-4)$
$F(0,\sqrt{15}$, $F(0,-\sqrt{15})$

19. $V(2,0)$, $V(-2,0)$

$F\left(\frac{2}{5}\sqrt{35},0\right)$, $F\left(-\frac{2}{5}\sqrt{35},0\right)$

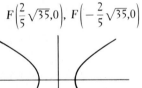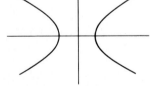

21. $V(4,-8)$, $F(4,-7)$

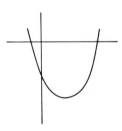

23. $(2,-1)$

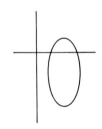

25. $(1.90, 1.55)$, $(1.90, -1.55)$, $(-1.90, 1.55)$, $(-1.90, -1.55)$

27. $m_1 = -\frac{12}{5}$, $m_2 = \frac{5}{12}$; $d_1^2 = 169$, $d_2^2 = 169$, $d_3^2 = 338$ **29.** $x^2 - 6x - 8y + 1 = 0$

31.   **33.**   **35.** $x^2 = -80y$ **37.**   **39.** $225y = 450x - x^2$

41. No   **43.** $\dfrac{x^2}{1.464 \times 10^6} + \dfrac{y^2}{1.461 \times 10^6} = 1$ **45.**   **47.** 13.5 ft

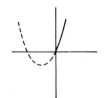

Exercises 2–1, p. 59

1. $A = s^2$ **3.** $V = \dfrac{8}{3}\pi r^2$ **5.** $p = 100(c - 3)$ **7.** $n_2 = -\dfrac{2}{3}n_1 + \dfrac{4000}{3}$ **9.** $-18, 70$ **11.** $15, 0$

13. $0, 2a^3$ **15.** $-5t + 7, 25t + 42$ **17.** $\sqrt{x^2 + 3}$ **19.** $\dfrac{\sqrt{x - 1}}{x^2 + 4}$ **21.** 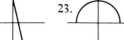   **23.**

25. $\sqrt[6]{x^6 + 1}$ **27.** $(4x - 5)^{7/2}$ **29.** $\dfrac{3x + 4}{\sqrt{2x + 1}}$ **31.** $\dfrac{1}{(x^2 + 1)^{3/2}}$ **33.** Yes, no   **35.** $y \neq 1$

37. $0.006T^2 + 4.000T + 340$ **39.** $\dfrac{(4 + \pi)p^2 - 480p + 14,400}{16\pi}$ **41.** $p = 2\left(x + \dfrac{A}{x}\right)$

Exercises 2–2, p. 67

1. cont. all x **3.** not cont. $x = -3$, div. by zero   **5.** cont. all x
7. not cont. $x = 1$, small change
9. 10.000, 8.500, 7.300, 7.030, 7.003, 4.000, 5.500, 6.700, 6.970, 6.997; $\lim\limits_{x \to 3} f(x) = 7$
11. 1.71, 1.9701, 1.997001, 2.31, 2.0301, 2.003001; $\lim\limits_{x \to 1} f(x) = 2$ **13.** 7   **15.** 1   **17.** 1   **19.** -2
21. 2   **23.** -4 **25.** does not exist   **27.** 0   **29.** 3   **31.** 0
33. 1.26×10^{-6}, 1.827×10^{-6}, 1.8837×10^{-6}, 1.8963×10^{-6}, 1.953×10^{-6}, 2.52×10^{-6};
$\lim\limits_{i \to 0.3} B = 1.89 \times 10^{-6}$ W
35. 34.8°C, 0°C

Exercises 2–3, p. 73

1. (slopes) 3.5, 3.9, 3.99, 3.999; $m = 4$

3. (slopes) $-3.5, -3.9, -3.99,$ -3.999; $m = -4$

5. 4   **7.** -4

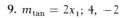

9. $m_{\tan} = 2x_1$; 4, -2 **11.** $m_{\tan} = 2x_1 + 2$; $-4, 4$ **13.** $m_{\tan} = 2x_1 + 4$; $-2, 8$

   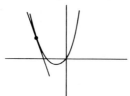   

15. $m_{\tan} = 6 - 2x_1$; 10, 0   **17.** $m_{\tan} = 3x_1^2 - 2$; 1, $-2, 1$ **19.** $m_{\tan} = 4x_1^3$; 0, 0.5, 4

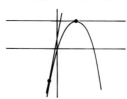   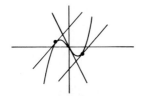   

Exercises 2–4, p. 77

1. 3   **3.** -2 **5.** $2x$ **7.** $10x$ **9.** $2x - 7$ **11.** $8 - 4x$ **13.** $3x^2 + 4$ **15.** $-\dfrac{1}{(x + 2)^2}$

17. $1 - \dfrac{1}{x^2}$ **19.** $-\dfrac{4}{x^3}$ **21.** $4x^3 + 3x^2 + 2x + 1$ **23.** $4x^3 + \dfrac{2}{x^2}$ **25.** $\dfrac{1}{2\sqrt{x+1}}$ **27.** $-\dfrac{3}{2\sqrt{1-3x}}$

Exercises 2−5, p. 81

1. $4.00, 4.00, 4.00, 4.00, 4.00$; $\lim\limits_{t \to 3} v = 4$ ft/s

3. $-14, -15, -15.8, -15.98, -15.998$; $\lim\limits_{t \to 4} v = -16$ ft/s

5. $4; 4$ **7.** $-4t; -16$ **9.** $3 + \dfrac{2}{t^2}$ **11.** $3t^2 - 6$ **13.** $12t - 4$ **15.** -4 **17.** 2 **19.** $2\pi r, 6\pi$

21. $0.285 \ \Omega/°C$ **23.** $\dfrac{25}{(5+R)^2}, \dfrac{25}{169}$ **25.** $4\pi r - \dfrac{200}{r^2}$ **27.** $-\dfrac{2}{r^3}, -2 \times 10^4$ lx/m, -2×10^2 lx/m

Exercises 2−6, p. 87

1. $5x^4$ **3.** $36x^8$ **5.** $4x^3$ **7.** $2x + 2$ **9.** $15x^2 - 1$ **11.** $8x^7 - 28x^6 - 1$ **13.** $42x^6 - 15x^2$
15. $x^2 + x$ **17.** 360 **19.** 14 **21.** $30t^4 - 5$ **23.** $120 - 32t$ **25.** 1 **27.** $3\pi r^2$ **29.** 6.0 V/°C
31. 300 m/km

Exercises 2−7, p. 90

1. $9x^2 + 4x$ **3.** $54x^2 - 60x$ **5.** $4x - 1$ **7.** $-14x^6 + 30x^4 + 4x^3 - 18x^2 - 6x$ **9.** $\dfrac{3}{(2x+3)^2}$

11. $\dfrac{-2x}{(x^2+1)^2}$ **13.** $\dfrac{6x - 2x^2}{(3-2x)^2}$ **15.** $\dfrac{-48x^5 - 12}{(4x^5 - 3x - 4)^2}$ **17.** $\dfrac{-12x^3 + 45x^2 - 14x}{(3x - 7)^2}$ **19.** 12 **21.** $1, -1$

23. $-\dfrac{1}{16}$ ft/s   **25.** 0.925 V/s   **27.** $\dfrac{E^2(R - r)}{(R + r)^3}$

Exercises 2−8, p. 96

1. $\dfrac{1}{2x^{1/2}}$ **3.** $-\dfrac{2}{x^3}$ **5.** $-\dfrac{1}{x^{4/3}}$ **7.** $\dfrac{3}{2}x^{1/2} + \dfrac{1}{x^2}$ **9.** $10x(x^2 + 1)^4$ **11.** $-192x^2(7 - 4x^3)^7$

13. $\dfrac{2x^2}{(2x^3 - 3)^{2/3}}$ **15.** $\dfrac{8x}{(1 - x^2)^5}$ **17.** $\dfrac{24x^3}{(2x^4 - 5)^{1/4}}$ **19.** $\dfrac{-4x}{(1 - 8x^2)^{3/4}}$ **21.** $\dfrac{3x - 2}{2(x - 1)^{1/2}}$

23. $\dfrac{-16x - 22}{(4x + 3)^{1/2}(8x + 1)^2}$ **25.** $x = 0$ **27.** 1 **29.** $\dfrac{188}{3}\sqrt[3]{60}$ **31.** $\dfrac{-450000}{V^{5/2}}, -4.50$ kPa/cm³

33. $\dfrac{-kqx}{(x^2 + b^2)^{3/2}}$ **35.** $\dfrac{-3kr}{[r^2 + (l/2)^2]^{5/2}}$

Exercises 2−9, p. 99

1. $-\dfrac{3}{2}$ **3.** $\dfrac{6x + 1}{4}$ **5.** $\dfrac{x}{y}$ **7.** $\dfrac{2x}{5y^4}$ **9.** $\dfrac{2x}{2y + 1}$ **11.** $\dfrac{-3y}{3x + 1}$ **13.** $\dfrac{-2x - y^3}{3xy^2 + 3}$

15. $\dfrac{-3x(x^2 + 1)^2}{y(y^2 + 1)}$ **17.** 1 **19.** $-\dfrac{x}{y}$

Exercises 2−10, p. 103

1.   **3.**   **5.**   **7.**

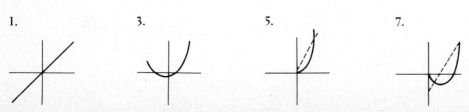

9.   **11.**   **13.**   **15.**

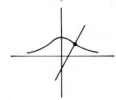

−0.2 lb/s

Review Exercises for Chapter 2, p. 105

1. −4   **3.** $\dfrac{1}{4}$ **5.** 1   **7.** $\dfrac{7}{3}$ **9.** $\dfrac{2}{3}$ **11.** −2   **13.** 5   **15.** −4x   **17.** $-\dfrac{4}{x^3}$ **19.** $\dfrac{1}{2\sqrt{x+5}}$

21. $14x^6 - 6x$ **23.** $\dfrac{2}{x^{1/2}} + \dfrac{1}{x^2}$ **25.** $\dfrac{1}{(1-x)^2}$ **27.** $-12(2-3x)^3$ **29.** $\dfrac{9x}{(5-2x^2)^{7/4}}$

31. $\dfrac{-15x^2 + 2x}{(1-6x)^{1/2}}$ **33.** $\dfrac{-2x-3}{2x^2(4x+3)^{1/2}}$ **35.** $\dfrac{2x - 6(2x-3y)^2}{1 - 9(2x-3y)^2}$ **37.** $\dfrac{1}{2}C_2, 0$ **39.** −31   **41.** 0.01 m/s

43. 0.04 V   **45.** $-\dfrac{2k}{r^3}$ **47.** $\dfrac{a}{\sqrt{v_0^2 + 2as}}$ **49.** $\dfrac{40V_2^{0.4}}{V_1^{1.4}}$ **51.** $4x - x^3, 4 - 3x^2$ **53.**

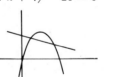

−5 kPa/cm³

Exercises 3−1, p. 110

1. $4x - y - 2 = 0$ **3.** $2y + x - 2 = 0$ **5.** $x + 4y - 21 = 0$ **7.** $8x - 4y - 7 = 0$

9. $\sqrt{3}x + 8y - 7 = 0, 16x - 2\sqrt{3}y - 15\sqrt{3} = 0$ **11.** $2x - 12y + 37 = 0, 72x + 12y + 37 = 0$

13. $-\dfrac{1}{4}$ **15.** $3x - 5y - 150 = 0$ **17.** $x - \sqrt{3}y + 4 = 0$

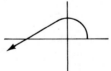

Exercises 3−2, p. 116

1. 3.16, 341.5°   **3.** 0.479, 211.5°   **5.** $a = 0$ **7.** 0.442, 58.1°   **9.** 301, 86.2°   **11.** 1.105
13. 136 ft/s, 28.1°; 124 ft/s, 345°   **15.** 32 ft/s²; 270.0° for both   **17.** 276 m/s, 43.5°; 2090 m/s, 16.7°
19. 21.2 mi/min, 296.6°

Exercises 3−3, p. 119

1. 0.0900 Ω/s   **3.** 1.22 $/week   **5.** 0.00401   **7.** 0.375 V/s   **9.** 100 ft²/min   **11.** 10.2 ft/s
13. −3.75 kPa/min   **15.** 2.51 × 10⁶ mm³/s   **17.** 128 km/h   **19.** 12.0 ft/s

Exercises 3−4, p. 126

1. Inc. $x > 1$, dec. $x < 1$ **3.** Inc. $-2 < x < 2$, dec. $x < -2, x > 2$ **5.** Min. (1, −1)
7. Min. (−2, −16), max. (2, 16)   **9.** Conc. up all x **11.** Conc. up $x < 0$, conc. down $x > 0$, infl. (0, 0)

13.

15.

17. Max.(3,18), conc. down all x

19. Max. $(-2,8)$, min.(0,0), infl.$(-1,4)$

21. Max.$(-1,4)$, min.$(1,-4)$, infl. $(0, 0)$

23. Max.(1,1), infl.(0,0), $\left(\dfrac{2}{3}, \dfrac{16}{37}\right)$

25.

27. $V = 4x^3 - 40x^2 + 96x$

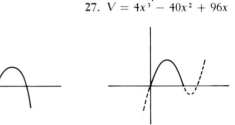

Exercises 3–5, p. 130

1. Inc. $x < 0$, dec. $x > 0$
Conc. up $x < 0$, $x > 0$
Asym. $x = 0$

3. Int. $(-\sqrt[3]{2},0)$, min. (1,3), infl. $(-\sqrt[3]{2},0)$, asym. $x = 0$

5. Int.(0,0), max.$(-2,-4)$, min.(0,0), asym. $x = -1$

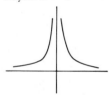

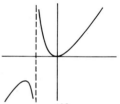

7. Int.$(0,-1)$, max.$(0,-1)$, asym. $x = 1$, $x = -1$

9. Asym. $R_T = 5$, inc. $R \geq 0$

11. $A = 2\pi r^2 + \dfrac{40}{r}$, min.(1.47,40.8)
asym. $r = 0$

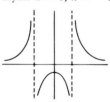

Exercises 3–6, p. 135

1. 196 ft   **3.** 0.250 ft/min   **5.** 8, 8   **7.** 5.00×10^5 ft²   **9.** $2\sqrt{2}$ by $2\sqrt{2}$ **11.** $5\sqrt{2}$, $5\sqrt{2}$

13. 12   **15.** 3 km   **17.** $\dfrac{4}{3}$ in.   **19.** 2 units   **21.** 0.58 L   **23.** Width $= \dfrac{16\sqrt{3}}{3}$ in., depth $= \dfrac{16\sqrt{6}}{3}$ in.

Review Exercises for Chapter 3, p. 137

1. $5x - y + 1 = 0$ **3.** $27x - 3y - 26 = 0$ **5.** 31.0, 359.5°   **7.** 48.0, 0.3°

9. Min.$(-2,-16)$
conc. up all x

11. Int: (0, 0), $(\pm 3\sqrt{3},0)$
Max:(3,54), Min:$(-3,-54)$
Infl:(0,0)

13. Min.$(2,-48)$
Conc. up $x < 0$, $x > 0$

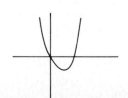

15. Int:(0,0) Asym: $x = -1$, $y = 1$ **17.** $2x - y + 1 = 0$ **19.** 3180 cm/s, 292.2°   **21.**

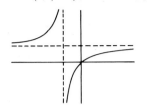

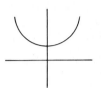

23. 3.27 in./min   **25.** $r = 1$, $\theta = 2$ **27.** Int:(1,0)   **29.** 13.0 km/h   **31.** 1.94 units

$\text{Max:}\left(2, \dfrac{1}{4}\right)$

$\text{Infl:}\left(3, \dfrac{2}{9}\right)$

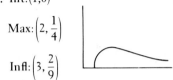

Asym: $P = 0$

$V \geq 1$ (only meaningful values)

Exercises 4–1, p. 143

1. $(5x^4 + 1)\,dx$ **3.** $8x(x^2 - 1)^3\,dx$ **5.** $x(1 - x)^2(-5x + 2)\,dx$ **7.** $\dfrac{dx}{(x + 1)^2}$ **9.** 12.28, 12

11. 1.71275, 1.675   **13.** -2.4, -2.473088 **15.** 0.6257, 0.62649   **17.** 16 in.²   **19.** 12.2 V

21. 2.64 in.³   **23.** $\dfrac{dA}{A} = \dfrac{2\,ds}{s}$

Exercises 4–2, p. 145

1. x^3 **3.** x^6 **5.** $\dfrac{3}{2}x^4 + x$ **7.** $\dfrac{2}{3}x^3 - \dfrac{1}{2}x^2$ **9.** $\dfrac{1}{x}$ **11.** $\dfrac{2}{x^3}$ **13.** $\dfrac{2}{5}x^5 + x$ **15.** $(2x + 1)^6$

17. $(x^2 - 1)^4$ **19.** $\dfrac{1}{40}(2x^4 + 1)^5$ **21.** $(6x + 1)^{3/2}$ **23.** $\dfrac{1}{4}(3x + 1)^{4/3}$

Exercises 4–3, p. 150

1. $x^2 + C$ **3.** $\dfrac{1}{8}x^8 + C$ **5.** $\dfrac{2}{5}x^{5/2} + C$ **7.** $-\dfrac{1}{3x^3} + C$ **9.** $\dfrac{1}{3}x^3 - \dfrac{1}{6}x^6 + C$ **11.** $\dfrac{1}{6}(1 + 2x)^3 + C$

13. $\dfrac{2}{5}x^{5/2} - \dfrac{5}{3}x^3 + C$ **15.** $\dfrac{2}{7}x^{7/2} - \dfrac{2}{5}x^{5/2} + C$ **17.** $\dfrac{1}{6}(x^2 - 1)^6 + C$ **19.** $\dfrac{1}{5}(x^4 + 3)^5 + C$

21. $\dfrac{1}{40}(x^5 + 4)^8 + C$ **23.** $\dfrac{1}{12}(8x + 1)^{3/2} + C$ **25.** $\dfrac{1}{6}\sqrt{6x^2 + 1} + C$ **27.** $\sqrt{x^2 - 2x} + C$

29. $y = 2x^3 + 2$ **31.** $y = 5 - \dfrac{1}{18}(1 - x^3)^6$ **33.** $12y = 83 + (1 - 4x^2)^{3/2}$ **35.** $y = 3x^2 + 2x - 3$

Exercises 4–4, p. 156

1. 9, 12.15   **3.** 1.92, 2.28   **5.** 7.625, 8.208   **7.** 0.464, 0.5995   **9.** 13.5   **11.** $\dfrac{8}{3}$ **13.** 9   **15.** 0.8

Exercises 4–5, p. 159

1. 1   **3.** $\dfrac{254}{7}$ **5.** $\dfrac{3}{2}\sqrt[3]{2}$ **7.** $\dfrac{747}{20}$ **9.** $\dfrac{4}{3}$ **11.** $\dfrac{81}{4}$ **13.** 2   **15.** 49   **17.** $\dfrac{364}{3}$ **19.** $\dfrac{110}{3}$

Exercises 4–6, p. 162

1.   3.    5.    7.    9. 6.8   11. 0.7

13.   15. 23.0

Exercises 4–7, p. 166

1. $\dfrac{11}{2} = 5.50$, $\dfrac{16}{3} = 5.33$ 3. 7.661, $\dfrac{23}{3} = 7.667$ 5. 0.2042   7. 2.996   9. 0.5205   11. 21.74

13. 10.58   15. 31.70

Exercises 4–8, p. 170

1. (a) 6   (b) 6   3. (a) 19.67   (b) 19.67   5. 0.2027   7. 3.084   9. 0.5114   11. 10.60

Review Exercises for Chapter 4, p. 171

1. $x^4 - \dfrac{1}{2}x^2 + C$ 3. $x^2 - \dfrac{4}{5}x^{5/2} + C$ 5. $\dfrac{20}{3}$ 7. $\dfrac{16}{3}$ 9. $\dfrac{1}{5(2 - 5x)} + C$

11. $-\dfrac{2}{7}(7 - 2x)^{7/4} + C$ 13. $\dfrac{9}{8}(3\sqrt[3]{3} - 1)$ 15. $-\dfrac{1}{30}(1 - 2x^3)^5 + C$ 17. $-\dfrac{1}{(2x - x^3)} + C$

19. $\dfrac{3350}{3}$ 21. $\dfrac{-6x\,dx}{(x^2 - 1)^4}$ 23. $\dfrac{(1 - 4x)\,dx}{(1 - 3x)^{2/3}}$ 25. 0.061   27. $y = 3x - \dfrac{1}{3}x^3 + \dfrac{17}{3}$ 29. 22

31. 0.842   33. 0.811   35. 1.01   37. 13.90   39. 4.02 mm³   41. $\dfrac{R\,dR}{R^2 + X^2}$

Exercises 5–1, p. 178

1. −96.0 ft/s   3. $v = 3t^2 + 5$ 5. 560 cm   7. 216 ft   9. 336 ft   11. 13.5 m   13. 1.0 C

15. 5.14 C   17. 667 V   19. 117 V   21. $\theta = 2t^2$ 23. 66.7 A   25. $\dfrac{k}{x_1}$

27. $m = 1002 - 2\sqrt{t + 1}$, 2.51×10^5 min

Exercises 5–2, p. 183

1. 2   3. $\dfrac{8}{3}$ 5. $\dfrac{3}{4}$ 7. $\dfrac{4}{3}\sqrt{2}$ 9. $\dfrac{1}{2}$ 11. $\dfrac{1}{6}$ 13. 12   15. $\dfrac{19}{3}$ 17. $\dfrac{15}{4}$ 19. $\dfrac{7}{6}$ 21. $\dfrac{256}{5}$

23. $\dfrac{2}{3}$ 25. $\dfrac{1}{3}$ 27. 1   29. 18.0 J   31. $\dfrac{128}{5}$ ft   33. 0.513 J

Exercises 5–3, p. 189

1. $\dfrac{1}{3}\pi$ 3. $\dfrac{1}{3}\pi$ 5. $\dfrac{348}{5}\pi$ 7. $\dfrac{768}{7}\pi$ 9. 72π 11. $\dfrac{16}{3}\pi$ 13. $\dfrac{1}{3}\pi$ 15. $\dfrac{1}{3}\pi$ 17. $\dfrac{2}{5}\pi$

19. $\dfrac{8}{3}\pi$ 21. $\dfrac{16}{15}\pi$ 23. $\dfrac{16}{3}\pi$ 25. $\dfrac{8}{3}\pi$ 27. $\dfrac{1}{3}\pi r^2 h$ 29. 3375π m³

Exercises 5−4, p. 196

1. $\left(\frac{10}{3}, 0\right)$ **3.** $\left(\frac{14}{15}, 0\right)$ **5.** $\left(-\frac{1}{2}, \frac{1}{2}\right)$ **7.** $\left(\frac{7}{22}, \frac{5}{22}\right)$ **9.** $\left(0, \frac{6}{5}\right)$ **11.** $\left(\frac{4}{3}, \frac{4}{3}\right)$ **13.** $\left(\frac{3}{5}, \frac{12}{35}\right)$

15. $\left(0, \frac{5}{6}\right)$ **17.** $\left(\frac{2}{3}, 0\right)$ **19.** $\left(\frac{9}{7}, 0\right)$ **21.** $\left(\frac{2}{3}a, \frac{1}{3}b\right)$ **23.** $\left(\frac{3}{8}a, 0\right)$

Exercises 5−5, p. 201

1. 128, 4   **3.** 214, 3.27   **5.** $\frac{64}{15}$ **7.** $\frac{2}{3}\sqrt{6}$ **9.** $\frac{1}{6}ma^2$ **11.** $\frac{4}{7}\sqrt{7}$ **13.** $\frac{8}{11}\sqrt{55}$ **15.** $\frac{64}{3}\pi$

17. $\frac{2}{5}\sqrt{10}$ **19.** $\frac{3}{10}mr^2$

Exercises 5−6, p. 204

1. 37.5 lb·in   **3.** 90.0 ft·lb   **5.** 2.40 J   **7.** 0.99 kJ   **9.** 10,000 ft·lb   **11.** 1800 N·m
13. 8.82×10^4 ft·lb   **15.** 1.26×10^6 N·m = 1.26 MJ

Exercises 5−7, p. 207

1. 499 lb   **3.** 196,000 N = 196 kN   **5.** 5320 lb   **7.** 35,900 N = 35.9 kN   **9.** 2560 lb
11. 11,700 lb

Exercises 5−8, p. 210

1. $\frac{38}{3}$ **3.** 21   **5.** $4\pi\sqrt{2}$ **7.** $\frac{56\pi}{3}$ **9.** 2.67 A   **11.** 10

Review Exercises for Chapter 5, p. 211

1. 160 ft/s   **3.** 4.22 s   **5.** 9.20 C   **7.** 3640 V   **9.** $y = 20x + \frac{1}{120}x^3$ **11.** $\frac{2}{3}$ **13.** 18

15. $\frac{48}{5}\pi$ **17.** $\frac{243}{10}\pi$ **19.** $\left(\frac{40}{21}, \frac{10}{3}\right)$ **21.** $\left(\frac{14}{5}, 0\right)$ **23.** $\frac{8}{5}$ **25.** $\frac{256}{3}\pi$ **27.** 9.00 ft-lb

29. 185,000 N·m = 185 kJ   **31.** 3990 lb   **33.** $\frac{46}{3}$ **35.** 0.286 Ω   **37.** 172,000 lb

Exercises 6−1, p. 220

1. 15.2°   **3.** 315°48′   **5.** $\frac{\pi}{12}, \frac{5\pi}{6}$ **7.** $\frac{5\pi}{12}, \frac{11\pi}{6}$ **9.** 72°, 270°   **11.** 10°, 315°   **13.** 0.401

15. 43.0°   **17.** 0.7071   **19.** 3.372   **21.** −1.732   **23.** −0.1221   **25.** $\frac{\pi}{9}, \frac{8\pi}{9}$ **27.** $\frac{13\pi}{18}, \frac{31\pi}{18}$

29.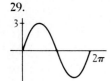

31.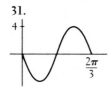

33.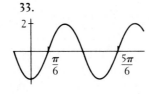

35.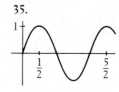

37.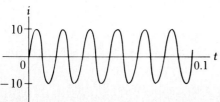

39.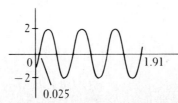

41.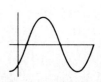

Exercises 6–2, p. 228
[*Note:* "Answers" to trigonometric identities are intermediate steps of suggested reductions of the left member.]

1. $\dfrac{\cos x}{\sin x}\cdot\dfrac{1}{\cos x}=\dfrac{1}{\sin x}$ 3. $\left(\dfrac{\sin y}{\cos y}\right)\left(\dfrac{1}{\sin y}\right)=\dfrac{1}{\cos y}$ 5. $(1-\sin^2 x)-\sin^2 x$

7. $\sin x\left(\dfrac{\sin x}{\cos x}\right)+\cos x=\dfrac{\sin^2 x+\cos^2 x}{\cos x}=\dfrac{1}{\cos x}$ 9. $\tan^2 x+\tan x\cot x=\tan^2 x+1$

11. $(\sin x\cos y+\cos x\sin y)(\sin x\cos y-\cos x\sin y)=\sin^2 x\cos^2 y-\cos^2 x\sin^2 y$
$$=\sin^2 x(1-\sin^2 y)-(1-\sin^2 x)\sin^2 y$$

13. $\cos x\cos y+\sin x\sin y+\sin x\cos y+\cos x\sin y=\cos x(\cos y+\sin y)+\sin x(\sin y+\cos y)$

15. $2\sin x+2\sin x\cos x=\dfrac{2\sin x(1+\cos x)(1-\cos x)}{1-\cos x}$

17. $\dfrac{\sin 3x\cos x+\cos 3x\sin x}{\sin x\cos x}=\dfrac{\sin 4x}{\frac12\sin 2x}=\dfrac{2\sin 2x\cos 2x}{\frac12\sin 2x}$ 19. $\dfrac{1-\cos\alpha}{2\sqrt{\dfrac12(1-\cos\alpha)}}=\sqrt{\dfrac{1-\cos\alpha}{2}}$

21. $2A\sin\dfrac{2\pi t}{T}\cos\dfrac{2\pi x}{\lambda}$ 23. $\cos\alpha=\dfrac{A}{C}$, $\sin\alpha=\dfrac{B}{C}$:

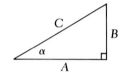

$$A\sin 2t+B\cos 2t=C\left(\dfrac{A}{C}\sin 2t+\dfrac{B}{C}\cos 2t\right)$$

25. 0.262 ft   27. 19.6%
$$=C(\cos\alpha\sin 2t+\sin\alpha\cos 2t)$$

Exercises 6–3, p. 233

1. $\cos(x+2)$ 3. $4\cos(2x-1)$ 5. $-10\sin 2x$ 7. $-6\sin(3x-1)$ 9. $8\sin 4x\cos 4x=4\sin 8x$
11. $-45\cos^2(5x+2)\sin(5x+2)$ 13. $\sin 3x+3x\cos 3x$ 15. $9x^2\cos 5x-15x^3\sin 5x$

17. $2x\cos x^2\cos 2x-2\sin x^2\sin 2x$ 19. $-2\sin 2x\sin x+4\cos 2x\cos x$ 21. $\dfrac{3x\cos 3x-\sin 3x}{x^2}$

23. $\dfrac{-4x^2\sin x^2-2\cos x^2}{3x^2}$ 25. $3\sin^2 x\cos x+2\sin 2x$ 27. $1-4\sin^2 4x\cos 4x$ 29. See table

31. $\dfrac{d\sin x}{dx}=\cos x$, $\dfrac{d^2\sin x}{dx^2}=-\sin x$, $\dfrac{d^3\sin x}{dx^3}=-\cos x$, $\dfrac{d^4\sin x}{dx^4}=\sin x$ 33. $3\sqrt2$ 35. 9.77 ft/s

Exercises 6–4, p. 235

1. $5\sec^2 5x$ 3. $2(1-x)\csc^2(1-x)^2$ 5. $6\sec 2x\tan 2x$ 7. $\dfrac{3\csc\sqrt x\cot\sqrt x}{2\sqrt x}$ 9. $30\tan 3x\sec^2 3x$

11. $-24\cot^3 3x\csc^2 3x$ 13. $2\tan 4x\sqrt{\sec 4x}$ 15. $-84\csc^4 7x\cot 7x$ 17. $x^2\sec^2 x+2x\tan x$

19. $-4\csc x^2(2x\cos x\cot x^2+\sin x)$ 21. $-\dfrac{\csc x(x\cot x+1)}{x^2}$

23. $\dfrac{-4\sin 4x-4\sin 4x\cot 3x+3\cos 4x\csc^2 3x}{(1+\cot 3x)^2}$ 25. $\sec^2 x(\tan^2 x-1)$ 27. $2\sec 2x(\sec 2x-\tan 2x)$

29. $24\tan 3x\sec^2 3x\,dx$ 31. $4\sec 4x(\tan^2 4x+\sec^2 4x)\,dx$ 33. -12 35. 140 ft/s

Exercises 6–5, p. 240

1. $\dfrac{\pi}{3}$ 3. 0   5. $-\dfrac{\pi}{3}$ 7. $\dfrac{\pi}{3}$ 9. $\dfrac{\pi}{6}$ 11. $-\dfrac{\pi}{4}$ 13. $\dfrac{\pi}{4}$ 15. -1.309 17. $\dfrac12\sqrt3$ 19. $\dfrac12\sqrt2$

21. -1.150 23. -1 25. $\dfrac{x}{\sqrt{1-x^2}}$ 27. $\dfrac{1}{x}$ 29. $\dfrac{3x}{\sqrt{9x^2-1}}$ 31. $2x\sqrt{1-x^2}$

33. $\sin\left[\text{Arcsin}\ \dfrac{3}{5} + \text{Arcsin}\ \dfrac{5}{13}\right] = \dfrac{3}{5}\cdot\dfrac{12}{13} + \dfrac{4}{5}\cdot\dfrac{5}{13} = \dfrac{56}{65}$

35. Let b = side opp. β; $\tan \beta = \dfrac{b}{1} = b$; $\tan \alpha = a + b = a + \tan \beta$; $\alpha = \text{Arctan}(a + \tan \beta)$

37. Let $\alpha = \text{Arctan}\dfrac{3}{6}$; $\theta + \alpha = \text{Arctan}\dfrac{3}{6-x}$; $\theta = \text{Arctan}\dfrac{3}{6-x} - \text{Arctan}\dfrac{1}{2}$

Exercises 6–6, p. 244

1. $\dfrac{2x}{\sqrt{1-x^4}}$ **3.** $\dfrac{18x^2}{\sqrt{1-9x^6}}$ **5.** $-\dfrac{1}{\sqrt{4-x^2}}$ **7.** $\dfrac{1}{\sqrt{(x-1)(2-x)}}$ **9.** $\dfrac{1}{2\sqrt{x}(1+x)}$ **11.** $-\dfrac{1}{x^2+1}$

13. $\dfrac{x}{\sqrt{1-x^2}} + \text{Arcsin}\ x$ **15.** $\dfrac{4x}{1+4x^2} + 2\ \text{Arctan}\ 2x$ **17.** $\dfrac{3\sqrt{1-4x^2}\ \text{Arcsin}\ 2x - 6x}{\sqrt{1-4x^2}\ \text{Arcsin}^2 2x}$

19. $\dfrac{8\ \text{Arcsin}\ 4x}{\sqrt{1-16x^2}}$ **21.** $\dfrac{9\ \text{Arctan}^2 x}{1+x^2}$ **23.** $\dfrac{-2(2x+1)^2}{(1+4x^2)^2}$ **25.** $\dfrac{3\ \text{Arcsin}^2 x\ dx}{\sqrt{1-x^2}}$ **27.** $\dfrac{-16x}{(1+4x^2)^2}$

29. Let $y = \text{Arcsec}\ u$; solve for u; take derivatives; substitute.   **31.** $-\dfrac{1}{\sqrt{R^2 - R_x^2}}\dfrac{dR_x}{dt}$

Exercises 6–7, p. 247

1. $d \sin x/dx = \cos x$ and $d \cos x/dx = -\sin x$, and $\sin x = \cos x$ at points of intersection.

3. $\dfrac{1}{x^2+1}$ is always positive

5. Dec., $x > 0$, $x < 0$;
Infl. (0,0)
Asym., $x = \pi/2$, $x = -\pi/2$

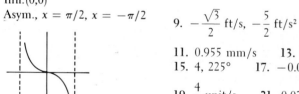

7. $8\sqrt{2}x + 8y + 4\sqrt{2} - 5\pi\sqrt{2} = 0$

9. $-\dfrac{\sqrt{3}}{2}$ ft/s, $-\dfrac{5}{2}$ ft/s²

11. 0.955 mm/s   **13.** 2π, 330°
15. 4, 225°   **17.** −0.0730 rad/s

19. $\dfrac{4}{3}$ unit/s   **21.** 0.02   **23.** 0.06 cm

25. 9.0 in.²   **27.** 14.05 ft

Review Exercises for Chapter 6, p. 249

1. $-12 \sin(4x - 1)$ **3.** $\dfrac{-\sec^2 \sqrt{3-x}}{2\sqrt{3-x}}$ **5.** $\dfrac{9}{9+x^2}$ **7.** $\dfrac{3 \sin^2 \sqrt{x}\ \cos \sqrt{x}}{\sqrt{x}}$ **9.** $\sec 2x(1 + 2x \tan 2x)$

11. $\dfrac{x - \sqrt{1-x^2}\ \text{Arcsin}\ x}{x^2\sqrt{1-x^2}}$ **13.** $-2 \csc 4x \sqrt{\cot 4x + \csc 4x}$ **15.** -1 **17.** $\text{Arccos}\ x$

19. $\dfrac{2y \sin 2x + \sin 2y}{\cos 2x - 2x \cos 2y}$ **21.** $-20 \sin 2t$ **23.** 2π, 330°   **25.** $7.27x + y - 9.44 = 0$

27. Infl. $\left(\dfrac{1}{2}\pi, \dfrac{1}{2}\pi\right), \left(\dfrac{3}{2}\pi, \dfrac{3}{2}\pi\right)$ **29.** 377 ft/s   **31.** −0.24 unit   **33.**   **35.** $5\sqrt{2} = 7.07$ cm.

Exercises 7–1, p. 256

1. $4 = \log_4 256$ **3.** $16 = 8^{4/3}$ **5.** $\dfrac{1}{3} = \log_8 2$ **7.** $16 = (0.5)^{-4}$ **9.** $0 = \log_{12} 1$ **11.** $\dfrac{1}{512} = 8^{-3}$

13. 343   **15.** 0.2   **17.**   **19.**   **21.** $\log_5 3$ **23.** $\log_6 x^{5/2}$ **25.** $\log_b 4y$

27. $\log_2\left(\dfrac{3}{5}x^2\right)$

29. 1.099   **31.** 1.921   **33.** $-\dfrac{1}{k}\ln\left(\dfrac{p}{p_0}\right) = \ln\left(\dfrac{p}{p_0}\right)^{-1/k}$ **35.** $-\dfrac{L}{R}\ln\left(\dfrac{E - iR}{E}\right)$

37. (a) 1.9021   (b) 4.2047   (c) 8.1316   (d) $8.9212 - 10$; 2.3×10^4

39. $\sinh 0.25 = \dfrac{1}{2}(e^{0.25} - e^{-0.25}) = 0.2526$ $1.0314^2 - 0.2526^2 = 1.0000$

$\cosh 0.25 = \dfrac{1}{2}(e^{0.25} + e^{-0.25}) = 1.0314$

Exercises 7–2, p. 260

1. $\dfrac{2 \log e}{x}$ **3.** $\dfrac{6 \log_5 e}{3x + 1}$ **5.** $\dfrac{-3}{1 - 3x}$ **7.** $\dfrac{4 \sec^2 2x}{\tan 2x} = 4 \sec 2x \csc 2x$ **9.** $\dfrac{1}{2x}$ **11.** $1 + \ln x$

13. $\dfrac{3(2x + 1)\ln(2x + 1) - 6x}{(2x + 1)[\ln(2x + 1)]^2}$ **15.** $\dfrac{1}{x \ln x}$ **17.** $\dfrac{1}{x^2 + x}$ **19.** $\dfrac{\cos \ln x}{x}$ **21.** $\dfrac{x \sec^2 x + \tan x}{x \tan x}$

23. $\dfrac{x + 4}{x(x + 2)}$ **25.** See table.   **27.** -1 **29.** $x^x(\ln x + 1)$ **31.** $\dfrac{10 \log e}{I}\dfrac{dI}{dt}$

Exercises 7–3, p. 263

1. $\dfrac{2(3^{2x})}{\log_3 e}$ **3.** $\dfrac{6(4^{6x})}{\log_4 e}$ **5.** $6e^{6x}$ **7.** $\dfrac{e^{\sqrt{x}}}{2\sqrt{x}}$ **9.** $e^x(x + 1)$ **11.** $e^{\sin x}(x \cos x + 1)$ **13.** $\dfrac{3e^{2x}(2x + 1)}{(x + 1)^2}$

15. $e^{-3x}(4 \cos 4x - 3 \sin 4x)$ **17.** $2(e^{2x} + e^{-2x})$ **19.** $\dfrac{e^{2x}}{x} + 2e^{2x}\ln x$ **21.** $12e^{6x}\cot 2e^{6x}$

23. $\dfrac{4e^{2x}}{\sqrt{1 - e^{4x}}}$ **25.** $(-xe^{-x} + e^{-x}) + (xe^{-x}) = e^{-x}$ **27.** $\dfrac{dy}{dx} = e^a; \; y = e^a$ **29.** $10e^{-0.1t}$

31. $\dfrac{d}{dx}\left[\dfrac{1}{2}(e^u - e^{-u})\right] = \dfrac{1}{2}(e^u + e^{-u})\dfrac{du}{dx} = \cosh u \dfrac{du}{dx}$

Exercises 7–4, p. 265

1. Int.(0,0), max.(0,0),
not defined for $\cos x < 0$,

asym. $x = -\dfrac{1}{2}\pi, \dfrac{1}{2}\pi, \ldots$

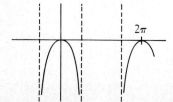

3. Int.(0,0), max. $\left(1, \dfrac{1}{e}\right)$,

infl. $\left(2, \dfrac{2}{e^2}\right)$, asym. $y = 0$

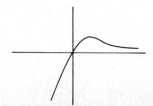

5. Int.(0,0), max.(0,0),
infl. $(-1, -\ln 2)$, $(1, -\ln 2)$

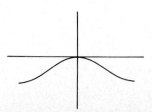

7. Int.(0,1), max.(0,1),

infl. $\left(\frac{1}{2}\sqrt{2}, \frac{1}{e}\sqrt{e}\right), \left(-\frac{1}{2}\sqrt{2}, \frac{1}{e}\sqrt{e}\right),$

asym. $y = 0$

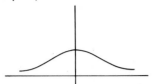

9. Max.$(1,-1)$, asym. $x = 0$

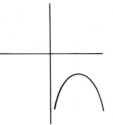

11. Int.(0,0), infl.(0,0),
inc. all x

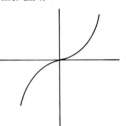

13. $y = x - 1$ **15.** -0.303 W/day   **17.** 1.21 **19.** $a = k^2 x$

21. $v = -e^{-0.5t}(1.4 \cos 6t + 2.3 \sin 6t), -2.02$ **23.** $d\left(\frac{T_2}{T_1}\right) = -\frac{1}{2}e^{k/\sin(\theta/2)}\csc\frac{\theta}{2}\cos\frac{\theta}{2}\,d\theta$

Review Exercises for Chapter 7, p. 267

1. $\dfrac{6x}{x^2 + 1}$ **3.** $2e^{2(x-3)}$ **5.** $-2x \cot x^2$ **7.** $\dfrac{2 \cos x \ln(3 + \sin x)}{3 + \sin x}$ **9.** $2(\cos 2x)e^{\sin 2x}$

11. $x^2(x + 3)e^x$ **13.** $x(1 + 2 \ln x)$ **15.** $\dfrac{2e^{2x}(x^2 - x + 1)}{(x^2 + 1)^2}$ **17.** $e^{-x}\sec x(\tan x - 1)$

19. $\dfrac{y(1 - 2x \ln y)}{x^2 - y}$ **21.** $1.73x + 3.00y - 0.47 = 0$ **23.**   **25.** $-0.15e^{-0.05t} = 0.05(3 - n)$

27. $-\$4.40$ **29.** -0.135 A   **31.** -16.0 J/min   **33.** 202 ft/s   **35.** 0.429
37. Derivative curve should be superimposed on function curve.   **39.** $2.28, 126.9°$

Exercises 8–1, p. 271

1. $\dfrac{1}{5}\sin^5 x + C$ **3.** $-\dfrac{2}{3}(\cos x)^{3/2} + C$ **5.** $\dfrac{1}{3}\tan^3 x + C$ **7.** $\dfrac{1}{8}$ **9.** $\dfrac{1}{4}(\text{Arcsin } x)^4 + C$

11. $\dfrac{1}{10}(\text{Arctan } 5x)^2 + C$ **13.** $\dfrac{1}{3}[\ln(x + 1)]^3 + C$ **15.** 0.179 **17.** $\dfrac{1}{4}(4 + e^x)^4 + C$

19. $\dfrac{3}{16}(2e^{2x} - 1)^{4/3} + C$ **21.** $\dfrac{1}{10}(1 + \sec^2 x)^5 + C$ **23.** $\dfrac{1}{6}$ **25.** $\dfrac{1}{2}$ **27.** $y = \dfrac{1}{3}(\ln x)^3 + 2$

Exercises 8–2, p. 274

1. $\dfrac{1}{4}\ln|1 + 4x| + C$ **3.** $-\dfrac{1}{6}\ln|4 - 3x^2| + C$ **5.** $-\dfrac{1}{2}\ln|\cot 2x| + C$ **7.** $\ln 2 = 0.693$

9. $\ln|1 - e^{-x}| + C$ **11.** $\ln|x + e^x| + C$ **13.** $\dfrac{1}{4}\ln|1 + 4 \sec x| + C$ **15.** $\dfrac{1}{4}\ln 5 = 0.402$

17. $\ln|\ln x| + C$ **19.** $\ln|2x + \tan x| + C$ **21.** $-\sqrt{1 - 2x} + C$ **23.** $\dfrac{1}{3}\ln\left(\dfrac{5}{4}\right) = 0.0744$ **25.** 1.10

27. $\pi \ln 2 = 2.18$ **29.** 0.549 C   **31.** 0.924 ft-lb

Exercises 8–3, p. 277

1. $e^{7x} + C$ **3.** $\dfrac{1}{2}e^{2x+5} + C$ **5.** $2(e - 1) = 3.44$ **7.** $\dfrac{1}{3}e^{x^3} + C$ **9.** $2e^{\sqrt{x}} + C$ **11.** $\dfrac{1}{2}e^{2 \sec x} + C$

13. $\dfrac{2e^x - 3}{2e^{2x}} + C$ **15.** $e^{\text{Arctan } x} + C$ **17.** $-\dfrac{1}{3} e^{\cos 3x} + C$ **19.** 0   **21.** 6.389   **23.** $\pi(e^4 - e) = 163$

25. $\dfrac{1}{8}(e^8 - 1) = 372$ **27.** $V = \dfrac{1}{3}(980 - 800e^{-30t})$

Exercises 8–4, p. 281

1. $\dfrac{1}{2} \sin 2x + C$ **3.** $\dfrac{1}{3} \tan 3x + C$ **5.** $2 \sec \dfrac{1}{2}x + C$ **7.** $\dfrac{1}{3} \ln |\sin x^3| + C$

9. $\dfrac{1}{2} \ln |\sec x^2 + \tan x^2| + C$ **11.** $\cos\left(\dfrac{1}{x}\right) + C$ **13.** $\dfrac{1}{2}\sqrt{3}$ **15.** $\dfrac{1}{5} \sec 5x + C$

17. $\dfrac{1}{2} \ln |\sec 2x + \tan 2x| + C$ **19.** $\dfrac{1}{9}\pi + \dfrac{1}{3} \ln 2 = 0.580$

21. $\csc x - \cot x - \ln |\csc x - \cot x| + \ln |\sin x| + C$ **23.** $\dfrac{1}{2}(\ln |\sin 2x| + \sin 2x) + C$ **25.** 0.347

27. $\pi \sqrt{3} = 5.44$ **29.** $s = \dfrac{4}{3} \sin 3t$ **31.** 0.742

Exercises 8–5, p. 286

1. $\dfrac{1}{3} \sin^3 x + C$ **3.** $-\dfrac{1}{2} \cos 2x + \dfrac{1}{6} \cos^3 2x + C$ **5.** $\dfrac{1}{3} \sin^3 x - \dfrac{1}{5} \sin^5 x + C$ **7.** $\dfrac{1}{120}(64 - 43\sqrt{2})$

9. $\dfrac{1}{2}x - \dfrac{1}{4} \sin 2x + C$ **11.** $\dfrac{1}{2}x + \dfrac{1}{12} \sin 6x + C$ **13.** $\dfrac{1}{2} \tan^2 x + \ln |\cos x| + C$ **15.** $\dfrac{1}{4} \sec^4 x + C$

17. $\dfrac{1}{6} \tan^3 2x - \dfrac{1}{2} \tan 2x + x + C$ **19.** $\dfrac{1}{15} \sec^5 3x - \dfrac{1}{9} \sec^3 3x + C$ **21.** $x - \dfrac{1}{2} \cos 2x + C$

23. $\dfrac{1}{4} \cot^4 x - \dfrac{1}{3} \cot^3 x + \dfrac{1}{2} \cot^2 x - \cot x + C$ **25.** $1 + \dfrac{1}{2} \ln 2 = 1.347$

27. $\dfrac{1}{5} \tan^5 x + \dfrac{2}{3} \tan^3 x + \tan x + C$ **29.** $\dfrac{1}{2}\pi^2 = 4.935$ **31.** $\dfrac{4}{3}$ **33.** $\sqrt{2}$

Exercises 8–6, p. 289

1. $\text{Arcsin } \dfrac{1}{2}x + C$ **3.** $\dfrac{1}{8} \text{Arctan } \dfrac{1}{8}x + C$ **5.** $\dfrac{1}{4} \text{Arcsin } 4x + C$ **7.** $\dfrac{1}{3} \text{Arctan } 6 = 0.469$

9. $\dfrac{2}{5}\sqrt{5} \text{Arcsin } \dfrac{1}{2}\sqrt{5}x + C$ **11.** $\dfrac{4}{9} \ln |9x^2 + 16| + C$

13. $\dfrac{1}{35}\sqrt{35}\left(\text{Arctan } \dfrac{2}{7}\sqrt{35} - \text{Arctan } \dfrac{1}{7}\sqrt{35}\right) = 0.057$ **15.** $\text{Arcsin } e^x + C$ **17.** $\text{Arctan}(x + 1) + C$

19. $4 \text{Arcsin } \dfrac{1}{2}(x + 2) + C$ **21.** -0.357 **23.** $2 \text{Arcsin}\left(\dfrac{1}{2}x\right) + \sqrt{4 - x^2} + C$ **25.** $\text{Arctan } 2 = 1.11$

27. $\dfrac{1}{4}\pi \text{Arctan } \dfrac{3}{4} = 0.505$ **29.** $\text{Arcsin } \dfrac{x}{A} = \sqrt{\dfrac{k}{m}}t + \text{Arcsin } \dfrac{x_0}{A}$ **31.** 0.22

Exercises 8–7, p. 292

1. $\cos x + x \sin x + C$ **3.** $\dfrac{1}{2}xe^{2x} - \dfrac{1}{4}e^{2x} + C$ **5.** $x \tan x \pm \ln |\cos x| + C$

7. $x \text{Arctan } x - \ln \sqrt{1 + x^2} + C$ **9.** $-2x\sqrt{1 - x} - \dfrac{4}{3}(1 - x)^{3/2} + C$ **11.** $\dfrac{1}{2}x^2 \ln x - \dfrac{1}{4}x^2 + C$

13. $\frac{1}{2}x \sin 2x - \frac{1}{4}(2x^2 - 1)\cos 2x + C$ 15. $\frac{1}{2}(e^{\pi/2} - 1) = 1.90$ 17. $1 - \frac{3}{e^2} = 0.594$

19. $\frac{1}{2}\pi - 1 = 0.571$ 21. $\frac{1}{2}\pi - 1 = 0.571$ 23. $q = \frac{1}{5}[e^{-2t}(\sin t - 2\cos t) + 2]$

Exercises 8–8, p. 296

1. $-\frac{\sqrt{1 - x^2}}{x} - \text{Arcsin } x + C$ 3. $\ln|x + \sqrt{x^2 - 4}| + C$ 5. $-\frac{\sqrt{x^2 + 9}}{9x} + C$ 7. $\frac{x}{4\sqrt{4 - x^2}} + C$

9. $\frac{16 - 9\sqrt{3}}{24} = 0.017$ 11. $\ln|\sqrt{x^2 + 2x + 2} + x + 1| + C$ 13. $\frac{1}{3}\text{Arcsec }\frac{2}{3}x + C$

15. $\text{Arcsec }e^x + C$ 17. π 19. $\frac{1}{4}ma^2$ 21. 2.68

Exercises 8–9, p. 298

1. $\frac{3}{25}[2 + 5x - 2\ln|2 + 5x|] + C$ 3. $-\frac{2}{27}(4 - 9x)(2 + 3x)^{3/2} + C$

5. $\frac{1}{2}x\sqrt{4 - x^2} + 2\text{ Arcsin }\frac{1}{2}x + C$ 7. $\frac{1}{2}\sin x - \frac{1}{10}\sin 5x + C$ 9. $\sqrt{4x^2 - 9} - 3\text{ Arcsec}\left(\frac{2x}{3}\right) + C$

11. $\frac{1}{20}\cos^4 4x \sin 4x + \frac{1}{5}\sin 4x - \frac{1}{15}\sin^3 4x + C$ 13. $\frac{1}{2}x^2\text{Arctan } x^2 - \frac{1}{4}\ln(1 + x^4) + C$

15. $\frac{1}{4}(8\pi - 9\sqrt{3}) = 2.386$ 17. $-\ln\left(\frac{1 + \sqrt{4x^2 + 1}}{2x}\right) + C$ 19. $-\ln\left(\frac{1 + \sqrt{1 - 4x^2}}{2x}\right) + C$

21. $\frac{1}{8}\cos 4x - \frac{1}{12}\cos 6x + C$ 23. $\frac{1}{3}(\cos x^3 + x^3\sin x^3) + C$ 25. $\frac{x^2}{2\sqrt{1 - x^4}} + C$ 27. 2.42

29. $\frac{1}{4}x^4\left(\ln x^2 - \frac{1}{2}\right) + C$ 31. $-\frac{x^3}{3\sqrt{x^6 - 1}} + C$ 33. $\frac{1}{4}[2\sqrt{5} + \ln(2 + \sqrt{5})] = 1.479$ 35. 208 lb

37. 210

Review Exercises for Chapter 8, p. 299

1. $-\frac{1}{2}e^{-2x} + C$ 3. $-\frac{1}{\ln 2x} + C$ 5. $\ln(1 + \sin x) + C$ 7. $\frac{2}{35}\text{Arctan }\frac{7}{5}x + C$ 9. 0

11. $\frac{1}{2}\ln 2 = 0.3466$ 13. $\frac{1}{12}\sec^4 3x + C$ 15. $-\frac{1}{3}\ln|\cos 3x| + C$ 17. $\frac{1}{9}\tan^3 3x + \frac{1}{3}\tan 3x + C$

19. $\frac{1}{2}\ln|2x + \sqrt{4x^2 - 9}| + C$ 21. $\sqrt{e^{2x} + 1} + C$ 23. $\frac{1}{6}(3x - \sin 3x \cos 3x) + C$

25. $-\frac{1}{2}x \cot 2x + \frac{1}{4}\ln|\sin 2x| + C$ 27. $\frac{1}{2}\sin e^{2x} + C$ 29. $\frac{1}{3}$ 31. $\frac{1}{2}x^2 - 2x + 3\ln|x + 2| + C$

33. $\Delta S = a \ln T + bT + \frac{1}{2}cT^2 + C$ 35. $v = 320(1 - e^{-t/10})$ 37. $y = \frac{1}{3}\tan^3 x + \tan x$ 39. 11.0

41. $4\pi(e^2 - 1) = 80.3$ 43. $(1.65, 0.24)$

Exercises 9–1, p. 304

1. $V = \pi r^2 h$ 3. $A = \frac{1}{2}bh$ 5. $A = \frac{2V}{r} + 2\pi r^2$ 7. $A = \frac{1}{4}ab$ 9. 24 11. -2

13. $2 - 3y + 4y^2$ 15. $6xt + xt^2 + t^3$ 17. $\frac{p^2 + pq + kp - p + 2q^2 + 4kq + 2k^2 + 5q + 5k}{p + q + k}$

19. $2hx - 2kx - 2hy + h^2 - 2hk - 4h$ **21.** 0 **23.** $81z^6 - 9z^5 - 2z^3$ **25.** $x \neq 0, y \not< 0$ **27.** $y \not> 1$

29. 18 V   **31.** 150 Pa   **33.** 6.67 cm   **35.** $L = \dfrac{(1.28 \times 10^5)r^4}{l^2}$

Exercises 9–2, p. 312

1.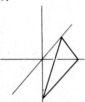

3.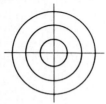

5.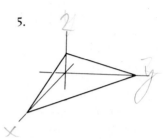

7.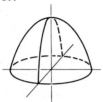

9.

11.

13.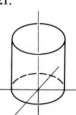

15.

17.

19.

21.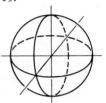

23.

25.

27.

29.

Exercises 9–3, p. 315

1. $\dfrac{\partial z}{\partial x} = 5 + 8xy, \dfrac{\partial z}{\partial y} = 4x^2$ **3.** $\dfrac{\partial z}{\partial x} = \dfrac{2x}{y} - 2y, \dfrac{\partial z}{\partial y} = -\dfrac{x^2}{y^2} - 2x$

5. $\dfrac{\partial \phi}{\partial r} = \dfrac{1 + 3rs}{\sqrt{1 + 2rs}}, \dfrac{\partial \phi}{\partial s} = \dfrac{r^2}{\sqrt{1 + 2rs}}$ **7.** $\dfrac{\partial z}{\partial x} = 4(2x + y^3)(x^2 + xy^3)^3, \dfrac{\partial z}{\partial y} = 12xy^2(x^2 + xy^3)^3$

9. $\dfrac{\partial z}{\partial x} = y \cos xy, \dfrac{\partial z}{\partial y} = x \cos xy$ **11.** $\dfrac{\partial y}{\partial r} = \dfrac{2r}{r^2 + s}, \dfrac{\partial y}{\partial s} = \dfrac{1}{r^2 + s}$

13. $\dfrac{\partial z}{\partial x} = \cos x - y \sin xy, \dfrac{\partial z}{\partial y} = -x \sin xy + \sin y$

15. $\dfrac{\partial f}{\partial x} = e^x(\cos xy - y \sin xy) - 2e^{-2x} \tan y, \dfrac{\partial f}{\partial y} = -xe^x \sin xy + e^{-2x} \sec^2 y$

17. -8 19. $2e$ 21. $-4, -4$ 23. $\left(\dfrac{R_2}{R_1 + R_2}\right)^2$ 25. $3.75 \times 10^{-3} \, 1/\Omega$

27. $\dfrac{k_2 + 2k_3FT}{L_0 + k_1F + k_2T + k_3FT^2}$ 29. $-\dfrac{nRT}{V^2}$

Exercises 9–4, p. 321

1. $dz = (4x + 3) \, dx - 2y \, dy$ 3. $dz = e^y dx + (xe^y - 2y) \, dy$
5. $dz = (y \cos xy + y \sin x) \, dx + (x \cos xy - \cos x) \, dy$

7. $dz = -\dfrac{2xy^2 \, dx}{x^4 + y^2} + \left(\dfrac{yx^2}{x^4 + y^2} + \text{Arctan} \dfrac{y}{x^2}\right) dy$ 9. Min $(1, 3, 0)$ 11. Min $(-1, 2, -1)$
13. $9.6 \times 10^4 \, \text{rad/min}^2$ 15. 1.3% 17. 4.1% 19. $40, 40, 40$ 21. 10 in., 10 in., 5 in.
23. $6xt - 6yt$

Exercises 9–5, p. 327

1. $\dfrac{1}{3}$ 3. $\dfrac{127}{14}$ 5. $\dfrac{1}{3}$ 7. $\dfrac{\pi - 6}{12}$ 9. 1   11. $\dfrac{74}{5}$ 13. $\dfrac{32}{3}$ 15. 8π 17. $\dfrac{28}{3}$

19. 18   21.

Exercises 9–6, p. 331

1. $\left(\frac{2}{3}, \frac{2}{3}\right)$ 3. $\left(\frac{9}{8}, \frac{18}{5}\right)$ 5. $\frac{4}{3}, \frac{4}{3}$ 7. 1.34   9. $\left(\frac{3}{4}, \frac{7}{10}\right), 0.775$ 11. $\left(\dfrac{a}{3}, \dfrac{b}{3}\right)$

13. $\dfrac{mb^2}{3}$ 15. $\dfrac{4}{3}\sqrt{3}$

Review Exercises for Chapter 9, p. 332

1. $\dfrac{\partial z}{\partial x} = 15x^2y^2 - 2y^4, \dfrac{\partial z}{\partial y} = 10x^3y - 8xy^3$ 3. $\dfrac{\partial z}{\partial x} = \dfrac{x}{\sqrt{x^2 - 3y^2}}, \dfrac{\partial z}{\partial y} = \dfrac{-3y}{\sqrt{x^2 - 3y^2}}$

5. $\dfrac{\partial u}{\partial x} = 2xy \cot(x^2 + 2y), \dfrac{\partial u}{\partial y} = 2y \cot(x^2 + 2y) + \ln \sin(x^2 + 2y)$ 7. $\dfrac{\partial z}{\partial x} = \dfrac{\partial z}{\partial y} = \dfrac{1}{2\sqrt{(x + y)(1 - x - y)}}$

9. $\dfrac{21}{2}$ 11. $\dfrac{e^2 - 3}{4}$ 13.   15. 0.113   17. $e^2x - y - e^2 + 1 = 0$ 19. 1.47

21. $\dfrac{\partial V}{\partial r} = \dfrac{ER}{(r + R)^2}, \dfrac{\partial V}{\partial R} = \dfrac{-rE}{(r + R)^2}$ **23.** 0.982   **25.** 1.25 cm   **27.** $\left(\dfrac{3}{2}, 3\right)$ **29.** $\dfrac{292}{15}$

31. $\left(\dfrac{14}{9}, \dfrac{8}{9}\right)$ **33.** 1.82   **35.** -12 W/min

Exercises 10–1, p. 338

1.   **3.**   **5.**   **7.**   **9.**   **11.**

13. $\left(2, \dfrac{\pi}{6}\right)$ **15.** $\left(1, \dfrac{7\pi}{6}\right)$ **17.** $(-4, -4\sqrt{3})$ **19.** $(2.77, -1.15)$ **21.** $r = 3 \sec \theta$ **23.** $r = a$

25. $r = 4 \cot \theta \csc \theta$ **27.** $r^2 = \dfrac{4}{1 + 3 \sin^2\theta}$ **29.** $x^2 + y^2 - y = 0$ **31.** $x = 4$

33. $x^4 + y^4 - 4x^3 + 2x^2y^2 - 4xy^2 - 4y^2 = 0$ **35.** $(x^2 + y^2)^2 = 2xy$

37. $s = \sqrt{x^2 + y^2}\,\arctan\left(\dfrac{y}{x}\right), A = \dfrac{1}{2}(x^2 + y^2)\arctan\left(\dfrac{y}{x}\right)$ **39.** $r = 100 \tan \theta \sec \theta$

Exercises 10–2, p. 341

1.   **3.**   **5.**   **7.**   **9.**

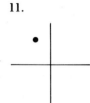
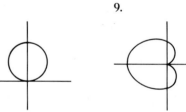

11.   **13.**   **15.**   **17.**   **19.**

21.   **23.**   **25.**   **27.**

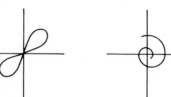

Exercises 10–3, p. 347

1. 9.49   **3.** 1.85   **5.** 4.0 ft/s   **7.** 2.0 ft/s   **9.** 8.25 ft/s   **11.** 1.54 ft/s   **13.** π **15.** $\dfrac{1}{2}$

17. π **19.** $\dfrac{1}{8}(e^{2\pi} - 1)$ **21.** π **23.** $\dfrac{1}{2}(\pi - 2)$ **25.** 5 ft/s, $5\sqrt{3}$ ft/s   **27.** 119 cm²

29. Use indicated procedure.   **31.** 4

Review Exercises for Chapter 10, p. 348

1. $\theta = \text{Arctan } 2 = 1.11$ **3.** $r^2\cos 2\theta = 16$ **5.** $(x^2 + y^2)^3 = 16x^2y^2$ **7.** $3x^2 + 4y^2 - 8x - 16 = 0$

9.   **11.**   **13.**   **15.**

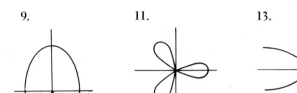

17.   **19.**   **21.** $\dfrac{9}{2}\pi$ **23.** $\dfrac{1}{48}\pi^3$ **25.** $\dfrac{13}{4}$ **27.** $r^2 = \dfrac{7.45 \times 10^7}{2.70 + 0.06\sin^2\theta}$

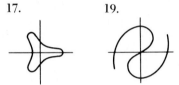

29. $v_r = \dfrac{2kx}{(x^2 + y^2)^2}$, $v_\theta = \dfrac{ky}{(x^2 + y^2)^2}$

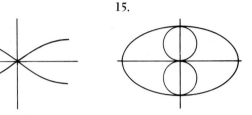

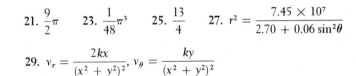

31. $(a^2 - b^2)x^2 + a^2y^2 + 2bx - 1 = 0$; $a = b$, parabola; $a^2 > b^2$, ellipse; $a^2 < b^2$, hyperbola
33. $v_r = 0$, $v_\theta = 5$ mi/s

Exercises 11–1, p. 356

1.

No.	2	3	4	5	6	7
Freq.	1	3	4	2	3	2

3.

No.	45	46	47	48	49	50	51	52	53	54	55	56	57
Freq.	1	1	1	2	2	0	1	0	1	0	2	0	1

5.

Int.	2–3	4–5	6–7
Freq.	4	6	5

7.

Int.	43–45	46–48	49–51	52–54	55–57
Freq.	1	4	3	1	3

9.   **11.**   **13.**   **15.**

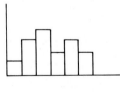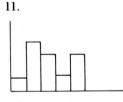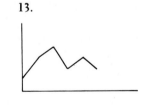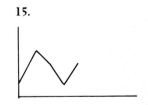

17. 4   **19.** 49   **21.** 4.6   **23.** 50.25

25.   **27.**   **29.** 3.435 A   **31.**

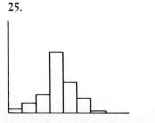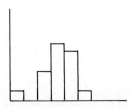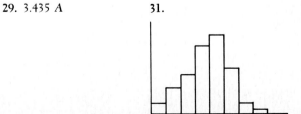

33. 20.9 mi/gal   **35.**   **37.** 4   **39.** 48, 49, 55

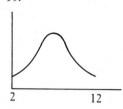

2   12

Exercises 11−2, p. 362

1. 1.50   **3.** 3.74   **5.** 1.50   **7.** 3.74   **9.** $60   **11.** 7   **13.** 60%   **15.** 58%   **17.** 64%
19. 75%

Exercises 11−3, p. 368

1. $y = (1,0)x - 2.6$ **3.** $y = -1.77x + 190$ **5.** $V = -0.6i + 11.3$

7. $V = 4.1 \times 10^{-15}f - 1.90$ **9.** 0.985   **11.** −0.90
    $f_0 = 0.460$ PHz

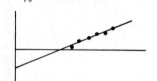

Exercises 11−4, p. 372

1. $y = 1.96x^2 + 5.2$ **3.** $y = 10.9/x$ **5.** $y = 5.95t^2 + 0.55$

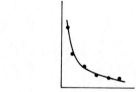

7. $y = 35.6(10^z) - 39.9$ **9.** $f = \dfrac{480}{\sqrt{L}} + 10$

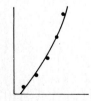

Review Exercises for Chapter 11, p. 373
1. 3.6   **3.** 0.8   **5.** 0.264 Pa · s   **7.** 0.014 Pa · s   **9.** 0.927 in.   **11.** 0.012 in.   **13.** 4

15.

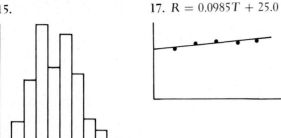

17. $R = 0.0985T + 25.0$ **19.** $y = 1.04 \tan x - 0.02$ **21.** $y = 1.10\sqrt{x} + 0.01$

Exercises 12–1, p. 380

1. $1 + x + \frac{1}{2}x^2 + \cdots$ **3.** $1 - \frac{1}{2}x^2 + \frac{1}{24}x^4 - \cdots$ **5.** $1 + \frac{1}{2}x - \frac{1}{8}x^2 + \cdots$

7. $1 - 2x + 2x^2 - \cdots$ **9.** $1 - 8x^2 + \frac{32}{3}x^4 - \cdots$ **11.** $1 + x + x^2 + \cdots$ **13.** $x - \frac{1}{3}x^3 + \cdots$

15. $x + \frac{1}{3}x^3 + \cdots$ **17.** $-\frac{1}{2}x^2 - \frac{1}{12}x^4 - \cdots$ **19.** $x^2 - \frac{1}{3}x^4 + \cdots$

21. (a), (c) functions are not defined at $x = 0$; (b) derivatives are not defined at $x = 0$.
23. $f^n(0) = 0$ except $f'''(0) = 6$

Exercises 12–2, p. 385

1. $1 + 3x + \frac{9}{2}x^2 + \frac{9}{2}x^3 + \cdots$ **3.** $\frac{x}{2} - \frac{x^3}{2^3 3!} + \frac{x^5}{2^5 5!} - \frac{x^7}{2^7 7!} + \cdots$ **5.** $1 - 8x^2 + \frac{32}{3}x^4 - \frac{256}{45}x^6 + \cdots$

7. $x^2 - \frac{1}{2}x^4 + \frac{1}{3}x^6 - \frac{1}{4}x^8 + \cdots$ **9.** 0.3103 **11.** 0.1901 **13.** $2.6 + 1.44j$ **15.** $6.08(\cos 9.5° + j\sin 9.5°)$

17. $1 + \frac{1}{2}x^2 + \frac{1}{24}x^4 + \frac{1}{720}x^6 + \cdots$ **19.** $x + x^2 + \frac{1}{3}x^3 + \cdots$

21. $\frac{d}{dx}\left(x - \frac{1}{6}x^3 + \frac{1}{120}x^5 - \cdots\right) = 1 - \frac{1}{2}x^2 + \frac{1}{24}x^4 - \cdots$ **23.** $\int \cos x\, dx = x - \frac{x^3}{3!} + \cdots$

25. 0.003099 **27.** 0.1249 **29.** Use indicated method.

Exercises 12–3, p. 388

1. 1.22 **3.** 0.09983 **5.** 2.718 **7.** 0.9986 **9.** 0.335 **11.** 0.8415 **13.** 1.0488 **15.** 1.032

17. 0.021 **19.** 2.4×10^{-10} **21.** $i = \frac{E}{L}\left(t - \frac{Rt^2}{2L}\right)$; small values of t **23.** 66 ft   **25.** 3.146

Exercises 12–4, p. 392

1. 3.32 **3.** 2.049 **5.** 0.5150 **7.** 0.49242 **9.** $e^{-2}\left[1 - (x - 2) + \frac{(x - 2)^2}{2!} - \cdots\right]$

11. $\frac{1}{2}\left[\sqrt{3} + \left(x - \frac{1}{3}\pi\right) - \frac{\sqrt{3}}{2!}\left(x - \frac{1}{3}\pi\right)^2 - \cdots\right]$ **13.** $2 + \frac{1}{12}(x - 8) - \frac{1}{288}(x - 8)^2 + \cdots$

15. $1 + 2\left(x - \frac{1}{4}\pi\right) + 2\left(x - \frac{1}{4}\pi\right)^2 + \cdots$ **17.** 0.111 **19.** 3.049 **21.** 2.03 **23.** 0.8746

25. Use indicated method.

Exercises 12−5, p. 399

1. $f(x) = \dfrac{1}{2} - \dfrac{2}{\pi}\sin x - \dfrac{2}{3\pi}\sin 3x - \cdots$

3. $f(x) = \dfrac{3}{2} + \dfrac{2}{\pi}\sin x + \dfrac{2}{3\pi}\sin 3x + \cdots$

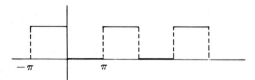

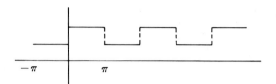

5. $f(x) = \dfrac{\pi}{4} - \dfrac{2}{\pi}\left(\cos x + \dfrac{1}{9}\cos 3x + \cdots\right) + \left(\sin x - \dfrac{1}{2}\sin 2x + \cdots\right)$

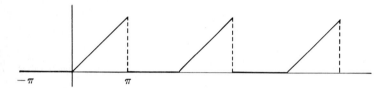

7. $f(x) = -\dfrac{1}{4} - \dfrac{1}{\pi}\cos x + \dfrac{1}{3\pi}\cos 3x - \cdots + \dfrac{3}{\pi}\sin x - \dfrac{1}{\pi}\sin 2x + \dfrac{1}{\pi}\sin 3x - \cdots$

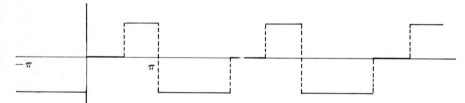

9. $f(x) = \dfrac{\pi}{2} - \dfrac{4}{\pi}\cos x - \dfrac{4}{9\pi}\cos 3x - \cdots$

11. $f(t) = \dfrac{2}{\pi} - \dfrac{4}{3\pi}\cos 2t - \dfrac{4}{15\pi}\cos 4t - \cdots$

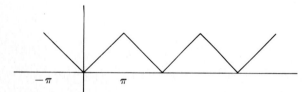

Review Exercises for Chapter 12, p. 400

1. $\dfrac{1}{2} - \dfrac{1}{4}x + \dfrac{1}{48}x^3 - \cdots$ **3.** $2x^2 - \dfrac{4}{3}x^6 + \dfrac{4}{15}x^{10} - \cdots$ **5.** $1 + \dfrac{1}{3}x - \dfrac{1}{9}x^2 + \cdots$

7. $x + \dfrac{1}{6}x^3 + \dfrac{3}{40}x^5 + \cdots$ **9.** 0.82   **11.** 1.09   **13.** 0.9325   **15.** 12.1655   **17.** 0.259

19. $\dfrac{1}{2} - \dfrac{1}{2}\sqrt{3}\left(x - \dfrac{1}{3}\pi\right) - \dfrac{1}{4}\left(x - \dfrac{1}{3}\pi\right)^2 + \cdots$

21. $f(x) = \dfrac{1}{2} + \dfrac{2}{\pi}\left(\cos x - \dfrac{1}{3}\cos 3x + \cdots\right)$

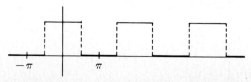

23. $(x + h) - \dfrac{(x + h)^3}{3!} + \cdots - (x - h) + \dfrac{(x - h)^3}{3!} - \cdots = 2h - \dfrac{2hx^2}{2!} + \cdots = 2h\left(1 - \dfrac{x^2}{2!} + \cdots\right)$

25.

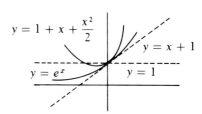

$y = 1 + x + \dfrac{x^2}{2}$

$y = x + 1$

$y = e^x$ $y = 1$

27. 0.0008331

29. $f(t) = \dfrac{1}{2\pi} + \dfrac{1}{\pi}\left(\dfrac{1}{2}\cos t - \dfrac{1}{3}\cos 2t + \cdots\right) + \dfrac{1}{4}\sin t + \dfrac{2}{3\pi}\sin 2t + \cdots$

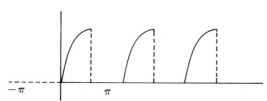

Exercises 13–1, p. 404

[*Note:* "Answers" in this section are the unsimplified expressions obtained by substituting functions and derivatives.]

1. $1 = 1$ **3.** $e^x - (e^x - 1) = 1$ **5.** $x(2cx) = 2(cx^2)$ **7.** $(-2ce^{-2x} + 1) + 2\left(ce^{-2x} + x - \dfrac{1}{2}\right) = 2x$

9. $(-12\cos 2x) + 4(3\cos 2x) = 0$

11. $[e^{2x}(1 + 4c_1 + 4c_2 + 4x + 4c_2x + 2x^2)] - 4[e^{2x}(2c_1 + c_2 + x + 2c_2x + x^2)]$
$$+ 4[e^{2x}(c_1 + c_2x + x^2/2)] = e^{2x}$$

13. $x^2\left[-\dfrac{c^2}{(x - c)^2}\right] + \left[\dfrac{cx}{(x - c)}\right]^2 = 0$ **15.** $x\left(-\dfrac{c_1}{x}\right) + \dfrac{c_1}{x} = 0$ **17.** $(\cos x - \sin x + e^{-x})$
$$+ (\sin x + \cos x - e^{-x}) = 2\cos x$$

19. $\left(e^{-x} + \dfrac{12}{5}\cos 2x + \dfrac{24}{5}\sin 2x\right) + \left(-e^{-x} + \dfrac{6}{5}\sin 2x - \dfrac{12}{5}\cos 2x\right) = 6\sin 2x$

21. $\cos x\left[\dfrac{(\sec x + \tan x) - (x + c)(\sec x \tan x + \sec^2 x)}{(\sec x + \tan x)^2}\right] + \sin x = 1 - \dfrac{x + c}{\sec x + \tan x}$ **23.** $c^2 + cx = cx + c^2$

Exercises 13–2, p. 407

1. $y = c - x^2$ **3.** $x - \dfrac{1}{y} = c$ **5.** $\ln(x^3 + 5) + 3y = c$ **7.** $4\sqrt{1 - y} = e^{-x^2} + c$ **9.** $e^x - e^{-y} = c$

11. $\tan^2 x + 2\ln y = c$ **13.** $x^2 + 1 + x\ln y + cx = 0$ **15.** $y^2 + 4\,\text{Arcsin }x = c$ **17.** $y = c - (\ln x)^2$

19. $\ln(e^x + 1) - \dfrac{1}{y} = c$ **21.** $3\ln y + x^3 = 0$ **23.** $2\ln(1 - y) = 1 - 2\sin x$

Exercises 13–3, p. 410

1. $2xy + x^2 = c$ **3.** $x^3 - 2y = cx - 4$ **5.** $x^2y - y = cx$ **7.** $(xy)^4 = 4\ln y + c$

9. $2\sqrt{x^2 + y^2} = x + c$ **11.** $y = c - \dfrac{1}{2}\ln\sin(x^2 + y^2)$ **13.** $\ln(y^2 - x^2) + 2x = c$

15. $5xy^2 + y^3 = c$ **17.** $2xy + x^3 = 5$ **19.** $2x = 2xy^2 - 15y$

Exercises 13–4, p. 412

1. $y = e^{-x}(x + c)$ **3.** $y = -\dfrac{1}{2}e^{-4x} + ce^{-2x}$ **5.** $y = x(3\ln x + c)$ **7.** $y = \dfrac{8}{7}x^3 + \dfrac{c}{\sqrt{x}}$

9. $y = -\cot x + c \csc x$ 11. $y = 3 + ce^{-x}$ 13. $2y = e^{4x}(x^2 + c)$ 15. $y = \frac{1}{4} + ce^{-x^4}$

17. $3y = x^4 - 6x^2 - 3 + cx$ 19. $xy = (x^3 + c)e^{3x}$ 21. $y = e^{-x}$ 23. $ye^{\sqrt{x}} = e^x + 2e$

Exercises 13–5, p. 418

1. $y^2 = 2x^2 + 1$ 3. $y = 2e^x - x - 1$ 5. $y^2 = c - 2x$ 7. $y^2 = c - 2\sin x$

9. $N = N_0(0.5)^{t/2}, 0.354N_0$ 11. \$1040.81   13. $\lim_{t \to \infty} \frac{E}{R}(1 - e^{-Rt/L}) = \frac{E}{R}$ 15. 0.952 A

17. $q = q_0 e^{-t/RC}$ 19. 11.04 lb   21. $v = 32(1 - e^{-t}), 32$ 23. 10.0 ft/s   25. $x = t^2 + 1, y = \frac{1}{t^2 + 1}$

27. $p = 15.0(0.667)^{10^{-4}h}$ 29. \$1260   31. 36.1°C

Review Exercises for Chapter 13, p. 420

1. $2 \ln(x^2 + 1) - \frac{1}{2y^2} = c$ 3. $ye^{2x} = x + c$ 5. $2x^2 + 4xy + y^4 = c$ 7. $y = cx^3 - x^2$

9. $y = c(y + 2)e^{2x}$ 11. $y = \frac{1}{2}(c - x^2)\csc x$ 13. $y = cxe^{x^3/3}$ 15. $(x^2 + y^2)(4x + c) + 1 = 0$

17. $y = 2\sin^2 x$ 19. $y = 2x - 1 - e^{-2x}$ 21. $r = r_0 + kt$ 23. 3.93 m/s   25. $5y^2 + x^2 = c$
27. $q = c_1 e^{-t/RC} + EC$ 29. $r = c \cos \theta$

Exercises 14–1, 14–2, p. 426

1. $y = c_1 e^{3x} + c_2 e^{-2x}$ 3. $y = c_1 e^{-x} + c_2 e^{-3x}$ 5. $y = c_1 + c_2 e^x$ 7. $y = c_1 e^{6x} + c_2 e^{2x/3}$
9. $y = c_1 e^{x/3} + c_2 e^{-3x}$ 11. $y = c_1 e^{x/3} + c_2 e^{-x}$ 13. $y = e^x(c_1 e^{x\sqrt{2}/2} + c_2 e^{-x\sqrt{2}/2})$
15. $y = e^{3x/8}(c_1 e^{x\sqrt{41}/8} + c_2 e^{-x\sqrt{41}/8})$ 17. $y = e^{3x/2}(c_1 e^{x\sqrt{13}/2} + c_2 e^{-x\sqrt{13}/2})$

19. $y = e^{-x/2}(c_1 e^{x\sqrt{33}/2} + c_2 e^{-x\sqrt{33}/2})$ 21. $y = \frac{1}{5}(3e^{7x} + 7e^{-3x})$ 23. $y = \frac{e^3}{e^7 - 1}(e^{4x} - e^{-3x})$

25. $y = c_1 + c_2 e^{-x} + c_3 e^{3x}$

Exercises 14–3, p. 430

1. $y = (c_1 + c_2 x)e^x$ 3. $y = (c_1 + c_2 x)e^{-6x}$ 5. $y = c_1 \sin 3x + c_2 \cos 3x$

7. $y = e^{-x/2}\left(c_1 \sin \frac{1}{2}\sqrt{7}x + c_2 \cos \frac{1}{2}\sqrt{7}x\right)$ 9. $y = c_1 + c_2 x$ 11. $y = c_1 \sin \frac{1}{2}x + c_2 \cos \frac{1}{2}x$

13. $y = (c_1 + c_2 x)e^{3x/4}$ 15. $y = c_1 \sin \frac{1}{5}\sqrt{2}x + c_2 \cos \frac{1}{5}\sqrt{2}x$ 17. $y = e^x\left(c_1 \cos \frac{1}{2}\sqrt{6}x + c_2 \sin \frac{1}{2}\sqrt{6}x\right)$

19. $(c_1 + c_2 x)e^{4x/5}$ 21. $y = e^{3x/4}(c_1 e^{x\sqrt{17}/4} + c_2 e^{-x\sqrt{17}/4})$ 23. $y = c_1 e^{x(-6+\sqrt{42})/3} + c_2 e^{x(-6-\sqrt{42})/3}$
25. $y = e^{(\pi/6-1-x)}\sin 3x$ 27. $y = (4 - 14x)e^{4x}$

Exercises 14–4, p. 433

1. $y = c_1 e^{2x} + c_2 e^{-x} - 2$ 3. $y = c_1 \sin x + c_2 \cos x + x^2 - 2$ 5. $y = c_1 e^{-x} + c_2 e^{-3x} + \frac{1}{8}e^x + \frac{2}{3}$

7. $y = c_1 e^{5x} + c_2 e^{-4x} - \frac{1}{18}e^{2x} - \frac{2}{5}$ 9. $y = c_1 e^{-5x} + c_2 e^{6x} - \frac{1}{3}$ 11. $y = c_1 e^{2x/3} + c_2 e^{-5x} + \frac{1}{4}e^{3x}$

13. $y = c_1 e^{2x} + c_2 e^{-2x} - \frac{1}{5}\sin x - \frac{2}{5}\cos x$ 15. $y = c_1 \sin x + c_2 \cos x - \frac{1}{3}\sin 2x + 4$

17. $y = c_1 e^{-x} + c_2 e^{-4x} - \frac{7}{100}e^x + \frac{1}{10}xe^x + 1$ 19. $y = (c_1 + c_2 x)e^{-3x} + \frac{1}{25}e^{2x} - e^{-2x}$

21. $y = \frac{1}{6}(11e^{3x} + 5e^{-2x} + e^x - 5)$ 23. $y = -\frac{2}{3}\sin x + \pi \cos x + x - \frac{1}{3}\sin 2x$

Exercises 14–5, p. 437

1. $x = 4 \cos 10t$ **3.** 20   **5.** $y = \frac{1}{4}\cos 16t$ **7.** $y = \frac{1}{4}\cos 16t + \frac{8}{63}\sin 2t - \frac{1}{63}\sin 16t$

9. $q = 2.23 \times 10^{-4}e^{-20t}\sin 2240t$ **11.** $q = \frac{1}{100}(1 - \cos 316t)$

13. $e^{-10t}(c_1 \sin 99.5t + c_2 \cos 99.5t) - \frac{1}{2290}(2 \cos 200t + 15 \sin 200t)$ **15.** $\frac{20}{229}(2 \sin 200t - 15 \cos 200t)$

17. $-[(6.5 \times 10^{-4})\sin 400t + (8.5 \times 10^{-6})\cos 400t]$

Review Exercises for Chapter 14, p. 438

1. $y = c_1 + c_2 e^{-x/2}$ **3.** $y = (c_1 + c_2 x)e^{-x}$ **5.** $y = e^x(c_1 \sin 2x + c_2 \cos 2x)$

7. $y = c_1 e^{8x} + c_2 e^{-7x} - \frac{1}{2}$ **9.** $y = (c_1 + c_2 x)e^{-3x} + \frac{1}{3}x - \frac{2}{9}$

11. $y = c_1 e^{2x/3} + c_2 e^{4x/3} + \frac{1}{2}x + \frac{25}{8}$ **13.** $y = e^{-x/2}(c_1 e^{x\sqrt{5}/2} + c_2 e^{-x\sqrt{5}/2}) + 2e^x$

15. $y = c_1 \sin 3x + c_2 \cos 3x + \frac{1}{8}\sin x$ **17.** $y = 2e^{-x/2}\sin(\sqrt{15}x/2)$

19. $y = \frac{1}{25}[16 \sin x + 12 \cos x - 3e^{-2x}(4 + 5x)]$ **21.** $y = (2 + 8t)e^{-4t}$; critically damped

23. $q = e^{-6t}(0.4 \cos 8t + 0.3 \sin 8t) - 0.4 \cos 10t$ **25.** $i = 0$ **27.** $y = \frac{10}{3EI}[100x^3 - x^4 + xL^2(L - 100)]$

Exercises 15–1, p. 444

1.

x	0.0	0.2	0.4	0.6	0.8	1.0
y	1.00	1.20	1.44	1.72	2.04	2.40

$y = \frac{1}{2}x^2 + x + 1$

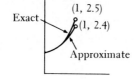

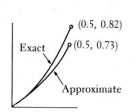

3.

x	0.0	0.1	0.2	0.3	0.4	0.5
y	0.0	0.10	0.22	0.36	0.53	0.73

$y = xe^x$

5.    **7.**    **9.**

5.

x	0.0	0.1	0.2	0.3	0.4	0.5	0.6	0.7	0.8	0.9	1.0
y	0.0	0.10	0.20	0.30	0.41	0.53	0.66	0.80	0.96	1.14	1.34

7.

x	0.0	0.2	0.4	0.6	0.8	1.0	1.2
y	0.0	0.20	0.41	0.64	0.93	1.35	2.12

9.

x	0.0	0.1	0.2	0.3	0.4	0.5	0.6	0.7	0.8	0.9	1.0
y	1.57	1.57	1.56	1.54	1.51	1.48	1.44	1.39	1.34	1.29	1.23

11.

x	0.100	0.200	0.300	0.400	0.500	0.600	(only tenths listed)
y	1.108	1.238	1.383	1.574	1.785	2.038	

13. 1.21   **15.** $i_{\text{approx}} = 0.08$ A, $i_{\text{exact}} = 0.09$ A

Exercises 15–2, p. 449

1. $y_2 = 1 + x + x^2 + \frac{1}{3}x^3$; $y = \dfrac{1}{1-x}$; $y_2(0.1) = 1.1103$; $y(0.1) = 1.1111$

3. $y_3 = x^2 + \frac{1}{2}x^4 + \frac{1}{6}x^6$; $y = e^{x^2} - 1$; $y_3(1) = 1.67$; $y(1) = 1.72$

5. $y_3 = \frac{1}{2}x^2 + \dfrac{1}{20}x^5 + \dfrac{1}{160}x^8 + \dfrac{1}{4400}x^{11}$ 7. $y_2 = \dfrac{3}{4} + x + \dfrac{1}{2}x^2 + \dfrac{1}{4}\cos 2x - \dfrac{1}{2}\sin 2x$

9. Maclaurin expansion: $y = 1 + x + x^2 + \dfrac{1}{3}x^3 + \dfrac{1}{12}x^4 + \cdots$

   Example A: $y_3 = 1 + x + x^2 + \dfrac{1}{3}x^3 + \dfrac{1}{24}x^4 + \cdots$

11. 0.58 ft/s

Exercises 15–3, p. 455

1. $F(s) = \displaystyle\int_0^\infty e^{-st}\, dt = -\dfrac{1}{s}e^{-st}\Big|_0^\infty = \dfrac{1}{s}$

3. $F(s) = \displaystyle\int_0^\infty e^{-st}\sin at\, dt = \dfrac{e^{-st}(-s\sin at - a\cos at)}{s^2 + a^2}\Big|_0^\infty = \dfrac{a}{s^2 + a^2}$ 5. $\dfrac{1}{s-3}$ 7. $\dfrac{6}{(s+2)^4}$

9. $\dfrac{s-2}{s^2+4}$ 11. $\dfrac{3}{s} + \dfrac{2(s^2-9)}{(s^2+9)^2}$ 13. $s^2 L(f) + s L(f)$ 15. $(2s^2 - s + 1)L(f) - 2s + 1$ 17. t^2

19. e^{-5t} 21. $\dfrac{1}{2}t^2 e^{-t}$ 23. $\dfrac{1}{54}(9t\sin 3t + 2\sin 3t - 6t\cos 3t)$

Exercises 15–4, p. 458

1. $y = e^{-t}$ 3. $y = -e^{3t/2}$ 5. $y = (1+t)e^{-3t}$ 7. $y = \dfrac{1}{2}\sin 2t$ 9. $y = 1 - e^{-2t}$

11. $y = e^{2t}\cos t$ 13. $y = 1 + \sin t$ 15. $y = e^{-t}\left(\dfrac{1}{2}t^2 + 3t + 1\right)$ 17. $v = 6(1 - e^{-t/2})$

19. $q = 1.6 \times 10^{-4}(1 - e^{-5\times 10^{-3}t})$ 21. $i = 5t\sin 50t$

Review Exercises for Chapter 15, p. 459

1.
x	0	0.1	0.2	0.3	0.4	0.5
y	1.00	0.90	0.82	0.76	0.71	0.68

3.
x	0	0.2	0.4	0.6	0.8	1.0	1.2	1.4	1.6	1.8	2.0
y	1.00	1.20	1.44	1.71	2.01	2.35	2.72	3.12	3.55	4.00	4.48

5.
x	0.0	0.1	0.2	0.3	0.4	0.5	0.6	0.7
y	0.50	0.56	0.63	0.72	0.84	0.99	1.19	1.50

7.
x	0.100	0.200	0.300	0.400	0.500	(only tenths listed)
y	0.908	0.833	0.774	0.728	0.697	

9. $y_3 = 1 + x + \dfrac{1}{2}x^2 + \dfrac{1}{2}x^3 + \dfrac{1}{12}x^4 + \dfrac{1}{60}x^5$ 11. $y_2 = -\dfrac{7}{36} + x - \dfrac{1}{9}x^3 + \dfrac{1}{4}x^4 + \dfrac{1}{18}x^6$

13. $y = e^{t/4}$ 15. $y = \dfrac{1}{2}(e^{3t} - e^t)$ 17. $y = -4\sin t$ 19. $y = \dfrac{1}{3}t - \dfrac{4}{9}\sin 3t$

21. $y_2 = \dfrac{7}{2} + x - (3+x)e^{-x} + \dfrac{1}{2}e^{-2x}$; $y_2(0.5) = 2.06$

23.
x	0.0	0.1	0.2	0.3	0.4	0.5	0.6	0.7	0.8	0.9	1.0
y	1.00	1.20	1.41	1.63	1.85	2.07	2.30	2.53	2.76	2.98	3.20

25. $i = 12(1 - e^{-t/2})$; $i(0.3) = 1.67$ A   27. $q = 10^{-4}e^{-8t}(4.0\cos 200t + 0.16\sin 200t)$

29.
x	0.00	0.05	0.10	0.15	0.20	0.25	0.30
y	0.00	0.30	0.59	0.88	1.16	1.43	1.69

31. $y = \dfrac{1}{4}t\sin 8t$

Index